THE HERETIC IN DARWIN'S COURT

THE HERETIC IN DARWIN'S COURT

The Life of Alfred Russel Wallace

ROSS A. SLOTTEN

COLUMBIA UNIVERSITY PRESS · NEW YORK

Columbia University Press

Publishers Since 1893

New York Chichester, West Sussex

Library of Congress Cataloging-in-Publication Data

Slotten, Ross A.

The heretic in Darwin's court : the life of Alfred Russel Wallace / Ross A. Slotten.

p. cm.

Includes bibliographical references (p.).

ISBN 0-231-13010-4 (cloth : alk. paper)

1. Wallace, Alfred Russel, 1823–1913. 2. Naturalists—England—Biography.
3. Spiritualists—England—Biography. 4. Socialists—England—Biography. I. Title.

QH31.W2S535 2004

508'.092—dc22

[B]

2003068833

Columbia University Press books are printed on permanent and durable acid-free paper.

Printed in the United States of America

Designed by Chang Jae Lee

c 10 9 8 7 6 5 4 3 2 1

Contents

Acknowledgments

First and foremost, I would like to thank Ben Williams and his staff at the library of the Field Museum of Natural History in Chicago for their early support of my project. As an independent scholar, I will always be indebted to Ben Williams, who provided me with an invaluable letter of introduction that opened doors that might otherwise have remained closed to me.

The writing of this book was indeed a journey. Along the way, I met or corresponded with a number of people who provided me with advice as well as access to original materials from which to work. I would like to single out Gina Douglas, librarian of the Linnean Society of London; A. Tatham, keeper of the collections of the Royal Geographical Society, London; Leslie Price, archivist of the Royal Botanic Gardens, Kew; Stella Brecknell, librarian of the Hope Entomological Collections at the Oxford University Museum of Natural History; John Handford, former archivist and librarian of Macmillan, London; Anne Barrett, college archivist of the Imperial College of Science, Technology and Medicine, London; Paul Cooper, assistant zoological librarian of the Natural History Museum, London; Robert W. O'Hara, an independent researcher who combed through the holdings of the Public Record Office for useful tidbits; Michael Palmer, archivist of the Zoological Society of London; Katharine Taylor, principal archivist of the Manchester Archives and Local Studies, Manchester Central Library; and Adam Perkins, Royal Greenwich Observatory archivist in the Department of Manuscripts and university archives at Cambridge University. All these individuals were unfailingly courteous when I repeatedly contacted them either by e-mail or in person for anything related to Wallace. I would also like to thank Lady de Bellaigue, at the library of Windsor Palace, for providing me with copies of letters to King Edward VII and Michele Minto, at the Wellcome Institute in London, for obtaining many of the photographs used in the text.

In the United States, librarians and archivists at the following libraries provided me with photocopies of letters to and from or relating to Wallace, not all of which were used in the final version of this biography: Dittrick

Medical History Center of Case Western Reserve University, Cleveland; Ernst Mayr and Houghton Libraries of Harvard University, Cambridge, Mass.; Milton S. Eisenhower Library of Johns Hopkins University, Baltimore; Yale University library, New Haven, Conn.; Harry Ransom Humanities Research Center of the University of Texas, Austin; Bookfellows Foundation of Knox College, Galesburg, Ill.; American Philosophical Society, Philadelphia; John Hay Library of Brown University, Providence, R.I.; and Smithsonian Institution Libraries, Washington, D.C.

I am especially grateful to Kenneth Parker, an indefatigable champion of Wallace, who gave me a tour of Wallace's haunts in Hertsford and who introduced me to Richard Wallace, one of Wallace's two grandsons, who kindly allowed me to examine memorabilia not yet in the hands of archivists or librarians. Throughout the research phase of this book, he encouraged me to forge ahead, delighted by the appearance of another (especially American) aficionado of his grandfather's life and works.

There were other Virgils who served as guides at critical junctures: Mark R. D. Seaward of the University of Bradford; Leonard G. Wilson, professor emeritus of the history of medicine at the University of Minnesota, Minneapolis; Richard Milner, senior editor of *Natural History Magazine*; Michael Pearson, who laboriously downloaded transcriptions of Wallace's Malay and American journals from an ancient computer; Bruce Evensen (my cousin), professor of journalism at De Paul University, Chicago; and Sam Fleishman, whose expert advice and guidance were indispensable to me. A number of friends helped me shape my book by reading it at various stages, among which I should mention John Davidson, Neel French, Matthew Lambert, Kevin Murphy, Mohamed Salem, and John Vranicar.

Last but not least, I must thank Robin Smith and Irene Pavitt at Columbia University Press, whose enthusiasm and attentiveness made publication possible; Henry Krawitz, for the unrewarding but Herculean task of reconciling the text, notes, and bibliography; and Sara Lippincott, a supreme editor, who meticulously scoured my manuscript for consistency and clarity and gave me the hope that this was a worthy undertaking.

THE HERETIC IN DARWIN'S COURT

Introduction

THE ANTHROPOLOGIST LOREN EISELEY, popularizer of the history of evolutionary thought, famously referred to the nineteenth century as "Darwin's Century." Although his book of that name is concerned with a good deal more than Darwin, the phrase has contributed to the myth of a lone scientist ultimately triumphing over universal opposition. Darwin was not alone, however. Another man also discovered the theory of natural selection, and he championed the theory as vigorously as did Darwin. His name was Alfred Russel Wallace. Although he was the century's greatest explorer-naturalist, few besides scholars know very much about him. For a time, Wallace shared the limelight with Darwin. It was Wallace who forced Darwin to publish *The Origin of Species*; indeed, were it not for Darwin, the nineteenth century probably would be known as "Wallace's Century."

Darwin, cautious to a fault, had been laboring for some twenty years on his theory, amassing what he hoped would be enough data to change the minds of the majority of his fellow scientists. At the time, most people believed that species had been separately (and divinely) created. Only two men knew the true nature and import of Darwin's work: his friends the botanist Joseph Hooker and the geologist Sir Charles Lyell. Darwin had begun to organize his material into a multivolume book to be entitled "Natural Selection." But an extraordinary thing happened. Sometime in June 1858, Darwin received a packet postmarked from the island of Ternate in the Dutch East Indies. The packet contained an essay by Alfred Russel Wallace, a thirty-five-year-old English naturalist with whom Darwin had struck up a correspondence three years earlier. The essay, "On the Tendency of Varieties to Depart Indefinitely from the Original Type," shocked Darwin. Wallace wrote:

> We believe that there is a tendency in nature to the continued progression of certain classes of varieties further and further from the original type—a progression to which there appears no reason to assign any definite limits. This progression, by minute steps, in various directions,

> but always checked and balanced by the necessary conditions, subject to which alone existence can be preserved, may be followed out so as to agree with all the phenomena presented by organized beings, their extinction and succession in past ages, and all the extraordinary modifications of form, instinct and habits which they exhibit.[1]

In fewer than a dozen pages, Wallace had outlined the theory of evolution by natural selection. "The struggle for existence," "the law of population of species," "the adaptation to the conditions of existence"—Wallace's terminology could have served as chapter headings for Darwin's book. Independently, Wallace had solved the riddle of the origin of species.

After a brief period of soul-searching, Darwin passed Wallace's essay on to Hooker and Lyell, not knowing how to handle the delicate issue of who should be credited with authorship of the theory. He left the decision to his two friends, who presented Wallace's essay along with a few of Darwin's notes on the theory at the July 1 meeting of the Linnean Society of London. Darwin's notes consisted of (1) the abstract of an essay on natural selection that he had written fourteen years earlier, in 1844, and that was to be published only if he died before completing the larger work, and (2) the abstract of a letter that he had penned in September 1857 to the American botanist Asa Gray describing his principle of divergence, or descent with modification from ancestral species. It was the same principle that Wallace had announced in 1855 in his essay "On the Law Which Has Regulated the Introduction of New Species," published in the *Annals and Magazine of Natural History*—which had passed almost unnoticed by English scientists (except Lyell, who had brought it to Darwin's attention) and was not acknowledged at the meeting.

By reading Darwin's notes and Wallace's essay in the order in which they were dated, Lyell and Hooker established Darwin's priority for the historical record; Wallace's essay was essentially presented as bolstering Darwin's conclusions. The doctrine was so radical, however, that it made little impact beyond eliciting a few unpublished comments from an audience that had too much respect for Lyell and Hooker to protest against it publicly. Thomas Bell, the president of the Linnean Society, would later pronounce the year 1858 to be one without any great breakthroughs in science.[2]

Still wary that Wallace would preempt him and encouraged by Hooker, Lyell, and a few others, Darwin abandoned his multivolume text and drew up a hurried four-hundred-page abstract, *On the Origin of Species by Means of Natural Selection, or the Preservation of Favoured Races in the Struggle for Life*. His book was published in November 1859, while Wallace was still traveling in the Malay Archipelago. Wallace never published his version of the theory and refused to challenge Darwin's claim of priority. In fact, he later declared himself grateful that it had not been left to him to present the theory to the world.[3] He gracefully deferred to Darwin—an act of abnegation that earned him Darwin's undying gratitude. It exemplified an ideal rarely realized: the promulgation of truth set above self-interest.

But the story of Wallace and Darwin does not fade away on this noble note. Wallace returned to England in 1862, after a harrowing eight-year journey in the eastern tropics, and continued to champion the theory of evolution by natural selection. It was Wallace who pioneered the field of biogeography, in which he applied his and Darwin's revolutionary idea to explain the past and present distribution of animals throughout the world. Another important contribution was the scientific investigation of the phenomenon of protective coloration (mimicry) in nature, which solved a number of problems that had long puzzled naturalists, whose thinking had been clouded by creationism. In fact, his indefatigable support of natural selection as the primary force driving evolution has since prompted the Harvard paleontologist and evolutionist Stephen Jay Gould to characterize Wallace's belief in a single, overarching principle of nature as "a fatal flaw."[4] Not even Darwin had such complete faith in his doctrine.

For three decades, Darwin and Wallace were rivals. At first their rivalry was amicable. Wallace did not share Darwin's conviction that sexual selection—that is, the kind of selection that depends on the reproductive advantage that certain individuals have over other individuals of the same sex and species—was a significant evolutionary force. Eventually he rejected Darwin's Lamarckian belief that characteristics acquired after birth could be transmitted to an organism's descendants, a concept that Darwin made one of the cornerstones of his theory of inheritance.

For a time, they worked together to shape modern evolutionary theory, but in the 1870s their rivalry inexorably divided them. To Darwin's horror, Wallace had become a committed spiritualist, having attended numerous séances at which he claimed to have witnessed communications with the dead. Spiritualism—which inevitably changed Wallace's mind about the natural world—was both a secular religion and a social movement. In the 1850s and 1860s, it had spread from upstate New York to England and Europe as far east as Russia, claiming millions of believers, including a number of prominent scientists. Sir William Crookes, president of the Royal Society (1913–1915), discoverer of thallium, and inventor of the cathode-ray tube; Darwin's cousin Francis Galton, a founder of modern statistical analysis and a notorious proponent of eugenics; the mathematician Augustus De Morgan; and the accomplished physicist and Nobel laureate Lord Rayleigh were only a few of the scientific men who believed in—or at least accepted as a possibility—the existence of spiritual phenomena.[5]

To many educated Victorians, traditional religious dogma was no longer acceptable, yet the new scientific materialism, with its faith in nothing but blind natural forces, was profoundly unsettling. Spiritualism's peculiar blend of science, philosophy, and concern with the supernatural offered a middle ground for agnostics, and it was to spiritualism that Wallace turned for answers to life's greatest mysteries. Natural selection could explain the origin of species, including the origin of humankind, but Wallace felt it could not explain the origin of our moral and intellectual nature, consciousness, life, or the origin of the universe. Wallace believed that he had found evidence of Mind or Intelligence behind natural laws, and of an Intelligent Designer manipulating those laws for a higher purpose: humanity's spiritual evolution after death. In 1870 he wrote:

> [A] superior intelligence has guided the development of man in a definite direction, and for a special purpose just as man guides the development of many animal and vegetable forms. . . . But even if my particular view should not be the true one, the difficulties I have put forward remain, and I think prove, that some more general and more fundamental law underlies that of "natural selection." . . . It is more probable, that the true law lies too deep for us to discover it; but there seems to me, to be ample indications that such a law does exist, and is probably connected with the absolute origin of life and organization.[6]

This was more than Darwin could bear. Wallace's beliefs opened a chasm between the two men that has survived them. Although Darwin did not dispute Wallace's co-origination of his theory, he began to doubt Wallace's scientific judgment. At times, he tried to undermine Wallace's credibility, most famously in rebuttals against him in *The Descent of Man*, but he did not have to try very hard: Wallace had undermined his own scientific credibility more effectively than any of his enemies—or friends—ever did. His fervent socialism and baffling stance against the efficacy of the smallpox vaccine did not enhance his reputation. By making scientific meetings and popular journals his bully pulpit for views that only a minority shared, Wallace alienated many of his fellow scientists. The result was that much of his later work was dismissed as the work of a crank, a man of science who had mysteriously been led astray, a view still held today.

Four major themes emerge from an analysis of Wallace's life: class relations, shifting paradigms, the nature of scientific research, and science versus religion. No examination of any important nineteenth-century English figure can ignore the issue of class. England was and to some degree still is a class-conscious society. Genealogy and education separated the select few from the lower classes. Without the proper pedigree, many people could never hope to advance materially or socially, at least in the early decades of the nineteenth century. Comportment, dress, how and where one was educated, avocations, and even word usage and the pronunciation of the mother tongue distinguished one person from another, leading to more subtle forms of discrimination. Wallace, who was not a member of the upper classes and had, moreover, been forced by economic circumstances to leave school at the age of thirteen, was viewed as an interloper by the majority of men who controlled English science. No matter what his position on any issue, his struggle would be an uphill one; that he often promulgated unorthodox ideas guaranteed that it would remain uphill. He made a name for himself not only because of his genius but also by force of his personality; yet he was the quintessential outsider both by birth and by choice. He never aspired to an elevated social position or bothered to conform to the rules of the social elite, instead empathizing with underdogs—especially the oppressed working classes—for whom he later became a passionate advocate.

As an outsider, he was perfectly positioned to promote a revolution. Thomas Kuhn has noted that revolutionaries are usually interlopers—mostly young people who have only recently entered a field and lack the preconceived notions and vested interests of those who subscribe to the old paradigm.[7] The old paradigm in nineteenth-century Western science was creationism. The theory of natural selection eventually replaced it with a radically new paradigm, a purely mechanistic one that dispensed with teleology and religious dogma. This was arguably the greatest intellectual revolution in modern Western history.

Yet Wallace, its co-originator, is not well known, at least among nonscientists. Introductory biology textbooks omit his name or quickly pass over him. Darwin is the sole originator, according to the contemporary popular view. Some have blamed Darwin for failing to give Wallace proper credit for his contributions. But this is not true. Darwin made plenty of allusions to Wallace. If anyone can be faulted, it is Wallace himself, who deferred to Darwin time and again throughout his long life, thus ensuring that posterity would forget him. It was also Darwin, not Wallace, who wrote the great book. Had Wallace completed "On the Law of Organic Change," his text on evolution, he might be celebrated today. Once *The Origin of Species* had been published, however, he saw no point in continuing to work on a book dealing with the same subject.

There are other reasons for Wallace's relative obscurity. He was a field biologist, not a laboratory scientist. Although Darwin spent five years as a traveling naturalist, his reputation was built mainly on his extraordinary experimental work. For nearly forty years he rarely left home, confining himself to the laboratories he had set up in his office and extensive gardens at Down House. His painstaking and methodical approach—the accumulation of facts and the obsessive testing of his hypotheses against those facts—elevated biology to a true science. In the twentieth century, laboratory science—at least in biology—is dominant. The naturalist is viewed as a quaint anachronism, like four-o'clock tea or formal dress in the hot sun. How many Nobel Prizes have gone to observers of nature, no matter how brilliant their observations?

Perhaps most important of all, Wallace tried to do the impossible in attempting to reconcile religion and science. Under the old paradigm, science and religion walked hand in hand. But the Darwin–Wallace theory of evolution irrevocably divided these two worldviews. Wallace hoped to create a new paradigm that transcended the old paradigm as well as the newer one, combining elements of both. He proposed a theory that unified the spirit hypothesis with the theory of natural selection, an intellectual shift that Darwin perceived as backtracking. And so it was, to a degree. Wallace was reacting to the excessive materialism that his revolution had wrought. He had omitted a soul from the monster he had created—an omission he discovered belatedly and hoped to remedy.

To us, Wallace's advocacy of spiritualism seems strange. In the nineteenth century, however, it was less strange. Although spiritualism captivated many in the upper classes—séances were common in Europe's royal courts and aristocratic salons—spiritualism was mainly a movement of the lower classes. The war that Wallace, Crookes, and a few other prominent scientist-spiritualists waged can be seen, in part, as a class war, for most of their detractors were upper-class scientists. One of their most formidable opponents, the physiologist and expert microscopist William Benjamin Carpenter, cited the lack of a proper scientific education (both Wallace and Crookes were self-educated) as a cause of their misguided application of the methods of inductive science—clearly an elitist assessment. Although Wallace can fairly be accused of excessive gullibility, he drew his fellow scientists' attention to phenomena now acknowledged to be real, even if their source was the human unconscious and not the spirit world in which Wallace so passionately believed. No one knew anything about the human unconscious in the 1860s and 1870s. Many scientists did not even believe that the brain was the seat of the mind. Wallace was not among them, being one of the first scientists to accept that fact. Not until the twentieth century did a new model of the mind arise that had its roots in the work of Wallace and Crookes, among others. It is not at first apparent that Wallace's legacy has been affected by his social class. But the views of his upper-class colleagues cast a long shadow on his reputation, and their reaction to his spiritualism was rather too extreme, whatever one may think of the spiritualist doctrine. Darwin, who was upper class, wealthy, and Cambridge-educated, invented a theory or two that seems as misguided as Wallace's spiritualism—his theory of pangenesis, based in part on the belief that characteristics acquired after birth could be transmitted, is one example—but they have been conveniently cast aside.

Some historians of science and evolutionary biologists continue to deprecate Wallace's nonbiological work in what often amounts to an attack on his character. H. L. McKinney, the foremost expert on Wallace, refers to his subject as a Jekyll and Hyde.[8] In his introduction to a recent reprint of Wallace's *Island Life*, the anthropologist H. James Birx writes that Wallace "continuously clung to his will to believe in pseudoscience and the ongoing spiritual progress of our species," adding, "To his lasting credit, Darwin gave priority to science and reason. On the other hand, Wallace's occultism and spiritualism represent not only his gullibility concerning the paranormal, but also his inability to come to grips with the naturalistic ramifications of organic evolution of our species. Surely, self-deception and a failure of nerve are no substitutes for truth and wisdom."[9] Like Birx, the distinguished evolutionist and biogeographer Ernst Mayr believes that Wallace "lost his nerve" when it came to applying the theory of natural selection to human beings, though his assessment is less harsh.[10] Even John Marsden, executive secretary of the Linnean Society—which has long celebrated its historical relationship with Wallace, honoring him at the semicentennial and centennial of the discovery of natural selection and recently commissioning a portrait of him to complement its portrait of Darwin—pokes fun at the spiritualists' belief in levitation, the materialization of spirits, and clairvoyance. "All these," Marsden writes about Wallace and other "distinguished, well-educated, and intelligent scientists" who belonged to the British Society for Psychical Research in the late nineteenth century, "believed in varying degrees in psychic powers and were prepared to testify to the veracity of the rogues who took them in. Some of the cases were laughable."[11]

But not everyone has so readily dismissed the early investigations of psychic phenomena. Carl Jung, who specialized in the study of psychology and the occult and spent a considerable amount of time at séances observing mediums, put the work of these pioneers into historical perspective:

> It is not a relapse into the darkness of superstition but an intense scientific interest, a need to direct the searchlight of truth on to the chaos of dubious facts. The names of Crookes . . . Wallace . . . and many other eminent men symbolize the rebirth and rehabilitation of the belief in spirits. Even if the real nature of their observations be disputed, even if they can be accused of errors and self-deception, these investigators have still earned themselves the undying moral merit of having thrown the full weight of their authority and of their great scientific name into these endeavours to shed fresh light on the darkness, regardless of all personal fears and considerations. They shrank neither from academic prejudice nor from the derision of the public, and at the very time when the thinking of educated people was more than ever spellbound by materialistic dogmas, they drew attention to phenomena of psychic provenience that seemed to be in complete contradiction to the materialism of their age.[12]

Nineteenth-century English class relations, shifting paradigms, the determinants of the nature of legitimate scientific research, and the perennial battle between science and religion generated a quagmire through which Wallace struggled and which influenced the direction of his life. This quagmire still surrounds him and explains why he remains unknown. The purpose of this study is to acknowledge these complex factors, place them in a broader framework, and produce a three-dimensional portrait of a man whose forays into spiritualism, socialism, antivaccinationism, and other unorthodox "isms" have been caricatured, overanalyzed, or ignored by specialists in the academic world. Although my methods adhere to scholarly standards, I am directing my book to a wider audience than historians of science and evolutionary biologists. My mission is to fill a gap in the popular imagination—one that Darwinian spin doctors have created at the expense of another great scientific thinker of the Victorian age.

CHAPTER 1

Origins of a Heretic

ALFRED RUSSEL WALLACE was born on January 8, 1823, in Usk, Wales, the eighth child of Thomas Vere and Mary Anne Wallace.[1] By a quirk in the registry, his middle name was misspelled "Russel" and was never corrected. The Wallace side traced itself back, like all Wallaces, to Sir William Wallace, the thirteenth-century hero who had led an unsuccessful revolt against Scotland's English overlords. His mother's relatives, French Huguenots, had fled to England at the end of the sixteenth century, anglicizing their name from Grenaille to Greenell. The Wallaces, devout Anglicans, were of the middle class, but Thomas Wallace, a lawyer who never practiced his profession, had squandered his inheritance of £500 a year as a result of a series of poor business decisions. To minimize expenses, in 1818 or 1819 he moved his family from Hertford, a town north of London, to Usk, a remote and picturesque market town in southeastern Wales, where rents were low and prices of goods half those of London. The Wallace family, which with Alfred's birth included five surviving children and a servant, occupied a spacious cottage along the Usk River. There Thomas Wallace learned the art of self-sufficiency; he grew his own fruits and vegetables, raised his own chickens, and tutored his children. It was the most satisfying period of his life.

In 1828 Mary Anne Wallace inherited money from her stepmother and the Wallaces returned to Hertford, where Alfred received his only formal education. At the local grammar school, he was taught elementary French but not enough Latin to make sense of Virgil, Cicero, or the other great Roman writers. Geography, which would become a lifelong fascination, was almost as tedious as Latin grammar, and science was not a part of the curriculum. In *My Life*, his autobiography, he writes that he acquired more knowledge from his father and older brothers than from his schoolmasters. From his father, who read the plays of Shakespeare and other classics to the family in the evenings, he developed a love of all types of literature. By the age of thirteen, he had read *Tom Jones*, *Don Quixote*, *Paradise Lost*, and the *Inferno*—demonstrating not only a precocious intellect but also a high degree of self-motivation. From his brother John, who was six years his senior and a talented mechanic, he learned to "appreciate the pleasure and utility of doing for one's self everything that one is able to do."[2]

But prosperity and tranquility were to be short-lived. In 1832 Alfred's twenty-two-year-old sister, Eliza, died from tuberculosis, a devastating blow to his parents, who already had endured the deaths of three other daughters. Four years later, Thomas Wallace lost the last of his personal savings as a result of bad real-estate investments. Shortly thereafter, Alfred's maternal uncle, the executor of the Greenell family's estate, declared bankruptcy, having (without anyone's knowledge or consent) borrowed against the Wallace children's small legacies and Mary Anne's modest inheritance to settle his debts. Once again driven to the brink of destitution, Thomas and Mary Anne Wallace and their youngest child, Herbert, moved from their comfortable house in Hertford to a small red-brick cottage in the village of Hoddesdon, a few miles south, to be near Fanny, their only remaining daughter, who at the age of twenty-four had taken a job as a governess to help support the family.[3] For thirteen-year-old Alfred, the reversal in the family fortunes altered the course of his life. His father could no longer afford to pay for his education. Moreover, there was no room for him in the new cottage. Around Christmas 1836, Thomas removed him from school and packed him off to London to board with John, an apprentice carpenter at a builder's yard, where he was expected to make his own way in the world.

The abrupt relocation to London seemed not to be too traumatic, however, and Alfred quickly adjusted. He shared a room and bed with John on Robert Street, off Hampstead Road, a five-minute walk from the workshop of a Mr. Webster, John's employer (and future father-in-law). At first he was not expected to work; he was merely an observer of working-class life. But it was an eye-opening experience. Here in London, the rudiments of his social conscience were awakened. At John's instigation, for the next six months he was exposed to the radical ideas of working-class men at the London Mechanics' Institute, one of several such institutions of higher learning scattered throughout the British Isles and established by forward-thinking entrepreneurs who needed skilled, educated men to manage their factories. Fired by the egalitarian teachings of the Welsh socialist and philanthropist Robert Owen, Alfred rejected the artificial constraints of the English class system, which pigeonholed every citizen and blocked the lower classes from sharing power and wealth with the ruling elite. Owen believed (and convinced the teenage Alfred) that one's character was formed by one's circumstances—that every child, from every class, was a blank slate who could be taught to behave in a morally correct or incorrect manner. In *A New View of Society*, published in 1813, he wrote that the supposition that every individual is accountable for all his sentiments and habits, and consequently deserves reward or punishment depending on his actions, was founded on "erroneous principles." According to Owen, society and its leaders—the clergy and politicians—were responsible for the criminal behavior that blighted the lower classes. He demonstrated in his mills at New Lanark, near Glasgow, that drunkenness, theft, lying, and violent quarrels—common events in the workplace—could be replaced by an improved work ethic and proper moral values if the work environment was more humane and children were sent to school and not to sweatshops. The promotion of the happiness of every individual and of society as a whole, known as utilitarianism, was Owen's goal, one that stirred the hearts of the workingmen (and their middle-class sympathizers) who came to hear his lectures at the Mechanics' Institutes.[4]

The antiecclesiastical diatribes of the deist Thomas Paine were also popular at the Mechanics' Institutes. In *The Age of Reason*, which Alfred read at this time, Paine stated that science and the mechanical arts were the engines of improvement; religion, he said, kept the citizenry ignorant of its natural rights and had become nothing more than a means of wealth for avaricious priests. Moreover, the "obscene stories, the voluptuous debaucheries, the cruel and torturous executions, the unrelenting vindictiveness" that filled half the Bible had taught people to persecute and revenge themselves on others instead of imitating the moral goodness and beneficence of God.[5] Paine transformed Alfred into an agnostic.

In the summer of 1837, Alfred joined his oldest brother, William, as an apprentice land surveyor in the countryside north of London. (Boarding with John had been a stopgap measure until William was ready to take him on.) William was twenty-seven years old, intelligent, well read, and worldly; his more radical political views clashed with those of their father, "a genuine Tory," Alfred reported many years later, who believed that political reform was "a sad giving-way to the ignorant clamour of the mob."[6] Alfred barely knew his brother when they began their travels, but he had great respect for William and was even a little afraid of him. In his autobiography, he portrays William as ascetic and humorless, an image that may or may not have reflected reality. William was saddled with considerable responsibilities, having inherited Alfred because his brother had nowhere else to go. His pecuniary circumstances were little better than their father's, and money was a constant source of anxiety—reflected in his reluctance to buy new clothes for Alfred, who was ashamed to be seen in the threadbare jackets and trousers he was rapidly outgrowing. Despite their fourteen-year age difference, however, the two brothers got on surprisingly well. They led a peripatetic existence, moving from town to town, never remaining anywhere long enough to develop lasting relationships. This gypsy life seemed to be in the Wallace blood; Alfred had never remained in any one place for more than a few years, nor would he for the rest of his long life.

The seven years that he spent with his brother were critical in terms of his intellectual and emotional development. Much of the time, William was his only steady companion; sometimes he was left alone for days or weeks, cut off from civilization, while William searched for work. This isolation allowed his keen mind to grow in its own direction, without guidance or pressure to follow a preconceived course, while his natural optimism kept him from succumbing to depression. His isolation also nurtured an idiosyncratic view of life. To survive in hard times—rural England in the early nineteenth century was difficult for all but the thin upper crust of society—one needed great inner strength and firm convictions, no matter how unorthodox, and the young Wallace had both to a remarkable degree. This unusual combination of emotional stability and intellectual fortitude was undoubtedly the foundation of his later unconventionality.

Alfred valued these periods of solitude, when he could roam the countryside on sturdy legs while his imagination was free to wander anywhere it liked. Both his legs and his brain covered a lot of territory in those years. When he began his apprenticeship with William, he knew virtually nothing about natural science; by the end of 1843, when he left William's service, he was beginning to grapple with the fundamental questions preoccupying the greatest philosophical naturalists of his era, including that of the origin of species.

Like most good land surveyors of the time, William was an amateur geologist. As the two brothers hiked through the parishes north of London, surveying the vast properties of the great landowners, William scoured hillsides and streambeds for unusual stones, which held in their mineral lattices the secrets of the past history of the earth. He dispelled some of the myths of Alfred's childhood, one of which was the nature of belemnites—cigar-shaped rocks known to English schoolboys as "thunderbolts"—which were abundant in the chalk and gravel deposits in the places they surveyed.

Despite their nickname, they had not been hurled from heaven by a raging God during a thunderstorm. William said that they were not rocks at all, but the fossilized internal shells of ancient squid that had thrived in some mysterious distant past, when the earth's flora and fauna differed from those existing today. With no previous instruction in natural history, Alfred also had assumed that chalk, a substance encountered everywhere around Hertford and reached in other places by digging ten to twenty feet below the surface, was the "natural and universal" substance of the earth, "the only question being how deep you must go to reach it."[7] William let him know that there was a science called geology and showed him that chalk was not found everywhere. Under William's tutelage—for William was often more like a teacher than a brother—a new world opened up to Alfred, one governed by laws and principles that his schoolmasters had neglected to include in their parochial curriculum.

Alfred was intrigued by fossils and geologic formations, but his enduring interest in geology and natural history would not develop until a few years later. In 1837 and 1838, he was more excited by surveying and mapping. Among William's cache of books and magazines, he had discovered a treatise on surveying, and he taught himself everything he could learn about his new occupation. His capacity for concentration and his predilection for the complete mastery of a subject showed themselves at an early age; now and for the rest of his life, he would throw himself into a field of study until he had exhausted all its possibilities. From the *Trigonometrical Survey of England*, he taught himself the fundamentals of mapmaking, and in his idle moments practiced calculating the distances between towns by measuring the angles and lengths of triangles formed by conspicuous landmarks, such as church spires and hilltop villages. The book familiarized him with the panoply of surveying instruments necessary to make more sophisticated calculations, but the only instrument he could afford was a pocket sextant, which he carried with him at all times, becoming expert in its use. He was learning the importance of boundaries and borders, concepts that would later emerge as crucial to his understanding of the geographic distribution of plants, animals, and the various races of *Homo sapiens*. He was always making maps, perhaps outmapping his brother, whose sketching abilities he could not match and deeply envied. Sibling rivalry may have motivated him as much as a genuine interest in the technical minutiae of surveying.[8]

In late 1839 or early 1840, Alfred and William moved to the small town of Kington, near the Welsh border. A few years earlier, Parliament had passed the Tithe Commutation Act, which established a monetary system for the payment of tithes to the Church of England. To improve government record keeping and augment the revenues of the church and its allies, the aristocratic landowners, every parish in England and Wales had to be surveyed and maps drawn listing the landowner, tenant, property, acreage, and tithe payable.[9] The Wallace brothers' job was to survey and make plans for the enclosure of the common lands, a euphemism for denying poor farmers access to pasture for their small herds of domestic animals and sources of fuel and building materials for their daily survival. Many of these previously open areas were the last remnants of pristine nature, where one could breathe clean air and enjoy the beauty of an unaltered landscape. Although in later years Wallace would express outrage at the loss of these lands, which he called an "all-embracing system of land-robbery," in his youth no such ideas ever occurred to him. He was absorbed in surveying and the pleasures of the wild, picturesque, boggy, and barren moors and felt no shame for his part in depriving his fellow citizens of their natural patrimony. For part of the time, he traveled through the counties of Shropshire and Radnorshire (now Powyshire), where he meticulously resurveyed a river called Senni Street, whose unusually twisted course had been erroneously mapped. In retrospect, he complained in his autobiography that "the numerous bends [had] been inserted at random as if of no importance."[10] He also took the opportunity to examine the boundaries of two distinct rock formations near Ludlow, sandstone on one side and Silurian shales on the other. By early 1841, when he was eighteen, he had acquired a good grasp of the basic principles of geology.

Later that year, the two brothers moved south to Neath, a Welsh town located on Swansea Bay, off the Bristol Channel. For the next two years, they helped design and superintend the erection of warehouses by means of powerful cranes. At night they taught themselves the fundamentals of civil engineering and the intricacies of making blueprints, guided by a book on Gothic architecture. They also learned methods for sounding rivers as they worked on a project to improve navigation in the Neath harbor. But for long periods they were idle, with William spending much of his time away, seeking employment or engaging in matters of business that did not interest his younger brother.

It was at this point in his life, when he had little else to do, that Wallace says he was first introduced to the variety, beauty, and mystery of the plant kingdom. But he already had an interest. Years earlier, while he and his father were walking through Hertford, they met a woman who remarked that she had found a rare saprophytic plant, the Indian pipe (*Monotropa uniflora*), which lives in a complex relationship with soil fungi that derive nourishment from the roots of forest trees. His father knew little about plants and could tell Alfred nothing about it, and William, his later mentor, would show no interest in native plants or animals unless they were fossils. At the time, Alfred hardly knew that there was a science of systematic botany, "that every flower and every meanest and most insignificant weed had been accurately described and classified and that there was any system or order in the endless variety of plants and animals which I knew existed."[11] But the chance remark of the Hertford woman stood out in his memory, and he admired anyone who knew the names of rare plants. On his ramblings through the rugged Welsh landscape during his brother's absence, he developed a passion for wildflowers, which he began to collect. He was more than a collector, however, for he possessed the scientist's drive to understand what he saw.

Alfred cultivated the friendship of the owner of a bookshop in Neath, a man named Charles Hayward, who directed him to numerous books, journals, and magazines on all aspects of science. From this shop, he acquired a small book containing good descriptions of a dozen or so of the most common natural orders of British plants. For the next year, he spent most of his free time wandering over the hills or along streams gathering flowers, becoming adept at plant identification. While learning to recognize the members of one order after another, he recollected in *My Life*, he grasped for the first time that there was a pattern underlying nature, a discovery that excited his curiosity. Confronted by numerous beautiful species that he could not categorize, he also discovered the limitations of his book, which was meant for readers with only a passing interest in botany. Obsessed by his new avocation, Alfred combed the advertising sections of scientific and educational publications that he found in the bookshop and among the various magazines of the proprietor of his lodging house. One day he was attracted by an advertisement for John Lindley's *Elements of Botany, Structural and Physiological*. Lindley, an expert on orchids, was one of England's foremost botanists, and his book was said to contain descriptions of all the natural orders of plants as well as numerous woodcut illustrations.[12] But the price of 10 shillings, 6 pence was steep for an eighteen-year-old apprentice surveyor. Alfred was paid less than £1 a week by his brother, who kept a small amount for himself and sent the rest of their salary to their parents. He had just enough money to buy the book, which he ordered impulsively from the bookseller without consulting William, who would certainly have disapproved.[13]

When *The Elements of Botany* finally arrived, however, Alfred was surprised to discover that he had not obtained the textbook on English wildflowers he thought he had ordered, but a treatise on systematic botany, which was useless for identifying British plants. "Not a single genus was described," he later wrote. "[It] was not even stated which orders contained any British species and which were wholly foreign, nor was any indication given of their general distribution or whether they comprised numerous or few genera or species."[14] Lindley's book, however, turned out to be well suited for a young man of raw intellect. It taught him how to apply scientific principles and methodology to the natural world. "Method, zeal, and perseverance" was Lindley's motto, and a careful study of his book grounded Wallace in the science of botany. "An extensive knowledge of structure . . . strengthens the perceptive powers and accustoms the mind to habits of careful generalisation," Lindley wrote. "It more especially leads to the consideration of the relationship one plant bears to another; and as plants which are most closely akin in structure are also most similar in their sensible properties, it often enables men to judge of the use of an unknown plant by the ascertained properties of those species in whose vicinity it takes its place by virtue of its natural affinities."[15]

Alfred borrowed a plant encyclopedia from Hayward and copied out the structural details of all genera and species of English plants, including ferns, mosses, and lichens, which he interleaved into the text of *The Elements of Botany*—an ambitious task. In the space of a few months, he became a highly competent amateur botanist, mastering the British flora. Method, zeal, and perseverance were virtues that the young Wallace readily embraced. Lindley's system also had practical ramifications that Wallace did not anticipate at the time: the roving botanist could take his principles to any part of the world and classify unknown species.

Alfred progressed rapidly beyond the stage of plant identification and began to collect and preserve, carefully drying his specimens between sheets of paper and pressing them beneath boards weighted with stones or books. Lindley encouraged the budding botanist to start a herbarium and provided specific instructions about the manner of preserving and protecting specimens. The herbarium consisted of sleeves separated by genus, with the species of each genus placed inside. Following Lindley, he made up tickets to label each specimen, giving the generic and specific name and the place where it was collected.[16] William thought that Alfred was wasting his time, but he remained undeterred. Every Sunday he took long walks through the mountains, filling his collecting box with botanical treasures. At night he puzzled out each species, categorizing, drying, and pressing as many as possible before going to bed. Step by step, he taught himself the abstruse science of taxonomy.

Taxonomy—the science, laws, and principles of classification—dates back to the ancient Greeks. Aristotle was the first to propose a classification system of the natural world, but not until the mid-eighteenth century did great advances in classification methods occur. The father of all taxonomists is the Swedish naturalist Carolus Linnaeus (Carl von Linné), who created the binomial system. Linnaeus assigned every known organism to a genus and species, establishing families, orders, and classes as well. He universalized the language of classification by assigning names in Latin; Latin had been used for short descriptions of plants, but not as a method for classification. However, the Linnaean method for classifying plants relied on a single characteristic—the sexual organs—and was considered artificial because it was not always based on natural relationships. Seeking simple ways for professional and amateur botanists to identify plants, Linnaeus aimed for the practical, recognizing that the flower and fruit are relatively stable characters, not subject to the great variation seen in other plant parts. The characters that Linnaeus used for his classification system included the number of pistils and stamens, the presence or absence of flowers, and the presence or absence of both sexual organs on the same plant.

Although Linnaeus's grouping of plants and animals into genera and species proved useful—and is still used—his system fell apart when it came to the broader categories of families, orders, and classes. All sorts of exceptions to his rules were found, especially as specimens poured in from different regions of the globe. Most botanists followed his system well into the nineteenth century, but the numerous exceptions to it prompted some naturalists—Lindley among them—to work out revisions. Linnaeus's maxim ("It is not the character which makes the genus but the genus which gives the character") did not reflect reality. Linnaeus himself occasionally cheated, classifying into a genus some plants that, in strict adherence to his own principles, should have been classified into another genus or family. Lindley and other advocates of what became known as the natural system disagreed with Linnaeus's reliance on a single character to classify a specimen: not only the pistil and stamen, but the root, the stem, the leaves, and other parts differentiated one group of plants from another. According to the natural system, plant species were grouped by affinities, or shared characteristics. A specimen was classified into a genus, a family, an order, or a class because of the resemblance of its various parts, properties, and qualities to those of other known plants, which made the natural system harder to learn but more scientific in principle. Reason, not memorization of a single prominent character, was the key to the natural system. The extensive index at the back of Lindley's book led the reader not only to the identification of a plant genus and species, but ultimately to its higher and broader taxonomic relationships.[17]

Linnaeus and Lindley assumed that each species had been created separately by God in the beginning. For example, the term "affinity" did not imply evolution or descent with modification from a common ancestor, but reflected the uniform and harmonious plan of God. The vast majority of naturalists at the beginning of the nineteenth century believed that all species had been separately created, with the Creator having placed them only in regions they were ideally suited for and nowhere else. A single pair of individuals had propagated the entire stock. Species were fixed and permanent entities, and whatever laws the Creator had used to regulate the geographic distribution of his creatures was something that would forever remain unknown. Although wedded to natural theology and its notion of a supreme, intentional designer, the natural system set the stage for modern evolutionary theory by arranging organisms based on true morphological relationships.

Alfred did not limit his studies to plants. On September 30, 1842, he opened a new book he had just bought—William Swainson's *Treatise on the Geography and Classification of Animals*, the starting point of his lifelong study of the geographic distribution of fauna.[18] How and why he was directed to this book is unknown, but perhaps Hayward, the Neath bookseller, made the initial providential suggestion. Swainson was a well-known English zoologist whose specialty was ornithology. His beautifully designed book, full of high-quality illustrations and plates, must have appealed to the young student of nature, whose experience of the planet's animal life was limited to a few of the species native to the British Isles.

Swainson was an ardent supporter of William Sharp Macleay, whose quinarian system of taxonomy enjoyed brief popularity in certain English scientific circles: Thomas Huxley was an early convert, and Charles Darwin seriously considered but eventually rejected it. In the era before the publication of *The Origin of Species*, Macleay's theory was one of several systematic treatments that attempted to explain the complex relationships among the earth's fauna. By applying a natural system to animal classification, Swainson hoped to show that animals were distributed according to a divine plan whose outlines were obvious, comprehensible, and definable. That plan, according to Swainson, harmonized best with Macleay's theory. The number 5 played a mystical role in the quinarian system: the earth was divided into five great biogeographic divisions—Europe, Asia, America, Africa, and Australia—that roughly paralleled the five varieties of the human species thought to exist. Moreover, among the five basic principles in Macleay's theory of the natural classification of animals, one stated that the animal kingdom could be subdivided into five circles, each formed from five smaller ones. The Vertebrata, for instance, formed a natural circle because "the reptiles (Reptilia) pass into the birds (Aves), these again into the quadrupeds (Mammalia), quadrupeds unite with the fishes (Pisces), these latter with the amphibious reptiles, and the Amphibia bring us back again to the reptiles, the point from which we started." According to Swainson's understanding of Macleay's theory, animals were related to one another by affinity or analogy. He gave as an example the goatsucker, the swallow, and the bat. The goatsucker and the swallow were related by affinity because both were birds, flew at the same hour of the day, and lived off insects, which were captured on the wing. The goatsucker and the swallow were related by analogy to the bat: all three had wings, flew at dusk, and fed in a similar manner. Circles touched at their nearest points, where a family in one class resembled a family in another, pictorially demonstrating the concepts of analogy and affinity.[19]

The quinarian system, though touted as "natural," was as contrived as that of Linnaeus, providing no clue as to the true relationships among organisms. Recognizing the difficulty of distinguishing varieties from species and believing that genera, orders, and classes were more natural assemblages than species, Swainson avoided the species question altogether by invoking creationism—a lapse that irritated Alfred, who had rejected the teleology of revealed religions. He dismissed Swainson's attempts to reconcile science and Scripture as "ridiculous" and made numerous annotations to that effect in the book's margins.[20]

Whatever its deficiencies, Swainson's book proved an invaluable resource. His clear and concise synopses of the ideas of past authorities as well as the leading scientific philosophers of the day—such as the evolutionary views of the French naturalist Jean-Baptiste Lamarck, the classification methods of the French paleontologist Georges Cuvier, the quinarianism of

Macleay, and the monogenic (single-origin) theory of the human species proposed by the English physician and ethnologist James Cowles Prichard—introduced Wallace to contemporary controversies in natural history. The deeper he dug into these controversies, the more fascinated he became.

In the autumn of 1843, Alfred composed what he called a "popular" lecture on botany, his reaction to a lecture on the Linnaean system of plant classification that he had heard at the Neath Mechanics' Institute one evening. The instructor, who upheld Linnaeus's system of classification as the most useful for determining the names of plants, considered the natural system of Lindley too impractical and complicated for the novice. In his essay, Alfred agreed that Linnaeus had made a great advance in classification when he substituted generic and specific names for the short Latin descriptions that had been used up to that point, but he asserted that the Linnaean system was only occasionally natural—accidental rather than logical. Alfred pointed out that Linnaeus grouped together plants that were unlike one another in essential characteristics, and, except for the naming of species, his method was useless when it came to understanding higher taxonomic relationships. The value of Lindley's observations, he said, was his detection of true affinities "under very diverse external forms."[21] But Alfred never intended to present his rebuttal before the students at the Mechanics' Institute. It was an exercise, one of a number in which he engaged, to write down, "more or less systematically," his ideas on various subjects.

In his autobiography, Wallace states that this essay is proof of his early interest in classification. It also demonstrates that he was no longer an amateur or a dilettante. The philosophical questions raised by Lindley and Swainson preoccupied him, and the breadth of his interests—botany, geology, zoology, and anthropology—suggests that he must have entertained greater aspirations than a career as a surveyor. Yet if he fantasized about devoting his life to the search for those laws that Swainson said the Creator would never reveal to mortal eyes, he did not record how he might transform that desire into a reality. The naturalists of his day were men of means; in the early 1840s, it was an unthinkable occupation for someone of his station in life and possessing such limited resources.

CHAPTER 2

The Struggle for Existence

In April 1843, Thomas Vere Wallace died at the age of seventy-two. The funeral and wake were held in Hoddesdon, and his body was removed to Hertford, where he was buried in the family vault at Saint Andrew's Church cemetery. He left his dependents without financial support. William, John, and Alfred were all working, but they had little money to spare for their mother, their sister Fanny, and their youngest brother, Herbert. Mary Anne Wallace, who was fifty-five years old, was forced to find work as a housekeeper. Fanny, still unmarried at thirty-one, was put on a boat to the United States, where she had accepted a teaching position at a small college in Macon, Georgia. Fourteen-year-old Herbert was pulled out of school and apprenticed to a trunk maker in London.

Alfred remained in William's service through the end of 1843. In January 1844, a decline in William's business forced Alfred to strike out on his own, and until he could find work he again lodged with John in London. His liberation had come at a difficult time: the country was in the throes of a depression. Unemployment was widespread, and political unrest compounded economic instability. One of the sources of this unrest was the price of dietary staples, still controlled by the 1815 Corn Law, which had been enacted to protect the landed gentry and large farmers from foreign competition. By remaining artificially high, the cost of basic food created hardship for all but the wealthiest people. The only solution seemed to be to relax laws against free trade so that less expensive food could be imported, a measure that the government, dominated by the aristocracy, stubbornly resisted. Groups like the Anti-Corn Law League, agitating for change, engaged the police in skirmishes throughout the country. Meanwhile, the rural British poor, having been displaced by the gradual enclosure of the lands, were migrating to the cities in search of new opportunities, hastening the country's transformation from an agrarian to an industrialized, urbanized society. Overcrowding and horrific sanitary conditions turned the largest cities in Britain into cesspools of disease. In the poorest neighborhoods, a fetid effluent of human waste flowed along open sewers, fed by common drains from the tightly packed tenements, setting the stage for the 1854 cholera epidemic, the worst such outbreak in London's recorded history.[1] Of all the beleaguered groups, the Irish suffered the most. In addition to the economic downturn, it was the time of the great potato famine, when millions of Irish Catholics either died or headed in droves to America, with barely a prick on the conscience of the Protestant English.

It was during this unpromising period that Alfred, unable to find work as a surveyor, applied for teaching positions through an agency in London. His inadequacy in Latin and algebra, however, led to numerous rejections. By a stroke of luck, a position that suited him perfectly turned up in the city of Leicester, where a clergyman needed someone to teach drawing, surveying, and mapping to students at his school. The salary, £30 or £40 a year, was paltry by most standards, but more than Alfred had ever earned before. Room and board were free, which provided him with a little extra cash, and his time would be his own when his duties ended in the late afternoon. Packing up his books, clothes, sextant, and herbarium, he headed to Leicester, a hundred miles north, to begin a new life.

Painfully aware of the deficiencies in his own education, Alfred struggled at first with his teaching duties. He was forced to relearn basic Latin and fine-tune his knowledge of algebra to keep one step ahead of his students, a task he applied himself to with his usual diligence. As a teacher, he was congenial and got on well with the boys, who for the most part were well behaved and attentive. The clergyman was a patient and kind-hearted man who had faith in Alfred's innate talents—even if Alfred did not share his religious convictions.

Alfred spent much of his spare time in the Leicester library. It was there that he picked up Thomas Malthus's pessimistic and controversial work *An Essay on the Principle of Population*, first published in 1798 and updated in 1830, a book whose harsh precepts would play a crucial role in the development of evolutionary theory. The depression of the 1840s must have made Malthus's words resonate with greater poignancy, while the utopian vision of the socialist Robert Owen, Alfred's early hero, seemed like a fairy tale, unrealizable in the foreseeable future. Owen, still a popular figure among the working class (his socialist paper the *New Moral World* had a weekly circulation of forty thousand), had disputed Malthus's bleak vision of humanity's fate in *A New View of Society*.[2] Never one to rely on anyone else's opinion—even that of a respected authority like Owen—Wallace hoped to discover what had aroused Owen's indignation.

Malthus's central thesis is well known: an unchecked human population increases geometrically, while the food supply increases arithmetically. Since the power of population growth was "indefinitely greater" than the power of the earth to provide sustenance, Malthus reasoned, the human population would inevitably outstrip its food supply. It already was happening in Ireland; the potato famine appeared to fulfill his prophecy, portending a catastrophe. In the natural world, the populations of plants and animals were kept within bounds by Necessity, "that imperious all-pervading law of nature," whose effects were wasted seed, sickness, and premature death. Humankind was not immune to this law and could not escape by any means, including the use of reason. The natural imbalance between population and food supply was an insurmountable obstacle to perfecting human society. The anxiety provoked by a dwindling means of existence was the root of humanity's misery and vice. Human history, Malthus believed, illustrated the incessant war between unbridled population growth and inadequate food supply. No community, state, or civilization could shield its people from this struggle: the highest and lowest cultures suffered equally. Shrinking resources led to territorial expansion, with the more pacific groups yielding to the more powerful. When might clashed with might, "the contest was a struggle for existence. . . . In these savage contests, many tribes must have been utterly exterminated. . . . The prodigious waste of human life occasioned by this perpetual struggle for room and food was more than supplied by the mighty power of population, acting, in some degree, unshackled from the constant habit of emigration."[3]

Malthus's dark, dystopian vision offended Owen. Malthus believed that fixed and unalterable laws of nature rather than political and social institutions were responsible for humanity's evil tendencies, a position that Owen rejected. The Malthusian point of view, a curious and novel version of the biblical assignment of original sin, conflicted with the antiestablishment thrust of the Enlightenment in general and Owen's philosophy in particular. Malthus's solution to the population problem was simple: abstinence from marriage and avoidance of premarital sex—recommendations in line with the teachings of the Anglican Church. (For the Irish and other undesirable elements, emigration conveniently fit the Malthusian bill.) All other checks—unwholesome occupations, harsh labor practices, unhealthy food and inadequate clothing, wars, infanticide—Malthus considered immoral, but social reform was never a part of his scheme. It was only by following "the ways of God," which required choosing virtue over vice and practicing the restraint of animal impulses instead of indulging them, that any improvement in the human condition would occur. His prescription was impossible to uphold, which only reinforced his thesis. The perpetual struggle for existence in the animal and plant worlds, as well as in the human world, was a concept that Malthus hammered into the heads of his readers, and it would haunt Wallace for the rest of his life.[4]

In Leicester, Alfred made one of his most important friendships. At the town library one evening, he met Henry Walter Bates. Bates was only nineteen in 1844, but he already was an accomplished entomologist and recently had published a paper on beetles ("coleopterous insects") in the *Zoologist*, an English journal dedicated to natural history.[5] The acceptance of an article for publication at such a young age must have impressed Alfred, who had not yet published anything. Bates's father owned a hosiery-dying factory, and although he made sure that his children were as well educated as their social position allowed, he expected Henry, the oldest of his four sons, to go into the family business. Henry, an excellent student, was profoundly disappointed when at the age of fourteen his studies were interrupted by his apprenticeship to another hosier. Inclined toward academia rather than trade, he was bored with sweeping out the warehouse and performing other menial chores thirteen hours a day. In his spare time, he attended classes at the Leicester Mechanics' Institute, where he was awarded prizes in Greek and Latin. Although nominally a Unitarian, he was influenced by the anticlerical spirit of the times and professed to be an agnostic, like his new friend Alfred Wallace.

On weekends, Henry and his younger brother, Frederick, searched the countryside for wildflowers, butterflies, and beetles. Henry was a meticulous collector, and his collections were in perfect order, every specimen accurately labeled and with a provenance describing its habitat and circumstance of capture. Alfred had scarcely paid any attention to insects while gathering plants. It was a revelation to him that hundreds, perhaps thousands, of species of beetle could be found within a few miles of town. Henry showed him a text containing descriptions of three thousand species of beetle inhabiting the British Isles alone, making the number in other regions of the world inconceivable. Henry also taught him how to collect and preserve insects, as well as the best places to look for them.[6] Through beetles, Wallace later acknowledged, he became conscious of the infinite variability of species—the almost endless modifications of structure, shape, color, and surface markings that distinguished one species from another, as well as their innumerable adaptations to diverse conditions. Such variability demanded explanation: fortuitously, beetles brought him to the heart of the problem of the origin of species. He took up his new hobby with characteristic zeal, buying a manual on British coleoptera, a collecting bottle, pins, and a storage box. Properly equipped, he made several excursions to the meadows and streams outside Leicester. One of his favorite haunts was a wild, neglected park that enclosed the picturesque ruins of a mansion five miles from the center of town, where the woods and bushy slopes were ideal collecting grounds. He walked to the park, either alone or in the company of a few of his students, taking a picnic lunch and spending the better part of the day engaged in this fascinating pursuit.[7]

The friendship of Alfred and Henry was temporarily interrupted in mid-March 1845 by the unexpected death of William Wallace from pneumonia at the age of thirty-six. Alfred buried his brother in the Llantwit-juxta-Neath church cemetery on March 15.[8] He returned to Leicester briefly to wind up his affairs at the school before reluctantly moving back to Neath in April to try to resurrect William's surveying business, which was heavily in debt. John came up from London to help out. The main difficulty the Wallace brothers faced was the collection of money, a task that Alfred detested. Impoverished by the depression, clients refused to pay, and Alfred was unable to muster the hard-nosed tactics required to squeeze the money out of them. By midsummer, the two brothers had dissolved the business after failing to turn a profit.

Alfred soon obtained a surveying job with a civil engineer in the Welsh coastal town of Swansea, a few miles west of Neath. The engineer badly needed surveyors who were good with a level and leveling staves to mark out land for a proposed railway through the Vale of Neath. It was the height of the great railroad mania; politicians and private investors seemed intent on transforming the landscape of the British Isles into a disfiguring grid of tracks and waystations. Suddenly Alfred's services were highly coveted, and he was offered the generous salary of 2 guineas a day. His occupation took him to places he had never seen before, and he could not resist bringing along his collecting gear. He was now free to do what he loved most: stroll along pleasant lanes and wooded paths near picturesque streams or climb over huge rocks and scramble up and down thick-forested bluffs in quiet contemplation of the wonders of nature. The region, rich in wildflowers, also abounded in insects. He must have been an odd sight as he tramped through the Welsh countryside—a tall, lanky figure in a wide-brimmed hat, one moment engaged in surveying the land, the next in ardently pursuing a lepidopterous treasure, which he deftly netted, killed, and stuffed into a pouch slung over his shoulder—though British citizens of the day probably would not have batted an eye, since eccentricity was a quality Victorians cultivated into a high art.

After completing the railroad survey, Alfred went to London to join other surveyors, draftsmen, and clerks in drawing up a lengthy report for Parliament, which required the railroad lines to be mapped out before the construction of track could proceed. In the end, their efforts came to nothing. The proposed capital outlay for 1,263 new railway lines was £500 million, a sum larger than the total amount of gold in the Bank of England! Many of the lines had been proposed merely for speculative purposes or to be sold to other companies with greater chances for success. Investors panicked, and the value of all railway shares plummeted. Not one-tenth of the planned railway system, Wallace later remarked, was ever realized.[9]

Alfred returned to Neath to try his hand at another business venture, this time persuading John to give up his job as a carpenter in order to set up an architectural, engineering, and construction firm. To save money as they hunted for projects, the two brothers boarded for several months with the Sims family, becoming friendly with one of the sons, Thomas, an aspiring professional photographer who eventually would marry their sister, Fanny. In the fall of 1846, after accumulating sufficient capital, John and Alfred bought a small cottage with a garden and chicken coop a mile from the center of town and moved in, along with their mother. The property bordered a river and looked across an unspoiled valley to the craggy peaks of distant mountains in provincial Wales. Their first commission was a survey of the adjoining parish of Llantwit-juxta-Neath. Unfortunately, one of Alfred's duties was to collect payment for the survey from farmers, most of whom were destitute and unable to speak or understand English. Once again, he found himself in a detestable position. Nevertheless, he finished the project and submitted a map of the area with the names and holdings of all the inhabitants.

There were to be happier assignments, the most personally satisfying among which was the design and construction of a building to house the Neath Mechanics' Institute, whose founder, William Jevons, a retired merchant and manufacturer from Liverpool, had been a friend of William's and thus gave the commission to the Wallace firm. The structure they built remains a minor landmark to this day. Jevons possessed a good library, from which Alfred borrowed books. When Jevons discovered Alfred's interest in science, he asked him to lecture at the institute, a prospect Alfred found terrifying. He was a dilettante, not an expert—an excuse he was to use throughout his life when showered with unwanted publicity or praise—but he was persuaded to begin with simple topics, like the principle of the lever and the pulley. His lack of self-confidence was not apparent, for he was considered a good lecturer and impressed his working-class audience.

John shared his brother's interest in natural history, and the two frequently wandered the countryside together, John searching for an uncommon bird or reptile, Alfred for an uncommon beetle or butterfly. Dismayed by the unsavory aspects of their business, the two brothers fantasized about going abroad to "more or less wild countries," as Wallace later wrote. In anticipation of making such a journey, one summer day they hiked through the mountains of Wales, camping in a cave. Neither had ever camped out before, and they spent an uncomfortable night sleeping among the rocks and shivering from the cold; they had not had the foresight to bring along appropriate clothes. Alfred reported that he enjoyed the wilderness experience.[10] However, it is not clear how serious the brothers were about traveling overseas together; after that brief, bone-chilling excursion, the subject seems not to have come up again.

During his time in Neath, Alfred kept in touch with Henry Bates, exchanging insect specimens and maintaining a lively correspondence on scientific issues. He told Henry that in "this little town" he felt intellectually isolated. He could find no one with more than a passing interest in any branch of natural history. "I quite envy you," he wrote, "who have friends near you attached to the same pursuit." Besides enumerating and describing the various species they collected, both men reflected on the scientific works they were reading at the time. The most contentious book of the day was *Vestiges of the Natural History of Creation*, which had been published anonymously because the author—who speculated on the origin of species, of life, of the universe, even of the human mind—rightly feared the consequences of his controversial and inflammatory hypotheses, which struck most Victorians as the height of blasphemy.[11] When it appeared in bookshops in 1844, it caused an immediate sensation.

The audacious author, whose identity Charles Darwin correctly guessed two years later, was Robert Chambers, editor of the popular *Chambers's Edinburgh Journal*, which advocated self-improvement, secular education, and political reform. Right from the start, Chambers set out a materialistic explanation for the creation of the universe and of life and rejected a literal interpretation of the Judeo-Christian conception as put forth in the Bible. The "whole of our firmament," Chambers wrote, was initially a diffused mass of "nebulous matter" filling the space it still occupies. Gradually, over some unspecified time period, the stars, the galaxies, our sun, and the earth formed. At first, the earth was too hot to support life: the absence of all traces of plants and animals from metamorphic rock suggested that "excessive temperatures prevailed" during the earth's early history. Citing the work of Sir Charles Lyell, Sir Roderick Impey Murchison, and Adam Sedgwick—the foremost geologists of the time—Chambers concluded that organic life began after the appearance of dry land, which emerged over aeons through a combination of sedimentation and an upward thrusting of rock by forces not clearly understood. Once dry land appeared, "there was now a theatre for the existence of plants and animals." On this theatrical stage, the simplest organisms were the earliest actors, followed by those that were more complex. But the atmosphere "had to go through some changes" before the earth was fit for terrestrial life. Chambers stated that "the Almighty" could not have brought forth each individual species through some sort of "immediate exertion." The construction of the earth and the solar system resulted not from divine intervention, but from natural laws that were the "expressions of his will." He wrote that "it is the narrowest of all views of the Deity, and characteristic of a humble class of intellects, to suppose him acting constantly in particular ways for particular occasions."[12]

Although the word "evolution" appears nowhere in Chambers's work (the term had not yet assumed its modern connotations), its author's belief in the interconnectedness of all organic beings implied something of the sort. Like most of his contemporaries, Chambers envisioned a great chain of being, a *scala naturae*, with the simplest organisms on the lowest rung of the ladder and the most complex (*Homo sapiens*) on the highest. Unlike them, however, he stressed its interconnectedness, recognizing that animals in the same class had descended from a common ancestor. "The giraffe has in its tall neck the same number of bones with the pig," Chambers pointed out. "The limbs of all the vertebrate animals are, in like manner, on one plan, however various they may appear."[13] To explain the linkage among the great classes of animals, he proposed transitional forms. For example, he posited an animal he called the Rynchosaurus, which united the body of a reptile with the beak and feet of a bird, as the predecessor of those two classes of animals. The marsupials linked egg-bearing vertebrates with live-bearing mammals. This ladder of organic life did not appear all at once, as described in the Bible, but developed gradually, step by step, over aeons of geologic time.

Anticipating the German evolutionist Ernst Haeckel's famous dictum that "ontogeny recapitulates phylogeny," Chambers perceived the embryologic record as a blueprint for the development of species. "[In] the gestation of mammals," he wrote, "the animalcule-like ovum of a few days is the parent, in a sense, of the chick-like form of a few weeks, and . . . in all the subsequent stages—fish, reptile, etc,—the one may . . . be said to be the progenitor of the other." He then presented a crude diagram, a vertical line with diagonal branches, demonstrating the stages through which a fetus passes, the lowest rung being the fish, the highest the mammal, each branch the point at which the adult form diverges from the fetus. "This diagram shews only main ramifications," Chambers said.

> [But] the reader must suppose minor ones, representing the subordinate differences of orders, tribes, families, genera, etc. if he wished to extend his views to the whole varieties of being in the animal kingdom. Limiting ourselves at present to the outline afforded by this diagram, it is apparent that the only thing required for an advance from one type to another in the generative process is that, for example, the fish embryo should not diverge at A, but go on to C before it diverges, in which case the progeny will be, not a fish, but a [reptile].[14]

Chambers asserted that the human species was not exempt from this law. If one examined the human embryo, one would see that the final stages exhibited features of the perfect ape, which were then suppressed before it became a "true human creature." He tried to head off his critics, who, he predicted, would object to his conception of the world as "degrading," a view he believed to be prejudiced and supercilious. If all creatures were a part of the Almighty Conception, he asked, why should they be held in contempt? Rather, "Let us regard them in a proper spirit, as parts of the grand plan."[15]

Chambers's work generated at least eighty reviews, most of which were unfavorable and some hostile. The majority of scientific men dismissed Chambers's theory as outlandish. Sedgwick, Woodwardian Professor of Geology at Cambridge University and president of the British Association for the Advancement of Science, criticized the book at the annual meeting of the association as a misconstrual of inductive science and an offense to morality.[16] Darwin made no public comments about *Vestiges*. His own work on the "mutability" of species was still in its infancy. In a letter to Thomas Huxley a few years after the publication of *Vestiges*, Darwin wrote, "I must think such a book, if it does no other good, spreads the taste of Natural Science. But I am perhaps no fair judge, for I am almost as unorthodox about species as the 'Vestiges' itself, though I hope not quite so unphilosophical." He disparaged the author's inaccurate knowledge of geology and zoology and found "the idea of a fish passing into a Reptile . . . monstrous." Huxley, never one to mince words, was merciless in his condemnation of the book. Not yet an evolutionist, he was not converted by Chambers's "unphilosophical" argument. He called the book an "attractive and . . . notorious work of fiction," concluding (in a review of the tenth edition in 1853) that "in the popular minds the foolish fancies of the 'Vestiges' are confounded with science, to the incalculable diminution of that reverence in which true philosophy should be held; and we should be unjust to our readers, and false to our own belief, if we commented upon them in any terms but those of the most unmitigated reprobation."[17]

When Alfred read Chambers's work in the fall of 1845, he was electrified by its novel arguments. It altered his perception of the natural world. In a letter dated November 9, he asked for Henry Bates's opinion; Henry's letter has not been preserved, but it would seem from Alfred's response that he was unimpressed. "I do not consider it [the theory of "progressive development" of plants and animals] a hasty generalisation," Alfred replied on December 28, "but rather as an ingenious hypothesis strongly supported by some striking facts and analogies, but which remains to be proved by more facts and the additional light which more research may throw upon the problem. It furnishes a subject for every observer of nature to attend to; every fact he observes will make either for or against it, and it thus serves both as an incitement to the collection of facts, and an object to which they can be applied when collected."[18]

It was this work that first convinced Wallace of the truth of organic evolution. Later in life, he proclaimed that Chambers's work had been undervalued; at the time of the book's publication, he was one of the few who placed it in its proper perspective, maintaining that Chambers's hypothesis of "progressive development" agreed more with geologic and natural facts than the idea of special creation, even if it erred in its details.[19] Chambers's work raised uncomfortable philosophical questions, provided few answers, and made no attempt to show how the various species of animals and plants had acquired their distinctive character. The "how" of the origin of species now became Alfred's intellectual holy grail.

CHAPTER 3

A Daring Plan

IN THE FIRST HALF of the nineteenth century, the most famous—or infamous—exponent of evolution was not the anonymous author of *Vestiges of the Natural History of Creation*, but the French naturalist Jean-Baptiste Lamarck, who died in obscurity in 1829. It has been stated that Lamarck's "volitional" explanation of organic development has been caricatured. However, his observations on the inheritance of acquired characteristics—that is, characteristics acquired as a consequence of the use or disuse of organs and parts—left an indelible stamp on the minds of naturalists, including Wallace. Lamarck's view of how the wading bird got its long legs, for example, reads like one of Aesop's fables: "[W]ishing to avoid immersing its body in the liquid, it acquired the habit of stretching and elongating its legs. The result of this . . . is that individuals will find themselves elevated as on stilts, on long naked legs."[1] Such "unphilosophical" statements left open to ridicule his more important theory of evolution, put forth at the turn of the century in his book *Philosophie zoologique*, in which he argued in favor of a "natural" process of organic change, the descendants of common parents deviating "indefinitely from [the] original type," in an inevitable progression from one form to another, up the ladder of complexity—the same idea espoused by the author of *Vestiges* forty years later.

Lamarck's speculations were one of the themes of Sir Charles Lyell's *Principles of Geology*, an influential three-volume study published in the early 1830s. Lyell, who had been born in 1797 into a family of Scottish gentry and studied at Oxford, was a barrister who chose to pursue his interests in geology rather than practice law. On scientific grounds, he rejected the tradition of using Scripture to support geologic theories. Without denying the truth of the Bible itself, he did not accept a literal interpretation of the story of Genesis.[2] Sometime in 1845 or 1846, Alfred read *Principles*, a book that would play a crucial role in the articulation of his own theory of evolution—as it would in Darwin's. But Lyell, despite his antiscriptural stance, was not an evolutionist. In fact, his book attacked the idea of evolution, especially the view advocated by Lamarck. Lyell perceived as dangerous Lamarck's denial of the part played by divine intervention in the creation of species and, worse, his suggestion that humans were derived from lower forms. "Henceforth his speculations know no bounds," Lyell remarks in *Principles*:

> He gives the rein to conjecture, and fancies that the outward form, internal structure, instinctive faculties, nay, that reason itself, may have been gradually developed from some of the simplest states of existence,—that all animals, that man himself, and the irrational beings, may have had one common origin . . . ; in fine, he renounces the belief in the high genealogy of his species, and looks forward, as if in compensation, to the future perfectibility of man in his physical, intellectual and moral attributes.[3]

Only *verae causae*—true causes—belonged in a scientific work, Lyell stated, and evolution was not a *vera causa* because it was not supported by factual evidence. Facts, he believed, supported the creationist theory. Species were fixed entities, derived from a single pair of individuals. Variations occurred "within certain limits," but "transmutation" never happened. Species lived out a natural life span, died, and were replaced by other species. He felt that embryology did not support the idea of gradual transmutation of one species into another; rather, it disclosed "the unity of plan" that runs through the animal kingdom, especially vertebrate animals. Lyell's concept of the species problem differed little from William Swainson's.[4]

At the time, Wallace made no comments about Lyell's ideas on species. But even as early as 1846, he would have rejected them in spite of Lyell's great powers of persuasion and his "philosophical," or inductive, method of reasoning; in 1846 Wallace was a disciple of *Vestiges*, and once he was converted to an idea he never let go. Wallace found Lyell's arguments about geologic processes persuasive, however, and his use of induction—the building of fact upon fact to support a hypothesis—appealed to Wallace's logical mind.

Lyell's signal contribution to the science of geology was his attack on catastrophism, a theory advanced by the renowned French paleontologist and anatomist Georges Cuvier. Catastrophists (in Lyell's interpretation) believed that the past history of the earth had been more violent than its present. Massive forces had uplifted the great mountain chains almost instantaneously; valleys had been created by sudden and devastating ruptures in the earth's crust; volcanic eruptions had spewed out lava, rock, and ash in volumes inconceivable now; periods of worldwide deluge had alternated with periods of quiescence. Whole floras and faunas had been wiped off the face of the earth during such catastrophic events and replaced with new creations, which, in turn, had been annihilated and replaced. Lyell dismissed the assertions of the catastrophists as fantasies, mere speculation with no basis in fact. He held that most geologic change was gradual and that where it appeared to be abrupt, the gaps in the story would be filled in as more evidence was gathered through the work of geologists. There was continuity between the geologic past and the present, and he noted many points of analogy between them. The present-day consequences of volcanoes and earthquakes, of floods and drought, were keys to the secrets of a past that catastrophists asserted was unknowable. Mountains and valleys formed over millions of years, Lyell said, uplifted and whittled down little by little through the action of wind, sand, sun, water, and other natural agencies, a process whose hallmarks were visible to any keen observer. Moreover, he showed that throughout the Tertiary period (which he subdivided into four successive epochs: the Eocene, Miocene, Older Pliocene, and Newer Pliocene), the assemblage of species had changed gradually—older species becoming extinct one by one as they were replaced by newer species. This uniformitarian doctrine, first formulated by the Scottish geologist James Hutton in 1785, emphasized the notion of deep time—time on an immense scale, almost beyond comprehension. And over such a vast amount of time, tremendous change was possible.[5] Lyell's uniformitarianism left a deep impression on Wallace—as deep as Thomas Malthus's idea of the struggle for existence and Robert Chambers's theory of organic evolution.

In September 1847, Wallace toured the extensive collections of specimens housed at the Muséum Nationale d'Histoire Naturelle of the Jardin des Plantes in Paris. His sister Fanny, who had returned from the United States, had offered to commemorate her reunion with her family by taking Alfred and John to Paris for a week. Fanny was fluent in French and acted as interpreter. A visit to the insect room of the British Museum on the journey home impressed him even more. The sheer number of beetles and butterflies assembled from remote regions of the globe overwhelmed him. He left feeling intoxicated by the allure of exotic places—especially the tropics, where species abounded in profusion.

In a letter to Henry Bates shortly after returning, Alfred reiterated his desire to travel abroad, a follow-up to a discussion the two men had had earlier in the summer about collecting somewhere in the tropics. "I begin to feel rather dissatisfied with a mere local collection," Alfred wrote that autumn. "Little is to be learnt by it. I should like to take some one family to study thoroughly, principally with a view to the theory of the origin of species. By that means I am strongly of [the] opinion that some definite results might be arrived at." Mindful of the vast number of entomological specimens at the British Museum, Alfred proposed that he and Henry go to some "unharvested" region of the Amazon to comb the banks of the world's mightiest river for botanical and zoological specimens, financing their way by selling duplicates of their collections in London and gathering as many facts as possible to solve that mystery of mysteries, the origin of species, thc great riddle that occupied the thoughts of the finest philosophical naturalists of the era.[6]

It was a daring plan, conceivable only by one with the arrogance and inexperience of youth. Swainson had warned the aspiring naturalist against such a proposition: "We recommend no one to make pecuniary speculations in subjects of natural history, under the idea that the prices to be obtained will repay them for the cost or trouble of their original acquisition."[7] Neither Bates nor Wallace had official sponsorship, and their middle-class status and inexperience would not persuade wealthy patrons to offer private support for an expedition with an uncertain outcome. Previous explorers had been independently wealthy or had been hired as naturalists by the Royal Navy. Alexander von Humboldt, a German aristocrat, had financed his five-year expedition to South America and spent his private fortune on his travels, subsequent publications, and scientific writings; Darwin's voyage around the world was underwritten by his father; Darwin's friends Thomas Huxley and Joseph Hooker doubled as medical officers and naturalists onboard naval ships. With no credentials except enthusiasm, Bates and Wallace could never hope to find work as traveling naturalists aboard an ocean-bound vessel.

But Wallace's plan was not as impulsive or ill-considered as it seemed. Since moving back to Neath in 1845, he had been reading extensively in natural history. He was indulging his passion for his "favorite subject—the variations, arrangements, distribution, etc., of species."[8] The Mechanics' Institute library contained the best books available for any eager student, including *Vestiges of the Natural History of Creation* and *Principles of Geology*. Wallace had by then tackled Humboldt's forbidding three-thousand-page chronicle, *Personal Narrative of a Journey to the Equinoctial Regions of the New Continent*, which was de rigueur for anyone contemplating travel as a naturalist. He also had read Darwin's *Voyage of the Beagle* at least twice. Wallace later said that more than any other published accounts, Darwin's *Voyage* and Humboldt's *Personal Narrative* had fired his imagination and filled him with a zeal to travel to some unexplored region of the earth for the purpose of making a contribution to natural science.[9]

Humboldt left for Spanish South America in 1799 and was given permission to explore the Amazonian and Andean regions of present-day Venezuela, Colombia, Ecuador, and Peru. One of his scientific missions was to demonstrate a connection between the Orinoco and Amazon Rivers, a problem that had puzzled geographers. But he and his traveling companion, Aimé Bonpland, never reached the sources of the Orinoco, which were in Brazil, an area off limits to private travelers at that time. Over a period of five years, however, he collected an immense amount of data. He was the first to create topographic maps of floral habitats. One of his most famous illustrations was a foreshortened and bisected view of Chimborazo, an Ecuadorian volcano, half of which was a realistic representation of the forested flanks and snow-capped peak, the other half a chart listing every plant collected on the volcano, with its precise location noted. Humboldt also invented an entirely new genre of travel literature by including objective descriptions of geology and natural history, details of his daily experiences, depictions of the customs and institutions of the indigenous peoples, and subjective impressions, judgments, and emotional reactions—all of which he combined into a seamless narrative with polymathic skill, aesthetic sensitivity, and literary craftsmanship. Humboldt's methodology, based on the precise and accurate recording of his observations and measurements, and the high quality of his written account set the standard for all future scientific travelers.[10]

Darwin also traveled for five years (1831–1836), most of which he spent sailing around the South American continent. As the *Beagle* sailed along the coast, Darwin, the ship's resident naturalist, made numerous excursions into the interior, collecting rocks and specimens of the local flora and fauna and examining fossils. In September 1835, the *Beagle* reached the Galápagos Islands, which would become famous as the epiphany of his travels. "It never occurred to me," Darwin later wrote, "that the productions of islands only a few miles apart, and placed under the same physical conditions, would be dissimilar."[11] He was also struck by the resemblance of the fauna of the islands to species on the nearby continent. Such "singular relationships" were among the observations that made an evolutionist of him, though there is no explicit statement of this in his travel book. By the time he returned to England in October 1836, Darwin had matured to such a striking degree both emotionally and intellectually that his father, who put little credence in phrenology, remarked to Darwin's sisters that "the shape of his head is quite altered!"[12]

Wallace told Bates that he admired Darwin's writing style, which was "so free from all labour, affectation, or egotism, and yet so full of interest and original thought." As another point in Darwin's favor, he noted that Darwin was "an ardent admirer and most able supporter of Mr Lyell's views"—referring to Lyell's uniformitarian ideas about geologic processes, not his thoughts on evolution.[13] (Darwin's ideas on evolution were still a secret.) However, Wallace clearly had more in mind than expressing admiration for Darwin's *Voyage*. He longed to produce a similar book, full of equal interest and original thought, about his own exploits in uncharted territories.

The decisive factor that set Alfred's sights on an expedition to South America was William Henry Edwards's book *A Voyage up the River Amazon: Including a Residence at Pará*.[14] Edwards, who had been born in 1822, was a respected American amateur naturalist who had written an account of his travels in Brazil in 1846 in the kind of purple prose that would sweep any headstrong young man off his feet—and Wallace's feet were well off the ground after he read about the wonders in store for a prospective traveling naturalist. Edwards portrayed Brazil as the naturalist's promised land, an Eden teeming with wildlife and exotic natives. Pacas and agoutis, motmots and chatterers, goatsuckers and manakins, and other birds and animals with strange names—all easily plucked like fruit from the bountiful branches of Brazilian trees—imbued that country with a magical aura that the two young men found irresistible. The region, populated by "Negroes of every shade of color" and beautiful Indian girls fluttering by like "visions," offered ample rewards for traveling anthropologists and ethnologists. The markets were cornucopias of exotic foods, the climate was one of the healthiest on earth, and the inhabitants were eager to assist the foreigner in every endeavor.[15]

Besides Edwards's earnest account, there were good reasons for selecting Brazil as a destination for scientific investigation. First, few places on earth boasted the biodiversity of the Amazon River basin, which included much of Brazil and parts of Peru, Ecuador, New Grenada (Colombia), Venezuela, and British Guiana (Guyana). Edwards's book testified to the richness of the fauna accessible to the amateur. Second, only a handful of explorer-naturalists had penetrated into the heart of the Amazon. Two Germans, Johannes Baptist von Spix and Carl Friedrich von Martius, spent eleven months (July 1819–

June 1820) ascending the Amazon River, beginning at Pará; and the Englishman Charles Waterton made four journeys between 1804 and 1824 to Guiana, an English colony on the northeastern border of Brazil. But their investigations were far from exhaustive.[16] Robert Schomburgk, a German-born English explorer, also limited his explorations to Guiana, while Darwin focused on the Spanish colonial states to the south and west of Brazil. Humboldt attempted to enter Brazil, but he was barred by the Portuguese colonial government because of his radical political views; his admiration for the French Revolution and his defense of abolitionism were well known to Portuguese authorities. After Dom Pedro I had declared independence from Portugal in 1822 and named himself emperor of Brazil, no foreigner could travel in Brazil. Civil unrest prevailed there for the next twenty years, until a constitutional monarchy was established in 1844. An opportunity thus existed for some enterprising young naturalist to do what Humboldt had been unable to do: explore scientifically the watershed between two of South America's greatest rivers, the Orinoco, which originated somewhere in Brazil, and the Amazon, whose important tributary, the Rio Negro, arose in Spanish territory.[17] The vast, biologically unexplored Amazon therefore excited great interest in the London scientific community and was the perfect place for two neophytes to make a name for themselves.

The third reason for traveling to Brazil was the relative inexpensiveness of a journey to South America. The cost of living in Pará was low and financial risk minimal as long as a steady stream of specimens could be shipped back to London. And, fourth, a network of foreign residents provided a solid infrastructure, a safety net for the collector unused to living in a remote land far from the comforts of home. Britain, now at the height of its imperialism, maintained a meddlesome presence in South America, even if territorial conquest was low on its agenda. Always looking for recipients of its exports, by threatening to use its military might Britain ensured that Latin America's doors never closed to its merchants (or other British citizens).[18]

There were considerable disadvantages, however. Humboldt and Bonpland had traveled on a swollen, raging, alligator-infested river in a small boat with a bottom so worn that it threatened to fall apart. In Argentina, Darwin had found himself in the midst of a revolution. Accounts by other travelers spoke of the stealth and malice of the Indians, who killed with silent darts dipped in curare. Above all other dangers loomed the threat to one's health. Plagued by mosquitoes, frequently soaked to the skin, emaciated as a result of inadequate food, and exhausted by their efforts, Humboldt and

Bonpland became severely ill—Bonpland to the point of death—on their return journey through Colombia to the Caribbean coast. "The dangers do not come from wild Indians or snakes or crocodiles or jaguars," Humboldt wrote, "but, as they naïvely say, from '*el sudar y las moscas*' (sweating and mosquitoes)."[19] More explorers died from tropical diseases than from shipwrecks or the poisoned darts of murderous indigenes. The prevailing opinion throughout the nineteenth century was that there was "something in the tropical climate inimical to Europeans." No distinction was made between the tropical climate and those diseases endemic to the tropics. Equatorial regions were associated with various forms of lethal "malarial fevers."[20]

If Bates hesitated, he was quickly persuaded by Wallace's passionate and eloquent arguments in favor of the vagabond life of an explorer-naturalist in the "torrid zone," as Humboldt called the equatorial rain forests of South America. No longer willing to maintain a foundering architectural business, Alfred had run out of ideas for a useful occupation, while Henry detested his latest job as a clerk in a brewery. Both men had little to lose and much to gain. Their hearts were now set on a journey to the Amazon.

Filled with all sorts of romantic notions about the American tropics, Bates and Wallace began to make arrangements for an expedition to Amazonia. Bates's father offered to lend the pair a small sum of money, which was added to their personal savings.[21] In preparation for his eventual departure for Brazil, Alfred and his brother John closed their architectural business. Wallace tells us nothing about what his family thought of his extraordinary decision to set out for the tropics. Since he always seemed to maintain good relations with his mother and siblings, they must have given his plan at least tacit approval. But his decision caused problems. John, who did not have enough surveying or other work to support the family, rented a small house and a few acres of pastureland near town. Mary Anne and Fanny remained with him to do the housework, while he went daily by small pony cart into town to sell milk from the cows he kept. But he made too little money to pay the rent, and in the spring of 1849 he decided to give it all up and sail for California during the mad rush for gold. Fortunately, Fanny married Thomas Sims that year and moved to London with her mother. Herbert, who was employed at the Neath Iron Works, stayed behind in Wales.[22]

Wallace and Bates agreed to meet in London in March 1848. Their primary object was to study the extensive collections of animals at the British Museum, the breadth of whose holdings symbolized England's ambitious imperialist agenda.[23] The treasure trove was analyzed by many of the most illustrious anatomists and taxonomists of the time, such as the paleontologist Richard Owen and the ornithologist John Gould. Here the two young naturalists familiarized themselves with the fauna of South America, noting the depth and breadth of the existing collection, which gave them an idea of the gaps that needed to be filled.

Many years later, in his essay "Museums for the People," Wallace would have some harsh things to say about the British Museum. Like most public museums of the mid-nineteenth century, it was as "unsuited for the amusement and instruction of the public" as it was for the purposes of "the scientific student." Many museums had begun as private collections, which only later were opened to the public. When these collections outgrew their space, Wallace complained, "a temple or palace" was designed to suit its patron's idiosyncratic taste, with little consideration given to lighting or the logical arrangement of objects. Any specimen was displayed, no matter how imperfect or badly prepared. Quantity, not quality, seemed to be the rule. As new specimens poured in from every region of the world, the task of naming, classifying, and cataloging them outstripped the abilities of curators. Disparate groups of animals, poorly labeled and with no attention paid to geographic distribution, might be exhibited together in narrow side galleries. Access to unexhibited specimens was not always easy, and even when accessible they were often in disarray and carelessly stored.[24] Despite all these difficulties, the British Museum was essential to Wallace and Bates's success in 1848, since its curators were interested in anything novel the two men could send back.

The British Museum was not the only institution interested in a naturalist's collections. The Royal Botanic Gardens at Kew hired its own legion of botanists to fill its conservatory with as many exotic plants as possible. In the mid-nineteenth century, the massive geometric glass conservatories like Kew were architectural and technical marvels, attracting visitors from all over Europe.[25] Wallace and Bates made the necessary pilgrimage to Kew, examining live plants in the humid hothouses and dried specimens preserved in herbaria beyond common view, plying the staff with questions, and feeling their way through the arcane world of academic botany. Although Kew did not hire them to make collections, Wallace and Bates agreed to send back unique botanical samples if the price was right.

By a fortunate coincidence, William Edwards, the author of *A Voyage up the River Amazon*, was in London at the same time and agreed to meet the two men one evening. No less effusive in person than in his prose and, at twenty-five, not much older than they were, Edwards encouraged Wallace and Bates to go to Brazil and gave them letters of introduction to some of his American acquaintances in Pará, home to the largest American and English expatriate population in northern Brazil.[26] On Edwards's advice, they accelerated their plans to leave England in order to reach Pará by late May, the beginning of the dry season.

There remained the issue of financial support. Convinced by Edwards's report of the plethora of wildlife, they consulted Edward Doubleday, keeper (curator) of the butterfly collections at the British Museum. Doubleday agreed that if certain conditions were met, two independent traveling naturalists could make sufficient money to underwrite their explorations—particularly since the whole of northern Brazil was very little known and their specimens were thus likely to be novel and attractive. A series of insects had to be as complete as possible, encompassing all known orders. As important as insects were land shells (for example, snails and other terrestrial mollusks), birds, and mammals. The rarer the species, the greater the revenue, with new species commanding the highest prices.[27]

Their experience limited to gathering plants and insects for their own private collections, Bates and Wallace had a lot to learn about the preparation, storage, and transport of large quantities of natural history specimens. For instructions on how to ship specimens to England, they were referred to the curator of the India Museum in London, Thomas Horsfield, one of the most famous naturalists of the day, who showed them the boxes in which he had shipped his butterflies from Java. Three feet long, two feet wide, and two feet deep, these boxes could hold a large number of cork-lined boards, fitted into vertical grooves about two inches apart, on which the insects were pinned. A large number of specimens could thus be safely packed into a small area, and any that came loose would fall to the bottom of the case and remain unharmed. As the two found not long after arriving in Brazil, such sturdy cases, though ideal for shipping pinned specimens, were far too cumbersome for the roving naturalist, who had to travel as lightly as possible. In the end, they made smaller storage boxes, substituting local soft woods or slices of the midribs of the ubiquitous palms for cork, which was not available.[28]

Knowing almost nothing about capturing and skinning birds and mammals, Wallace and Bates also needed a crash course in taxidermy. But with so little time left before their planned departure, they realized that most of their education would have to take place in the field. Other explorer-naturalists had been better prepared. While studying medicine in Edinburgh, Darwin had learned the art of taxidermy from a former slave who earned his living stuffing birds, having traveled to Guiana with Charles Waterton in the 1820s. An expert marksman with the zeal and leisure time for hunting and bird shooting, Darwin spent his free summers backpacking, collecting, and chasing wild game.[29] Wallace, who seems to have had little experience with a gun, would have to practice shooting before journeying to the tropical forests, whose denizens would be obscured by dense foliage and deep shade. No doubt a large number of specimens would elude him as he progressed along the learning curve.

There were other drawbacks to collecting in a foreign country, however. Preservation posed the greatest challenge: a badly preserved organism bearing little resemblance to its living counterpart was useless. Hard tissues such as shells, bones, beaks, and some skins formed a disproportionate part of what was shipped back to England and the Continent. Soft internal structures, equally important to the taxonomist, would be destroyed if improperly stored. Whole animals placed in "spirits" (alcohol) often became distorted, though the more delicate organs were protected. If a glass jar was not airtight, the alcohol evaporated, destroying the specimen—and a ruined specimen meant lost revenue. An independent naturalist, living hand-to-mouth in hostile conditions, counted every item as a contribution toward his survival and elected to amass the more profitable objects. The philosophical naturalist, though, had to strike a balance between economic necessity and the advancement of scientific knowledge. To make his mark in the world, he needed to obtain specimens of value to the scientist at the university or natural history museum and to procure plants and animals for the shelves of the dilettante, both tasks demanding expertise in taxonomy and classification. In order to identify and classify a specimen properly, it had to be examined with an eye as critical and knowledgeable as that of an expert jeweler examining a diamond.[30] The audacity of Wallace and Bates's expedition lay not in its conception, but in its execution.

The two men were put in touch with Samuel Stevens, an enthusiastic collector of British coleoptera and lepidoptera and brother of J. C. Stevens, a well-known natural history auctioneer. Samuel Stevens was the natural history equivalent of an impresario, acting as agent for many wayfaring naturalists, advertising their work at the leading scientific societies, disposing of their collections, providing a steady supply of cash, and sending his clients in the field the latest information on matters of general scientific interest. He was honest, reliable, and endowed with both common sense and practical knowledge, qualities indispensable to Wallace and Bates. He requested a commission of 20 percent plus 5 percent for insurance and freight. Each insect would bring in 4 pence, leaving 3 pence for the collector—a pittance, but the going rate at the time.[31] There was no question that quantity as well as quality was the key to making a profit; this meant not hundreds but thousands of specimens.

Stevens was doubtless instrumental in educating Wallace and Bates in the methods of collecting and preserving specimens, though Swainson had included several chapters on preservation techniques in *A Treatise on the Geography and Classification of Animals.* Stevens later wrote a treatise on the subject. As a natural history agent, he was the perfect educator, with one eye on science and the other on the bank. His advice was pragmatic. Echoing Humboldt, he advised his readers that every specimen, "dry or in spirit," should have a number attached to it corresponding to a number in the collector's notebook. The notebook should also contain information regarding the locale and the season in which the specimen was found, as well as its habits, habitat, and local name. The importance of a field notebook could not be overemphasized. The notebooks of Humboldt and Darwin were invaluable for their later work and the work of their fellow scientists. Meticulous records were essential for anyone hoping to make lasting contributions to science, whether in the physical sciences or natural history.[32]

The basic equipment of the collector included knives, scissors, scalpels, pliers, a large assortment of pins of various sizes, needles, a hammer, a small hatchet, cotton, paper, a folding net, a hoop net, a water net, forceps, a digger, glass phials, large and small packing cases, and a great number of pillboxes, all of which could be purchased from Stevens's shop on Bloomsbury Street near the British Museum. Stevens also recommended having a good supply of arsenic soap, which was used as a pesticide to protect killed specimens against scavenging insects and putrefaction. Since the soap had to be made up in the field, he provided the recipe. Before 1750, salt and alum had been used to preserve bird and mammal skins, but after a few years skins thus "preserved" fell apart. In the 1750s, a Frenchman named Jean-Baptiste Becoeur invented arsenic soap, a milestone in the history of preservation and an advance responsible for the existence of vast modern bird and mammal collections.[33] Although a small quantity may be absorbed through the skin, arsenic soap is not particularly toxic, which was fortunate for nineteenth-century naturalists since every specimen had to be washed thoroughly by hand.

In the field, freshly killed mammals and birds had to be treated within a few hours of death. Once putrefaction set in, the feathers of birds fell off. Before skinning an animal, its dimensions and the color of its eyes and soft parts had to be noted because, once dead, its colors faded quickly and shrinkage after skinning and preservation prevented accurate measurements of body size. Smaller mammals could be preserved in gin, or a "proof spirit" half diluted with water, after a slit was made in the underside. Larger mammals were skinned by making an incision in a straight line from the anus to the throat, with the skin then carefully detached with a knife. The skull and skeleton were left intact, but the eyes, brain, and tongue were scooped out and disposed of, as were the muscle tissue and internal organs. All adherent fat was scraped away; if any remained, the skin was strewn with powdered tan, made from willow or oak bark, or another potent astringent that dissolved fat without penetrating other tissues, before the application of the arsenic soap. The ears, lips, and feet of the larger mammals were doused with turpentine to accelerate drying and destroy potentially destructive insects. When completely dry, the skin was rolled up, hair innermost, beginning with the head. To prevent damage from abrasion, dried grass or moss was inserted during the rolling process. The skin had to be unrolled periodically and checked for moisture. If possible, it was further exposed to the sun and sprinkled once again with turpentine. If insects were detected, strong tobacco or, better, aromatic spices were added.[34]

Birds were treated in a similar fashion, but the incisions had to be as small as possible and in the least visible parts. Any blood was immediately wiped away to prevent staining of the plumage. A blunter instrument, like a wooden style, which was similar to a surgical probe, was preferable to a knife in separating the skin and feathers from the skeleton. The only bone left in place was the os coccyx (tailbone); without it, the tail feathers were likely to drop out. In the smaller birds, a fragment of the skull was cut away to extract the brain and eyes; in the larger birds, the brain and eyes were removed from the skull through multiple incisions in the back of the neck, the skin of which was then stitched back together. Stevens recommended that, if at all possible, a second specimen of the same species with its internal organs intact be preserved in spirits for the benefit of the taxonomist. His advice covered all faunal groups—insects, arachnids, crustaceans, sponges, starfish, land shells, freshwater and marine shells, reptiles, and fish—and was indispensable for Wallace and Bates.

The final hurdle that the two would-be explorers had to overcome was the acquisition of passports, which were dispensed by the Foreign Office only through an official recommendation. That recommendation came from Sir William Hooker, the curator of the Royal Botanic Gardens at Kew (and father of Darwin's friend Joseph Hooker). Hooker was impressed with the enthusiasm and competence of the two young men, whom he had met during their visits to the conservatory and who had promised to do everything in their power to obtain interesting specimens for the museum and to communicate any information they thought worthwhile.[35] Hooker wrote each man two recommendations—one for the Foreign Office and the other for the Brazilian authorities. The letter to the Brazilian authorities was meant to confirm their identities and facilitate their progress into the interior of the country.[36]

Wallace left London to spend a few days with the Bates family in Leicester, where he had a brief opportunity to practice shooting and skinning birds while he made last-minute preparations for his departure. From Leicester, he and Bates proceeded to Liverpool, stopping first at the greenhouses of Chatsworth, the quality of whose tropical collections rivaled that of Kew. Chatsworth also possessed the finest orchidarium in the world—the main attraction for Wallace, who had become enchanted with orchids ever since reading an article years earlier by John Lindley. Lindley, who described one species as "too delicate and beautiful for a flower of the earth," gave orchids such a weird and mysterious charm that on the basis of these descriptions alone Wallace longed to go to the tropics, where some of the most exotic specimens flourished.[37] On the evening of April 24, Wallace and Bates arrived at Liverpool harbor after a cold and miserable ride atop the stagecoach from Chatsworth. The next day, they called on a man who had collected butterflies in the Brazilian town of Pernambuco, near Pará, and who regaled them at dinner with stories of the country, the people, and the beauties of nature, as reflected in his impressive collections. The following morning, April 26, they were ready to board their ship, the merchant vessel *Mischief*, and sail for Brazil.[38]

A daguerreotype of Wallace, taken a few months before he sailed and published in his autobiography, depicts a rather severe-looking young man, seated with hands folded, dressed formally in a black suit and black ascot, his full head of hair parted on the right and swept sideways over the left ear. His clean-shaven face is not unhandsome; his intelligent eyes, with their penetrating gaze, are partly obscured by round, wire-rimmed glasses. Behind this faintly studious mien, however, was a steely strength. By the time of his departure, Wallace was well suited to the rigors of travel in one of the wildest unexplored regions on earth. With no firm commitments to anyone except his mother and sister, adapted to living in the most primitive accommodations of his own society, grounded in the sophisticated science of surveying and mapping, possessed of a foundation of knowledge astonishingly wide for a young man of his era who lacked a formal education, experienced as a botanist and beetle collector, imbued with the deistic and humanistic philosophies of Thomas Paine and Robert Owen, and inspired by the bold adventures and original discoveries of his idols Humboldt and Darwin, at the age of twenty-five Alfred Russel Wallace was ready to make his mark in the world. He had become accustomed to the life of a wanderer, and now he was embarking on a mission that was the stuff of many young men's dreams. Here was the quintessential Victorian explorer—eccentric, independent, intrepid, brilliant, and already harboring the seeds of iconoclasm.

CHAPTER 4

Travels on the Amazon . . .

ON MAY 26, 1848, twenty-nine days after leaving England, the *Mischief*, with its two naturalists aboard, crossed the equator and anchored off the coast of Brazil, six miles from the small village of Salinas, the only port of entry to the vast Amazonian watershed. It had been a rough journey, at least initially. After reaching the Bay of Biscay, just off the southwestern coast of France, the *Mischief* encountered gale-force winds that nearly swamped it. For five days, Wallace was confined to his cabin, laid low by seasickness (a malady he shared with Darwin). Mercifully, the seas calmed and he was able to venture on deck to admire the vistas and ponder the wonders of the Sargasso Sea, teeming with seaweed and marine life. Shortly before the ship's arrival at Salinas, the water had begun to lose its blue color, changing from green to olive to olive-yellow as a consequence of the immense quantity of silt carried thousands of miles out to sea from the interior of South America. Although land was still not visible, the water was too shallow to permit a safe approach. A signal was hoisted for a local pilot to guide the *Mischief*. Pará, its ultimate destination, lay seventy miles inland on the banks of the Rio do Pará, which was not a true mouth of the Amazon—despite what maps at the time indicated—but was connected to it by an intricate network of deep and narrow channels.[1]

The next morning, the vessel carefully maneuvered southward through the Baia de Marajó, propelled by a light breeze and the tide. To the naked eye, the view was unremarkable. Even with the aid of the captain's telescope, the southeastern shore to the left revealed nothing more than an undulating terrain of bare sandhills and scattered trees. But to the right, on the northwestern shore some ten miles away, was a mass of densely packed trees, the front of a forest that clothed the surface of the country for two thousand miles, from the coast to the foot of the Andes, within whose recesses many wonders were waiting to be revealed. The air was humid and the sky overcast; lightning flashed on the horizon. In the evening, the ship glided noiselessly up the Pará River past two fishing villages and many native canoes, which looked like toy boats beneath the lofty walls of the jungle. "And when the sun rose in a cloudless sky," Wallace wrote rapturously about his first morning in the tropics, "the city of Pará, surrounded by dense forest, and overtopped by palms and plantains, greeted our sight, appearing doubly beautiful from the presence of those luxuriant tropical productions in a state of nature, which we had so often admired in the conservatories of Kew and Chatsworth."[2]

As Wallace and Bates awaited clearance from the customs officers, they observed canoes conveying the black and Indian passengers to and from town, while clanging church bells and exploding rockets heralded one of the innumerable festivals. Vultures soared overhead, and swallows crowded steeples and housetops. Pará, a city of fifteen thousand inhabitants, was picturesque: low whitewashed houses with red-tiled roofs and numerous towers and cupolas, all interspersed with clusters of feathery palms. It was the largest city in the Amazon basin, the capital of the least known province of a vast empire, a province equal in extent to all of western Europe.

Their first two weeks in Brazil were a profound disappointment. When they got a little inland, Pará was not quite so picturesque. The handsome but ruinous public buildings, the city squares overgrown with grass and weeds, the slivers of gardens and patches of garbage dumps between houses, and the narrow and "horribly rough" streets depressed them. "The general impression of the city to a person fresh from England is not very favourable," Wallace complained to his boyhood friend George Silk. "There is such a want of neatness and order, such an appearance of neglect and decay, such evidences of apathy and indolence, as to be at first absolutely painful."[3] Bates wrote that although rents were low, people seemed content with lodging "of a quality which would be spurned by paupers in England."[4] To get anything done in Brazil, they would soon discover, one had to cultivate the trait of *paciencia*, so foreign to the restless Anglo-Saxon mentality.

Most disturbing of all was the scarcity of visible wildlife. On their first walk into the forest, Wallace and Bates looked about, expecting to see monkeys as plentiful as at the London Zoo and hummingbirds and parrots in profusion. But for several days, they did not see a single monkey, and the number and beauty of birds and insects fell far short of their expectations. The majority of birds were small and dull-colored, no more exotic than the birds of England, and few had pleasing songs. The only vertebrates they encountered in great numbers were lizards, and then not in the forest but in Pará, camouflaged by the dilapidated stone and mud walls they seemed to favor. Wallace realized too late that travelers crowded into one description all the wonders and novelties they had observed over weeks and months, inadvertently giving their readers an erroneous impression of abundant variety (though animals of all sorts were indeed plentiful when he knew how and where to look for them). He summarized his initial disenchantment in his travel narrative, noting that the weather was not as hot, the people not as strange, and the vegetation not as striking as the picture he had conjured in his imagination.[5]

In part, William Edwards could be blamed for this disenchantment. Having arrived at the end of the wet season (on Edwards's advice), Wallace missed the "glories" of the vegetation. Moreover—and this was his own fault—he mistook secondary growth for primary rain forest, which in the environs of Pará had been hacked and burned long before to make way for human habitation and cultivation. Gradually—like eyes adjusting to the darkness—the beauty and majesty of the rain forest became apparent to him. Even in damaged areas overrun by weeds as commonplace as any back home there were many beautiful flowers and flowering vines, though not in the profusion he had anticipated from the accounts of other travelers. It was the relative paucity of flowers, however, that amazed Wallace. Later he would remark that he had never seen in the dank, shady forests of Brazil anything to match the beauty of the expansive, colorful English countryside, where blooming crabapples and fields of buttercups and daisies abounded. The greater proportion of plants in the Amazon, he learned, had inconspicuous green or white flowers or were like the Melastomas, whose magnificent purple blooms fell to the ground within hours of opening, while the tree remained flowerless for the rest of the year. It was not uncommon for him to travel for a week upriver without seeing a single impressive flowering tree or shrub.[6]

When he finally experienced true virgin equatorial forest several weeks after his arrival, Wallace was astonished; as his analytical powers began to sharpen, his disappointment faded. He wrote that "the observer new to the scene would perhaps be first struck by the varied yet symmetrical trunks, which rise up with perfect straightness to a great height without a branch, and which, being placed at a considerable average distance apart, give an impression similar to that produced by the columns of some enormous building." Overhead, perhaps a hundred feet up, was an unbroken canopy of leaves and branches, dense enough to block out the sun. "There is a weird gloom and a solemn silence, which combine to produce a sense of the vast—

the primeval—almost of the infinite," he continued in a more philosophical vein. "It is a world in which man seems an intruder, and where he feels overwhelmed by the contemplation of the ever-acting forces, which, from the simple elements of the atmosphere, build up the great mass of vegetation which overshadows, and almost seems to oppress the earth."[7] In botanical diversity, the Amazonian forests surpassed those of northern Europe, even if the flowery show did not. Instead of extensive tracts of pine, oak, or beech, one rarely found two trees of the same species in close proximity, with the exception of palms. Despite a painstaking search along a road extending ten miles into the forest through unchanging terrain, Wallace identified only two specimens of the massaranduba—referred to by the locals as the cow tree because its bark produced a copious supply of a substance that to the natives was as pleasant to drink as cow's milk but that to Wallace and Bates was rank and thick as glue. Such a phenomenon was difficult to explain according to the creationist doctrine of geographic distribution and speciation, but it was a tantalizing fact that Wallace recorded in an appendix to his travel narrative, though without proposing any alternative hypothesis.[8]

Wallace and Bates spent the first few weeks at the country home of Daniel Miller, the British vice consul in Pará. After many inquiries, they found a house of their own in the nearby village of Nazaré. What convinced them to take this house was its proximity to virgin forest. On their initial inspection, they approached Nazaré by a circuitous jungle route three miles in length. Beyond the swampy suburbs of Pará lay the lofty forest, with its hundred-foot trees whose trunks were enveloped in the dense greenery of clambering plants. Creepers and climbers hung in long festoons, sometimes curling and twisting on the ground like monstrous serpents. The forest floor was a tangled mass of bushes and shrubs, brightened by the rare flower and numerous brilliantly colored butterflies. Repeatedly they stopped "to examine and admire." As they forged ahead, the landscape abruptly changed: the gloom and shade of the forest retreated, and they could feel the powerful sun beating down on their heads from a cloudless sky. The ground was warm beneath their feet, and the intense tropical heat silenced the cries of birds and animals. Then they entered forest terrain once again, the cool air and the moldy smell of rotting vegetation a welcome balm for the senses. Soon the track was swampy and difficult to negotiate. Wild banana palms with sword-like blades some eight feet long obstructed their path, along with other exotic trees and dense shrubs. The mucky terrain alarmed them; with each step, they knew that they might tread on a "venomous reptile." Despite the hazards, they would spend a lot of time in this part of the forest, which became one of their best hunting grounds for birds and insects.

The house they rented in Nazaré consisted of four rooms of moderate size, with a roofed veranda extending around it that provided shade from the punishing afternoon sun and an escape from the sweltering interior. They worked and ate their meals on the veranda; at night they slept inside, native style, in comfortable hammocks. The cultivated grounds contained orange, banana, and other fruit trees, as well as small plots of coffee and cassava. On three sides of their property the forest beckoned, with its promise of zoological and botanical rewards. Settling in for a few months' work, they quickly established a schedule. At dawn they drank a cup of coffee and then immediately set off into the forest to search for birds, which were most active at that time of day. While they were thus engaged, the servant they had hired headed to Pará to buy provisions, returning in time to prepare their breakfast. Between ten and three o'clock, they reentered the forest, this time dedicating themselves to entomology. As the temperature rose into the nineties and it became too hot for outdoor work, the two men returned home for a midafternoon siesta, like everyone else. In June and July, heavy afternoon showers brought refreshing coolness. First there was a dramatic gathering of black clouds and a rush of wind through the trees that tore off leaves and flowers, followed by flashes of lightning and crashing thunder; then came torrents of rain. At four they dined. Then they attended to the preservation of whatever specimens they had collected that day. After tea at seven, they sometimes walked into Pará to observe the Brazilian street life or socialize with the European and American expatriates, who gave frequent parties.

The greatest treasure they captured in those early weeks was a specimen of *Haetera esmeralda* (now *Cithaerias esmeralda*), a butterfly whose wings were transparent except for a bright violet patch on the hind wing. (Samuel Stevens would show it off at the August 6, 1849, meeting of the Entomological Society.) They were less successful in capturing the gigantic blue Morpho butterflies—the jewels in any lepidopterist's cabinet—which floated by frequently, but whose flight tactics were maddeningly evasive. In one of his journals Bates writes, "On Tuesday, collected 46 specimens [of butterflies] of 39 species. On Wednesday, 37 specimens of 33 species, 27 of which are different from those taken the preceding day."[9] At a profit of only 3 pence a specimen, the risk of failure nagged at them constantly.

Although their adventure had begun inauspiciously, the first two months ended with great success. Together they forwarded 3,635 insects packed into a large chest—553 species of butterflies and moths, 450 beetles, and 400 belonging to other orders—in addition to twelve chests of native plants. Among the flora were very few orchids, but Wallace sent 100 dried plants—principally ferns—to Kew. Ferns were "tolerably abundant," he told Sir William Hooker in a letter dated August 20, 1848. "There are many minute species. There may probably be one hundred species altogether found near Pará." He offered to send the base of the stem of an unusual palm he had discovered, as well as living specimens of any plants that Hooker desired. To their agent, Stevens, he wrote that what had most impressed him thus far was the immense diversity of life and the "strange forms and beautiful colours" of the insects and birds, whose numbers were "truly astonishing."[10]

In late August 1848, Wallace and Bates left Nazaré for an expedition up the Tocantíns River. Joining the Pará River forty-five miles southwest of Pará, the Tocantíns was the third largest river in the Amazon system and reached sixteen hundred miles into the interior of Brazil. The vessel they sailed in was called a *vigilinga* (country canoe). It had two masts and was less than thirty feet long. Over the stern, creating a cabin, was a palm-thatched roof that reminded Wallace of a gypsy tent; over the prow was a smaller and lower thatched roof, under which provisions and baggage could be stowed for a journey of up to three months. The *vigilinga* had to be sturdy enough to survive in heavy seas—the Tocantíns, though a river, had "vast sealike expanses of water to traverse."[11] Their captain was an American expatriate named Leavens who managed some rice mills near Pará. An excellent marksman, amateur naturalist, expert taxidermist, and fluent in Portuguese, Leavens was the perfect guide and no doubt helped his passengers to find, capture, and preserve their specimens.

As they moved upriver in their country canoe, they noted the difference between the fauna here and that near Pará, including two new species of butterfly and a different species of sloth, which was caught swimming across the river and stewed for dinner. Difficulty finding a crew prevented them from sailing as far as they would have liked, and after less than three weeks on the river they were forced to turn back. Before returning to Pará, they convinced Leavens to proceed a little farther to some spectacular waterfalls. On their way to and from the falls, they passed through the habitat of the hyacinthine macaw (*Macrocercus hyacinthinus*, now *Anodorhynchus hyacinthus*), or blue macaw, as Wallace referred to it. Three feet long from the tip of its beak to the end of its tail and covered with feathers of a soft indigo blue, it was the largest and one of the most beautiful species of the parrot family—and it was rare, which made it a prize catch for the lucky naturalist. In the morning and evening, Wallace caught sight of the birds flying off in pairs or feeding on the rocklike nuts of the mucuja palm, but he failed to get a single specimen. Beyond the cataracts, a region he would never visit, these macaws were said to be common. The sharply restricted territory of this elegant macaw raised a question: What limited the range of such a strong flier? According to principles laid down by men like William Swainson and Sir Charles Lyell, the macaw should have been found everywhere in the Amazon basin. Why would the Almighty so "exactly" restrict the macaw's geographic distribution? "It appears with the rock," Wallace concluded, "and with this there is no doubt a corresponding change in the fruits on which the birds feed." It was another one of those puzzles that required some scientific explanation, not the invocation of creationism.[12]

Not long after returning to Nazaré, Wallace was incapacitated by an infection of his hand, the result of an accidental gunshot wound he had received when his gun misfired. Advised by a doctor to put his arm in a sling, he was unable to do anything for two weeks—not even pin an insect. Once he had recovered, he assisted Bates in packing their specimens for England. Accompanying this shipment was a letter to Stevens describing their adventure up the Tocantíns, portions of which Stevens published in the *Annals and Magazine of Natural History*, prefaced by an advertisement to collectors proclaiming the two naturalists as "enterprising and deserving young men" who had set out on an expedition to South America to explore "the vast and unexamined regions of the province of Pará."[13] Stevens exhibited the collections at the January 1, 1849, meeting of the Entomological Society, declaring that Wallace and Bates had sent a box containing "many rare and valuable specimens."[14]

The September 1848 shipment, however, would be their last together. After only four months, the partnership abruptly ended in an argument over an unspecified matter. If they had envisioned themselves as a latter-day Alexander von Humboldt and Aimé Bonpland, that dream also ended. The details of the argument were never discussed publicly, and no clues to its exact nature can be found in extant letters or memoirs. The rancor was so deep that Wallace barely mentions Bates in his travel narrative; Bates, for his part, glosses over the incident in his own account. In later years, the episode was denied completely, no doubt to preserve a Victorian veneer of civility. But in a letter to Hooker, Richard Spruce, an explorer who arrived in the Amazon in the summer of 1849, confirmed that a serious disagreement between the two men had occurred.[15] Whatever the cause, whether scientific or personal, it was serious enough that Wallace and Bates avoided each other for more than a year.

Wallace now faced the prospect of exploring an unknown country alone. Someone with less self-assurance would have returned home, but he was committed to his original plan, with or without Bates. Leaving Bates to fend for himself, he temporarily lodged in the *rocinha* (country house) of the Swiss consul, whom he had met in June during one of his "exploratory rambles." There he remained for several weeks, waiting for transportation to the island of Marajó, the place he had selected as his next destination. Marajó was said to harbor rare and unusual waterbirds, which was just what Wallace needed to offset the unforeseen expenses he would incur as an independent traveler.

A delay in his departure to Marajó gave Wallace an opportunity to observe and collect the many small birds that inhabited the region around Pará. Among the most common was the bush shrike, an attractive bird with long silky feathers banded or spotted in black and white that hid in thick and impenetrable vegetation, hopping from branch to branch as it picked off insects. Shotgun in hand, and with the stealth of a native, Wallace crept up within a few yards of the bird, but only after several attempts did he succeed in bagging one instead of blowing it to pieces. The ant thrush was another common bird he had difficulty procuring. Each time he tried to retrieve his downed specimen, he was invariably forced to beat a hasty retreat while powerful-jawed ants, the bird's chief prey, attacked him in ferocious swarms, biting and stinging as they crawled up his arms and legs with a rapidity that amazed him.

But there were revelations that compensated for his troubles. Goatsuckers, swallows, tyrant flycatchers, and jacamars—birds with scarcely any resemblance to one another—all competed for the same food, seizing their prey on the wing. Ibis, spoonbills, and herons foraged side by side, selecting the same food from the shallow water along the beach. Pigeons, parrots, toucans, and chatterers—all fruit-eating birds from distinct and widely separated avian families—often fed on the same tree, vying for the same fruits, while avoiding the fruit of certain trees that no bird fed on. These facts he discovered during his dissections, when he found the same insect parts or the same small crustacea or the same seeds in the stomachs of his specimens. Such findings contradicted the beliefs of most naturalists of the day, who held that the varied forms and structures of birds' bills, for example, were specially created to accommodate the food each species was appointed to eat. "In all works on Natural History, we constantly find details of the marvelous adaptation of animals to their food, their habits, and the localities in which they are found," Wallace remarked five years later as he reflected on these early experiences, "but . . . it must strike every one, that the numbers of birds and insects of different groups, having scarcely any resemblance to each other, which yet feed on the same food and inhabit the same localities, cannot have been so differently constructed and adorned for that purpose alone." Some other principle seemed to be regulating the infinitely varied forms of animal life, Wallace thought, but that principle was still a mystery to him in 1853.[16]

In early November 1848, Wallace finally sailed for Marajó, which was wedged like an enormous stone in the mouth of the Amazon delta, but he returned to Pará in February 1849 with little that was new or rare. Bates, meanwhile, had left for a large estate in Caripi, twenty-three miles northeast, where, having better luck than Wallace, he collected nearly twelve hundred species of insects and numerous birds, reptiles, and shells that local people caught for him. He arrived back in Pará about the same time as Wallace, but the two avoided each other even though they rented neighboring houses in Nazaré. Wallace did better there. Aided by an experienced hunter who was once the slave of the German naturalist Johann Natterer, he acquired a number of novel specimens, including toucans and trogons. Boys from the neighborhood quickly learned that he would pay for all kinds of *bichos*, as his specimens were called by the locals, and nearly every day they brought him snakes, which he preserved in spirits.

By July 1849, Wallace was ready to set out for the heart of the Amazon. Based on information he had obtained from the locals and expatriates, he was determined to go as far as the trading town of Santarém, about five hundred miles upriver at the confluence of the Amazon and the Rio Tapajós. The year in Pará had been an apprenticeship, a time for honing his craft. Now adept at shooting, skinning, and preserving his specimens, and fluent in the lingua franca of the country, he had acquired the skills necessary for making a successful foray into the unexplored interior. But without Bates, he lacked a trusted English-speaking companion. It was perhaps for this reason that he—or perhaps his mother—made arrangements for his younger brother, twenty-one-year-old Herbert, who had been somewhat adrift for the past seven years, to come to Brazil as his assistant. Eager or not, Herbert, who now preferred to be called by his middle name of Edward, agreed and arrived in Pará in June 1849.

Wallace barely knew him, having left Hertford when Edward was only seven and a half. Edward had been educated at home by their father and then sent for a year or two to a cheap boarding school in Essex. The London trunk maker to whom he was apprenticed at the age of fourteen gave him a job as a shop assistant, but he had little interest in making or selling trunks and he lasted for only a year. He then moved to Neath after William found him employment as a pattern maker at the Neath Abbey Iron Works, a job that he found equally deplorable but endured for the next four years until being invited to Brazil. Edward had no mechanical aptitude or inclination for the kinds of work available in the nineteenth century to a young man without skills, advanced education, or money. He was a self-styled intellectual, with a love of poetry and a distaste for manual labor. His dream was to be a poet; indeed, his talents were recognized by local newspapers, which published his verses. But a poet could not survive on verses alone. Edward's misfortune, Wallace wrote in his autobiography, "was that he . . . was not possessed of sufficient energy to overcome these deficiencies of nature and nurture."[17]

After Edward's arrival in Pará, Wallace could not leave at once. One of the items he awaited from England was a letter from the Foreign Office in response to a request he had made in May to facilitate his journey into the interior. Johannes Baptist von Spix and Carl Friedrich von Martius had penetrated the interior with just such government endorsements, he reminded Stevens, his agent. Stevens wrote a letter to the Foreign Office stating that Wallace had sent "several valuable consignments containing numbers of novelties, especially in the Insect tribe, portions of which have been purchased by Mr. Gray for our National Collection at the British Museum." With backing from John Edward Gray, the director of the museum, Stevens appealed to the Foreign Office for assistance, and a memorandum was finally drafted under the direction of Viscount (Lord) Palmerston, the foreign secretary, instructing the British consulate in Pará to obtain the necessary documents for Wallace from the Brazilian authorities.[18]

Anxious to leave Pará for Santarém before the onset of the rains and already vexed by the unnecessary delay, Wallace decided to take the chance that the papers would be awaiting him at Santarém or farther upriver. He and Edward departed in early August, a month later than he had planned. Their craft, a small sailing vessel with just enough room for their equipment and food stores, reeked of rotting fish and decaying animal hides. The deck was uneven and leaky, and water drenched their clothes and hammocks. But they soon adapted to the inconveniences and unpleasantness of river travel in a primitive country, passing the time with books borrowed from their English and American friends in Pará.

For the first ten days, the Wallace brothers followed a labyrinth of narrow channels, a slow and tedious journey dependent on the fluxes of the tide (no wind ever filled the sails), before joining the swifter waters of the mighty Amazon. It was the first time that Wallace had actually seen the main body of the river itself, and its immensity awed him. The distance from shore to shore was so great that the river's banks were often invisible. He let his imagination wander to the sources of the Amazon in the distant Andes, to the ancient Incas who once reigned over half the continent, to the silver-laden mountains of Potosí, to the gold-seeking Spaniards and wild Indian inhabitants of Ecuador and Peru. "What a grand idea it was to think that we now saw the accumulated waters of a course of three thousand miles," Wallace wrote in a moment of poetic abandon, "that all the streams that for a length of twelve hundred miles drained from the snow-clad Andes and were here congregated in the wide extent of ochre-coloured water spread out before us!"[19]

In 1849 the general course of the river, its important bends and main tributaries, were accurately laid down, a fact Wallace noted in the appendix of his travel narrative. Its extent, in a straight line from east to west, was approximately 2,050 miles; from north to south, its tributaries covered another 1,720 miles. From four degrees north latitude to twenty degrees south, every stream flowing down from the Andes emptied into the Amazon. To convey its massive dimensions, he imagined uniting every river from Saint Petersburg to Madrid into a single mighty flood. But he discovered that the details of the Amazon's course were incorrect. The numerous islands and parallel channels, the deep bays and great lakes, and the varying widths of the river were unknown. Geographers at the time even disputed whether the Pará River, on which Wallace had set sail, was or was not a tributary of the Amazon. Wallace therefore set out to survey as much of the Amazon as he could while slowly making his way to Santarém.

Over the next two and a half weeks, thunderstorms and violent squalls pushed them swiftly toward their destination. The Amazon was a vast, silt-choked expanse, littered with floating islands of vegetation detached from the shoreline and enormous tree trunks ripped up by the roots and driven by the currents like twigs in a stream, sometimes carrying with them birds and animals. Flocks of parrots, macaws, herons, and ducks ascended in bursts of color from the thick forests along the banks into the gray, cloudy skies. Twenty-eight days and 474 miles after leaving Pará, Wallace and Edward arrived in Santarém, at the mouth of the blue, transparent waters of the Rio Tapajós.

Life moved at a languorous pace in Santarém. In the evening, the local potentates and world travelers assembled at the house of Captain Hislop, an eccentric Scotsman who had lived on the Amazon for forty-five years, where they smoked, took snuff, and talked politics, science, and philosophy well into the night. Wildlife around Santarém was plentiful, and a few excursions shortly after their arrival were highly productive. On September 12, Wallace wrote another letter to Stevens in which he complained that the main difficulty was getting men to accompany him short distances and mentioned that he had not yet received any papers from the Foreign Office. He also wanted some guidance in choosing future destinations. Monte Alegre, teeming with beetles in its thousand-foot-high hills, was to be his next stop, but after that he was open to suggestions. He could go up the Rio Negro, which branched off from the main body of the Amazon at the town of Barra (Manaus), toward the sources of the Orinoco—though he feared that the northernmost region of Brazil was poor in coleoptera—or he could go up the Rio Madeira, another major tributary of the Amazon, into Bolivia. In a letter written two days later, Wallace pleaded: "Pray write whenever you can, and give me all the information you may be able to obtain, both as to what things are wanted in any class or order and as to localities."[20]

In mid-September, Wallace and Edward sailed to Monte Alegre, a town several miles east of Santarém on the northern bank of the Amazon. Santarém was on the southern bank, and from a natural history perspective Wallace would soon discover that the distinction was important. Monte Alegre had been built on a hill one-quarter of a mile from a tributary and was reached by an arduous walk through deep sand. Cacti twenty to thirty feet tall and branching like candelabra grew in a desert landscape that strikingly contrasted with the humid opulence of the rain forest everywhere else along the Amazon. In the distance and not easily accessible were the lofty hills, the home of beetles. But the best collecting grounds for insects, he found, were closer to town, in the shady groves on rocky slopes above the river, where numerous springs gushed and ferns, mosses, and creeping plants grew on the wet rocks. The trip was marred when Edward sprained his leg, which swelled and abscessed above the knee, incapacitating him for two weeks.

While Wallace and Edward languished in Monte Alegre, Bates was making his way upriver. He passed the village in early October, unaware that the

Wallaces were so close by. He arrived in Santarém on October 9, but remained only briefly before proceeding to Obydos, fifty miles upstream. A few days later, Wallace and Edward returned to Santarém. On October 27, Richard Spruce disembarked from one of Captain Hislop's vessels and occupied a house near the Wallaces, though he initially kept his distance. The Amazon buzzed with the activity of English naturalists, who seemed to be avoiding one another like bees belonging to competing hives. Spruce had met Wallace earlier in Pará, having sailed to Brazil with an assistant on the same ship that brought Edward from Liverpool. But Wallace was wary of other traveling naturalists following his estrangement from Bates. "I forgot to mention that we have several times seen Mr. Wallace senior—when we arrived he had made up his mind to go to the Rio Negro, but hearing me talk of Monte Alegre put it into his head to go thither and he is preparing to start in a few days," Spruce had written to Hooker on August 3, 1849. "He does not appear to wish for our company either now or at the time when we ourselves propose starting."[21] Between August and October, Wallace's Anglophobia had softened. Perhaps it was the effect of Captain Hislop's wine-soaked evening gatherings, which elevated everyone's spirits and broke down barriers. On one such evening Wallace and Spruce struck up what was to become a lifelong friendship.

Spruce was born in 1817 near the city of York. Although trained to be a schoolmaster like his father, he felt unsuited to the stresses of teaching and decided to devote himself to his great passion, botany. Despite suffering from a number of unspecified but debilitating illnesses, he wandered for several months in the Pyrenees, sometimes at elevations greater than ten thousand feet. He gathered some of the rarest flowers known from that region and a larger number of mosses than he had been led to believe existed, doubling the quantity of species previously reported. His work came to the attention of Sir William Hooker and George Bentham, another botanist at Kew, and he determined—perhaps hearing of the success of Wallace and Bates in Brazil—to undertake the botanical exploration of the Amazon valley with Kew's support. Bentham agreed to act as his agent, identifying his specimens and selling his collections, and Spruce set off for Brazil on the same vessel that carried Edward Wallace.

Spruce was a first-rate botanist, with a keen eye, and became an invaluable resource for Wallace, who later regarded him as one of the greatest but least known naturalists of the nineteenth century. Few have surpassed him in terms of accurate and vivid description, while his accounts of the pharmacological properties of many of the Amazon's indigenous plants anticipated research more than a century later. Spruce, however, lacked the inclination to theorize, a quality his new friend had in abundance. Their missions also differed. Wallace's primary purpose in coming to South America was to find a solution to the most important scientific problem of the time: the origin of species. Spruce's goal was narrower. As an emissary of the Royal Botanic Gardens, he was in Brazil to acquire an intimate knowledge of the flora of the Amazon valley and the Andes. His greatest contribution to the British Empire would be his procurement of seedlings and cuttings from the cinchona tree, the source of quinine and the only treatment at the time for malaria, which he shipped to India, where the Raj established extensive plantations for harvesting the bark containing the life-saving drug.

Spruce's comportment reflected his intellectual acumen. Neatness was a virtue he prized. His clothes were neat; his handwriting was impeccable; and his writing materials, books, herbaria, microscope, stores of food, and clothing all had their appointed places. Meticulous and exacting, he was a model scientist. But he had other virtues that cemented the friendship between the two men. Like Wallace, he was fond of literature and was an "advanced Liberal" in politics, empathizing with the struggles of the working class. "Nothing more excited his personal indignation," Wallace reported, "than to hear of the petty, but cruel, persecutions to which they [the working classes] are often subjected."[22]

Spruce enjoyed Wallace's company, especially the "animated and thoughtful conversation in the evenings," which would not last long, as both found it difficult to keep their eyes open after eight o'clock, fatigued from their long investigations during the day.[23] Wallace acted as guide, acquainting Spruce with the paths across the *campo* (savanna). Wallace told Spruce about the famed Amazon water lily (*Victoria amazonica*), which he had seen near Monte Alegre, and produced a fragment of its leaf; the whole was more than four feet in diameter. This plant, which grew in Amazonian lakes and side channels but not in the main river, was still a novelty in 1849, though a specimen had been successfully cultivated at Chatsworth in 1848; a famous photograph of the time shows the daughter of Chatsworth's ingenious gardener standing on a colossal leafy platform in the middle of a pond. From the native Indians, Spruce had heard of a water plant near Santarém that they nicknamed *forno* (oven) because of its resemblance to the circular oven used for baking *farinha*, the chief item of the native diet. To confirm that *forno* and *V. amazonica* were one and the same, Wallace and Spruce took a boat across the Amazon to a large island. After walking two miles, they found a patch of these enormous water lilies, growing in barely two feet of water, with deep, tenacious roots and the classic floating leaves.[24]

The geography and vegetation of Santarém and Monte Alegre were unique to Amazonia. Instead of dense, impenetrable forest, the landscape was open, punctuated by rocky hills dotted with shrubs and stunted trees. Insects—especially butterflies—differed from almost every species Wallace had found near Pará. Spruce was similarly puzzled by the marked difference between the flora of Santarém and that of the rest of the Amazon. Nearly every plant he collected was a species new to him, though he had examined numerous specimens in the Kew herbarium before leaving England. It was in Santarém that Wallace first obtained direct evidence that a great river could limit the range of a species, a fact that was to influence his thinking about the relationship of geographic distribution and evolution. The phenomenon of an unexpected restriction of habitat was not limited to the blue macaw. On one side of the Amazon, for instance, he found *Callithea sapphirina*, a butterfly with wings of a velvety sky-blue; on the opposite bank was the closely allied species *C. leprieurii*, indigo-hued and with different markings on its underside. Neither species ever seemed to commingle with its sister species across the water. Several months would pass before he truly grasped the significance of this finding, for he continued to send home specimens to England without clearly denoting on which side of the river they had been captured.

In *A Narrative of Travels on the Amazon and Rio Negro*, written four years later, he speculated on this interesting finding. Places no more than fifty or a hundred miles apart often had distinctly different species of insects and birds. It was understandable, he said, that Europe and North America or tropical America and equatorial Africa—continents widely separated by seas—might have few species in common. But in all parts of the world could be found animals distributed among smaller local groups, with almost every district having peculiar species found nowhere else. Great mountain chains might serve as barriers to the mingling of species; for example, the Rocky Mountains and the Andes separate two distinct climatic and zoological zones. But what if no such obvious barriers existed? He reasoned that there must be many other kinds of boundaries that limited the range of animals despite similarities of climate and soil type. Small rivers rarely prevented animals from passing, especially birds and insects. Very large rivers, like the Amazon and the Tocantíns, however, seemed to affect the range of animals of many orders. And it was this natural cause, not creationism, that he later believed accounted for the separation of two species of butterflies. Only after discovering that the Amazon and its larger tributaries served as barriers to the ranges of various orders of animals of all classes—especially birds, mammals, and insects—did he begin to keep an accurate record of the precise geographic location of every specimen he collected.[25] As meticulous as other naturalists had been, few had ever been so precise, and virtually no one (as far as he knew) comprehended the contribution of such isolating mechanisms to the formation of species.

CHAPTER 5

. . . *And the Rio Negro*

THE RAINS CAME EARLY to the Amazon in 1849, roughly two months ahead of schedule. It was therefore with much anxiety that the Wallace brothers pushed off in late November for Barra. On the final leg of their journey, the winds mysteriously vanished and they made slow progress, while they suffered day and night from either drenching rain or swarming mosquitoes. Christmas was dismal, celebrated aboard their canoe with a *farinha* pudding, fish, and coffee. Wallace records the following in *A Narrative of Travels on the Amazon and Rio Negro*: "[While] eating . . . our thoughts turned to our distant home, and to dear friends who at their more luxurious tables would think of us far away upon the Amazon."[1]

They arrived in Barra on the morning of December 31, their spirits improved by the sudden change in the color of the water: the yellow Amazon had given way to the inky black Rio Negro. The Rio Negro differed strikingly from the Amazon in other ways. There were no floating islands or logs and uprooted trees "with their cargoes of gulls."[2] In fact, there was scarcely any sign of life on the stygian, sluggish waters. After disembarking, they presented letters to Senhor Henrique Antony, an Italian gentleman who was the chief merchant and leading dignitary of the city. For more than forty years, Senhor Henrique, as he liked to be called, had been "the travelers' friend at Barra." He was beloved by foreigners and locals alike. Richard Spruce later honored him by naming "the finest new genus of plants" that he found on the Rio Negro *Henriquezia*. One species, *H. verticillata*, Spruce described as "a noble tree of eighty to a hundred feet high, having its branches and leaves in whorls, and bearing a profusion of magnificent foxglove-like flowers."[3] Senhor Henrique offered the young men the use of two large rooms in his house and extended an invitation to share his dinner. Wallace's initial impression of Barra, a city of five or six thousand people mostly of mixed Portuguese and Indian descent, was unfavorable. It was little more than a third-rate trading post, importing European cotton goods of inferior quality and quantities of cheap cutlery, beads, mirrors, and other trinkets that the citizens traded with the Indians for Brazil nuts, *salsasparilha* (a native plant whose roots were believed to have medicinal properties), and fish. The town lacked any amusements, "unless drinking and gambling on a small scale can be so considered." Most people never opened a book or had any "mental occupation." On Sundays, everyone made the social rounds after Mass to review the various scandals of the week, which were too outrageous for him to detail in his travel journal. He judged that morality was at the lowest ebb possible in any civilized community.[4]

In Barra, Wallace found the long-awaited documents from the British Foreign Office. He therefore made plans to leave as quickly as possible to search for the "celebrated umbrella chatterers" (*Cephalopterus ornatus*), a bird species that was said to be most abundant during the wet season. He set off for three days without his brother on a journey up the Rio Negro to the bird's reputed haunts. Within days, he quickly acquired twenty-five specimens of the umbrella bird, in addition to other species endemic to that part of the river basin. Without its striking crest and neck plume, Wallace later mused, the male umbrella bird, averaging eighteen inches in length, would in size and overall color be nothing more than a crow. But its crest, composed of long, slender feathers that arose from the contractile skin of the head, was in his opinion the most fully developed and beautiful of any bird known. The shafts of the feathers were white, while the plumes, which curved outward at the tip, were glossy blue and hairlike. When erect, the crest formed an elongated shiny blue dome of radiating feathers—hence the umbrella appellation—reaching below the beak, which was obscured by the elegant ornamental mass. The other unusual feature was a long cylindrical plume of scaly feathers that hung from the neck and occasionally puffed out in an impressive display. The Indians called the bird *ueramimbé* (trumpet bird) because its loud, hoarse cry resembled the sound of a deep, resonant musical instrument.[5]

The habits and geographic distribution of this unusual bird were the subject of the first article that Wallace published in a scientific journal, appearing in the *Proceedings of the Zoological Society of London* in 1850. It already contained the most characteristic elements of his writing style, which was straightforward and economical, full of vivid and accurate description, and marked by the self-confidence and ease of an expert naturalist. This earliest publication lacked only the tendency to speculate, which would creep into and leave an unmistakable stamp on his later writing. Wallace had now begun to pay particular attention to geographic distribution, unlike most other naturalists at the time. This species, he said, was confined predominantly to the Rio Negro, where it was most commonly found in the numerous islands of the river. He did not believe that it occurred in any of the other great tributaries of the Amazon, though a hunter had informed him that an allied species—a white umbrella bird—inhabited the forests of the upper Uaupés, a tributary of the Rio Negro. It was his intention, he announced to his readers, to verify this hunter's report.[6]

When Wallace returned to Barra, it was raining daily, sometimes incessantly, making the collecting and preservation of specimens almost impossible. Feathers and hair dropped out; insects grew moldy. The dreariness was improved by a little society of German, English, and American expatriates, all hosted by the ebullient Senhor Henrique. Henry Bates had arrived in Wallace's absence, and the two men put aside their differences. In the brief intervals between showers, Bates and the Wallace brothers made forays into the neighboring forest and flooded lowlands. These rambles filled in the time for two or three months, but Wallace longed to escape. For him, the rains brought an "unexciting monotony" and created an aura of "desolation and death." Bates, less restless than Wallace, enjoyed the noiseless solitude of the place, its astonishing diversity, and the luxuriance of trees, foliage, and flowers.[7]

In February or early March 1850, an English bird collector arrived in Barra with a young Indian assistant who was expert in the use of the *zarabatana* (blowpipe), the principal weapon of many of the indigenous tribes. Nine or ten feet long, the blowpipe was artfully constructed from the stem of the palm *Iriartea setigera*, which varied in diameter from the thickness of a finger to two inches. After these stems were carefully dried, the soft pith was bored out and the bore polished to near-perfect smoothness. Its craftsman then fitted a conical mouthpiece at one end, tightly wrapped the pipe with long, flat strips of the Jacitára palm, and smeared the whole with black beeswax. Darts were fashioned from the hard rind of palm-leaf stalks, thin strips being cut and made sharp as needles by scraping the ends with a knife or an animal's tooth. They were then steeped in a freshly made poison the natives called *urari* (curare), concocted from the bark of the vine *Strychnos toxifera*.[8] To aid in its trajectory, each dart was furnished with a little oval tuft of a silky material from the seeds of the silk-cotton tree (*Eriodendron samaüma*). In an inexperienced man's hands, this contraption was unwieldy, but adult Indians in the upper Amazon used the blowpipe with remarkable adeptness and could take deadly aim at a bird or another animal fifty or sixty yards away. It was far more useful than a gun, bringing down an animal in silence without scattering the troop or flock, and far less destructive. The Indian assistant gave Wallace and Bates lessons in the use of this instrument, but they gave it up in despair well before they could attain mastery and reverted to their noisy and ruinous guns. Of the blowpipe Bates concluded, "None but the stealthy Indian can use it effectively."[9]

In late March, Wallace and Bates once more parted company—this time amicably—and would not meet again for another twelve years. To minimize the duplication of specimens—and perhaps to temper any perception of rivalry—the two men agreed to explore different parts of the Amazon River basin. Bates thus left on March 26 for Ega, a village four hundred miles west of Barra on the Solimões, the main body of the Amazon upriver from the Rio Negro to the Peruvian border. Wallace planned to head up the Rio Negro once a boat arrived to transport him, leaving Edward behind in Barra. After a year in Amazonia, Edward knew that he was not suited for the life of a naturalist. Having little interest in birds and insects, he accompanied his older brother without enthusiasm, a necessary quality to succeed as a collector. He did not share Wallace's passion for discovering the origin of species. His main passion was poetry, and at every opportunity he put his thoughts and experiences into verse. Seasickness and "Forests high and gloomy / Where never a ray of sun / Can pierce its way to enter / Those shades so thick and dun" were not worth enduring, he wrote, even though "Lasses darkly delicate / With eyes that ever kill" breathed to him "in whispers / That we are in Brazil."[10]

At the end of August, the boat finally arrived from Pará, carrying money from the sale of Wallace's specimens and a large parcel of letters that included some from home and from his brother John in California, some dating back more than a year. Wallace spent most of the night responding to the more important letters and packing up his collections. Edward planned to return to Pará in six months, when a vessel would be making the trip downriver, and from there sail to England, where he hoped to launch a literary career. On August 31, Wallace left his younger brother to contemplate the "killing eyes" of the senhoras and set off on the most ambitious journey of his career.

For the next five months, Wallace traveled up the Rio Negro in a vessel that seemed like a yacht compared with the precarious crafts he had become accustomed to on the Amazon. Measuring thirty-five feet long and seven feet wide, it was sturdy and spacious. There were two areas covered with semicircular roofs, called *toldas*, one fore and the other aft. Indian paddlers stood at the bow, rowing with long-poled oars. The captain, Senhor João Antonio de Lima, a Portuguese trader who had for years resided on the upper Rio Negro, was well prepared to deal with its "semi-civilised and savage" inhabitants. As gifts or for trade, he had brought bales of fabric—coarse cotton cloth, flimsy but brilliantly colored prints, checked and striped cottons, and the commonest calico—and a quantity of blue and red handkerchiefs. He also had brought axes and cutlasses and knives "in profusion," thousands of fishhooks, shot, beads, buttons, and magnifying glasses—everything the indigenous population wanted or thought it needed. There was plenty of *caxaça* (native rum) and wine for Lima's own use, as well as cooking supplies sufficient to last a small family for six months, for Lima's wife, children, and an elderly servant named Old Jeronymo accompanied him.

They traveled at a steady rate, which contrasted pleasantly with the unpredictable and anxiety-provoking pace of Wallace's earlier journeys. Lima, a "grizzly" man of average build whose face reminded Wallace of a portrait of some banished lord displayed in the National Gallery of London, was a good companion. Fairly well educated, he was a great raconteur, with an inexhaustible fund of anecdotes from his early life in Portugal; he enjoyed entertaining his English passenger with stories "of the grossest kind," which Wallace ranked with those of Rabelais or Chaucer. At daybreak they would have coffee and biscuits and then stop again for a breakfast of chicken or fresh fish in the late morning. Along the way, they visited desolate and half-deserted villages, passing the time with one of Lima's old friends or loading and unloading cargo. Now and then, they bought live fowl, eggs, bananas, and oranges from the owner of a solitary hut in the middle of nowhere. At six o'-clock, they had dinner and then sipped coffee on top of a *tolda* while watching goatsuckers and hummingbirds dart into and out of the forest in search of insects and listening to the twilight sounds of chanting tree frogs and the cries of howler monkeys. It was the most pleasant hour, and this stretch of the river was mercifully free of mosquitoes. Two hours later, they usually moored in a place where they could hang their hammocks and sleep until four or five the next morning. In mid-October, they reached the major falls of the Rio Negro, and for three days they traversed treacherous rapids or portaged around them, keeping one step ahead of death and destruction in the powerful eddies, whirlpools, and foaming breakers. The last set of rapids gave way to the smooth waters of the "great and unknown river" Uaupés.

At the village of Nossa Senhora de Guía, Senhor Lima's home, Wallace was given an unoccupied house, where he deposited his boxes and hung up his hammock for a brief stay. While the crew reveled day and night with their families in the village, he set out into the surrounding woods to shoot birds and catch insects. Lima dispatched a couple of unenthusiastic Indians to hunt for birds with their blowpipes and poisoned darts, but they returned empty-handed or with birds that Wallace dismissed as common and worthless, which anyone could have found around the huts of every village along the Rio Negro. He had no better luck augmenting his insect collection. Instead, he focused his efforts on the river, where new species of fish were hauled in every day. Amazed by their variety and the strangeness of their forms, he drew and described each one in great detail before turning them over to the cook. However, one species was too tiny to turn over to the cook. *Vandellia cirrhosa* attached itself to other fish and aquatic animals, feeding off the blood of its host. "This minute fish," he wrote in his notebook, "enters the urinary passage of men and women, wounds and extracts blood within, and all efforts to extract it are usually unavailing. Effusion of blood, inflammation, and death have in several instances occurred."[11] Another, the electric eel (*Gymnotus electricus*), was not highly esteemed as food because of the number of forked or branched bones in every part of its body. It belonged to a strange family or subfamily of fish with unique features—an elongated form lacking dorsal and ventral fins but having a long posterior fin and an anus located in the anterior, an anatomical peculiarity that occurred as a result of the intestine passing forward instead of backward. In one long-snouted species, the anus budded in front of the eyes.[12]

From Nossa Senhora de Guía, Wallace headed up a small stream to the haunts of the cock-of-the-rock (*Rupicola crocea*), which the English naturalist Charles Waterton had described as an "elegant crested bird . . . of gloomy disposition."[13] Accompanied by two Indian hunters and traveling in a canoe provided by Lima, Wallace set off into the wilderness toward the lonely granite mountain that the birds were known to inhabit. Thick virgin rain forest soon gave way to flat bush and scrub normally flooded in the rainy season but now sandy and almost open. The stream, winding in sharp curves, flowed with a rapid current, making progress tedious and difficult. After a few days of paddling through this dreary landscape, they once again entered luxuriant rain forest. Arriving at a village, he set up his workshop in one of its half dozen ramshackle and windowless huts, whose earthen floor and thatch roof still smelled of smoke. The village's inhabitants lived in squalor and spoke no Portuguese. Using the few words of the local language he knew, he managed to hire some boys to shoot birds with blowpipes. Others with remarkably good eyesight would creep along by his side and point out small animals before he could see them himself or bring back some specimen, like a minute hummingbird, which he never would have found in the dense thicket.

The habitat of the cock-of-the-rock was ten to twelve miles from the village; what was most striking to Wallace was the species's limited range, "an example of a bird having its range defined by a geological formation and by the physical character of the country," he noted in his travel narrative.[14] Amid the damp boulders of the granite mountain, male birds gathered in groups of forty or fifty, producing a fantastic spectacle as they displayed to attract the dark, dusky-brown females. Characterized by an intense orange color, with a narrow maroon band along its crest feathers and a partly concealed black tail, the male cock-of-the rock was among the most beautiful birds of South America, one highly coveted by collectors in England.[15]

After promising a good payment for every captured *gallo*, as the bird was called by the inhabitants, Wallace persuaded almost the entire male population of the village to accompany him. Loaded with the barest necessities and forgoing his one luxury, coffee, he proceeded with his half-savage entourage through the dense jungle, each man carrying a blowpipe, bow and arrows, reed mat, and supply of *farinha*. He provided the salt and put his faith in the bounty of the forest for the rest. The path, initially easy to negotiate, soon became a tangle of creepers, vines, and thick bush. Hooked spines caught his shirtsleeves, while the naked Indians walked on with ease; he mused that they no doubt regarded him "as a good illustration of the uselessness and bad consequences of wearing clothes upon a forest journey."[16] After many hours of hard walking, they set up camp at the foot of the small mountain, which rose to a height of several hundred feet. A few of the men pursued a herd of wild pigs and staggered back to camp with one fat specimen, which provided a feast for several days.

The true geographic features of the granite mountain were deceptive: from a distance, it appeared to be a smooth, forest-covered hill of gradual slope; up close, it was a nearly perpendicular mass of granite of "extraordinary jaggedness," with great caverns and ridges forming deep sloping gutters or vertical channels with sharp edges, which Wallace discovered to his dismay the following day when he took a path as far as possible up the mountain. Solitary rocks varying in size from a wheelbarrow to a house obstructed his way. After struggling over these boulders for hours, staring over "fearful ridges," and making hazardous descents into gloomy chasms, his reward was a single bird. But what a specimen! "I caught a glimpse of the magnificent bird sitting amidst the gloom," he later wrote, "shining out like a mass of brilliant flame." At his appearance the bird flew off in alarm, but Wallace brought it down with his shotgun. He returned to the campsite and skinned his prize before sunset. Over the next four days, he obtained a dozen more specimens, two of which he personally shot—in one case waiting patiently for hours under a tree in driving rain until the "brilliant flame" flew within firing range.[17]

Wallace stayed in the remote Indian village for two more weeks before rejoining Lima's boat. In a letter home, he outlined his plans for the next part of his journey. No longer indecisive or encumbered by Edward or anyone else, he had a clear idea of his objectives:

> My canoe is now getting ready for a further journey up to near the sources of the Rio Negro in Venezuela, where I have reason to believe I shall find insects more plentiful, and at least as many birds as here. On my return from there I shall take a voyage up the great river Uaupés, and another up the Isanna, not so much for my collections, which I do not expect to be very profitable there, but because I am so much interested in the country and the people that I am determined to see and know more of it and them than any other European traveler. If I do not get profit, I hope at least to get some credit as an industrious and persevering traveler.[18]

At the end of January 1851, after some delay, Wallace gathered together his bulky baggage and headed northwest with four Indians in one of Lima's canoes. What impressed him along this stretch of the Rio Negro was the river's enormous width; in some places it was 15 or 20 miles across and crowded with large, forest-clad islands. For the first 450 miles, dense forest extended back from the water's edge as far as the eye could see. Gradually the features changed as the alluvial floodplain gave way to isolated mountains, rock pillars, and a vast granitic plateau. By the eighth day of his journey, he had reached the border town of San Carlos, the principal Venezuelan village on the Rio Negro and the farthest point that Humboldt had attained coming from the opposite direction. Along the way, he had diligently and accurately surveyed the river with the aid of a thermometer, calculating the elevation of various stages above sea level by measuring the water's boiling point. Unfortunately, he lost that thermometer and a second one broke before he had completed his measurements. Nevertheless, it was with a sense of pride that he disembarked at San Carlos, for he was now following in the footsteps of his illustrious predecessor.

From San Carlos, Wallace continued north on the Rio Negro into Venezuela, to the village of Tómo. In Tómo he was taken to the only European inhabitant, who agreed to guide him to the next village in an *oba*, a craft hewn from a single log and just large enough to hold the two of them plus Wallace's cumbersome luggage. He changed *obas* at this village and replaced his European guide with two Indian ones. The journey now became more treacherous. After floating over a succession of miniature cascades and down a sunken channel whose interminable snaking course made mapping with his remaining instruments almost impossible, they arrived at Pimichín, a village consisting of only two huts. That evening, he had a chance encounter with a jaguar, which ignored him but sent him packing back to his hut with a deeper respect for the hidden hazards of the Amazonian rain forest.[19]

A few days later, an entourage of porters he had hired from a local Indian trader appeared in his camp. One man and eight women carried his gear and supplies—a hundred-pound basket of salt, four baskets of *farinha*, and several other baskets and boxes containing miscellaneous items. His two assistants accompanied them to learn the route and then returned to guide him. It was a distance of only ten miles from Pimichín to his destination, Javíta, but they left in the midafternoon, hours later than he desired. This meant trekking over trackless landscape in the dark. Despite his fear of treading on cold, slithering, venomous snakes or being caught in the jaws of a second jaguar, he plodded onward, vaguely confident that nothing bad would happen to him or his two men. At last they came to a cleared field, but they had to cross at least a dozen streams before reaching his hut; they stumbled into water in the darkness and then thrashed about until they could find a bridge. These bridges—rarely little more than a log as narrow as four inches, with barely a handhold—were of unknown length and height above the water, and the streams below were of uncertain depth. Crossing "under such circumstances," he wrote, "was rather a nervous matter." Twice he lost his balance and fell; luckily, the distance was not far and the water not very deep. Finally, miragelike, the lights of the village appeared. When he entered the camp, he was directed to a shed, where he dined on smoked turtle's eggs, salt fish, and *farinha*, after which repast he fell fast asleep in his hammock. The next morning, he moved to larger quarters in an abandoned convent.

The risky journey, he believed, had been well worth making. It was a great moment, for he had crossed the boundary of the Amazon valley and was among the streams of another great waterway, the Orinoco, where no other naturalist had ever been—a journey that placed him in the company of the great explorer-naturalists of the South American continent. In so remote a district of the Amazon, he felt certain that he could make some important discoveries and was determined to stay for a month. He sent his Indian assistants into the forest to bring back, in his words, "splendid trogons, monkeys, and other curious birds and animals."[20]

But . . . The sagas of explorers are filled with many "buts," except for the expurgated versions so prevalent in the English bookshops of the time. After months of clear skies and near-drought conditions, it began to rain in thick, drenching sheets on the very night when Wallace unloaded his paraphernalia at Javíta and began to dream of the zoological riches he would send back to England and to ponder his attendant fame. Day after day for the next four weeks, it poured intermittently from morning until midafternoon, the tranquil river swelling into a raging torrent. Insects dwindled in numbers. It taxed his ingenuity to discover new methods of drying and preserving his specimens.

Through sheer determination, Wallace amassed a considerable number of butterflies, beetles, and other insect orders, many of which did not inhabit the lower Amazon and Rio Negro and were new to science. During his forty-day stay in Javíta, he procured at least forty new species of butterfly alone. He obtained fewer bird and mammal species, but amassed a good collection of fish. Pressured for time—he could begin his examinations only in the late afternoon, after returning from a forest ramble—and exposed to the torments of sandflies "in swarms of millions," which bit every exposed surface of his skin, he meticulously recorded his observations in his notebook, going over each specimen of fish in painstaking detail, measuring its length and fin sizes, studying its external character from mouth to tail, and making accurate sketches. Resisting speculation about the cause or causes of differences and similarities, he made comparisons within and between species. About one specimen, he wrote that it was "20 to 30 in. long—very closely allied to [another species] but the body thicker, the size larger, the first ray of p[roximal] fin and of d[orsal] smaller in proportion. The p[roximal] spine and d[orsal] fin plates striated—furrowed—the colour as in [an allied species] purplish olive."[21]

Evenings in Javíta were dull, with no one to talk to and nothing to read. He missed the joy of books and the companionship of friends with similar intellectual interests. Occasionally he would attend an Indian *festa,* where everyone drank freely and danced. Yet he was not unhappy. He noted that he was the only white man among two hundred peaceful, "half-wild" inhabitants who, when they were not reveling, were piously occupied in cleaning their churches, streets, and homes. He admired the young girls—"far superior in their graceful forms" to English village girls of the same age, whom they resembled "save in their dusky skin"—and envied the nakedness of the boys, their heads, bodies, and limbs unrestricted by the burdensome clothes of the modern world. The villagers of Javíta lived simply, their only luxury being salt. He reflected on the miseries he had left behind in England, where many who were worse off "in physical and moral health" than the Brazilian Indians longed for nothing but material wealth. While in a state of "excited indignation against civilised life in general," he put his thoughts into verse:

> Rather than live a man like one of these,
> I'd be an Indian here, and live content
> To fish, and hunt, and paddle my canoe,
> And see my children grow, like young wild fawns,
> In health of body and in peace of mind,
> Rich without wealth, and happy without gold![22]

He wrote to Spruce from Javíta that he was enjoying himself "amazingly" in a romantic and unexplored country far beyond the region of the ague.[23] But not long afterward, his two Indian assistants, who felt uncomfortable living among people whose language they did not know, ran away and he was unable to find anyone in the village to work for him. For the next two weeks, he was forced to be his own cook and housemaid. But he had managed to persuade the local boys to catch beetles for him, paying a fishhook for each specimen. Every evening, three or four of them would come into his quarters at the convent with their treasures imprisoned in pieces of bamboo or carefully tied up in leaves. Despite all the wetness, it was a harvest that surpassed his collections from the forests of the Amazon and Rio Negro combined. It was with much regret that he prepared to leave Javíta to rendezvous with Senhor Lima, who had arranged to go with him up the unexplored Uaupés River.

Arriving back in Nossa Senhora de Guía at the end of April 1851, he expected to leave immediately for the Uaupés but was forced to exercise *paciencia* once again as Lima's boat made its way from Barra to meet him.

While he waited, Wallace passed the time fishing and sketching. The number of species of fish he discovered astonished him. From the Rio Negro alone, he identified 160 distinct forms, representing a fraction of the total he believed inhabited the river and its tributaries—and the Rio Negro was only a part of the vast Amazonian watershed. His collections clearly showed that the diversity of life was not limited to terrestrial organisms. Every *igaripé* (path of the canoe), as the Indians referred to the small streams of the Amazon and its tributaries, yielded surprises in apparent violation of the divine law of geographic distribution, which predicted an identity of species in similar habitats. In a letter home, Wallace mentioned three literary goals he hoped to achieve after returning to England from Brazil. The first was to write a book on the fish of the Amazon and Rio Negro. The second was to write a book on the palms of the Amazon valley. The third was to write a book on the physical geography of the Amazon. There was a fourth as well, which he did not mention because it was probably taken for granted: to write a travel narrative in the manner of Darwin and Humboldt.[24]

On June 3, 1851, Wallace finally began his journey up the Uaupés River, accompanied by Lima and two Indian assistants. While Lima hunted down *farinha* and *salsasparilha*, Wallace sought fish, insects, birds, and Indian artifacts. River travel was predictably miserable. As they grasped bushes, creepers, and tree branches to pull the boat along, their canoe was invaded by biting and stinging ants, which crawled over their bodies and under their clothes, becoming entangled in hair and beard. River levels were high and water flooded the banks, making landing difficult and cooking impossible. *Farinha*, water, and a piece of dried fish were often their only daily fare.

Wallace first encountered what he termed the "true [human] denizens of the forest" along this stretch of the river and was as fascinated as Darwin had been during his first meeting with the primitive inhabitants of Tierra del Fuego nearly twenty years earlier.[25] Wallace spent several days among these people, sleeping overnight in a native lodge a hundred feet long, forty feet wide, and thirty feet high, which had been partitioned into apartments for separate families. From the accounts of other travelers and his own observations, he estimated that there were at least thirty different tribes along the Uaupés and its tributaries. He viewed them with the eye of an anthropologist (or ethnologist, as such an investigator identified himself at the time), noting details of their dress, language, comportment, eating habits, and celebrations. Except when in the presence of *brancos* (white men), when they slipped on petticoats, the women went about naked and relatively unadorned. The men, however, presented the most unusual appearance and were different from all the other indigenes he had encountered; it was, he wrote, as if he had been "suddenly transported to another quarter of the globe." Their hair was parted in the middle, combed behind the ears, and tied into a long ponytail that trailed a yard down their backs. On top of their heads, a comb made of palm and grass and ornamented with colorful tufts of toucan feathers obtruded; in their pierced ears were small pieces of straw. This feminine appearance was further enhanced by their habit of wearing large necklaces and bracelets consisting of beads and their custom of plucking the hair of the eyebrows and removing all body hair. With their genitals hidden behind shields, they seemed to Wallace to embody the myth of the famed Amazon women warriors, which he now interpreted as a case of mistaken identity.[26]

He was fascinated by the Indian diet. Along with the usual yams, sweet potatoes, peppers, cassava bread, and various fruits, they regularly ate insects. Six orders were deemed edible: hymenoptera, neuroptera, homoptera, coleoptera, aptera, and annelida (or segmented worms—including earthworms and leeches—now no longer classified as an insect order). One delicacy was the great-headed red ant, which at certain seasons swarmed in the thousands from enormous mounds and created a stir of excitement in the villages; everyone rushed out and gathered as many as they could into baskets and calabashes. The ants were eaten alive: they were grasped by the head, as one might hold a strawberry, and the egg-rich abdomen was bitten off and eaten along with *farinha.* Wallace found that the great-headed red ant was more to his liking when roasted or smoked and sprinkled with salt. Termites, which the natives chased out of their nests with long blades of grass, were also considered a delicacy.[27]

Wallace, in turn, fascinated the natives. Over six feet tall, he towered above his diminutive hosts, who—clothed in nothing but paint, feathers, and ornaments—crowded around him to examine his strange pale skin, thick beard, light hair, blue eyes, and spectacles, which no one had ever seen before and could not understand. A hundred pairs of bright eyes would watch him eat, walk, and talk. It was unnerving, but he carried on as if unconcerned.

At one of these villages, he was afflicted with dysentery and grew alarmed when he could not eat or drink anything without precipitating cramps and bloody diarrhea; after five days, he recovered and pushed on westward to São Jeronymo, a mile below the first and most dangerous of the falls of the Uaupés. The next day, he and a Brazilian man he had met ascended the river to get *farinha.* Crossing the boiling, foaming, eddying river, they entered a small *igaripé*, unloaded their *oba*, and had their supplies carried through the rugged forest while the little boat was portaged around the falls. While he waited for Lima's pilot to appear from an unsavory mission to take Indian slaves, for the next five days he skinned birds, sketched fish, and amassed an extensive collection of orchids. He was punished by myriad *piums*, or tiny buffalo gnats, which covered his feet and hands with swollen and inflamed purplish bites and displaced as a source of misery the *chigoes*, a type of flea that burrowed under his toenails and swelled with eggs and blood. At night a vampire bat (*Phyllostoma hastatum*) might enter through any aperture it could find in his sleeping quarters and attack him. Vampire bats were common on the Amazon and its tributaries, especially the deeper one traveled into the interior. Swooping silently over its slumbering victim, this bat would administer a painless bite or scratch, sucking the blood until it was shooed away. The bleeding was hard to stop. Lima's young son presented a ghastly sight one morning when he appeared with both of his legs thickly smeared and blotched with blood. One of Wallace's assistants was a particular favorite of the bats; despite all precautions, he was even bitten on "the most prominent part of his person" through a hole in his trousers that the bat had managed to reach by working its way through the netting of his hammock. "We could not help laughing at the catastrophe," Wallace wrote, "but to him it was no laughing matter." Wallace usually escaped by tightly wrapping himself in his blanket, though once he was bitten on the nose and another time on the tip of his toe.

Wallace's ultimate destination was Juruparí (Devil) cataract, near the Colombian border, about a month's voyage from the mouth of the Uaupés. Its prize of prizes, the vaunted white species of umbrella bird, had never before been seen or collected by a European, though its presence there had been confirmed by several independent accounts. Wallace's intention had always been to distinguish himself in some manner in English scientific circles, and so he hoped to bring home a menagerie of exotic animals for the admiration of scientists and public alike. But to thoroughly explore Juruparí required returning to Barra, now fifteen hundred miles away, to make preparations for a journey of several months, for which he would need additional supplies and articles of barter. The best time to continue upriver to the Juruparí cataract was November to February, when the river was low and the weather relatively dry, so he had no choice but to turn back immediately.

On July 24, he arrived in Nossa Senhora de Guía on his return journey. Although he was in a hurry, no one else was. His Indian assistants spent two weeks preparing his canoe when two days would have sufficed, and after that was done they found other reasons for delaying their departure. When Wallace finally arrived in Barra in mid-September, he found his friend Spruce stranded there, unable to assemble a crew for a journey upriver. Spruce invited him to share his living quarters during what Wallace hoped would be a short stay before resuming his trip up the Uaupés. But he was stunned when he opened one of the many letters that awaited him. Postmarked in early June, it was from Daniel Miller, now the British consul in Pará, who conveyed the shocking news that Edward had contracted a severe case of yellow fever and was unlikely to recover. Wallace had assumed that Edward had left Brazil in February or March and was safely home. Yet there were no other letters explaining Edward's reasons for being in Pará in June, nor were there letters from the consulate, from acquaintances in Pará, or from home about Edward's condition. The only subsequent letter he received reported the death of Miller from an attack of "brain fever." He was left in the dark about the outcome of his brother's illness.

In his travel narrative, Wallace tells us nothing more about his reaction to Edward's probable death. He reports that he spent two weeks buying supplies, packing his miscellaneous collections, and arranging for his expedition up the Uaupés. At night he luxuriated in the enjoyments of rational conversation with Spruce, "to me, at least, the greatest, and here the rarest of pleasures." He outwardly appeared to carry on as usual, as though his brother's illness were no more serious than a cold or a broken arm. But what to do about Edward must have posed a horrible dilemma for him. It apparently was resolved in favor of science. He opted not to go to Pará. No letters describing his thoughts or feelings on the subject have been preserved or discovered, and he provides no further clues in his autobiography. In part, his choice may have been dictated by economic necessity: he had to continue his explorations if he hoped to make enough money to support himself and his family. Vessels coming from or going to Pará were infrequent—he would later learn that Edward's ship had been delayed for weeks for necessary repairs—and once Wallace had reached Pará, he almost certainly would have had to leave for England because of a shortage of funds.

Given Wallace's reticence on the subject, it is not clear exactly when he learned of Edward's death. The details of Edward's final days come from a surprising source: Henry Bates. Bates had arrived in Pará not long after Edward, having all but given up a career as a traveling naturalist. His personal servant at Ega, on the Solimões, had robbed him of everything, including his shoes. For nearly a year, he had received no money from the sale of his specimens, and life in such abject poverty had been unbearable. Miraculously, a boat from Pará arrived with £40 for him, and two months later another £30 was delivered, along with letters from his father begging him to come home. "Considering the unsatisfactory nature of my future prospects in this profession, I think I do better in returning to a more certain prospect of establishing myself. I have now, therefore, only one idea, that of returning to England," he wrote to Samuel Stevens in a letter dated December 23, 1850, and reprinted in the *Zoologist*.[28] In the same letter, he half-heartedly proposed that he might venture to the Amazon sometime again in the future, perhaps going to Peru or Venezuela after making better arrangements. At the end of the letter, there is a hint of jealousy: "Mr. Wallace, I suppose, will follow up the profession, and probably will adopt the track I have planned for Peru; he is now in a glorious country, and you must expect great things from him. In perseverence and real knowledge of the subject, he goes ahead of me, and is worthy of all success."[29] Not until May, when he arrived in Pará depressed and determined to return to Leicester, did he learn from Stevens that his collections were selling very well indeed. A sizable amount of money awaited him. He was also acquiring a coterie of admirers as a result of the extracts of his letters in the *Zoologist*. One man had written that "these letters appear to me to be distinguished by a devotion to science, and by an enthusiasm which are seldom exhibited. . . . His descriptions depict the primeval forests of South America, where, in not a few instances, no European foot but his own seems ever to have trod, with a glowing freshness and vivacity which brings everything in the clearest manner before the eye of the mind."[30] To top it off, Stevens reported that a new species of butterfly that Bates had discovered had been named *Callithea batesii*. All this news changed his mind, and he canceled his plans to leave Brazil.[31]

On June 2, when Wallace was beginning his first journey up the Uaupés River, Bates and Edward were having tea with Consul Miller. Immediately afterward, Edward began to shiver, and the shivering was followed by a high fever and vomiting. Too sick to go back to the house he had rented in Nazaré, he was taken by Bates to the nearby home of a friend. A wave of yellow fever had struck Pará in May; in the previous year, yellow fever—the first such epidemic since the late eighteenth century—had afflicted 75 percent of the population, and 5 percent of those infected had died. On the heels of both epidemics came an outbreak of smallpox; doctors were in short supply, attending to the myriad sick from two potentially lethal diseases. As yellow fever ravaged the white population, smallpox afflicted the Indians and blacks. Bates got the best medical care available, but he was horrified when the doctor began treating Edward for yellow fever instead of "merely constipation," which had been Bates's diagnosis. The next day, Edward seemed to be progressing well. "But he committed the imprudence of getting up and walking barefoot about a cold brick floor after mustard plasters had just been taken from his feet," Bates wrote in a letter to Mary Anne Wallace not long afterward. The next day Edward's fever recurred, and he developed the "black vomit," a sign of imminent death. Despite the skillful ministrations of the physician, he continued to deteriorate. For the next four days, he "suffered fearfully." Bates bravely slept by his side in a vain effort to nurse him back to health, with Miller visiting periodically for support. Edward died at 2:00 A.M. on Sunday, June 8, and was interred somewhere in Pará, in a burial ground set aside for victims of the epidemic, whose corpses were believed to be contagious.[32]

Not long after Bates returned to his accommodations in Nazaré, he, too, became ill with fever and began vomiting. He sent a servant to town to buy medicine, while he spent several nervous hours pacing back and forth on the veranda, wrapped in a blanket. He drank various teas and then a "good draught" of a concoction of elder blossoms to induce sweating, before falling "insensible" into his hammock. He did not wake until midnight, by which time he was very weak, with every bone in his body aching. He purged himself with Epsom salts. Forty-eight hours later his fever broke, and in eight days he had completely recovered. The unpleasant task of informing Edward's mother of her son's death fell to him, since Edward's older brother was somewhere in the wilderness and thus out of reach. In addition to describing the details of Edward's illness, Bates noted Edward's "genial temper and kind heart." The letter stated that "he did not converse freely after being first taken [by the fever], but felt regret at being taken thus when on the eve of departure for England."[33]

In Barra that autumn, before starting again up the Uaupés River, Wallace was suddenly informed by the local authorities that he needed a passport, "an annoyance I had quite forgotten." Loaded and ready to go, he found the red tape a "bore." Senhor Henrique, always friendly to travelers, advised him to leave without it—he would send it on when it was ready—and with Henrique's blessing Wallace departed at the end of September 1851. Spruce accompanied him for a day to collect plants, but Wallace's canoe was not large enough to accommodate two traveling naturalists on a longer journey. The next morning, they bade farewell to each other. Not long after his departure, Wallace developed a severe headache, joint pains, and fever, which left him weak and apathetic. His Indian assistants paid no attention to his sufferings, and he was truly alone, in the heart of a wild country where he might disappear from the face of the earth. In his feverish dreams he reviewed his past life and aspirations and fantasized about an ignominious end on the Rio Negro. He wondered about the fate of Edward and about John in California, where an epidemic of cholera was raging. But these gloomy thoughts vanished when he had recovered, and he vowed never again to travel in such remote, uninhabited lands without a civilized companion.

Dressed in light cotton clothes, his head protected from the blistering sun by a black felt hat, Wallace and his half-naked Indian crew made their way slowly upriver through intermittent storms, some with hurricane-force winds that would have capsized a less well balanced vessel. In November, one of his best assistants fell ill; a few days later, another became sick. Both had to be left behind. Because of the difficulty of finding more help, Wallace simply bought a smaller canoe and continued his ascent of the Rio Negro, stopping at the town of São Gabriel before proceeding to São Joaquim, at the mouth of the Uaupés, where Senhor Lima awaited him.

Upon arriving at São Joaquim, he was stricken again with what turned out to be malaria. While he was laid up in his hammock suffering from violent fevers, his Indians took the opportunity to steal some of the *caxaça* that he had brought along for preserving his fishes—as well as anything else they could get their hands on. During a brief remission in his symptoms, he decided to dismiss them before he lost everything. For a few days, he thought he was better, but then the fevers returned, this time at night, followed by periods of depression and vertigo. His feverish nights alternated with nights of profuse sweating. He had little appetite and suffered from extreme lassitude. Fortunately, Lima and Old Jeronymo were there to attend him.

Word of his illness reached Spruce, who had finally found a boat to take him upriver and had set up camp in São Gabriel. "I had sad news two days ago from my friend Wallace," Spruce wrote to Sir William Hooker at Kew. "[He] writes me by another hand that he is almost at the point of death from a malignant fever, which has reduced him to such a state of weakness that he cannot rise from his hammock or even feed himself. The person who brought me the letter told me that he had taken no nourishment for some days except the juice of oranges and cashews."[34]

For days, Wallace was unable to speak intelligibly and could barely turn over in his hammock. He took quinine for some time without any apparent benefit. After two weeks, his symptoms ceased temporarily, leaving him emaciated and feeble, but then they returned, occurring daily. The fevers and sweats, which lasted from noon until night, prevented him from getting any rest. He remained in this state of misery until late January 1852, when the symptoms gradually diminished in severity. He began to eat heartily, yet gained little strength at first; only with difficulty, and aided by two walking sticks, could he walk across the room of his hut. Then the "ague" left him completely, and he could walk with one stick to the river. A week later, he felt strong enough to travel downriver to São Gabriel to visit Spruce. Undeterred by three months of illness, he was determined to make his pioneering expedition to the upper Uaupés. But the best time for collecting in this region, when the level of the river was at its lowest and the rains had not yet come, had passed while he lay "half-dead" in São Joaquim.

Nevertheless, Wallace left on February 16, this time well supported by a crew of seven, courtesy of Lima. Five days later, he was back in São Jeronymo. Fever and debilitation forced him to rest there for a short time before pushing off again to the village of Juarité, on the border between Brazil and Colombia. After Juarité, Lima's right-hand man refused to go any farther, though he had been paid for the whole voyage, and another crewman also deserted. Wallace does not state the reason for their behavior, which he condemned as deliberate deception, but the crew had just portaged around two sets of waterfalls and had had to unload all of his cargo and carry it a considerable distance through the forest and up a steep path. They may well have had to carry Wallace, too, because he reports that he was still very weak and feverish. Undoubtedly knowing that worse lay ahead, the two men had decided to quit rather than suffer. But nothing could stop Wallace now, and he forged on with those brave (or foolish) enough to keep him company. By March 1, they were negotiating rapids, which weakened the canoe. For an ax, a shirt, trousers, two knives, and some beads, he secured a larger *oba*. There were more rapids to overcome—some of them "furious cataracts" and others nearly perpendicular falls. Twelve were so steep and turbulent that the canoe had to be entirely unloaded. The loading and unloading every few hours was a great annoyance. Bird and insect specimens left out to dry were overturned, drenched with spray, and blown about by the wind, as were his notebooks and other papers. "It was an excellent lesson in patience to bear all with philosophical serenity," Wallace recalled.[35]

On March 10, his heart leaped into his throat when his entire crew "sprang like otters" into the water, swam to shore, and disappeared into the forest. But they had bolted only at the sight of a tree with edible fruit. When a good stock of fruit was on board, they pushed off again. That night they passed safely over the most dangerous rapids, and at last they entered "the country of the painted turtle and the white umbrella bird," where he was determined to remain for two weeks until he could procure those rarities. Once again, Wallace was elated. He had gone where no European had gone before. But he decided that he had reached the limit of his expedition. They had been lucky thus far, having negotiated without serious accident a river famous for its difficulties and the dangers inherent in its navigation. He only regretted the loss of his instruments for determining the latitude, longitude, and height above sea level of the distant reaches of the Uaupés River. But he felt that ascertainment of the true course and sources of "this little-known but interesting and important river" by some other adventurer was an object worth the danger and expense of a voyage.[36]

Wallace's quest for the elusive white umbrella bird ended in disappointment. A native hunter he had hired killed a few of the black species but never a white one. Concerning the white species, he was given many contradictory statements. Some people knew nothing about it, while others said that it was very rarely seen. He was therefore inclined to believe that if it existed at all, this white "species" was simply a sport, like the occasional white blackbirds or starlings he saw in England. It was also possible, he later concluded, that his local informants had mistaken the white bell-bird for a white umbrella bird, since the bell-bird has a fleshy erectile appendage on the base of its upper mandible similar to that of the umbrella bird.[37] The *anambé de catinga*, another bird he had long been searching for, as well as one or two small birds new to him, a dozen new species of fish, and two or three rare butterflies constituted his natural history specimens from the virgin territory of the upper Uaupés.[38] He also collected the weapons, implements, and ornaments of the natives—an object of his travels in the region—and compiled a vocabulary of the language of one of the tribes. As for the other rare species, the painted turtle, a single specimen had allegedly been saved for him by a Brazilian trader he had met some weeks earlier; it had managed to escape from its corral in a small stream before Wallace could lay eyes on it. Although his collections had fallen short of his expectations, his heroic accomplishment had not.

At the end of March 1852, he headed back to Barra, collecting live animals along the way. Late in April he stopped off to see Spruce, who was still at São

Gabriel. Both had spent so many months in the company of native peoples that they could not string together an English sentence without Portuguese interpolations. He continued downstream to Barra with his collections, both living and dead. Losing one live animal a day, he found himself caught up in a race against time. Of one hundred live animals he had bought or been given, only thirty-four remained by the time he reached Barra on May 17: five monkeys, two macaws, twenty parrots and parakeets of twelve species, five small birds, a white-crested Brazilian pheasant, and a toucan. At Barra he was greeted by Senhor Henrique, who found him a small, mud-floored, leaky-roofed room in town.

As badly as he needed the rest, Wallace could not stay long. When the vessel that would take him to Pará finally arrived, he gathered up his collections from the upper Uaupés as well as everything that had remained in storage during his long absence: a great number of cases and boxes, including six large containers that Senhor Henrique had been unable to ship to England because "the great men of Barra" suspected that the crates contained contraband material. Then there was the absurd requirement of the passport to leave Barra, which could be obtained only after passing through a great number of bureaucratic hoops, testing the limits of his *patiencia*. Added to this was the miserable condition of the city, which had not seen a single vessel from Pará in five months. All supplies were exhausted, and staples were scarce. The little society of foreigners who had relieved the monotony of life had long since disbanded. Only Senhor Henrique remained, a pillar of stability in a world that verged on anarchy.

On June 10, Wallace boarded his canoe for Pará, hauling with him his specimens amassed over almost three years and his menagerie of live animals. The toucan, now full grown and tame—a special prize—escaped and drowned. In his collections were nearly ten thousand birds, all skinned and carefully prepared for mounting, an extensive herbarium of Brazilian plants, and a collection of birds' eggs that surpassed the holdings of any English museum.[39] Wallace continued to suffer from periodic bouts of malaria during the trip. He arrived in Pará on July 2, still very weak and unable to exert himself, and not long afterward obtained passage to England on the brig *Helen*. Setting sail with a heavy heart on the morning of July 12, 1852, he took leave of "the white houses and waving palm trees" of the city that had been the gateway to his heaven and hell.

CHAPTER 6

Disaster at Sea . . . and a Civilized Interlude

ABOARD THE *HELEN*, Wallace's menagerie and specimen cases were stowed in the hold, sharing space with other cargo, including 120 tons of India rubber, twenty small casks of balsam and rice chaff, and a large quantity of palm oil. After a week at sea, Wallace had a recurrence of fever— slighter this time—that soon abated. For the next few weeks, he suffered from seasickness and stayed in his cabin, which he shared with the ship's captain, John Turner.[1]

On Friday, August 6, Captain Turner entered their quarters shortly after breakfast and said calmly, "I'm afraid the ship's on fire. Come and see what you think of it." Going on deck, Wallace and Turner saw thick smoke emanating from the forecastle. The crew opened the hatchway and began to throw out casks of the highly flammable palm oil, but was forced to retreat when the smoke became too dense. Buckets of water were hurled into the hold—to no effect. Rushing back to their cabin, Wallace and the captain found smoke issuing from the floorboards. Turner grabbed his instruments, books, and charts. Wallace took his tin box, which contained, along with a few shirts, his watch, and a little money, his drawings of fish and palms, the diary he had kept on the Rio Negro, plus some notes and observations of the Rio Negro and the Uaupés; most of his clothes and other items, including the rest of his notebooks, had to be abandoned.

Turner ordered his men to lower the two lifeboats, which were outfitted with bread, water, and other necessities. The lifeboats leaked, their planks having shrunk from long exposure to the sun. Blankets, rugs, pillows, and clothes, all tossed in indiscriminately by the crew, became soaked and heavy and overloaded the boats. Two men were assigned to each boat to bail and stop up the leaks; the others continued their futile efforts to put out the fire. A furious boiling and hissing rumbled through the bowels of the ship, and at noon flames burst into the cabin and deck. The quarterdeck was scorching hot; Turner saw that there was no hope of saving the ship and ordered everyone into the lifeboats. Wallace had to let himself down by rope, but his grip weakened and the rope slipped through his hands, flaying skin off his fingers. He tumbled into a pile of wet clothes and blankets in a lifeboat half filled with water and immediately set to bailing, the saltwater stinging his excoriated hands.

All afternoon, they floated near the burning ship, as if in a dream, watching the progress of the flames, which rushed up the shrouds and sails "in a most magnificent conflagration." In the rolling of the waves, the weakened masts broke and fell into the sea. The deck burned away next, its ironwork glowing red-hot. Most of the parrots, monkeys, and other animals either were incinerated or died of suffocation. A few had retreated to the bowsprit, but the men were unable to coax them into the boats and they were immolated with the rest of the ship, except for a lone parrot that fell into the water, perched on a piece of debris, and was retrieved. By nightfall, the ship was a cauldron of fire, with the cargo of India rubber "forming a liquid burning mass" at the bottom. It was only now that they felt hunger and broke open tins of biscuits and raw ham. There was a considerable swell, and the wreckage, some of it still on fire, threatened the lifeboats. All night, the boats tossed on the billowing ocean, unsettling Wallace's stomach; they continued to leak and had to be bailed out, one man bailing with a bucket, another with a mug. Hoping that the flames would act as a beacon to attract other ships in the area, Turner kept the lifeboats near the burning carcass, but by morning the *Helen* was nothing more than a crescent of charred timber and no ship had appeared. Wallace recorded their location: lat. 30°30' N; long. 52° W. They were stranded in the Sargasso Sea, seven hundred miles from Bermuda, the nearest land, but fortunately they were in the track of vessels bound to and from the West Indies.

Day after day, scorched by the sun and drenched by the sea, they languished in their boats, which were tied together, one behind the other, masts up, sails unfurled. One night, two boobies circled them; on another, a flock of unidentified birds flew by. When the evening was clear, Wallace stretched out and stared at the stars, watching showers of meteors. Their supplies of food and water were dwindling. Sometimes it rained, and they hoped to catch freshwater, but all they had for receptacles were their clothes and the sails, which were soaked with saltwater and therefore useless. The captain imposed strict rationing, and everyone suffered from thirst in the blistering sun. The one bit of luck was that the boats had stopped leaking.

The full force of his loss did not strike Wallace immediately. Stunned by their harrowing ordeal, he was convinced that he would not survive. He thought about the absurdity of having taken his money with him, which he had done instinctively. As the days passed, however, he began to believe that he might live and regretted not having grabbed his shoes, coat, and trousers. Shock and denial gave way to profound depression when he considered the irretrievable loss of four years' work. He later wrote:

> All my private collection of insects and birds since I left Pará was with me [on the *Helen*], and comprised hundreds of new and beautiful species, which would have rendered (I had fondly hoped) my cabinet, as far as regards American species, one of the finest in Europe. But besides this, I have lost a number of sketches, drawings, notes, and observations on natural history, besides the three most interesting years of my journal, the whole of which, unlike any pecuniary loss, can never be replaced. How many times, when almost overcome by the ague, had I crawled into the forest and been rewarded by some unknown and beautiful species! How many places, which no European foot but my own had [trod]! How many weary days and weeks had I passed, upheld only by the fond hope of bringing home many new and beautiful forms from those wild regions; every one of which would be endeared to me by the recollections they would call up![2]

One special prize lost in the fire was a palm leaf, fifty feet long, that he had intended for the botanical room of the British Museum. But he realized the futility of such thoughts and sank within himself to find the strength to bear his fate "with patience and equanimity."

At 5:00 P.M on August 15, after ten miserable days adrift, they were spotted by a passing ship, which reached them within three hours. To celebrate their imminent rescue, Captain Turner gave them permission to drink the rest of the water. Their savior was the *Jordeson*, a merchant vessel traveling from Cuba to London. Wallace discovered that he could barely stand when he got on board, and he was nearly knocked over by the mad rush to the water casks. The lucky survivors were then treated to tea.

But the *Jordeson* was no guarantee of safety. It was a slow-going ship, encumbered by its unexpected cargo of people. On August 22, they survived an encounter with a waterspout, which failed to swamp them. September was hurricane season, and they did not escape. One morning, the wind shredded four or five sails before they could be hauled down. The ship was buffeted in the roiling, foamy sea and colossal waves crested over its decks.

By night, it was pitching fearfully, with water pouring over the cabin skylights. The compass spun around in all directions. The boat creaked under the strain. Water leaked in. The pumps clacked in a way that made everyone nervous. It was so wet that one day a fire could not be started to cook food, and passengers and crew were forced to content themselves with a stale biscuit for dinner.

Unprepared for the sudden doubling of its crew, the *Jordeson* was short of provisions. From the first day, food was strictly rationed. Several times, the sailors caught dolphins, which were not bad as a meal. The cheese and ham were depleted first, followed by the peas. The beef and pork were of the worst quality—worse than Wallace had thought possible—but he ate them anyway. The men were at the point of starvation when they were finally able to replenish their supplies from a passing ship; these rations also disappeared. In the end, they were reduced to catching rats, which the men cooked and devoured in desperation, consuming even the contents of the grease pot. On September 29, as they reached the English Channel, they were intercepted by a violent gale that destroyed many vessels much more seaworthy than the *Jordeson*, which limped toward its destination with four feet of water in its hold. The eighty-day nightmare voyage came to a peaceful end on October 1, 1852, when they landed at Deal, on the southeastern coast of England. Emaciated and emotionally battered but thankful to be alive, Wallace was finally back on English soil.

After four years of isolation in the pristine wilderness of the Amazon and nearly three months at sea, Wallace must have been both comforted and appalled by the sight of London—filthy, polluted, and sprawling, its majestic skyline belying the overcrowded slums and open sewers. His travails had come to an end. For that he was grateful. But his journey to the Amazon had been a failure. The death of a brother, his own near death from acute malaria, the loss of most of his specimens and his precious journals, and another near-death experience at sea were indelible stains on an expedition that had begun with so much promise. A single paper on the umbrella bird, a few novel birds and insects that had arrived in England before his first voyage up the Rio Negro, numerous drawings and descriptions of fish but no specimens for taxonomists to examine, a set of "careful drawings" of all the species of palms he had encountered, some notes and measurements for a map of the Uaupés, his first journal noting his observations on the Rio Negro, and a handful of letters to his family, Sir William Hooker, and Samuel Stevens documenting his discoveries and describing his experiences were all that he had to show for his four years of effort. And he was no closer to a solution to the origin of species, the mystery that had drawn him to South America in the first place. In similar circumstances, an ordinary man would have sunk into a deep depression, wallowed in self-pity, or at least interpreted the mad twist of events as a warning to adopt some humbler occupation. But Wallace was not an ordinary man.

He arrived in London looking like the castaway he was, clothed in rags and without any luggage, unable to tolerate standing for long periods because of so many days at sea. He was not prepared to face his mother and sister in such a state; his wasted, disheveled appearance would have shocked them. But his pitiable condition did not deter him from showing up at the Entomological Society of London on October 4, only three days after landing at Deal. With ankles swollen from malnutrition and immobility, he must have been a pitiful sight as he hobbled into the meeting. He was the object of everyone's attention as he recounted the glories of his exploration and the magnitude of his losses—the embodiment of all that was great in the English character. The tactic was at least effective: no one could doubt his devotion to the cause of science.

Accompanying Wallace to the meeting was Stevens, who was as much his savior as the captain of the *Jordeson* had been. Stevens had insured his collections for £200 (though Wallace felt they were worth at least £500) and presented him with a badly needed check. He had also taken Wallace to the nearest shop to buy a warm suit to replace his own of flimsy calico—all that remained of his Brazilian wardrobe—and then to the tailor and haberdasher for a new set of clothes. For a week Wallace, roomed with Stevens's mother, who fed him nourishing food to add weight to his skeletal frame and make him presentable to his friends and family.

Although Wallace had sworn never to travel again, he changed his mind after only four days ashore. In a letter to Richard Spruce, he spoke of undertaking another expedition. "How I begin to envy you in that glorious country where 'the sun shines for ever unchangeably bright,'" he wrote. "Fifty times since I left Pará have I vowed . . . never to trust myself more on the ocean. But good resolutions soon fade, and I am already doubtful whether the Andes or the Philippines are to be the scene of my next wanderings."[3] Humboldt once remarked that the joy of finding oneself back in civilization would be short-lived if one had learned "to feel deeply the marvels of tropical nature."[4] The memory of what one had endured quickly faded, he observed, and soon the weary traveler began to plan another journey. Humboldt was right.

The £200 from the insurance company was considerably less than his collections were worth, but sufficient for a young man to live frugally. In the 1850s, £60 a year was the lower limit of income for a member of the middle class; for example, a tutor earned approximately £150 annually. At the opposite end of the economic spectrum, an aristocrat's annual income ranged from £3,000 to £50,000 or more. The sum of £300 was enough for a "gentleman" to marry, a gentleman being defined in 1850 as a member of the aristocracy or the upwardly mobile middle class who did not have to work and could lead a life of leisure.[5] Although Wallace was not a gentleman in material terms, he certainly was in the moral sense (despite his radical political and scientific ideas), defined by such Victorian middle-class spokesmen as Samuel Smiles, who stereotyped the true gentleman as "honest, truthful, upright, polite, temperate, courageous, self-respecting, and self-helping."[6]

In his autobiography, Wallace says nothing about his reunion with his mother and sister. It must have been an emotional experience. In the space of a decade, the Wallace clan had been greatly reduced. Two sons and a father had died, and a third son had left to seek his fortune in America. Over a period of forty years, Wallace's mother had lost six of her nine children, a mortality rate exceptional even in the nineteenth century. It was therefore left to Wallace to take charge. In December, he rented a house on Upper Albany Street, near the Zoological Gardens, and consolidated his family's expenses by moving in with his mother, sister, and brother-in-law, Thomas Sims, who was one of England's first professional photographers. (Fanny, a good watercolorist, colorized her husband's photographic portraits.) The house was also close to the offices of the Zoological Society in Hanover Square, with easy access to Stevens, who worked near the British Museum.

By Christmas, Wallace had regained enough strength to make preparations for the next phase of his life. With the disintegration of four years' worth of work in the flames of a burning ship, he had lost considerable ground. To establish scientific credibility, he had to publish something, which spurred him to write up his travels with the help of George Silk—a boyhood friend who was now private secretary to the archdeacon of the Anglican church in Kensington—and publish a number of scientific papers based on the few surviving notes, sketches, and letters. He also began to circulate in the London scientific community, making important professional contacts.

In 1852 there were at least thirty-six major scientific and literary societies in London. Wallace narrowed his attendance to three—the Entomological, Zoological, and Royal Geographical Societies—that met in the afternoon or evening once or twice a month.[7] Not yet a member of any society, he could attend their meetings only as a guest and use their libraries only by prior arrangement, but within a short time he became a familiar presence. He also visited the Linnean Society and the Kew Herbarium to consult works of botany. Having made his name well known to the officers of the Zoological and Entomological Societies, he was given permission to visit the zoo—which was not yet open to the public—and was welcomed at the scientific meetings of both societies, where he met most of London's zoologists and entomologists. Membership in a scientific society required endorsement by several of its members, acceptance by a vote at one of the general meetings, and the payment of dues.[8] With one exception, as we shall see, Wallace made no effort to join any society, principally to save money. After his departure from England in 1854, he would be elected a "corresponding" member of the Zoological and Entomological Societies, which exempted him from annual dues and entitled him to receive a journal of their proceedings at a reduced price.

Wallace attended a lecture at the Zoological Society one evening in December 1852 on some echinococci (parasites) found in the liver of a zebra that had died in the zoo. The lecture was given by Thomas Huxley, a young anatomist who had made a name for himself. Huxley was already a fellow of the Royal Society, England's most prestigious scientific organization, and not long after his lecture he was appointed professor (chairman) of natural history and paleontology at the Royal School of Mines. He did not read his paper, but with the aid of diagrams and sketches he presented a lucid description of the echinococcus's structure and development and of the "strange transformations" it underwent when the parent worm migrated from the intestines to other internal organs. Given Huxley's formidable erudition, Wallace thought that he projected a mature image, though at twenty-seven he was actually two years younger than Wallace. "I was particularly struck with his wonderful power of making a difficult and rather complex subject perfectly intelligible and extremely interesting to persons who, like myself, were absolutely ignorant," Wallace later wrote. "I was amazed, too, at his complete mastery of the subject, and his great amount of technical knowledge of a kind to which I [had] never given any attention, the structure and development of the lower forms of animal life." From that time forward, he looked up to Huxley as a scientist "immeasurably superior" to him, though the two would not become friends for another decade.[9]

Wallace spent a considerable amount of time at the British Museum examining the zoological collections—especially birds and insects—in an attempt to interpret his own incomplete series of specimens and identify major gaps in the zoological record, which would help him make a decision about his next field of operations. Sometime in late 1853 or early 1854, he actually met Darwin in the insect room, where they conversed for a few minutes. Darwin had long since retreated to Down House and had stopped attending the monthly gatherings of the scientific societies on a regular basis, but he occasionally visited the museum. In this brief initial encounter, Wallace apparently made no impression—Darwin does not mention the meeting in his autobiography—and Wallace himself recollected that nothing "of any importance" passed between them.[10]

The activities of the scientific societies were a crucial part of Wallace's education. He learned a great deal about the interests, ideas, and perceptions of his fellow naturalists. Papers deemed worthy of attention by the president and council members of each society were read at the meetings, and they gave Wallace a sense of the style, content, and format that members expected. Communications from working naturalists flowed in from different parts of the world on a regular basis, including letters from Henry Bates in Amazonia, excerpts of which were frequently read at the general meetings of the Entomological Society. Such a plethora of information and data poured in that the society's president, Edward Newman, the founding editor of the *Zoologist* and a former president of the Zoological Society, would express his satisfaction "that exotic entomology is claiming a greater share of attention than at any previous time."[11]

Newman admired men like Wallace and made a point of encouraging his efforts. Born in Hampstead in 1801, he was a Quaker by birth and the son of a successful manufacturer. For a time he assisted his father in the wool trade, but he spent most of his free hours pursuing his interest in natural history, especially entomology. In 1833 he helped to establish the Entomological Society of London, and ten years later he founded the *Zoologist.* His personal mission was to make "natural history a pursuit for the shopkeeper and the mechanic"—British science being almost exclusively the province of upper-class amateurs—and to that end he went into business as a printer of scientific books. Among the first contributors to the *Zoologist* was Henry Bates, the nineteen-year-old son of a hosiery dyer. The study of nature ennobled the student, Newman believed; it elevated his "moral dignity" and improved his mind, for what pursuit was more dignified or more instructive than the study of God's creations—a rhetorical question he asked in his 1850 presidential address to the Zoological Society. The *Zoologist* aimed to "combine scientific truths with readable English." Until that time, an article was considered to have a "scientific character" only if Latin descriptions were used, but Newman preferred "pure, plain, intelligible English," which would make the subject of natural history accessible to those "not deeply versed in Science." This was almost heresy, and he labored unceasingly against the prejudices of the scientific elite ("the merely scientific," he called them), who resisted his attempts to democratize their specialties. These men had scoffed at the idea of a journal aimed at the masses, but by the time of Newman's death in 1876 no one was scoffing and the list of contributors to the *Zoologist* included almost every British naturalist of note.[12]

This process of democratization, led by Newman and others, encouraged self-made men like Bates, Wallace, and Spruce to pursue their groundbreaking journeys and devote "the prime of their lives" to the cause of science. But Newman and like-minded colleagues performed a more important, perhaps unintended, function. By creating a forum for the exchange of ideas that was open to interested people from all social classes (including radicals like Wallace and Huxley), they were instrumental in preparing the educated public for the general acceptance of the imminent revolution in scientific thought. To paraphrase Lenin, Newman would hand the enemy—which, ironically, included himself since he never accepted the Darwin–Wallace theory of evolution—the noose with which it would hang itself.

Given the little material that he had salvaged from his collections, Wallace was astonishingly productive. In the space of a year, he wrote two books and four important papers, an indication of what he could have accomplished had his collections and journals survived. Two of his papers dealt with the broader issue of geographic distribution. He read the first, "On the Habits of the Butterflies of the Amazon Valley," at the Entomological Society on November 7 and December 5, 1853. He read the second, "On the Monkeys of the

Amazon," at the Zoological Society on December 14, and it was published in the society's *Proceedings* the following year. It also was published in the *Annals and Magazine of Natural History*, a journal founded in 1837, whose mission was to be a "ready medium of communication for the lovers of Natural History in all parts of the world." At the time, the *Annals* was the only journal that did not limit its subject matter to a single field of natural history but welcomed contributions from investigators in various fields, including geology and botany. Its circulation extended to the European continent and the United States, giving it a wider audience than that of the *Zoologist*.[14]

The paper on butterflies focused on the "habits, manners and peculiari ties" of the various families, genera, and species that Wallace had observed and collected in Pará and along the Amazon and Rio Negro. In addition to accurate and detailed descriptions, it contained precise information on their geographic ranges and distribution. Some species were found only along the margins of rivers, Wallace wrote, while others could be found in second-growth woods and shady paths or, more rarely, in the gloomy recesses of the virgin forest. He occasionally speculates on some aspect of their behavior; referring to butterflies of two genera, for example, he states that "they always keep within a few inches of the ground, like the Satyridae, which in their peculiar ocellated markings they both so much resemble; almost leading us to suppose that the colour and marking of an insect affects its habits, or vice versa." One passage indicates the direction of his thinking at the time:

> All these groups [the *Heliconia* species] are exceedingly productive in closely allied species and varieties of the most interesting description, and often having a very limited range. . . . [A]s there is every reason to believe that the banks of the lower Amazon are among the most recently formed part of South America, we may fairly regard those insects, which are peculiar to that district, as among the youngest of species, the latest in the long series of modifications which the forms of animal life have undergone.[15]

If Wallace's paper on butterflies nibbled at the issue of the origin of species, his paper on monkeys took larger bites. In his analysis of the distribution of monkeys, Wallace reported an unusual discovery: that the Amazon and its two great tributaries, the Rio Negro in the north and the Rio Madeira in the south, formed the natural limits beyond which certain species never passed. It was an interesting piece of deductive work, since he had to rely partly on the reports of other naturalists, who were perhaps less rigorous in identifying the precise ranges of their specimens. He had seen twenty-one species himself. The rare woolly monkey (*Lagothrix humboldtii*, now *L. lagotricha*), which he had not observed, inhabited the district between the Rio Negro and the Andes but was unknown east of the river. A species of spider monkey (*Ateles paniscus*) was found in the Guiana portion of the Amazon watershed up to the Rio Negro but not beyond it. The short-tailed monkey (*Brachiurus couxiu*, now *Cacajao melanocephalus*) had a similar range, while a distinct species of that genus (*C. calvus*) lived in the upper Amazon toward Ecuador and the upper Rio Negro toward Guiana. One species of "sloth-monkey" (*Pithecia*) was found to the north of the upper Amazon, while another species of the same genus was found to the south, and he was able to assign several other species to distinct zones. Wallace therefore envisioned four "great divisions" of eastern equatorial South America. These zones he termed the Guiana (north of the Amazon and Rio Negro and including parts of Venezuela), the Ecuador (southwest of the Rio Negro toward the Andes, a region now part of Peru, and parts of Colombia), the Peru (bounded on the north by the upper Amazon and on the east by the Rio Madeira), and the Brazil (east of the Rio Negro and the Rio Madeira).

In his article, Wallace did not discuss the geologic history of these regions, but in *A Narrative of Travels on the Amazon and Rio Negro*, which he published in 1853, he ventured a theory. He imagined that the vast central plains of South America, now barely above sea level, were at one time a gulf or large inland sea. Gradually that sea filled with alluvial deposits from three great rivers coursing down from the surrounding highlands, and the resultant basins became stocked with species from the three large landmasses that he hypothesized had existed before the sea disappeared. Those rivers—the Amazon, the Rio Negro, and the Rio Madeira—would eventually form natural dividing lines, beyond which certain species never passed. But as one approached the sources of these rivers, he observed, the rivers ceased to be boundaries and the range of species overlapped—not just monkeys but also birds and larger mammals. This phenomenon raised four questions in his mind. First, were closely allied species ever separated by a wide interval of country? Second, what topographical features determined the boundaries of species and genera? Third, did the "isothermal lines" described by Humboldt to define botanical zones accurately encompass the range of animal species, or were the lines and the species distribution independent of each other? Fourth, why were certain rivers and mountain ranges geographic barriers while others were not?[16]

Wallace's questions clearly went to the heart of the problem of the origin of species. He provided no answers, merely concluding that certain large barriers, like great rivers and mountain chains, somehow determined the range of allied orders and families of animals. However, he implied that some law other than special creation determined the distribution of allied genera and species of animals. Since reading *Vestiges of the Natural History of Creation* in 1845, Wallace had pondered the "great secret" of the actual steps by which each species was produced, with its special adaptations to the conditions of existence. His views in the intervening years had increased in sophistication, and he had come a long way toward furnishing facts in favor of the "ingenious hypothesis" of the anonymous author of that notorious book. Even if the supporting specimens were irretrievably lost, those facts were indelibly imprinted on his mind. Wallace had already accumulated enough evidence at this early date to convince him that allied species were related by common descent and not independently created. Otherwise, he would not have posed his four questions.

During his presentation to the Zoological Society, Wallace complained about the sloppiness of previous travelers, who had failed to give precise locations of the places where their specimens were taken, noting that "there is scarcely an animal whose exact geographical limits we can mark out on the map." Johannes Baptist von Spix, for instance, had given the "banks of the river Amazon" as a locality, not realizing that a species on one side did not appear on the other. Wallace's disdain annoyed John Edward Gray, curator of the zoological section of the British Museum, who noted pointedly that "we have specimens collected by Mr. Wallace himself marked 'Rio Negro' only." Hoist with his own petard, Wallace stammered an inadequate response, recalling too late that only after several months of collecting had he begun to understand the relationship of geographic boundaries and species distribution.[17]

The paper on monkeys also demonstrated Wallace's refusal to accept a statement merely because it was made by a known authority. Humboldt believed that multiple individuals produced the unearthly vocalization of the howler monkey, but the Indians insisted that a single animal produced the noise. Wallace settled the question by dissecting a monkey and discovering in the throat a bellowslike muscular apparatus that caused air to reverberate in the animal's body cavity and was responsible for the eponymous howl.

Wallace also made a major scientific contribution in economic or ethnobotany. Among the few papers he had rescued were his notes and drawings of Amazonian palms, which served as the basis of his other book published in 1853: *Palm Trees of the Amazon and Their Uses*. When he examined his first palms in the Amazon, he was unable to distinguish closely allied species. But, like the Spanish inhabitants of the Galápagos Islands, who could identify the island a tortoise came from, the Indian guides knew the differences, which were subtle and "permanent." In his appendix, Wallace created a useful table, listing the various palms, giving their botanical and colloquial names, and summarizing their uses. Few naturalists before him had paid much attention to such native lore. Along with Henry Bates and Richard Spruce, Wallace pioneered this ethnographic field, which bridged the gap between the biological and anthropological sciences.

Although *Palm Trees of the Amazon* received a favorable review from the *Annals and Magazine of Natural History*, and its section on the economic value of palms was singled out for praise, Wallace derived very little money from it.[18] He had hired an artist from Kew to give "life and variety" to his crude sketches and then found a publisher and printed 250 copies at his own expense, money that he barely recovered through the book's limited sale. But botanists were less impressed. William Hooker was taken aback by Wallace's audacity in ascribing his own generic and specific names to allegedly new species, especially to a commercial palm that "the immortal [Carl Friedrich von] Martius" had identified as *Attalea funifera* but that Wallace believed was a new species (*piassaba*) of the genus *Leopoldinia*. Wallace had also erred in the number of new species he identified; it was later shown that three of the fourteen already had been classified.[19] "I wish he would keep to his Zoology, where he is at home," Hooker complained in a letter to Spruce.[20] Spruce replied that when he had seen some of Wallace's sketches, he pointed out which palms were new and which were already described—thus shifting part of the blame to himself. He had proposed making the book a joint effort, in which he would take the literary part and Wallace the pictorial, but Wallace, who never collaborated with anyone on any project during his long life, refused Spruce's proposition. Disappointed with Wallace's final product, Spruce told Hooker that while most of the illustrations were "very pretty" and the accounts of the palms' uses were good, the descriptions were "worse than nothing" and without much to interest the botanist—though, in defense of his friend, he argued that Wallace had correctly placed the erstwhile *A. funifera* in the proper genus.[21]

Wallace's aggressive posturing rankled other members of the scientific community. Newman came to his defense, addressing the general tendency of society members to deprecate the work of others. "Such terms as 'species-man,' 'mere collector,' 'theorist,' and many others, need but be quoted in explanation of my meaning," Newman said in 1854, in his presidential address to the Entomological Society, continuing:

> Now the sentiment implied by such expressions is unphilosophical; it is not in accordance with the catholicity of science; it implies that we are drawing a comparison between ourselves and others disadvantageous to those others, and it therefore has a direct tendency to foster, if not to create, feelings that ought not to exist. . . . [T]he man who, in whatever station of life, devotes his time, by night and by day; at all seasons, in all weathers; at home and abroad, to the positive capture and preservation of those specimens which serve as the objects for all our observations: he is the real labourer in the field, and if we would keep the lamp of our science constantly burning, it is to him alone that we can look for fuel to feed its flame.[22]

In his autobiography, Wallace says that he would have returned to tropical America had Bates given up his work in the Amazon. He briefly considered going to Central Africa, but he eventually chose Asia as the best place to continue his work in zoogeography:

> During my constant attendance at the meetings of the Zoological and Entomological Societies, and visits to the insect and bird departments of the British Museum, I had obtained sufficient information to satisfy me that the very finest field for an exploring and collecting naturalist was to be found in the great Malayan Archipelago, of which just sufficient was known to prove its wonderful richness, while no part of it, with the one exception of the island of Java, had been well explored as regards its natural history.[23]

There were practical reasons for going to the Malay Archipelago as well. Singapore was an English colony, and part of the northern coast of Borneo recently had been ceded by the sultan of Brunei to an enterprising but controversial Englishman, Sir James Brooke, known as the White Rajah of Sarawak. The numerous Dutch settlements on some of the other islands offered good facilities for the traveler. Finally, the success of generations of white men demonstrated that the country was not an unhealthy place for Europeans.

At this point, Wallace's knowledge of ornithology was surprisingly deficient. He had gone to the Amazon as an entomologist and a botanist, not as an ornithologist. But without greater knowledge of this important branch of natural history, he realized, he would not advance scientifically. He believed that a study of birds was more likely to provide answers to key questions about geographic distribution than a study of insects.[24] An examination of mammals might have been equally fruitful, but birds were more numerous, less bulky, and easier to capture. One mystery he had not entirely solved concerned the factors that prevented the strong-flying hyacinthine macaw from inhabiting a contiguous region in the Amazon, where the only barrier seemed to be a large but not impassable river.

In preparation for his journey to the Far East, therefore, Wallace immersed himself in avian taxonomy. One of his greatest difficulties was finding a compact summary with brief descriptions of the more important known bird species. The most comprehensive work was Prince Charles-Lucien Bonaparte's *Conspectus Generum Avium*, the first two volumes of which were published in 1849. Bonaparte, a nephew of Napoleon, was one of the leading ornithologists of his day. Inspired by the evolutionary ideas of Jean-Baptiste Lamarck, he had set for himself the monumental task of publishing a natural arrangement of all known genera and species of birds (the species numbering more than eight thousand), segregated by subclasses, orders, families, and subfamilies—an undertaking that no naturalist had yet accomplished or even attempted.

The system that Bonaparte adopted is closer to contemporary conceptions of phylogenetic arrangements than to any ornithologist's arrangement in the pre-Darwinian–Wallacean era, though Wallace would later find himself "at a loss to understand by what principles [Bonaparte] had been guided," for it was sometimes impossible to tell "whether two families follow each other because [Bonaparte] thinks them allied, or merely because the exigencies of a consecutive series compels him so to place them."[25] Bonaparte's two volumes amounted to eight hundred pages and catalogued 7 orders, 102 families, 230 subfamilies, and 1,290 genera—all the birds identified up to that time except for waterfowl, grouse, and pheasants, groups that he did not have time to incorporate into later volumes (he died in 1857) but that were relatively unimportant for Wallace's purposes. For each species, Bonaparte listed the known geographic distribution. For almost half of those 8,000 species, he provided short and "excellent" descriptions in Latin by which each bird could be easily identified. In many families, like the cuckoos and woodpeckers, every known species was described; in others, a large number were described. However, from some regions—especially the Malay Archipelago—Bonaparte's lists were incomplete and contained few descriptions of known species.[26]

After purchasing Bonaparte's massive volumes, Wallace set himself the equally monumental task of adding descriptions of every bird previously identified from the Malay Archipelago but not noted, or incompletely noted, in Bonaparte's text. He also listed those species not represented in the collections of the British Museum. Laboriously culling information from a variety of sources, he copied out in abbreviated form in the book's wide margins those characters that he thought would most easily help him identify every bird already identified by Bonaparte or by others. If his specimen did not fit any description, then he could be fairly certain that he had obtained a new or an undescribed species.[27]

For butterflies, Wallace consulted a book by the French naturalist Jean-Alphonse Boisduval, who had described all known species of two families: Papilionidae and Pieridae. Although the book dealt with only the two families, Wallace found its descriptions "so clear and precise" that it was of "immense" interest to him. For other families of butterflies and beetles, he made notes and sketches from the collection at the British Museum, but, as he later wrote, "I soon found that so many of the species I collected were new or very rare, that in the less known groups I could safely collect all as of equal importance."[28]

On June 13, 1853, Wallace read a paper on the Rio Negro and Uaupés River at the Royal Geographical Society, impressing a number of its influential members. He had made observations not only of the course of his canoe but also of as many visible points, hills, inter-island channels, and houses as possible. For distances, he had timed the journey with a good watch, making estimates of the rate of travel up- and downriver. With his pocket sextant, he had determined several latitudes by altitudes of the sun or, at night, by the positions of some of the fixed stars. His object, he said, had been to give a "tolerable idea" of the course and width of the Rio Negro between Barra and the mouth of the Cassiquiare—a tributary that joined the Rio Negro just north of San Carlos, the border town between Brazil and Venezuela—and to map the first four hundred miles of the virtually unknown Uaupés, though the loss of his thermometers had prevented him from determining its altitude. From his data, he had created a large map, which was reduced and lithographed for his presentation and contained much useful information, including the names of the Indian tribes that inhabited the region, the more important botanical products of the surrounding forests, and the isolated granitic peaks and mountains.[29]

Wallace's presence at this bastion of scientific and political conservatism amused Spruce. "You, who go of nights to Geographical Societies' [*sic*] meetings and other long-faced reunions, will perhaps despise our mode of passing the time," he wrote from San Carlos, "and yet I daresay would have liked now and then to listen to tales of *frades* and *moças* [friars and girls], and of men who could turn themselves into *boas* and *cobras grandes*."[30]

But the society was the perfect venue for enlisting the sympathy of men who could help him get to the Far East. The cost of a passage to Singapore or anywhere in Asia was prohibitive—far more than to South America. It may have been Brooke, whom he met at one of the society's meetings, who suggested that Wallace formally contact Sir Roderick Impey Murchison, the president of the society.

The Royal Geographical Society, established in 1830, had as its mission "the promotion and diffusion of the most important and entertaining branch of knowledge—geography." Since England was a great maritime power with a large number of foreign possessions regularly visited by government officials, scientists, scholars, and adventurers, the society would serve as a centralized repository for geographic information, both classified and unclassified. Its members included the leading statesmen of the day as well as naval and military officers on whose reports those statesmen depended.[31] The society's library was an excellent source of information for Wallace, but at its meetings he was able to mingle with the British establishment.

Murchison was one of the original founders of the Royal Geographical Society. Born in Scotland in 1792 and educated at a military college, he eventually studied geology, giving up his career with the army to devote the rest of his life to science, like his close friend Sir Charles Lyell, who also had changed careers. In 1831 Murchison was elected president of the society, a position he held until his death in 1872. His most important scientific contribution was the discovery of a major system of geologic strata, distinguished by the complete absence of fossil vertebrates and land plants, that he named the Silurian. As one went deeper into the strata (that is, farther back in time), there was a gradual disappearance of fossils of any sort, which Murchison took as evidence that one could identify the true beginnings of life on earth—thus contradicting Lyell's assertion that "we shall never be permitted to behold the signs of the earth's origin, or the evidences of the first introduction into it of organic beings [because] so vast a scheme [does not] lie within the reach of our philosophical inquiries, or even of our speculations." Although (unlike Lyell) Murchison believed in the progressive appearance of more complex forms of life (*scala naturae*), he was not and never would be an evolutionist.[32] From various accounts, Murchison was something of a character. His unofficial biographer, Sir Clements Markham, describes him as having "a fine presence, and old world manner. No one ever filled the presidential chair with so much grace and dignity." But he was also pompous in manner and fussy about his evening dress, which included a frilly shirt, white hat, and spencer. He loved to mix with the aristocracy and was a frequent name dropper, though his vanity, Markham felt, was harmless and hurt no one.[33]

Although Wallace liked Murchison, calling him "one of the most accessible and kindly men of science," Murchison's affected demeanor explains the somewhat odd and uncharacteristically formal letter that Wallace composed to solicit some kind of backing—preferably financial—from the society.[34] Writing in the third person on June 30, he proposed leaving England in the autumn or winter of 1853, making Singapore his headquarters, and visiting, in succession, Borneo, the Philippines, Celebes, Timor, the Moluccas, and New Guinea, remaining for one or two years in each place. Wallace wrote: "His chief object is the investigation of the Natural History of the Eastern Archipelago in a more complete manner than has hitherto been attempted; but he will also pay much attention to Geography, and hopes to add considerably to our knowledge of such of the islands as he may visit." He requested from the council of the society a recommendation to Her Majesty's Government for a free passage to the archipelago and aid in obtaining permission from Spain and Holland to visit parts of their territories. "As some guarantee of his capabilities as a traveller," he continued, "he may perhaps be excused for referring to his recent travels for nearly five years, in South America, where alone and unassisted he penetrated several hundred miles beyond any former European traveller, as shewn by the Map and description of the Rio Negro, which he has had the honour to lay before the Society at its last meeting." He then summarized his tragic losses on the return journey, which necessitated his application to the Royal Geographical Society.[35]

Murchison agreed to make the recommendation to the government, but he did not act immediately because he had to make a trip to the Continent. In the council minute book of the Royal Geographical Society for July 22, 1853, is a note acknowledging the receipt of Wallace's letter and the recommendation that Murchison contact the government about a free passage to Singapore and procure letters of introduction from the governments of Spain and Holland to their East Indian colonies. In late August, the Foreign Office informed the society that the Spanish and Dutch governments had given their consent to Wallace's "projected scientific expedition" and had promised to assist and protect him during his visit to their colonies.[36]

But this information apparently was not given to Wallace. Nearly two months passed with no word from the society, so he decided to leave for a two-week hiking trip in Switzerland with his friend George Silk. When he arrived in Paris on his way to Switzerland, a communiqué from Norton Shaw, the secretary of the Royal Geographical Society, awaited him at his hotel, informing him that he was to sail at once. He was offered free passage on a ship as far as Ceylon, but no farther. Dismayed, Wallace replied that a passage to Ceylon would not do because it was inconvenient and the cost of travel from there to Singapore was more than he could afford. He was also in the midst of completing *A Narrative of Travels on the Amazon and Rio Negro* and did not want to leave until it had been accepted for publication. He would take his chances and wait for another opportunity.[37]

After returning from Switzerland and receiving no other offers from the Royal Geographical Society, Wallace again flirted with the idea of going to the snowy mountains of East Africa if transportation to Borneo was not feasible or if the society had no interest in the Malay Archipelago. No doubt regretting his curt response to Shaw, he feared that he might never leave England. It is a measure of his desperation that he was anxious to go somewhere—anywhere—to carry on his work and replenish his dwindling bank account, but Shaw reassured him that the Far East was a suitable destination. All Wallace had to do was notify him when he was ready to go. Africa was never mentioned again.

After Wallace completed his *Travels* manuscript in early November 1853, he sent a letter to Shaw informing him that the book was finished. He excerpted a directive from the Foreign Office, dated September 27, stating that "Mr. Wallace should apply to the admiralty when he is ready to proceed in order that he may be informed whether an opportunity then exists for a passage in one of Her Majesty's ships to Singapore." He said that he was ready anytime to accept whatever passage "the Lords Commissioners of the Admiralty may be pleased" to grant him.[38]

Yet another month passed without any word from Shaw. Wallace wrote again, this time with greater insistence. The problem was not with Shaw, who had twice sent letters to the Foreign Office on Wallace's behalf, but with the Foreign Office, which did not respond to Shaw until January 11, 1854. Shaw had requested that Wallace be granted passage on the *Juno*, which was sailing to Australia, whence he could proceed on his own to Singapore. In anticipation of imminent departure, Wallace sent off his heavy baggage by merchant ship to Singapore, but then the *Juno*'s destination was inexplicably changed. Wallace was distraught. Two days later, he found another ship, the *Frolic*, bound for Australia, which would sail in a few days.[39] He arranged his passage at once and proceeded to Portsmouth, where the *Frolic* was anchored.

The *Frolic* lay alongside a number of other warships and carried twelve guns. Wallace was introduced to the crew and made himself at home. He found the captain to be a small, nervous man, but congenial and possessed of scientific and literary tastes. The ship's doctor also promised to be a pleasant sailing companion. Days passed, and still the ship received no orders to sail. Wallace amused himself by reading a Spanish edition of *Don Quixote* in his cabin. The crew was fully occupied, as though the ship were out at sea. Signals between the *Frolic* and the Admiralty confused Wallace, who waited anxiously for any information about departure. After more than two weeks on board and still no word, he sent a letter to Shaw on February 6, venting his frustration: "We are still at anchor. . . . Day before yesterday 'Annul Sailing' was telegraphed from the Admiralty. So we wait further orders. The Captain knows nothing about it, is anxious to know whether he is delayed for a time or is countermanded all together. . . . Captain Nolloth and the officers are all very pleasant and I anticipate an agreeable voyage so should not now like to be turned off."[40]

On February 8, Wallace was given the bad news that the *Frolic* would not be sailing for Australia, but he was offered passage on the *Juno* once again. He lugged his baggage ashore and searched for the captain of the *Juno*, feeling left "to exercise patience here as much as in the Amazon."[41] But the *Juno* never sailed for the Far East.

Wallace did not leave until March 4, when he was given free passage on a Peninsular & Oriental steamer bound for Cairo and beyond. Murchison had mollified him by obtaining a first-class ticket. Recalling his difficulties in Brazil, at the last minute—almost as an afterthought—Wallace asked to be admitted to the Royal Geographical Society. He evidently hoped that the prestige of the initials FRGS after his name would obviate the challenges of travel in foreign territories. It was the first and last time he ever solicited admission to any society. His request was accepted, and on February 27 he joined the illustrious ranks of the other great English explorers of the nineteenth century as a fellow of the Royal Geographical Society.[42]

Wallace's voyage coincided with the onset of the Crimean War, which explains the comedy of errors he experienced. England did not actually declare war on Russia until March 12, but tensions had been mounting for almost a year. England and France had come to the aid of the Ottoman Empire, which verged on dismemberment by Russia. The Ottomans had declared war on October 4, 1853, inciting the Russians to attack and destroy Turkish ships on the Black Sea. On January 3, 1854, England and France sent ships to the Black Sea, which further antagonized the Russians. Wallace was caught in the chaotic interlude when cabinet members were bickering and diplomacy was spinning out of control. Troops and ships were mobilized, and new crises materialized daily.[43] It was a fortunate accident that Wallace proceeded by merchant vessel and not on the armed *Frolic*; otherwise, he might have stumbled into a situation fully as perilous as his worst misadventures in the tropics.

Wallace's ship sailed by way of Gibraltar and Malta before entering the harbor at Alexandria. By barge he traveled up the Nile, past the pyramids, and disembarked at Cairo, the dirty but picturesque "city of romance." By horse-drawn coach, he continued overland to Suez—this was before the railroad was built or the dream of a canal had been realized. Along the way, he passed the endless trains of camels known as the Indian & Australian Mails, an expensive mode of transporting goods from east to west that he would have to forgo in favor of the less expensive and longer route around the Cape of Good Hope. He luxuriated in his first-class accommodations aboard the steamer *Bengal* and arrived in grand style at Singapore on April 20, 1854, "to begin the eight years of wandering throughout the Malay Archipelago, which constituted the central and controlling incident of my life."[44]

CHAPTER 7

The Malay Archipelago

WHEN WALLACE DISEMBARKED with his "chattels" in Singapore in April 1854, he found himself in an exotic world. Singapore was like few other cities on earth and certainly like no other city he had ever seen. The harbor was crowded with massive war and merchant vessels from Europe, dwarfing the hundreds of colorful praus and junks of the Malays and Chinese moored nearby. Throngs of people pushed past him on the street, including the Portuguese of Malacca, East Indians of various sects, native Malays, Javanese and other peoples from the outer islands, the occasional Englishman, and the ubiquitous Chinese. The town was an anthropologist's paradise.

Wallace had not come to Singapore alone. As he vowed when he was languishing in the Amazon in the grip of malaria, he brought with him an assistant and companion, a young apprentice collector named Charles Allen, the son of a London carpenter who had done some work for Fanny and Thomas Sims. After locating equipment and supplies sent ahead a few months earlier, the first task of the single-minded thirty-one-year-old naturalist and his inexperienced adolescent apprentice was to find a place to live. Accommodations in Singapore were not inexpensive; Wallace and Charles had no choice but to spend the first week in a costly hotel.

From the perspective of a naturalist, the town of Singapore had little to offer. Over the preceding half century, the expanding population had denuded the once thick-forested island. It was not much better beyond the city limits, where no insect or animal worth collecting was to be found along the barren, dusty roads leading to plantations of nutmeg and Oreca palm. Eventually, Wallace and Charles rented a room in a French Catholic mission, nestled among foothills eight miles from town. Atop of these hills was intact forest, which Chinese woodcutters were assiduously clearing, but the wood, sawdust, and decaying leaves they left behind proved to be ideal habitats for insects and their larvae. In a matter of days, Wallace made a remarkable collection.[1]

Wallace quickly settled into a routine. At half past five, he would awaken, bathe, and have coffee, and then arrange and put away the insects collected the day before, setting them out in a safe place to dry. While he was doing this, Charles mended nets, filled pincushions, and packed up the supplies for the day's outing. At eight, they would have breakfast and within an hour leave for the jungle. By the time they reached the top of the mission's hill, they were drenched in sweat. For four or five hours, they would wander amid the felled trees and tangled vines gathering up insects—mainly beetles, some of which Wallace judged "very rare and beautiful"—and then return to bathe, change clothes, and kill and pin insects until dinnertime at four o'clock. After another hour or so of work, which included meticulously recording what he had captured in "registries"—two separate notebooks, one for insects, the other for birds and mammals—Wallace would relax with a cup of coffee, read, and then go to bed, though occasionally his collections were so numerous that he would continue killing and pinning well into the night.

By the end of May, he had shipped to Samuel Stevens 700 species of beetles, among which were 130 highly esteemed species of longicorns, prized for their attractive colors and antennae, which were long, thin, and playful, like a Dalí mustache—totaling approximately 1,000 specimens. Perhaps inspired by Henry Bates's example, he began to document his experiences by sending regular reports to the various journals and societies, beginning with a letter dated May 9, 1854, printed in the *Zoologist*. For the next five years, the English zoological world was kept abreast of both men's exploits, the one on the South American continent, and the other halfway around the globe in some of the remotest places on earth.

Sixteen years old and looking a few years younger, Charles would remain in Wallace's service for only a year and a half. Wallace was a perfectionist, with a clear vision of his objectives, and he placed extraordinary demands on himself and others in order to achieve them. He was a tough taskmaster who hoped to elevate Charles to his own impossibly high standards. Wallace notes in a letter to Fanny that what struck him most about Charles in their first few weeks together was his untidiness: Charles had arrived in Singapore with his clothes somehow in tatters, looking like a homeless urchin shanghaied off the streets of London. At first, Wallace indulged the boy's foibles and held out hope for improvement, allowing Charles to kill and pin flies, bugs, and wasps but not trusting him with the more delicate and valuable beetles. But by the following summer, when they were in Borneo, Wallace would be at his wit's end. He had reported his travails to his mother, sister, and Stevens, all of whom were scouring London for a better prospect. In response to his sister's report that they had found a suitable replacement, Wallace let out a torrent of frustration:

> Do not tell me merely that [the new candidate] is "a very nice young man." Of course he is. So is Charles a very nice boy, but I could not be troubled with another like him for any consideration whatever. . . . From you I should like to know whether he is quiet or boisterous, forward or shy, talkative or silent, sensible or frivolous, delicate or strong. Ask him whether he can live on rice and salt fish for a week on any occasion—whether he can do without wine or beer, and sometimes without tea,—coffee or sugar—whether he can sleep on a board—whether he likes the hottest weather in England—whether he is too delicate to skin a stinking animal—whether he can walk twenty miles a day—whether he can work, for there is sometimes as hard work in collecting as in anything. Can he draw (not copy)? Can he speak French? Does he write a good hand? Can he make anything? Can he saw a piece of board straight? (Charles cannot, and every bit of carpenter work I have to do myself.) Ask him to make anything—a little cardboard box, a wooden peg or bottle stopper, and see if he makes them neat, straight and square. Charles never does anything the one or the other. Charles has now been with me more than a year, and every day some such conversation ensues:
>
> "Charles, look at these butterflies that you set out yesterday."
>
> "Yes, sir."
>
> "Look at that one—is it set out evenly?"
>
> "No, sir."
>
> "Put it right then, and all the others that want it."
>
> In five minutes he brings me the box to look at.
>
> "Have you put them all right?"
>
> "Yes, sir."
>
> "There's one with the wings uneven, there's another with the body on one side, then another with the pin crooked. Put them all right this time."
>
> It most frequently happens that they have to go back a third time. Then all is right. If he puts up a bird, the head is on one side, there is a great lump of cotton on one side of the neck like a wen, the feet are

> twisted soles uppermost, or something else. In everything it is the same, what ought to be straight is always put crooked. This after twelve months' constant practice and constant teaching! Day after day I have to look over everything he does and tell him of the same faults. Another with a similar incapacity would drive me mad.[2]

Singapore was poor in birds and mammals, and Wallace had no intention of limiting his collections to insects. In July, he and Charles left for Malacca, sixty miles north of Singapore, on the coast of the Malay Peninsula facing Sumatra. Malacca was an old city that had changed rulers many times over the preceding 300 years. In 1511 it was seized by the Portuguese; 130 years later, it was taken by the Dutch; 150 years after that, the Dutch entrusted it to the British for protection against an imminent invasion by Napoleon's forces. Although England returned most of the East Indies to Holland, it negotiated hard to retain Singapore and Malacca, and the Dutch yielded reluctantly. In the case of Singapore, they made an enormous mistake—almost as great as their trade of Manhattan for one of the Spice Islands. Within thirty years, Singapore grew from a small fishing village to the largest free port in the Far East. England further expanded its foothold on the Malay Peninsula through a treaty with Siam; eventually, British Malaya would become Singapore and Malaysia, and Dutch Malaya would become Indonesia.

Wallace and Charles traveled for nine weeks in and around Malacca, but Wallace spent two of those weeks in bed recovering from a return bout of malaria. After a government doctor had "killed" his malady with liberal doses of quinine, Wallace walked to a government bungalow fifteen miles in the interior, accompanied by Charles, six Malay porters, and a local guide, a "young gentleman of Malacca" who had a "taste" for natural history and a familiarity with the Malays and their language. They brought enough provisions to last for a month, since food was difficult to obtain in the interior. A few weeks later, after making a good collection of birds and insects and observing the people, who were devout Muslims, he set out on his first true adventure to Mount Ophir, fifty miles east of town in the heart of the peninsula. With his native crew and English assistant, he passed through forested country that grew wilder and hillier by the day. Knee deep in mud, they were annoyed by leeches, which lay in wait on the leaves of plants, standing erect and moving their heads right and left in search of some mammal to attach themselves to. These Wallace and his men avoided; but others they did not see attached by the half dozen to their exposed skin and turned their clothes into a bloody mess. In the evening, while bathing, he found several on his body sucking their fill. One latched onto his neck a fraction of an inch from the jugular vein. How he extracted it he does not say, but like a true scientist he overcame his revulsion and described several of these undesirable companions as "beautifully marked with stripes of bright yellow."[3]

Along the way, Wallace took measurements of the altitude for a topographic map of the region, and for a week he roamed the forests, filling his boxes with specimens that were eventually shipped to England. He hiked to the summit of Mount Ophir, ascending precariously by grabbing roots and creepers and refreshing himself with the water he found in unopened pitcher plants, a kind of "insect soup too strongly flavoured with formic acid."[4] Now and then, between the rolling clouds below, he had magnificent panoramic views.

In a letter to his mother written in July, Wallace had mentioned Cambodia as a possible destination, but when he returned to Singapore from Malacca in September, he contacted the "White Rajah," Sir James Brooke, governor of Sarawak, who persuaded him to go to Sarawak instead. Sarawak was a sliver of territory on the northern coast of Borneo, the interior of which was terra incognita to exploring naturalists. Brooke offered Wallace every assistance for exploring the land under his rule. He invited him to join a small group of English residents and practice Malay, which he had to master if he was to expand his operations beyond the Malay Peninsula into the vast, unknown archipelago itself.

Brooke had lived in the Malay Archipelago for fifteen years. An aristocrat by birth, he led an aimless life as a youth and never completed his formal education. He enlisted in the British army in 1819 at the age of sixteen, but his military career was cut short six years later, when he was severely wounded in Assam during the Anglo-Burmese War. He returned to England, and in five years, when he was fully recovered, he rejoined the army. After arriving in Madras, he changed his mind about army life and boarded a navy vessel to the Far East. For the next year, he traveled in Malacca, Singapore, and China, an excursion that fired his imagination. Back in England, he convinced his father to buy a schooner so that he and some friends could embark on a "wild adventure" somewhere as traders. They sailed for China, their schooner laden with merchandise, but the poorly planned scheme was a fiasco and they were forced to sell their ship and cargo at a loss. In 1835 Brooke's father died, leaving him a considerable sum of money. He bought another vessel and sailed around the Mediterranean, becoming expert in commanding a ship and more artful as a trader. By 1838 he was ready for a longer voyage. Once again, he found himself drawn to the Far East, and this time he chose as his principal destination the Malay Archipelago, most of whose islands were unexplored except for those parts the Dutch occupied. These islands were reputed to be richer in mineral wealth than the Americas and unrivaled in natural beauty; moreover, current political conditions offered the British a chance to gain a stronger foothold in Southeast Asia. Brooke shared the vision of Stamford Raffles, the founder of modern Singapore, of establishing a chain of commercial outposts from India to Australia that would dominate trade in the region. Both Holland and the Malay states were weak, and the northern coast of Borneo, with no significant settlements and a location close to China and Singapore, seemed the ideal place to seek one's fortune. In August 1839, Brooke landed in northern Borneo at Kuching, Sarawak's nominal capital, a small town of fifteen hundred inhabitants. The rest of the country was even more sparsely populated. In the interior were the pagan Dyaks, who did not cultivate enough to sustain themselves and depended on trade for survival; the scattered Islamic Malays, who lived off the proceeds of their small gardens and from fishing; and two or three hundred Chinese, who prospected for gold or mined antimony. Two years later, Brooke was awarded control of Sarawak by the sultan of Brunei in gratitude for his assistance in subduing insurgents who threatened the royal court.

Brooke established a benign, paternalistic dictatorship, convinced that "no Asiatic is fitted to govern a country" without enlightened European guidance. Hoping to avoid the usual corrupting influence of Western civilization on its colonized subjects, he drew up a series of regulations and rules of conduct that he had printed in Singapore in the Malay language. Murder, robbery, "and other heinous crimes," he wrote, were to be punished according to harsh Bruneian law. All men—Malays, Chinese, Dyaks, and Europeans—were free to trade and enjoy their profits. He imposed a tax on every citizen, "more nominal than onerous," guaranteeing protection for those who "act rightly" but vowing not to tolerate those who disturbed the public peace or committed crimes. With only a small staff consisting of four European and eight native assistants, in less than a year he brought peace to his little state, and soon an influx of Chinese merchants and traders brought prosperity. Brooke was pleased with his novel experiment in enlightened government, informing the British public that it "ensures the independence of native princes, and will advance the inhabitants further in the scale of civilisation by means of this very independence."[5]

By all accounts, he was a dashing and charismatic figure. His private secretary and biographer, Spenser St. John, describes him as "handsome, elegant in look as well as in manner, fond of the lighter accomplishments of music and poetry and full of ability, and with his friends, brilliant in talk." He was uncommonly brave and fearless. But when Wallace met him in 1854, he was a changed man, at least physically. A year earlier, he had been stricken with smallpox. His recovery was long and slow, and the disease left him fatigued and permanently disfigured.

Wallace sailed from Singapore on October 16 and arrived in Kuching on November 1. He would remain in Borneo until January 1856, a crucial period in his eight-year sojourn in the Far East, during which he formulated the details of his evolutionary theory. For a part of that time, he was a guest of the rajah, who acted as mentor, colleague, friend, and sounding board. Although Brooke served as the inspiration for Rudyard Kipling's *The Man Who Would Be King*, a tale of greed and imperial arrogance, he was a more complex figure, vilified by some yet admired by many. Wallace was among his most loyal admirers and named a new species of butterfly he discovered in Sarawak after him: *Ornithoptera brookiana*. With a curved band of brilliant metallic green spots across its velvety black wings from tip to tip, it is among the most elegant species of birdwings, so called because of their enormous wingspan—some females of the species *O. victoria* can reach ten inches—and graceful flight.

Wallace's visit to Sarawak came during a peaceful interlude in Brooke's life. Only a few years earlier, he had vanquished a community of pirates who terrorized the northern coast of Borneo, but the manner in which he had achieved this success inflamed passions in Parliament and triggered an investigation. The English treasury had been tapped to provide money for every pirate head captured in battle, and Brooke's little army of headhunters delivered more than five hundred heads; whether all of them had belonged to pirates was another question. The investigating commission had just left Singapore at the time of Wallace's arrival; whatever the outcome of their findings, Brooke was enjoying a quiet life for the first time in more than a decade.

Brooke had built his residence on a knoll overlooking the Sarawak River. According to St. John, it was originally a four-room, lofty house surrounded by broad verandas, with a library in front, a "splendid" dining room in the middle, and two bedrooms in back. By the time of Wallace's visit, Brooke had added a wing for his private use, leaving the other rooms for his staff and visitors. Fruit trees surrounded the house, and fragrant jasmine bordered the lawn and paths. He paid particular attention to his garden, especially the roses, which were his favorite flower. Partly hidden by dense foliage were pigeon houses, kitchens, and servants' quarters. Rooms were comfortably furnished with matted floors, easy chairs, pictures, and books—all arranged, one visitor observed, "with more taste than bachelors usually display." The walls of the large central room were adorned with every type of firearm. Over the years, Brooke had filled the library with all the classics, books by the best historians and essayists, travel narratives, theological works, and various maps, encyclopedias, and other reference books. "I well remember a sneaking parson from Singapore who came on a visit, examining the library, " St. John writes, "and when he found works of Priestley and Channing alongside those of Horsely and Pye Smith, going away and privately denouncing the Rajah as an infidel and an atheist, or, worse still, a Unitarian." In this room, Brooke would spend part of his day reading or playing chess.[6] At 11:00 A.M., he went to the Kuching courthouse to rule on pending cases, then returned home at 2:00 P.M. What Spenser St. John called "the great feeding-time" began at sunset. Brooke took his seat at the head of the table, and his staff and guests were seated according to their rank. After dinner, they all retired to the salon to smoke cigars and talk. At this time, the Chinese dropped in. They crept up to touch their rajah's hand, and then retired to a corner of the room and squatted in silence for a couple of hours to listen to the proceedings before creeping silently out again. The conversation usually centered on religious or philosophical topics and would last until 11:00 P.M., at which point the rajah would send his guests off to bed.

Wallace enjoyed these lively nightly discussions. St. John had fond recollections of Wallace and the debates he stimulated. "If he could not convince us that our ugly neighbours, the orang-outangs, were our ancestors," St. John recalls, "he pleased, delighted and instructed us by his clever and inexhaustible flow of talk—really good talk. The Rajah was pleased to have so clever a man with him, as it excited his mind, and brought out his brilliant ideas." The next day, everyone would catch one another in Brooke's extensive library, checking for authorities to bolster their arguments in the evening's discussions. Among the many books and journals in the library was a copy of *Vestiges of the Natural History of Creation*, which Brooke had read in 1850 and was fond of discussing. With Wallace present, the content of the book aroused heated debate.[7]

Wallace and Charles spent their first four months exploring the various parts of the Sarawak River, from its mouth to the picturesque limestone mountains that it wound through inland. In March 1855, they ventured into Borneo, arriving at a coal mine managed by an Englishman, who put them up temporarily. Wallace was so pleased with the site that he had a small house with two rooms and a veranda built for himself and remained there for the next nine months.

In the untouched virgin forest of Borneo, insects were widely scattered, but the felled trees and large clearings near coal mines attracted a wide array of insects. Wallace had arrived just as the rains were diminishing; the sun shone every day, and butterflies and wasps ventured into open, sunny places. In less than two weeks, he had collected twice the number of beetles he had amassed in the preceding four months, on average taking in twenty-four new species a day. By the end of his nine months, he had collected two thousand distinct species in one square mile, aided by local Dyaks and Chinese workmen, who were paid a penny for each insect they brought him.

On March 12, 1855, Wallace began a special journal—separate from his travel diary and the insect, mammal, and bird registries he kept—to track his collections. In it, he made intermittent entries until sometime in 1859. Less polished than his travel journal, it contains a mixture of observations, anecdotes, extracts from scientific texts, and musings. He intended to use the notes in this new journal for a book called "On the Law of Organic Change," and it is the only surviving document that gives a clue to the ideas on speciation that were brewing in his mind during his first five years in the Malay Archipelago.[8]

One of Wallace's chief reasons for coming to Borneo was to observe the orangutan in its natural habitat and obtain good specimens of the different varieties and species of both sexes thought to exist. So impressed was he by this great anthropoid ape that he recorded five separate accounts of his observations.[9] After a thorough examination of sixteen specimens (nine males and seven females), all but one of which he shot himself, he published a treatise on the orangutan in the *Annals and Magazine of Natural History*—a series of three articles intended to dispel previous misconceptions and inaccuracies. The Dyaks had told him that there were at least three types of orangutan. The first and most abundant the Dyaks called *mias chappan*; it was large, and the face expanded laterally into fatty protuberances or ridges over the temporal muscles—mistermed "callosities," Wallace said. He believed that the *mias chappan* corresponded to the orangutan classified by taxonomists as *Simia satyrus*. The second and smaller type the Dyaks called *mias kassu*, corresponding to *S. morio* and described by the anatomist and paleontologist Richard Owen. As for the third species, Wallace could never get the Dyaks to define the differences with precision. Brooke—also something of a specialist in the orangutan, having presented his observations in a letter addressed to the Zoological Society on March 25, 1841 (and published in the *Annals* in 1842)—agreed initially with the general Dyak belief in the existence of three species. Edward Blyth, curator of the Museum of the Royal Asiatic Society of Bengal in Calcutta, imagined yet a fourth species, based on a few skeletons that he had received from Sarawak, distinguished from the other three by its shorter and more robust limbs and slightly projecting jaw. (In a letter to Darwin dated August 5, 1855, he names this species *Pongo owenii*.) But all these conclusions, Wallace said, had been made from a small number of specimens. By comparing the skulls of several specimens, he showed that there was only one species—with "great individual variation in form and proportion of scull [*sic*] and skeleton [as] decided as those existing between the most strongly marked forms of the Caucasian and African crania in the human species."[10] He believed that the error of the Dyaks had originated from an incorrect determination of sex: they thought that large, ridged skulls belonged to females as well as males, and that small, smooth skulls with short canines belonged to males as well as females. After returning to Sarawak in December, he presented his skins and skeletons to Brooke, who changed his mind after hearing Wallace's argument and studying the facial anatomy. (Today taxonomists agree with Wallace, recognizing only one species—reclassified as *P. pygmaeus*—along with two subspecies: *P. pygmaeus abelii* of Sumatra and *P. pygmaeus pygmaeus* of Borneo.)

The articles that Wallace composed for the *Annals* are progressively daring. Emboldened by his subversion of previous authoritative accounts and with growing confidence in the validity of evolution, he was audacious. The springboard for his most radical speculation was, of all things, orangutan dentition. The purpose of the orangutan's huge canine teeth, Wallace pointed out, was usually ascribed to defense against larger carnivorous animals, whereas he had observed that the orangutan never used its teeth in defending itself but instead relied on the brute strength of its arms. Nor were the canines necessary for acquiring or tearing food, since the orangutan's diet consisted of fruits and soft vegetable matter: "Do you mean to assert, then, some of my readers will indignantly ask, that this animal, or any animal, is provided with organs and appendages which serve no material or physical purpose?" Without offering any further explanation, he added, echoing Sir Charles Lyell, "Naturalists are too apt to imagine, when they cannot discover, a use for everything in nature." But he saved his most provocative comment for the end of the article. Why, he asked, would an animal of such a "high type"—one that so closely approximated a human being in physical structure, though with marked external differences—be confined to so limited a district in the world? If almost all other animals had been represented in previous ages by "allied yet distinct" forms, so must the orangutan, perhaps an indication of the former existence of allied species even more gigantic and more or less human in form: "Every class and every order has furnished some examples, from which we may conclude, that all isolations in nature are apparent only, and that whether we discover their remains or no, every animal now existing has had its representatives in past geological epochs."[11]

The immense variability within the same species, the remarkable similarity in structure and behavior to humans, and the presence of apparently useless characteristics were all due to some other law not yet discovered by the community of scientists. Even humankind, he implied, was represented by allied forms in the past. Evolution—or transmutation—was stamped all over his article. This allusion to ideas set forth in Robert Chambers's *Vestiges*, with its revolutionary implications, and the jab at his fellow naturalists were both arrogant and heretical. Yet no naturalist who had ever contemplated such phenomena could deny that Wallace was raising necessary questions.

The *Zoologist* published parts of these three articles, omitting the offensive theoretical sections and preserving the more innocuous commentary.[12] Wallace also contributed anonymously an article entitled "A New Kind of Baby" to the November 22, 1856, issue of *Chambers's Journal*. For three months, he had raised an infant orangutan that he had rescued from a swamp. Its uncanny resemblance to a human infant was a source of both amusement and serious reflection. The baby "must be a descendant of some very primitive people," he boldly stated, suggesting to his audience that orangutans and human beings were derived from the same ancestral stock.

Wallace was in a mischievous mood. Some months before writing these four extraordinary articles, he had composed the even more extraordinary essay "On the Law Which Has Regulated the Introduction of New Species," which was published in the September 1855 issue of the *Annals and Magazine of Natural History*. It was written at the end of the wet season in February 1855, when he was staying in a little house on the mouth of the Sarawak River at the foot of a mountain. The incessant rains had kept him indoors much of the time, and he had little else to do but "ponder over the problem which was rarely absent from my thoughts."[13] His head was filled with a vast amount of information, culled from his reading of William Swainson, Humboldt, Darwin, and the British Museum catalogues of insects and reptiles, which he knew by heart. At his side were Charles-Lucien Bonaparte's annotated *Conspectus Generum Avium* and Lyell's *Principles of Geology*. The nine months he had spent in the eastern tropics, the four years in the Amazon, and his readings, conversations, and attendance at scientific meetings had all provided him with a vivid impression of the diversity of life on the planet. He was unsatisfied with the manner in which naturalists had utilized these facts. Wallace's essential premise—his Sarawak Law—was that "every species has come into existence coincident both in space and time with a preexisting closely-allied species." As he explained it fifty years later in his autobiography, "[This] clearly pointed to some kind of evolution. It suggested the when and where of its occurrence, and that it could only be through natural generation, as was also suggested in the *Vestiges*; but the how was still a secret."[14]

The immediate stimulus for writing "On the Law Which Has Regulated the Introduction of New Species" was an article by Edward Forbes, a noted biogeographer, marine biologist, and professor of botany at King's College in London, published in the *Proceedings of the Royal Institution of London* in October 1854. Wallace may have seen this article in Brooke's library, or he may have received a copy from Stevens in January or February. The article, "On the Manifestation of Polarity in the Distribution of Organized Beings in Time," struck him as ludicrous. "I was annoyed to see such an ideal absurdity put forth when such a simple hypothesis will explain all the facts," he later observed to Bates.[15] Forbes was a disciple of William Sharp Macleay and his quinarian theory. His abstruse concept of "polarity" was a law or an attribute "of the divinely originating scheme of creation."[16] He attributed the radical differences between the fauna of the distant past and the fauna of recent times—with scanty numbers of species in the intermediate epochs—to two separate creations, and he saw a balance, a divine harmony, in the fossil record, which in his diagram resembled a dumbbell. With considerably less passion than Forbes's article evoked, Wallace wrote, "It is only in consequence of some views having been lately promulgated, in a wrong direction, that [I] now venture to present [my] ideas to the public, with only such obvious illustrations of the arguments and results as occur to [me] in a place far removed from all means of reference and exact information."[17]

He opened his article by advocating Lyell's uniformitarian premise that changes to the earth's surface had occurred over an immense but unknown period of time and operated continuously. As the earth's topography slowly changed, so had the "whole series" of organic life. "[The] present condition of the organic world," he wrote, "is clearly derived by a natural process of gradual extinction and creation of species from that of the latest geological periods. We may therefore safely infer a like gradation and natural sequence from one geological epoch to another." Wallace's introduction concluded with an allusion to the influence of the *Vestiges* on his thinking. He felt that he had gathered enough facts over a decade to provide structural support for an otherwise "wildly" untenable hypothesis: that some form of evolution (a word he did not use in his essay) was responsible for the past and present distribution of plants and animals. His reasoning was simple and elegant. Organizing his observations into several propositions that any naturalist would have to admit were true since they were based on known phenomena, he led his readers by inductive reasoning to his inevitable law.

The broadest categories of organisms (classes and orders), Wallace said, were spread over the entire earth, while families and genera were more limited in range. Similarly, genera of widely disseminated families and species of widely disseminated genera were also more limited. In regions rich in species, the most closely allied species either shared habitats or occupied adjoining habitats. (For example, birds as a class exist on every continent, while the Trochlidae, the hummingbird family, inhabit only the Americas. Throughout the Americas, the various genera and species of Trochlidae occupy overlapping or contiguous territories.)

Wallace also noted that in geologic formations could be found evidence of extinct species related to extant ones. Whether in geologic time or geographic space, a continuity of organisms suggested modifications of some unidentified ancestral forms, with gradual radiation of the new forms from their origins into different and distant zoological regions. Furthermore, no species had ever come into existence twice, a fact consistent with some theory other than the reigning dogma. If each species had been specially created, he concluded, no logical reason should prevent its repeat appearances over the course of the earth's history.

Wallace believed that his law explained four phenomena: the relationships of families, genera, and species to one another; the geographic distribution of plants and animals; the temporal distribution of plants and animals (that is, the present and past "arrangements" of life on earth); and the existence of rudimentary organs. Wallace borrowed from *Vestiges* and transformed into a brilliant metaphor the crude diagram of a vertical line with its diagonal branches demonstrating the manner in which orders, families, genera, and species had arisen, as implied by embryologic development. He suggested that two or three distinct species may have been derived from a common "antitype [*sic*]," or ancestral form, and each of them may have served as an "antitype" for other closely allied species: "If we consider that we have only fragments of this vast system, the stem and main branches being represented by extinct species of which we have no knowledge, while a vast mass of limbs and boughs and minute twigs and scattered leaves is what we have to place in order, and determine the true position originally occupied with regard to the others, the whole difficulty of the true Natural System of classification becomes apparent to us."

With this metaphor, Wallace cast aside more than two thousand years of thinking. He had advanced beyond the author of *Vestiges* by rejecting the concept of a great chain of being. Theories like Macleay's and Forbes's, which arranged species or groups in circles, not only were illogical but were not confirmed by the actual relationships of organic beings in nature. His arboreal metaphor illustrated two principles of nature: organisms were related by common descent, and they became modified and diverged into separate species.

In light of his theory, Wallace introduced the subject of the geographic distribution of organisms, citing Darwin's investigations in the Galápagos Islands. To the best of his knowledge, no one had yet explained the phenomenon of the Galápagos, which contained "little groups of plants and animals peculiar to themselves, but most nearly allied to those of South America." These ancient islands, he said, originally had been colonized by castaways from the nearby mainland, carried to the islands by the action of winds and currents. Over an unknown period of time, the original inhabitants became extinct and were replaced by closely allied or modified "prototypes." But the Galápagos were only the most famous illustration of a more widespread biological phenomenon. By examining specimens from every region of the earth and fossils that had been recovered from geologic formations, one could demonstrate the temporal and spatial relationships of present-day animal and plant species to closely allied extant or extinct species. Intermittent geologic changes must have modified environmental conditions, he reasoned, leading to the extinction of some species and the survival and success of others. Arguing against catastrophic events on a global scale, he suggested that such extinctions were gradual events, played out over long periods of time. "To discover how the extinct species have from time to time been replaced by new ones down to the very latest geological period," Wallace wrote, "is the most difficult, and at the same time the most interesting problem in the natural history of the earth."

Although his theory appeared to support the notion of a progressive development from lower to higher forms of life, retrogression, Wallace said, was not inconsistent with this theory. According to his metaphor, the outermost twigs of a branch might be lost, while others, closer to the trunk and therefore less complex, might develop into different species. Hence it was possible for an existing species to be more "primitive" than its predecessors. Forbes's paucity of species in the intermediate epochs was an illusion, he argued, caused by the imperfect fossil record. During periods of "geological repose," species proliferated; during periods of geologic activity, extinctions occurred. Extinctions sometimes exceeded creations, and vice versa, but there was continuity throughout the biological record. His theory did not depend on the completeness of the fossil record, but was "founded upon isolated groups of facts, recogniz[ed] their isolation, and endeavour[ed] to deduce from them the intervening portions."

In the final section of his article, Wallace stated that his theory offered an explanation for rudimentary organs, anomalies that baffled naturalists. The striking similarity between, for example, the jointed finger bones of the paddle of the manatee and the more developed hands and feet of other mammals raised an important question. "If each species has been created independently, and without any necessary relations with preexisting species," Wallace asked, "what do these rudiments, these apparent imperfections mean?"

Wallace sent "On the Law Which Has Regulated the Introduction of New Species" to Stevens, who presented it to the editors of the *Annals*. In the September 1855 issue, they placed it in an unassuming position between "On the Genus *Assimia*" by John Edward Gray, director of the zoological section of the British Museum, and "On Some New Species of *Hemipedina* from the Oolites" by a naturalist named Thomas Wright. Nothing happened. Edward Forbes, one person who might have responded, had died unexpectedly on November 18, 1854, one month after the publication of his article, which Wallace would not have known when he wrote his essay. The apparent lack of response doubtless precipitated Wallace's provocative commentary in the orangutan articles. Hoping to generate a scientific dialogue, he was becoming increasingly aggressive.

But Wallace was partly to blame for his colleagues' silence. By omitting the word "transmutation" from his essay, a word that made the hair stand up on the back of the necks of most English naturalists, he failed to pull in his readers, or at least those who did not grasp the paper's implications at first glance. His omission introduced an unintentional ambiguity in an otherwise clear exposition. It may have cost him greater recognition.

Wallace wrote to Brooke from the field to express his disappointment and enclosed a copy of the article with his letter. Brooke remained unconverted to Wallace's ideas, but he was indignant about the state of English science. "My great surprise is . . . at the bigotry and intolerance with which views or facts apparently adverse to received systems and doctrines are received," he wrote. "You say your little pamphlet is to feel the pulse of scientific men in regard to this hypothesis! What a reign of intolerance to need such caution! It is this which makes me despair of advance. What harm can [it] do us? What good can it not do us? And yet the inquiry is as beset with bristles as a porcupine's back."[18]

Eventually Stevens sent word of the effects of his article. What naturalists wanted him to do, Stevens relayed, was to "collect more facts," not theorize.[19] But Wallace's Sarawak paper was not ignored by everyone. Lyell had read it, and he was struck to such a degree that he felt compelled to begin his own species notebook.

Since its publication beginning in 1830, Lyell had been making additions and corrections to his *Principles of Geology* to keep pace with the paleontological record, which was expanding at an unprecedented rate. The fourth edition, which Wallace owned, would soon be out of date, and Lyell struggled to maintain his primacy in a field that all acknowledged he had almost single-handedly elevated to a true science.[20] The new fossil discoveries seemed to support the doctrine of progressive development, and, in the view of some geologists and naturalists. this progressive development pointed to a divinely inspired plan, with humanity as the capstone of creation. But Lyell felt that the evidence also could be used to support Jean Baptiste Lamarck's heinous idea about "transmutation," or common descent, and the appalling conclusion that human beings and orangutans were related.[21] Lyell remained adamantly opposed to Lamarck. Species were stable entities, he insisted, not "plastic," and somehow appeared steadily throughout the course of geologic history. After their creation by an invisible guiding force, they lived out their natural species life history, became extinct, and were replaced by others newly created. On November 26, 1855, shortly after having read

Wallace's article in the *Annals*, Lyell began a journal to index books and articles on the species question. Wallace's use of the uniformitarian principle to argue persuasively against Lyell's firmly held beliefs was unsettling, and Lyell's first entry included extracts and notes from Wallace's paper. Two days later, he also recorded his private thoughts about Wallace's theory in a second journal. Cracks were forming in the edifice of his grand scheme about the history of the earth.[22]

Another individual who read "On the Law Which Has Regulated the Introduction of New Species" was Charles Darwin. "Nothing very new," he wrote in the margins. "Uses my simile of tree. It seems all creation with him. . . . It is all creation, but why does . . . his law hold good; he puts the facts in striking point of view. Argues against our supposed geological perfect knowledge. Explains Rudimentary organs on same idea (I sh[oul]d state that put generation for creation and I quite agree)."[23] Around this time, in December 1855, Darwin was in the midst of an intensive and exhaustive study of variation in domesticated species and had drawn up a memorandum, addressed to a score of naturalists around the world, requesting specimens of any domestic breed of poultry, pigeons, rabbits, cats, and dogs bred for many generations in remote parts of the globe. One of the men he contacted was "R Wallace," and he read Wallace's paper shortly before or shortly afterward.[24] Darwin would have known something about Wallace's explorations and discoveries from commentaries in the *Zoologist* and other journals or through the grapevine in the scientific community—one of the men he corresponded with was Brooke—but whether Darwin knew anything at the time about Wallace's work in South America is unclear.

The request from Darwin flattered the younger naturalist, and Wallace was eager to comply. He obliged with specimens of not only the domesticated duck—a peculiar breed with a long, flat body and erect, penguinlike gait—but also the jungle cock, which had been domesticated on the island of Lombok, where Wallace would go that spring; it probably was one of the original species, he explained, from which the domestic breeds of poultry were derived.[25] Darwin was delighted, for he was amassing enough material to test his great theory; his only complaint was that the acquisition of all these specimens from naturalists around the world was costing him a small fortune.[26]

In mid-April 1856, Lyell traveled to Darwin's house to discuss Lamarck's theory of "transmutation." After three days of study, relaxation, and conversation, Darwin explained to Lyell for the first time his theory of natural selection as the evolutionary mechanism. With Wallace's article still fresh in his memory, Lyell unexpectedly urged Darwin to publish at once, sensing that Wallace and Darwin were on the same trail. Claiming that he had not yet gathered enough facts, Darwin told Lyell that he was not ready.

"I wish you would publish some small fragment of your data, pigeons if you please & so out with the theory & let it take date & be cited & understood," Lyell wrote a few weeks after his visit, reiterating his concern about Wallace.[27] Darwin thanked him for his sympathy but remained opposed to publishing simply for the sake of priority. "To give a fair sketch would be absolutely impossible, for every proposition requires such an array of facts," he replied with some distress. "I do not know what to think. . . . I certainly should be vexed if any one were to publish my doctrines before me."[28]

A few days after Lyell's return to London, Darwin met with Joseph Hooker, Thomas Huxley, and another naturalist to discuss different aspects of the species question. Darwin was sending out a trial balloon, as Wallace had done with the publication of his paper. Apparently, the trio was not prepared to embrace the whole "Lamarckian" doctrine. But Lyell's advice disturbed Darwin. He wrote to Hooker to ask his opinion on the matter, stating that he feared abuse from the journal editor or scientific council members for a mere sketch of his views. He did not wish to suffer the fate of the author of *Vestiges*. "It [would be] dreadfully unphilosophical to give a résumé, without exact references, of an unpublished work," Darwin wrote.[29] Hooker agreed with Darwin that his ideas were too preliminary to publish, but Darwin could not shake off Lyell's letter. Rejecting the idea of a short, unannotated sketch or outline, he decided to write a more extensive work, which would organize eighteen years of data into a compelling argument to support the theory of evolution. Another five months passed before he wrote the first chapter of his proposed book, which he entitled "Natural Selection."

Few men reacted appropriately to Wallace's *Annals* paper, which required too radical a break with traditional thinking. Even Darwin did not fully appreciate its originality, and he was slow to grasp that Wallace was the serious threat Lyell believed him to be. Many years later, Huxley reflected on Wallace's essay:

> After a reperusal . . . I cannot confess myself very much surprised that this remarkable paper should have attracted so much less attention than it deserved and that it failed to exert a decisive influence upon the course of biological speculation. . . . But the acceptance of this statement as a law of nature carries certain obvious consequences. For ei-

> ther species have been produced independently of one another by separate acts of creation, or the later species have proceeded from the earlier by the modification of the latter. No other alternative than one of these two is, so far as I am aware, conceivable now. I think it may be said without fear of contradiction, that if the book of Genesis had never existed no sane man would hesitate to prefer the hypothesis of modification to that of independent creation.[30]

In the meantime, Wallace had spent a week in early December 1855 at Brooke's private mountain retreat, twenty miles upriver from Kuching and perched on a summit that was reached by clambering over a rocky slope and up a rudimentary staircase made of notched tree trunks. Brooke's redoubt was surrounded by a tangled mass of luxuriant vegetation. In the cool, fresh air hung the stench of durians, a fruit that was ambrosia to the Malay palate but repugnant to the uninitiated Westerner.[31] Here the two men relaxed, bathing in a nearby spring under an overhanging rock and enjoying the mangosteens and lansats—"two of the most delicious of the subacid tropical fruits," in Wallace's opinion—brought to them in heaping basketfuls by the local Dyaks, who worshiped Brooke as some sort of deity. Brooke gave Wallace permission to use this cottage for a few weeks after Christmas, but not before Wallace returned with him to celebrate the holidays at his palace in Kuching. Every Christmas, all the Europeans in Sarawak were invited to enjoy the hospitality of the rajah, who, as St. John noted, "possessed in a pre-eminent degree the art of making every one around him comfortable and happy." When Wallace returned to the mountain cottage after Christmas, he took with him Charles and a Malay boy named Ali, intended as Charles's replacement. Only fourteen years old, Ali surpassed Charles in every way, and for the next seven years he would be Wallace's trusted companion, servant, and assistant.

Brooke's refuge was ideal for collecting moths. In his eight years of travel in the Far East, Wallace would never find a spot as entomologically productive. At night he sat out on the veranda, reading beside a lamp placed on a table—his pins, forceps, net, and collecting boxes at hand—and waited to see how many moths were drawn to the solitary light. The best times, he discovered, were the rainiest, but only when the moon was completely shrouded by clouds. On four occasions, he captured more than 100 moths, most of them separate species. Some settled on the wall or the table; others flew up to the rafters as he chased them all over the veranda. In the next four weeks, he collected 1,386 specimens. He owed his success to the house itself, with its low, sloping roof and white walls, which both attracted moths and prevented their easy escape, prompting him to recommend—not entirely in jest—that entomologists traveling in the Malay Archipelago take along a white canvas veranda-shaped tent.[32]

Leaving Sarawak at the end of January or early February 1856, Wallace was determined to investigate the lesser-known islands of the eastern archipelago: Celebes, the Moluccas, Timor, and New Guinea. His two years in Singapore and Sarawak had been a necessary training ground for excursions farther afield. Having obtained a good sense of the flora, fauna, and geography of Borneo and the Malay Peninsula, he was prepared to fulfill his true mission. Wallace wrote to his London friends:

> I look forward with unmixed satisfaction to my visit to the rich and almost unexplored Spice Islands,—the land of the Lories, the cockatoos and the birds of paradise, the country of tortoise-shell and pearls, of beautiful shells and rare insects. I look forward with expectation and awe to visiting lands exposed to destruction from the sleeping volcano and its kindred earthquake; and not less do I anticipate the pleasures of observing the varied races of mankind, and of becoming familiar with the manners, customs and modes of thought of people so far removed from the European races and European civilisation.[33]

The port of entry to this region was Macassar, the modern city of Ujung Pandang, on the southern tip of Celebes. His boat from Kuching, however, arrived in Singapore on the very day that a vessel was leaving for Macassar, and he was now stranded for months in Singapore. For the next few weeks, he arranged his specimens for shipment back to England; they included two casks with five orangutan hides in spirits, a box containing sixteen orangutan skulls and two skeletons, six boxes holding fifteen hundred moths and five thousand other insects, a box of dried ferns, and a large case of bird and mammal skins, shells, and reptiles. For the orangutan hides alone, he expected to earn at least £250, and Stevens had found an enthusiastic insect collector, William Wilson Saunders, who became his entomological patron and ensured him a steady source of income.

Wallace summed up his collecting experiences in Borneo in an article for the *Zoologist*. He had acquired few novel birds and mammals except for the orangutans, but he had gathered enough "presumptive evidence" to hypothesize that there had been a land connection linking Sumatra, Borneo, and the Malay Peninsula in the not too distant geologic past. For an ornithologist seeking new species, Borneo thus was not the best collecting ground. He also addressed the complaints of some of his armchair naturalist colleagues, who were disappointed with his insect collections, "almost as if I made as well as collected them." Where were all the big beautiful insects they were accustomed to seeing from the tropics? Why was he wasting their time with small, dull specimens? He reminded his readers of the true rewards of the dedicated naturalist. Despite illness and privation, despite the craving for intellectual and "congenial" society, he was more than compensated by the pleasures of looking forward to a time when the specimens he had amassed would furnish him with the "inexhaustible food for study and reflection" and remind him of the strange and beautiful scenes in which they were obtained.[34] To Stevens, Wallace privately berated one of his critics; "[His] remarks are very amusing, but he is no entomologist to despise small insects. He errs as most people do in believing that the tropical insects are generally large and beautiful. . . . Before I left London the constant cry was 'Do not neglect the small things.' 'The small things are what we want because they have never been collected in the tropics.'"[35]

The complaints from home exacerbated his frustration at being trapped in Singapore. But he was not too bored or frustrated to compose another essay, "Attempts at a Natural Arrangement of Birds," in which he applied the principle of common descent, or his Sarawak Law, to bird classification.[36] Although he had not yet discovered the mechanism driving evolution, he already viewed the world through Darwinian eyes two years before Darwin announced an identical theory and long before other naturalists had abandoned creationism. Some of the branches on the metaphorical tree had been lost or had not yet been recovered, but absence of proof was not proof of absence. Where other authorities saw unbridgeable gaps between species, Wallace saw potential connections.

Wallace set sail from Singapore in late May 1856 on a vessel bound for Bali and Lombok, islands off Java that he would not have visited had he found a more direct passage to Macassar. From his perspective as an exploring naturalist, Bali and Lombok were inconvenient detours from his primary destination. Java itself had been thoroughly worked over by the English naturalist Thomas Horsfield, as well as by the Dutch earlier in the century; like other naturalists, Wallace regarded Bali and Lombok as zoological extensions of their larger neighbor. But his accidental voyage was to lead to one of the most important discoveries of his scientific career.

Before his departure, he wrote to Bates, summarizing two years of observations. Between Santarém and Pará and Santarém and Barra, he said, there were more differences in the species of all classes than among Malacca, Java, Sumatra, and Borneo combined. In comparison with the Amazon, the Malay Archipelago suffered from an "excessive poverty" of diurnal lepidoptera. Only in species of the genus *Papilio* were the two countries approximately equal. But in other orders, like the coleoptera, he believed that the archipelago surpassed what Bates had collected around Ega on the Amazon. His own collecting had suffered from six months of voyages and illness and six months of excessive rain; otherwise, he might have been more productive. He then listed for Bates what he had obtained thus far. In the space of two years, he had amassed approximately thirty thousand specimens, of which some six thousand were distinct species. The birds of the islands he had so far explored were well known and not worth collecting, but he had settled the question of the number of orangutan species. He liked the people and customs of the Malay Archipelago less than those of Brazil. Provisions and labor were expensive, traveling was tedious and expensive, and the high wages of servants prevented him from living in the "free-and-easy-style" of Amazonia.[37]

In mid-June, Wallace arrived on the lush island of Bali, which astonished and delighted him. Never had he seen so beautiful and well cultivated a region outside Europe. As he rode on horseback into the interior, he marveled at the terraced rice paddies and the complex but efficient system of irrigation, which ensured a perpetual harvest of rice and other staples. Much to his regret, he had budgeted too little time to add anything significant to his collections, but he was pleased to have caught a glimpse of the only existing remnant of the great Hindu civilization in the Far East. On June 16, he crossed the fifteen-mile strait to the island of Lombok, a two-day cruise with spectacular views of the mist-enshrouded volcanoes of both islands. After anchoring a quarter of a mile offshore, he boarded a small boat to the black volcanic sand beach. For the next several days, he wandered around Ampanam, Lombok's main town, shooting birds. Among his first specimens was a bird related to an Australian species. Although this was a novel and surprising acquisition, the scarcity of birds disappointed him, so he arranged to visit another district where they were said to be more plentiful. With a small quantity of luggage, he was rowed to his destination, a one-day journey from Ampanam.

There the surrounding landscape pleased Wallace: volcanic cinder cones, which had eroded into smooth hills over the long period of geologic time that Lyell had posited; enclosed valleys; and open plains. The slopes were covered with dense, scrubby bamboos, prickly trees, and shrubs; in the intervening basins were groves of majestic palm trees teeming with birds, the most conspicuous of which were flocks of screaming white cockatoos. On his treks through the region, he came across enormous mounds that no one could explain, but he soon discovered that they had been constructed by the orange-footed scrubfowl (*Megapodius reinwardt*), a bird the size of a hen, with large feet and long, curved claws that it employed to rake together all kinds of rubbish—dead leaves, sticks, stones, rotten wood—and heap it into mounds six feet high and twelve feet wide, in the middle of which its brick-red eggs were buried. Wallace told the local people that birds made the mounds, but they refused to believe him. More difficult to explain to himself was the discovery of this *Megapodius*, an "Australian" genus, on an island so close to Java, which ought to be a zoological extension of Asia.[38]

The most beautiful bird he encountered was a species of *Pitta*, or ground thrush. It was elusive, hiding in dense thickets and requiring great patience for capture. Wallace's practice was to walk cautiously along the narrow pathways near its habitat. When he detected a ground thrush, he stood motionless, imitating the bird's whistle and waiting for a half hour or more until it came within shooting range. In addition to the ground thrush, there were green doves, little crimson and black flower-peckers, large black cuckoos, metallic king crows, golden orioles, and fine jungle cocks. The last he obtained for "Mr Darwin."[39]

But he paid a price for this diversion. He missed his connection to Celebes; he would have to wait for two and a half months for the next boat. The delay, though aggravating and a blow to his finances, gave him time to reflect on a remarkable and unexpected discovery. Lombok was inhabited by species allied to the fauna of Australia, while Bali contained species allied to the fauna of Asia. The short, deep strait between these two islands was a faunal divide, partitioning Asia from Australia. He had stumbled on what appeared to be the exact division between two distinct zoogeographic regions. "The islands of Baly and Lombock," he observed in a letter to Stevens, "though of nearly the same size, of the same soil aspect, elevation and climate and within sight of each other, yet differ considerably in their productions, and in fact belong to two quite distinct Zoological provinces, of which they form the extreme limits."[40]

Wallace was impressed not only by the absence on Lombok of the most common birds found in the great "Indian" islands of Borneo, Java, and Bali but also by the absence of important groups of mammals, such as monkeys, large cats, and rodents. He was also impressed by the presence there of cockatoos and birds of the family Megapodiidae, which did not exist on Bali. "South America and Africa, separated by the Atlantic, do not differ as widely as Asia and Australia," he would write in a paper two years later. "[In] a few hours we may experience an amount of zoological difference which only weeks or even months of travel will give us in any other part of the world!"[41] Wallace's discovery contradicted everything he had ever read about the kinds of fauna that should occupy neighboring islands that otherwise appeared nearly identical. That a mere fifteen-mile gap could be a chasm between two fundamentally different faunal groups was astonishing, and it bolstered his conviction that a mechanism other than a divine plan explained the phenomenon of animal distribution.

At the end of August, he boarded a little schooner and finally left for Macassar, which proved a disappointment. After four months there, he had little to show for his efforts. By December, when dark clouds appeared on the horizon portending the onset of the rainy season, he longed for better hunting grounds. He therefore jumped at the chance to go to the Aru Islands, an opportunity that had turned up unexpectedly.

Macassar was the only Dutch port that traded regularly with Aru. A thousand miles to the east, near the coast of New Guinea, the Aru Islands were one of the great objects of Wallace's journey to the Malay Archipelago. Because of the monsoons, native vessels made the voyage only once a year, leaving Macassar in December or January and returning in July or August. He would travel by native prau, owned and commanded by a Eurasian captain, half Dutch and half Malay, who offered to take him there and bring him back six months later. The captain assured him that two types of bird of paradise were abundant in the Aru Islands—the large yellow great bird of paradise (*Paradisea apoda*) and the smaller red king bird of paradise (*P. regia*)—both of which he thought could easily be shot or bought from the natives. Thus encouraged, Wallace agreed to the passage, whose cost the captain left up to him. He dashed off a letter to Stevens: "What I shall get there it is impossible to say. Being a group of small islands, the immense diversity and richness of the productions of New Guinea will of course be wanting; yet I think I may expect some approach to the strange and beautiful natural productions of that unexplored country. Very few naturalists have visited

Arru [*sic*]. . . . I suppose not twenty specimens of its birds and insects are positively known."[42] But it was birds of paradise that Wallace coveted most. No European naturalist had ever seen a specimen in its native habitat. Ornamental feathers from these magnificent birds were the rage in Europe, adorning women's hats and adding a dash of exoticism to their wardrobe, and the native New Guineans were glad to oblige them. Trade in the feathers escalated during the nineteenth century, and the supply seemed endless until the threat of extinction brought people to their senses—but that was no one's concern in 1856. The first skins of one species had begun to arrive in Europe from New Guinea (Papua) in the eighteenth century. Native birdcatchers, interested more in spectacular ornamental feathers than in scientific accuracy, amputated the feet and sometimes the wings for convenience before trading with merchants. *Paradisea apoda*, as Linnaeus christened the bird, thus gave rise to the myth of a footless bird, perpetually on the wing. Papuan ornithology remained a mystery until the 1770s, when a French traveler named Pierre Sonnerat returned home from the Moluccas and the Philippines with several unusual specimens. Others followed Sonnerat, most significantly René-Primevère Lesson, a resident naturalist aboard the vessel *Coquille*, which set sail from France in 1822 on a voyage of scientific exploration and discovery. Lesson brought back to Paris skins from more than a dozen species of bird of paradise, all obtained from the natives and therefore not accurately localized. Today the family Paradisaeidae is known to contain twenty genera, subdivided further into forty distinct species, and geographically limited to New Guinea, the Moluccas, and northern Australia. They are among the most beautiful and bizarre birds on earth, the orchids of the world's avifauna.

The trip to Aru was Wallace's riskiest journey to date. In all his travels, he never had ventured beyond the confines of European power. He felt as he had when, after leaving school at the age of thirteen, he was packed off to "that scene of all that is strange and new and wonderful to young imaginations—London!" Aru, separated into three distinct parts by narrow channels and surrounded by numerous islets—on a map greater Aru resembles a cracked egg—was inhabited by "lawless traders and ferocious savages." He did not share this observation with his mother and sister, who were beseeching him to return home in every letter they posted to him. On December 10, he wrote euphemistically that he was "going out of reach of letters for six months." He was well and in good spirits, dining in luxury on rich cow's milk, good bread, and magnificent mangoes. The mangoes were his reason for leaving, he added lightly, as he feared he would overindulge.[43]

After a smooth and enjoyable voyage, he arrived at Aru in early January 1857. On March 10, he sent a letter to the Entomological Society summing up his six weeks on the islet of Wamma in the town of Dobbo, a village jerry-built on a spit of land wide enough for three rows of houses. He shared with his colleagues an interesting observation that he had made about the zoogeography of the region: no two islands were exactly alike despite their proximity and similarity of habitat. For example, the Ke Islands (another island cluster, which he had passed before reaching Aru) and Wamma, only sixty miles apart, differed markedly in insect species.[44] Wallace did not elaborate on the implications of that difference in this letter, but in his species notebook he had noted the resemblance of the Malay Archipelago to the Galápagos Islands. "Here we must suppose special creations in each island of peculiar species," he observed, "though the islands are all exactly similar in structure, soil and climate and some of them within sight of each other. It may be said it is a mystery which we cannot explain, but do we not thus make unnecessary mysteries and difficulties by supposing special creations contrary to the present course of nature?"[45] Those of his entomological friends who might have been made uncomfortable by his theorizing were doubtless pleased with the news that nine-tenths of his specimens were new to English collections.

Shortly after his arrival in Dobbo, Wallace inquired about birds of paradise. To find them, he would have to travel deeper into the interior of Aru, reachable only by boat; but he was delayed for weeks because of pirates, who had just attacked a prau and stripped its men of everything on board, including their clothes. The pirates had escaped into the labyrinth of Aru's channels and islands, where they could continue to harass the local population without fear of capture. Not a soul would take him anywhere for any price, but in mid-March, when the danger of piracy abated, he finally secured a boat and a willing crew. Setting sail early one morning, he and his crew skirted the western coast of Wokan, the largest and northernmost of the three main Aru Islands, and entered the mouth of a small river. They made their way through a mangrove swamp until late afternoon, when they disembarked at a small village. Here, for the price of a chopping knife, he rented a five-foot sliver of space in a shed "of the most miserable description" that was already occupied by a dozen people. He ordered his crew to sleep in the prau to guard his provisions and gear. On the day of his arrival, he found some men who assured him that they could obtain specimens of birds of paradise, but after they set off into the humid forest he never heard from them again. On the second day, however, Ali unexpectedly returned to the hut with a perfect king bird of paradise. As Wallace examined the exquisite creature, which was not much larger than a thrush, admiring its gorgeous colors and extraordinary tail ornamented with two wirelike feathers hanging downward and spiraling into two glittering buttons, he could barely contain himself. "Thus one of the great objects of my coming so far was accomplished!" he later wrote. "My admiration and delight over this exquisite winged form quite amused my Aru hosts, who saw nothing more . . . than we do in the robin or chaffinch."[46] What excited Wallace more was the knowledge that he was sending back a specimen that far surpassed in quality anything ever seen before in Europe:

> Here I am an established fact, the first European I believe who has ever resided in a Papuan island. Here I am in forests where white and black cockatoos, brilliant lories of scarlet, blue and green, the racquet-tailed kingfishers and the birds of paradise . . . make the air musical with their shrill discordant cries. The emotions excited in the minds of a naturalist who has long desired to see the actual thing which he has hitherto known only by description, drawing or badly-preserved external covering, especially when that thing is of surpassing rarity and beauty, require the poetic faculty fully to express them. The remote island in which I found myself situated, in an almost unvisited sea, far from the tracks of merchant-fleets and navies; the wild luxuriant tropical forest, which stretched far away on every side; the rude uncultured savages who gathered around me—all had their influence in determining the emotions which I gazed upon this "thing of beauty."[47]

This early success was not quickly matched. It rained incessantly. Wallace felt the need to penetrate farther into the Aru jungles, but he could not persuade any of the locals to guide him, pirates being the usual excuse. Eventually, he cajoled one native to captain his prau, and he and his crew set off at the end of March, sailing upriver until they reached Wanumbai, a village consisting of two houses, one of which he occupied after hard bargaining, the landlord yielding for the price of ten yards of cotton cloth. As in the Amazon, Wallace was the object of great curiosity. During his stay, there was an endless parade of onlookers. "I found the tables turned upon me," he noted with amusement, "& I was become even as the Zulus or Aztecs which I had been one of the gazers at in London. I was to the Arru [*sic*] Islanders a new & strange variety of man, & had the pleasure of affording them in my own person an instructive lesson in comparative Ethnology."[48]

From his base in Wanumbai, he trekked into the jungle along paths of black mud that made his walks miserable. He was rewarded with a good collection of birds and beautiful butterflies. He obtained a specimen of the strange marsupial *Cuscus maculatus*, which had a woolly coat like a lamb's, a long and powerful prehensile tail, and stumpy legs with long-clawed feet that resembled hands.

Wallace's delight with the people and satisfaction with his collections were counterbalanced by various physical torments. Insect bites caused the most misery, and he jokingly wrote that insects were taking revenge for his persecution of them. Sand flies attacked him at night, and during his daily walks through banana plantations and black mud, mosquitoes assaulted his legs. In the moist tropical heat, these bites soon became ulcerated, which made walking impossible and confined him to his house for a week; there the incessant chatter of his hosts further tortured him. In the early-morning hours, just before sunrise, he lay pensively in bed, listening to the raucous calls of the birds of paradise as they searched for food; the cries resounded through the forest, changing direction rapidly and repeatedly. At the crack of dawn came the screeches of lories and parakeets and the screams of cockatoos. He was a pathetic sight, crawling down to the river to wash, his agony reaching its highest pitch as he helplessly watched a magnificent *Papilio ulysses* or another beautiful butterfly waft by. There was nothing to do but be patient and attend to one or another necessary task, like skinning or cleaning specimens. The bites, the stings, the pain, the constant irritation of insect pests of the tropics—all these he could bear without complaint. "But to be kept prisoner by them in such an unknown country as Arru [*sic*]," he lamented, "where rare and beautiful creatures are to found in every forest ramble, a country reached by such a long and tedious voyage and which may never be visited again, is a punishment too severe for a Naturalist to pass by in silence."[49]

Wallace was consoled by Ali's enterprise. He brought him one bird of paradise after another, capturing excellent specimens of both the king and the great species and enabling Wallace to send a biblical note of triumph back home: "Rejoice with me, for I have found what I sought. . . . I have a few specimens absolutely perfect. . . . I believe I am the only Englishman who has ever shot and skinned (and ate) birds of Paradise, and the first European who has done so alive, and at his own risk and expense; and I deserve to reap the reward, if any reward is ever to be reaped by the exploring collector."[50]

Wallace's ultimate goal was the vast and mysterious island of New Guinea, where he would seek the other thirteen or so known species of birds of paradise. New Guinea was one of the most dangerous places on earth for a white man; headhunters and cannibalistic tribes murdered anyone who trespassed on their territory. He had learned from his contacts in Dobbo which places were safe and which were not. Their accounts of the abundance of birds excited him so much that he could think of nothing else. His planned route would take him to the island of Ternate; from there, he could reach the westernmost arm of New Guinea, where a naturalist could work in relative safety, the inhabitants having been tamed by missionaries.

While recuperating from his skin infections in Wanumbai, Wallace wrote a treatise on the great bird of paradise for the *Annals*, intending to correct the many erroneous statements published by Lesson and other naturalists. He had discovered that its range was limited to one part of Aru and the southernmost peninsula of New Guinea, while an allied species inhabited the northern peninsula of New Guinea and one or two nearby islands. It was interesting, he said, that although the Ke Islands were closer to New Guinea than were the Aru Islands, no species of bird of paradise was found at Ke. He reemphasized the critical importance of pinning down as precisely as possible the range of a species, relying on one's own observations or those that were verifiable. Lesson, the previous authority on the bird of paradise, was led to inaccuracies by failing to state where and how his specimens had been obtained. Knowing nothing about the birds' habits, he was also unable to describe the most spectacular feature of their behavior: the courtship display of the males, during which vertical plumes behind the wings were spread in a magnificent fan as they performed a ritualized dance on the branches of trees.[51]

The bird of paradise was not the only avian family that stimulated Wallace to reflect on zoogeography. Once he had established himself in the central forests of Aru, new species "burst upon him," forcefully revealing Aru's intimate zoological connection with New Guinea. Although separated by at least 150 miles, Aru and New Guinea had many birds in common. The only other islands in the world known to possess species identical with those of the proximal mainland were Great Britain and Sicily. Ceylon and Tasmania had species—even genera—unique to them, though they were closer to India and Australia, respectively, than Aru was to New Guinea. From a zoological perspective, Aru seemed to be an outlying part of New Guinea, from which it had been separated during a recent geologic epoch.

Wallace's conclusion was further supported by the geologic evidence: soundings of the surrounding seas demonstrated the relative shallowness of the ocean bed between Aru and New Guinea, while the depth westward was "fathomless." The most crucial evidence, however, was the presence of three broad channels traversing the three portions of greater Aru, which could be explained only if they had been true rivers with their source in the mountains of New Guinea and had been reduced to their present condition by the subsidence of the intervening land.[52]

Wallace remained in Wanumbai for another six weeks. At night little marsupial rats ran about and nibbled at everything. Four or five types of ants attacked anything not isolated by water-filled moats, one variety even swimming across them. Huge spiders lurked in baskets and hid in the folds of his mosquito net. Centipedes and millipedes nestled under his pillow and ventured out to crawl on his head. Scorpions sought refuge under every board. "We need hardly to mention bugs and fleas & to tell the truth," he confided to one of his journals, "I would not exchange all the others in my list for these last if as abundant as they often are at home. They are a constant & unceasing source of actual torment & disgust;—spiders centipedes & scorpions are large ugly and dangerous, but you may live among them a long time & get no hurt but fright."[53]

On May 6, with his bird box nearly overflowing and his legs still a mess, Wallace made the fifteen-hour trip back to Dobbo, arriving at dusk. He occupied the only place available, the courthouse of the Dutch commissioners, which was nothing more than a shed at the end of the village. Since his departure in March, traders had arrived in great numbers and more houses had been thrown up along the crowded little spit of land. Fleets of boats were lined up on shore to be repaired and repainted for the homeward voyage. An Australian menagerie picked through the refuse. Parrots, lories, and fruit doves cooed and chattered in cages hung from every doorway. An occasional small kangaroo wandered around the houses. During the day, there was at least one cockfight, with a circle of spectators cheering and betting as the birds tore each other to pieces. At night, there was singing and music. On Fridays, Dobbo's Muslims prayed at the town's mosque, the most distant from Mecca of any in the world.

By the end of June, threatening clouds had gathered on the horizon. The shops were hastily boarded up, and the traders prepared for departure; in a week, Dobbo would be a ghost town. The captain of Wallace's prau arrived, as promised, and on July 2 they departed for Macassar along with fifteen other praus, forming a grand regatta. With fair weather and a strong, steady wind at their backs, they arrived nine days later. Awaiting him were seven months' worth of letters, which he spent the night opening and reading. One was from Bates, congratulating him on the neglected article in the September 1855 *Annals*:

> I was startled at first to see you already ripe for the enunciation of the theory. You can imagine with what interest I read and studied it, and I must say that it is perfectly well done. The idea is like truth itself, so simple and obvious that those who read and understand it will be struck by its simplicity; and yet it is perfectly original. The reasoning is close and clear, and although so brief an essay, it is quite complete, embraces the whole difficulty, and anticipates and annihilates all objections. Few men will be in a condition to comprehend and appreciate the paper, but it will infallibly create for you a high and sound reputation. A new method of investigating and propounding zoology and botany inductively is necessitated. New libraries will have to be written.[54]

Wallace would later confess to Bates that he reread his letter at least twenty times.[55]

There was also a letter dated May 1, 1857, from Darwin on the subject of Wallace's article; he was equally positive, though more reserved:

> I can plainly see that we have thought much alike and to a certain extent have come to similar conclusions. In regard to your paper in the Annals, I agree to the truth of almost every word . . . and I daresay that you will agree with me that it is very rare to find oneself agreeing pretty closely with any theoretical paper. This summer will make the twentieth year (!) since I opened my first notebook on the question how and in what way species and varieties differ from each other. I am now preparing my work for publication, but I find the subject so very large, that though I have written many chapters, I do not suppose I shall go to press for two years.[56]

If Darwin's revelation about the duration and magnitude of his work on the species problem was meant to intimidate Wallace, it did the opposite, acting like a spur.[57] Further emboldened, Wallace produced several original papers in a prolific outburst: "On the Natural History of the Aru Islands" for the *Annals*, "Note on the Theory of Permanent and Geographical Varieties" and "On the Entomology of the Aru Islands" for the *Zoologist*, and "On the Habits and Transformations of a Species of *Ornithoptera*, Allied to *O. Priamus*, Inhabiting the Aru Islands, near New Guinea," for the *Transactions of the Entomological Society*. All were written in Macassar, when he had the leisure to examine and contemplate his specimens before shipping them off to England. For a month, he was occupied with analyzing, notating, and preparing a collection of nine thousand specimens comprising sixteen hundred species, more than three-quarters of them distinct species of insects. His labors exhausted him more than usual. In no other part of the tropics was so much care required to protect his collections. Ants posed a formidable threat to both his birds and his insects; two of these pestiferous species were minute and impossible to eradicate. Poor Ali received a severe scolding on two occasions for uncharacteristic carelessness in having allowed the edge of a palm mat and a slender bit of rattan to touch Wallace's shelves, which were suspended from the ceiling. "One would think the ants must every night explore and wander everywhere," Wallace complained, "for they never fail to discover even a hanging thread by which to ascend." Ants established colonies in bird skins, devouring the eyelids and the bases of the beaks and thus ruining days of work and the efforts of weeks of exploration. Spiders, larvae, and mites were other pests that seemed immune to arsenic, camphor, and any other poison he could think of. Added to these were roaches and rats, which could get at anything not tightly sealed in a box.[58]

But Wallace's meticulousness paid off. The quality of his specimens impressed his colleagues, who pronounced his collections a very important addition to the knowledge of New Guinea fauna and the distribution of organisms in the Eastern Hemisphere. At the February 23, 1858, meeting of the Zoological Society, John Gould, England's leading ornithologist, presented "a highly interesting series of birds collected by Mr. A. R. Wallace in the Aru Islands. Among them were two species of birds of paradise. . . . Hitherto these magnificent birds have only been sent to this country in mutilated condition, their skins having been prepared and dried by the Papuans, frequently without their wings, and almost always without their legs. Mr. Wallace's skins, however, are perfect, and in the highest possible condition." Wallace had also sent 121 species of birds, "showing great perseverance and energy on [his] part."[59] One mammal, the ratlike marsupial—which was one of Wallace's tormentors and subsequently was captured, skinned, and preserved—Gray of the British Museum honored with his name: *Myoictis wallacii.*[60] The Entomological Society was also pleased. At the March 1, 1858, meeting, Stevens exhibited

> some beautiful Lepidoptera and Coleoptera taken by Mr. Wallace in [the] Ke and Aru Islands. . . . Mr. Westwood observed that it was extremely interesting to see the fine Papilios, &c. which had been found seventy years ago by the Dutch in the Islands of the Indian Archipelago, and since almost forgotten, were now being rediscovered and sent to this country in such admirable preservation: the best of thanks were due to Mr. Wallace and those who, like him hazarded their lives in unhealthy tropical climates to collect objects of Natural History, and he trusted they would receive the pecuniary reward they so well merited.[61]

By now, Wallace was working hard on the species problem, trying to assemble the facts he had amassed into some greater schema. He was on the verge of a great discovery—of that he was convinced—a belief confirmed by the letters from Darwin and Bates. In his essay "On the Natural History of the Aru Islands," published in the December 1857 supplement of the *Annals,* he trod on dangerous ground, targeting Lyell, whose version of the theory of special creation he picked apart. How, for example, could the geographic distribution of Vertebrata be accounted for? Why were the same species not found in the same climates all over the world? How did this theory explain the striking entomological, ornithological, and mammalian differences between New Guinea and Borneo, both virtually identical in climate and geography—differences not only at the species level but spanning the zoological spectrum of genera, families, and whole orders? Conversely, how could the remarkable zoological congruence between Australia and New Guinea—the former a desert, the latter a tropical rain forest—be explained? "[We] can hardly imagine that the great variety of monkeys, of squirrels, of Insectivora, and of Felidae, were created in Borneo because the country was adapted to them, and not one single species given to another country exactly similar, and at no great distance." Wallace concluded that "some other law has regulated the distribution of existing species than the physical countries in which they are found, or we should not see countries the most opposite in character with similar productions, while others almost exactly alike as respects climate and general aspect, yet differ totally in their forms of organic life."[62] That law was his Sarawak Law, which he applied to the geographic distribution of species in the Malay Archipelago.

At one time, Wallace hypothesized, New Guinea and Australia were joined, and during this period the climate and physical features of each resembled those of the other. A separation occurred, leading to a change in climate and the extinction of certain groups of fauna. New species were then somehow "gradually introduced" into each, closely allied to the preexisting species, many of which were common to the two places. As time passed, these allied species became peculiar to their particular region, with few identical but many very similar, which explained the relative congruence between the fauna of Australia and that of New Guinea. The law of close affinity would not permit the appearance of groups not closely related to the original inhabitants. For example, a tiger would not suddenly materialize in New Guinea or a kangaroo in Borneo. At a later geologic period, the island complex of Aru separated from New Guinea. "Its separation must have occurred at a very recent period," he concluded, "the number of species common to the two showing that scarcely any extinctions have since taken place, and probably as few introductions of new species."

Wallace asked his readers to imagine the following scenario. Suppose that the Aru Islands were to remain undisturbed for a period "perhaps equal to about one division of the Tertiary epoch." It is reasonable to believe that the change of vertebrate species over such a long time would be complete and that an entirely new "race" of vertebrates—all closely allied to those now existing—somehow would be introduced. Simultaneously, a new fauna gradually would have appeared in New Guinea. This future Aru and New Guinea would then resemble present-day New Guinea and Australia, with their faunal similarities and differences. Let the same uniformitarian process then continue for another geologic period. In one country, some species would have become extinct and not replaced with allied species, while in the other a series of modified but closely allied species somehow would have been introduced. Thus the faunas would come to differ not only at the species level but also at the genus level. If one imagined further geologic changes—say, the elevation of Aru into a mountainous country and its expansion by alluvial plains, and the flattening of New Guinea with a corresponding reduction of its surface area, leading to the extinction of many species—new species might then be more rapidly introduced into the modified and enlarged territory of Aru. Some genera and families that had gone extinct in New Guinea but still were present in Aru could proliferate there, echoing a phenomenon seen in Madagascar, where the families and some of the genera belonged to existing African groups, whereas others were peculiar at the species, genus, and even family level, though retaining a general resemblance to African forms. Wallace recorded in his species notebook: "It is quite unnecessary to suppose that new species have ever been created 'perfectly dissimilar in forms, habits and organisation' from those which have preceded them; neither do 'centres of creation' [here he quotes Lyell], which have been advocated by some, appear either necessary or accordant with facts, unless we suppose a 'centre' in every island and in every district which possesses a peculiar species."[63]

Wallace addressed one final question about the word "species" itself: What is the difference between a "species" and a "variety"? He presented the conundrum in his article "Note on the Theory of Permanent and Geographical Varieties"—it would be published in the *Zoologist* in January 1858—which sought to bring the evolutionists and creationists head to head. "As this subject is now attracting much attention among naturalists and particularly among entomologists," Wallace wrote, almost certainly with Darwin in mind, "I venture to offer the following observations, which, without advocating either side of the question, are intended to point out a difficulty, or rather a dilemma, its advocates do not appear to have perceived."

When did variation end and speciation begin, Wallace wondered. The generally adopted opinion was that species were independent creations that varied only within certain limits during their whole existence, while "permanent" varieties were not independent creations but produced by ordinary generation from species. The term "permanent variety" was an oxymoron: according to the old system of belief, boundaries were not supposed to be blurred. But in a state of nature, boundaries sometimes did blur. He admonished naturalists to stop straddling the fence: either reject special creation altogether and agree that species differed from varieties only by degree, as a result of common descent, or call any group with "permanent characters, however slight," a species.[64]

Shortly after completing "Note on the Theory of Permanent and Geographical Varieties," Wallace sent a paper to the Entomological Society in order to inflame even further the controversy over the origin of species. In this paper, "On the Habits and Transformations of a Species of *Ornithoptera*," he demonstrated that not only was it sometimes difficult to distinguish two species from each other but even genera might be difficult to differentiate. His specimen of a species of *Ornithoptera* was a case in point: the adult form so closely resembled an allied species that it might be considered a variety, yet its caterpillar almost exactly resembled that of a third species, from which it could be distinguished only by subtle changes in the color and size of certain appendages. Furthermore, the caterpillars and pupae of the genus *Ornithoptera* itself almost exactly resembled those of the genus *Papilio*![65]

Privately, in his species notebook, Wallace was more blunt: "Lyell says, that varieties of some species may differ more than other species do from each other without shaking our confidence in the reality of species. But why should we have that confidence? Is it not mere prepossession or prejudice like that in favour of the stability of the earth which he has so ably argued against? In fact what positive evidence have we that species only vary within certain limits?"[66]

Every new fact that Wallace discovered reinforced his belief in evolution. His Sarawak Law had wide applications. But descent with modification was not the mechanism of evolution; it merely refashioned into a new paradigm what was readily observable and already known. The mechanism of evolution remained as elusive as ever. Without that critical element, his arguments would remain unconvincing to the majority of his fellow scientists, whose instinct was to reject a hypothesis based on scientific principles for one justifiable on a priori grounds alone.

CHAPTER 8

The Mechanism Revealed

IN LATE SEPTEMBER 1857, Wallace sent a rambling letter to Darwin that has not survived intact. He must have reported the difficulties of life as a traveling naturalist, because in his reply Darwin offers sympathy and encouragement for his "laborious undertaking." Wallace also seems to have alluded to his two-year-old paper, "On the Law Which Has Regulated the Introduction of New Species," in the *Annals and Magazine of Natural History* and asked for feedback, providing additional data that he had gathered to support his hypothesis. From Darwin's answers to this and other concerns, it is clear that Wallace elaborated on his belief in a former land connection between New Guinea and Australia in the distant past. Moreover, he wondered whether Darwin was planning to discuss the origin of humans in his upcoming book.[1]

Wallace did not have time to wait for Darwin's response. On November 19, he boarded a Dutch steamer bound for Amboyna (Ambon) and Ternate. The boat stopped first at Coupang (Kupang) in Timor and then at Banda, in the heart of the Spice Islands. Once again, Wallace was struck by the marked contrast of the archipelago's islands, which differed in their vegetation despite the fact that all were volcanic in origin. East of Bali, they were almost barren, with scanty and scrubby plants, while the Moluccas, including Banda, Ke, and Aru, were covered with dense, verdant forests. He left Amboyna in early January 1858, but not before writing a long-overdue reply to Henry Bates's letter congratulating him on his September 1855 *Annals* paper. "To persons who have not thought much on the subject, I fear my paper on the succession of species will not appear so clear as it does to you," he told Bates. "That paper is, of course, only the announcement of the theory, not its development. I have prepared and written portions of an extensive work embracing the subject in all its bearings and endeavouring to prove what in the paper I have only indicated."[2]

It was the first time that he had admitted to anyone his intention to publish a book on the origin of species, though in his species notebook he had mentioned this goal as early as 1856. In the book's final chapter, he planned to disprove Sir Charles Lyell's arguments against evolution.[3] He confided to Bates that Darwin's letter of May 1, 1857, had "much gratified" him. He felt that Darwin might "save him the trouble" of writing the second part of his hypothesis by proving that there was no difference between the origin of varieties and of species, or that Darwin might give him trouble by arriving at another conclusion. "Your collections and my own," he told Bates, "will furnish valuable material to illustrate and prove the universal applicability of the hypothesis."

Wallace reached the island of Ternate on January 8, 1858, his thirty-fifth birthday. Ternate was midway between Celebes and New Guinea and approximately 250 miles from each. Although only 65 square miles in extent, it had a past worthy of Timbuktu. Ternate and its tiny neighbor, Tidore, had been the center of the world's spice trade when the Portuguese and Spanish arrived in the early sixteenth century. But by the time of Wallace's visit, they were dismal backwaters under Dutch authority, with magnificent ruins the only evidence of their splendid past.

Wallace gave introductory letters to a man named Duivenboden, known as the King of Ternate because of his immense wealth and excellent connections in the region. Descended from an old expatriate Dutch family, Duivenboden owned half the town, many ships, and more than a hundred slaves. He had been educated in Europe and was well versed in science and literature, and yet, like the White Rajah of Sarawak, he preferred to preside over an insignificant kingdom in one of the remotest places on earth rather than live in the heart of European society as an ordinary citizen.[4] The house that Duivenboden found for Wallace was a wreck, but Wallace saw potential in the solidity of its foundations and ample size. Its setting was perfect: there was a "wilderness" of fruit trees, with the market and beach only five minutes away, and towering over everything was the island's four-thousand-foot-high volcano, with virgin forest extending up to the summit. But the feature that sealed the contract was the deep well, with its "inexhaustible supply" of cold water, a luxury in that part of the world. After minimal renovation and the purchase of a set of "cheap but elegant" bamboo furniture, Wallace established himself as "an inhabitant of the earthquake-tortured island of Ternate" for the next two and a half years.

In his first month on the island, Wallace added little to his collections. He was too busy fixing up his house and preparing for a visit to the zoologically unknown island of Gilolo (Halmahera), only ten miles east of Ternate's harbor. He noted to Bates that "not a single insect has ever been collected there." Before mailing the letter, he enclosed a memorandum with an estimate of the total number of distinct species of insects he had collected over three and a half years, which clearly demonstrated the archipelago's impressive entomological diversity: 620 species of butterflies, 2,000 of moths, 3,700 of beetles, 750 of bees and wasps, 660 of flies, 500 of bugs and cicadas, 160 of locusts, 110 of dragonflies, and 40 of earwigs, totaling 8,540 species (the number of specimens is unrecorded).[5]

Wallace left for Gilolo sometime in late January or early February on a boat owned by a Dutchman, captained by a Chinese, and rowed by a crew of Papuan slaves—a human diversity that did not escape his notice. They landed at the village of Sedingole, in the center of the island. Gilolo, a miniaturized version of Celebes, was shaped like a distorted starfish (minus one appendage), with four mountainous, densely forested arms. Only the coast was settled; the interior remained as wild as in the days before the Dutch arrived. After two unproductive days in Sedingole, Wallace set sail for Dodinga, navigating smoothly between small islands along the coast and reaching the village at dawn the next day. Dodinga, located on a narrow isthmus separating the north and south peninsulas, was guarded by a Dutch corporal and four Javanese soldiers, who occupied a small ancient Portuguese fort that was in shambles from having been battered by earthquakes. The luxuriant forest, brightened by masses of large scarlet flowers, promised zoological treasures, but Wallace was unable to do much because of an attack of malaria. On March 1, he returned to Ternate with a small number of insects and birds to await the return of Duivenboden's schooner from Macassar, which would take him to New Guinea, the land of the bird of paradise.

From January 8 until March 1, his documentation is scanty: a "fortnight" in Ternate and "a month" in Gilolo; a list of insects and birds in his insect and bird registries; a letter to Henry Bates and another to Henry's brother, Frederick; a few notes in his private journal about the islands, his difficulties, and the behavioral and physical characteristics of the people. And yet at some time during those two months, he found the time to pen an essay that opened up a new chapter in the history of science.

For reasons that will soon become clear, Wallace omitted from his highly acclaimed travel narrative *The Malay Archipelago*, published in 1869, not only any allusion to that seminal essay but also any credit for the independent discovery of the theory of natural selection. Ambiguities still surround the composition of the paper. His journals mention nothing about it, the original was lost after its publication in England, and no drafts were saved. It was allegedly sent from Ternate in February, but his insect, bird, and mammal registries all indicate that he was on the island of Gilolo, not Ternate, in February 1858, an incongruity that Wallace never clarified and that remains a minor mystery.[6] Eventually, prodded by friends, he told a tale that was full of romance and drama.[7]

According to his recollections, he was lying in bed one day in his cottage in Ternate, suffering from a "rather severe" attack of malaria. Never far from his mind, even in such trying moments, was the main impetus for traveling to the ends of the earth, the problem that had obsessed him for a decade. Up to this point, he was convinced that changes in species had taken place by natural succession and common descent, either slowly or rapidly, depending on conditions. Almost fifty years later, in *My Life*, his autobiography, he reconstructed his thinking at the time: "The problem then was not only how and why do species change, but how and why do they change into new and well-defined species, distinguished from each other in so many ways; why and how do they become so exactly adapted to distinct modes of life; and why do all the intermediate grades die out (as geology shows they have died out) and leave only clearly defined and well-marked species, genera, and higher groups of animals?"[8]

Wallace had not limited his thinking to animal species. His examinations of plants led to similar conclusions. He was also acutely aware of human diversity—mental, moral, and physical. He had not exempted the human species from natural law; this seems clear from what we know of his September 1857 letter to Darwin and from the tenor of "A New Kind of Baby," a paper on the orangutan that had been published in the November 22, 1856, issue of *Chambers's Journal*. With such thoughts tormenting him—and in a "cold fit," wrapped up in a blanket despite an ambient temperature of eighty-eight degrees Fahrenheit—he stated that he somehow was led to think of the "positive checks" described by Thomas Malthus, whose work he had not recalled until that moment. Malthusian checks on human population growth—war, famine, disease, infertility—also applied to animal populations. In the space of two hours that had elapsed between the onset of chills and their subsidence in a pool of sweat, Wallace said that he had devised the entire theory of natural selection, which, despite physical exhaustion, he sketched out that same evening. It was a spark of inspiration that brought together years of experience and contemplation. Over the next two evenings, he wrote the theory out in full.

Wallace called his essay "On the Tendency of Varieties to Depart Indefinitely from the Original Type; Instability of Varieties Supposed to Prove the Permanent Distinctness of Species," an allusion to both Lamarck and Lyell.[9] In his opening remarks he states that "there is a general principle in nature which will cause many varieties to survive the parent species, and to give rise to successive variations departing further and further from the original type and which also produces, in domesticated animals, the tendency of varieties to return to the parent form." He interweaves two major themes: divergence from a common ancestor, or "antitype," the unusual word he ascribes to the progenitor; and survival of the fittest, leading to the formation of new species. His purpose is to strike at the very foundation of the notion of the "fixity" of species. He already had introduced his principle of divergence and descent with modification—the Sarawak Law—in "On the Law Which Has Regulated the Introduction of New Species" and in "Attempts at a Natural Arrangement of Birds," in which he applied the principle. His revelation was the mechanism, the long-sought-after causation, which came to him in a flash of insight.

As in his 1855 essay, Wallace constructed his argument inductively, elaborating on commonly known facts. The life of wild animals is a struggle for existence, he reminded his audience, and in that struggle animals are concerned with two things: self-preservation and the survival of their infant offspring. The survival not only of the individual but of the entire species depends on the successful procurement of food during hard times and the avoidance of enemies; otherwise, large numbers of animals will die, and the species will verge on extinction. "The struggle for existence" was not a novel phrase. Malthus, Robert Owen, and Lyell had used it. So had the Swiss botanist Alphonse de Candolle. Malthus had applied the term to human populations; Lyell, to animal populations; and Candolle, to plant populations. Wallace extended it much further. The struggle for existence explained what before seemed inexplicable: why some species were abundant and others, closely allied, were rare. The "law of population" and not fecundity, he said, determined the relative abundance or rarity of a species. Any animal, large or small, would increase in a geometric ratio if permitted. By means of a simple calculation, it could be shown that one pair of birds would generate 10 million offspring in fifteen years. Malthus had demonstrated this phenomenon in human populations, but populations in the animal world were generally stable. While hunting birds of paradise, for example, Wallace had noted a superabundance of immature males and relatively few adults:

"It is evident, therefore, that each year an immense number of birds must perish—as many in fact as are born. It would therefore appear that, as far as the continuance of the species and keeping up the average number of individuals concerned, large broods are superfluous."

What controlled population levels? Availability of food, but also predation, disease, and impaired fertility. If food supplies increased arithmetically while animal populations increased geometrically, animals would quickly outstrip their food supply. As long as the environment remained physically unchanged, the numbers of its animal population could not increase materially. The numbers dying annually were immense, and those that died usually were the weakest—the very young, the very old, the diseased—while those that survived were the most healthy and vigorous. In the struggle for existence, the weakest and most defective must die.

At this point, Wallace introduced the concept of "utility" of variations. Any variation, however slight, would affect the population either positively or adversely. All varieties fell into two classes: those that remained in the minority under stable conditions and those that achieved numerical superiority over the parent population. A drought, a plague of locusts, a new predator—in short, any change that made conditions difficult for a species—would tax its "utmost power" to avoid extermination. Clearly, the least numerous and most "feebly organised" variety would suffer first and become extinct if the pressure was great enough. If the parent species was less fit in any way, the stronger variety would surpass it in numbers.

Once a variety replaced a species, it could never return to the "original" form. This new, improved, and populous race might in time give rise to new varieties that exhibited "several diverging modifications of form," any of which, if improving fitness, could become dominant under the right conditions. "Here, then," Wallace wrote, "we have progression and continued divergence deduced from the general laws which regulate the existence of animals in a state of nature." But he did not advocate an inevitable forward progression. If conditions changed again, the parent form and the other failed varieties—assuming they still existed—might suddenly flourish at the expense of the formerly fittest variety, no longer superior and now in danger of extinction.

Domesticated species were generally inferior forms, sustained and propagated artificially; in a state of nature, most of these animals would die. If they survived in the wild, their descendants would tend to "return" toward the fitter and better adapted wild form. Lamarck's hypothesis—that acquired characteristics are passed on to one's progeny—was unnecessary. In the case of the giraffe, for instance, those of its ancestors that had longer necks survived in times when food was scarce because of their ability to reach the loftiest foliage; thus their offspring were more numerous, while the competing, shorter-necked variety gradually became extinct.

Wallace's principle had wide applications. As Bates had predicted, new avenues of study suddenly opened up for scientists. As Wallace noted:

> Even the peculiar colours of many animals, especially insects, so closely resembling the soil or the leaves or the trunks on which they habitually reside, are explained on the same principle. An origin such as is here advocated will also agree with the peculiar character of the modifications of form and structure which obtain in organized beings—the many lines of divergence from a central type, the increasing efficiency and power of a particular organ through a succession of allied species, and the remarkable persistence of unimportant parts such as colour, texture of plumage and hair, form of horns or crests, through a series of species differing considerably in more essential characters.

In amassing his collections, Wallace had proved to himself that variation was a fact of nature. Now he held that those variations tended to be inherited—though he did not use that term—and that the most successful variations proliferated, while those that were less successful tended to lead to extinction, a dynamic process constantly regulated by checks and balances in the surrounding environment. Changes were not sudden (as Lamarck would have it) but gradual, occurring over immense periods of time and progressing "by minute steps in various directions." His theory agreed with all the phenomena of life: extinction, derivation of present-day species from past species, physical form, instincts, and habits. He had discovered a true natural system, one without a predetermined balance, teleology, or divine plan.

Once Wallace had written his essay in its final form, he decided to send it to Darwin, who he hoped would find the ideas in it helpful in his own work on the origin of species. In the accompanying letter—which has been lost, though he recounts the gist of it in his autobiography—he asked Darwin to show the essay to Lyell for his opinion if he thought it "sufficiently important."[10] It was a bold request—one that makes sense only if Wallace already knew of their friendship and was aware of Lyell's interest in his theoretical work. It is therefore likely that the mail steamer from Singapore that arrived in Ternate on March 9 carried Darwin's response to Wallace's letter of the previous September.[11] In it Darwin writes, "You say that you have been somewhat surprised at no notice having been taken of your paper in the Annals. I cannot say that I am; for so very few naturalists care for anything beyond the mere description of species. But you must not suppose that your paper had not been attended to: two very good men, Sir C Lyell and Mr E Blyth at Calcutta, specially called my attention to it. . . . [T]hough agreeing with you on your conclusions in the paper, I believe I go much further than you; but it is too long a subject to enter on my speculative notions."[12]

What Wallace did not know, and what he could not have known at the time, was that his discovery of the theory of natural selection marked another turning point in his life—but for reasons that (he later alleged, with characteristic understatement) had nothing to do with Darwin. In *My Life*, he remarks that his solution to the origin of species caused him to branch out into a different intellectual direction. Before his discovery, he had devoted his time to describing, cataloguing, and working out the distribution of his birds and insects. If he had not hit on the mechanism of evolution, he believed, he would have spent "the best years" of his life in what he came to consider "comparatively profitless work." His new discovery "swept all this away," and he decided to let others complete the tedious task of identifying and naming his specimens. He could now dedicate himself to "the great generalizations" that his labors and that of others had made possible.[13]

Wallace arrived in Dorey, on New Guinea, on April 10. He had left Ternate on March 25 on a prau belonging to his friend Duivenboden, accompanied by Ali and three other servants. As the rugged mountains of the New Guinea coast loomed ahead, he gazed with intense interest at the place he had longed to visit, the succeeding ridges of the interior, "where the foot of civilized man had never trod."[14] Dorey itself, however, had seen previous European visitors. The French naturalist René-Primevère Lesson had stopped there forty years earlier, and now German missionaries were headquartered in the town. Wallace presented a letter from the Dutch governor to one of the missionaries, who facilitated his contact with the native chief. The chief built him a hut facing the sea.

Wallace remained for ten days in this impoverished village, ingratiating himself with the chief by offering him gifts and promising payment for any specimens—particularly of the vaunted birds of paradise—that his people could provide, but even with these incentives they brought him no birds of paradise and very little of anything else. On May 5, a Dutch steamer anchored off the beach near his hut, and the little luck he had enjoyed came to an end. The steamer carried the prince of Tidore, son of the sultan, and his retinue, who monopolized both food and wildlife. The prince, who remained on board his vessel and sent everyone else out as his emissaries, had come to Dorey to acquire the skins of birds of paradise, ruining Wallace's arrangements (however unrewarding) with the locals. Even worse, all fish and vegetables were sold to the Tidoreans, leaving Wallace and his men with almost nothing to eat. The only benefit the ship brought was its German doctor, whose services Wallace badly needed. After clambering over some fallen trunks and branches, he had cut his ankle, which then swelled and ulcerated. When the ulcer healed, the Achilles tendon just above its insertion on the heel became inflamed. The doctor advised poulticing the ankle for several days, which only made the inflammation and swelling worse. The area then had to be leeched and covered with ointments and more poultices for several weeks, driving Wallace to the brink of despair. Once again, he was languishing in a hut in a remote part of the world that he might never be able to revisit, forfeiting who knew how many new species a day because of a simple mishap. A month would pass before Wallace dared venture out. Not long after recovering from his ankle ulcer, he was afflicted with fever and a severe soreness of his mouth, tongue, and gums, which prevented him taking anything but liquid meals. By turns, everyone in his party became ill with malaria or dysentery. One of his men—a quiet but reliable eighteen-year-old named Jumaat, from Boutong, an island south of Celebes—developed a fever, lingered for a couple of weeks, and died in late June despite Wallace's ministrations with the limited stock of medicines he had at his disposal. A Muslim, like all of Wallace's assistants, Jumaat was buried in the Islamic fashion, enshrouded in cotton cloth that Wallace provided for the occasion.

The trip to New Guinea was a disaster for Wallace, culminating in the tragic death of one of his assistants. He did not see a single bird of paradise, the sole reason for having risked his life and the lives of his men. He caught nothing rare or novel. The most abundant insects were ants, which descended on his hut with a vengeance, carrying off specimens as he worked and even tearing the labels off the cards the specimens were attached to. Flies attacked his bird skins, laying masses of sticky eggs that hatched ravenous maggots. Wallace sent a summary of his visit to Norton Shaw at the Royal

Geographical Society, suggesting that it contained nothing worthy of insertion in the society's transactions, but Shaw and Sir Roderick Murchison disagreed. "Mr. Wallace was, as he himself truly observes, the sole European inhabitant, and we may safely add, the sole civilised being, on New Guinea for three months," Murchison said in his anniversary address of 1859. "The researches of this skillful naturalist were necessarily confined to a small portion of the island, Doree. . . . Mr. Wallace's paper supplies us with by far the best account of the geology and geography of the place he visited, while other Societies have properly received his contributions to botany and zoology."[15]

Wallace returned to Ternate in mid-August, grateful to experience again the simple joys of life—milk in his tea and coffee and variety in his diet. He found five months' worth of mail and journals awaiting him, including news of the safe arrival of his Aru collections and the "admiration they excited." The news infused in him a new spirit of determination. More cheering was the £1,000 that these collections had earned him. "This makes me hope I may soon realise enough to live upon and carry out my long cherished plans of a country life in Old England," he wrote to his "dear old" mother.[16] Apparently the world wanted more, as though he were a machine manufacturing new species. Samuel Stevens received the brunt of his anger. "You ask me if I go out to collect at night; certainly not," Wallace wrote.

> [A] man who works, with hardly half an hour's intermission, from 6 A.M. until 6 P.M., four or five of the hottest hours being spent entirely out of doors, is very glad to spend his evenings with a book (if he has one) and a cup of coffee and be in bed soon after 8 o'clock. Night work may be very well for amateurs, but not for the man who works twelve hours every day at his collection. . . . You and Dr Gray seem to imagine that I neglect the mammals, or I should send more specimens, but you do not know how difficult it is to get them: at Dorey I could not get a single specimen.[17]

But the news that Wallace hoped to hear most of all had not arrived: there was nothing from Darwin or Lyell. He had mailed his essay to Darwin in early March; the earliest he could have expected a response was twenty weeks later, toward the end of July. Although he must have been disappointed, he was probably not surprised; no important paper he had written thus far had elicited an immediate reaction from anyone.

In mid-September, he made a second expedition to Gilolo, which was cut short when he discovered that his path to the slopes of the distant summits, reputed to be teeming with birds, was barred by impassable grass eight to ten feet high. He next prepared for a journey to Batchian, 150 miles to the south. The unpacking from one unsuccessful voyage; the repacking for another trip, which might be equally unfruitful; and the frustrating search for a boat and crew tried his patience. As Wallace occupied himself with these annoyances, the Dutch mail steamer arrived with more mail from home. This time, his packet included two letters, one from Darwin and the other, oddly, from Joseph Hooker, one of England's finest botanists and a man he knew only by reputation. The letters from Darwin and Hooker told an extraordinary story. By a coincidence almost beyond belief, Wallace's essay outlined the same doctrine that Darwin had discovered and had been laboring over for almost two decades but had not yet published![18]

The Darwinian theory, as it is known today, is a two-part theory, embodied simply and elegantly in Wallace's essay: evolution from a common origin through the agency of natural selection. The great power of the Darwin–Wallace theory is the unification of two independent ideas—the survival of the fittest and the common origin and divergence of species—to explain, better than any other previous theory, the distribution and diversity of life on the planet. Yet how did two men, working on opposite sides of the globe, arrive at similar conclusions independently and virtually simultaneously?

Not long after returning from the Galápagos in 1836, Darwin became convinced of the probability of evolution, but the theory he originally conceived was not the one he published some twenty years later. In the late 1830s, he believed that gradual changes in the movements of the earth's crust somehow encouraged species to vary, become modified, and evolve into different species. As it had for Wallace, Lyell's uniformitarian doctrine provided Darwin with a scientific framework for examining the organic world. But an intensive investigation of barnacles, which occupied eight wearying years of his life, altered his views on the cause of variations, though he confessed that he doubted whether the work was worth the amount of time he had devoted to their study. "[It] was . . . evident that neither the action of the surrounding conditions, nor the will of the organisms . . . could account for the innumerable cases in which organisms of every kind are beautifully adapted to their habits of life," he wrote in his autobiography.[19] Barnacles taught Darwin what Wallace would learn from his experiences as a field biologist: that variations occurred naturally and spontaneously. Variations were the material on which natural selection acted. Moreover, Darwin concluded that variations appeared as a natural outcome of sexual relations. Internal factors (which were not yet defined, for genetic theory lay in the future) and sex, not external factors, played a fundamental role in the development of variation. As Darwin scholar Janet Browne has noted, this was one of the great intellectual achievements predating the publication of *The Origin of Species*.[20]

In 1842, when the nineteen-year-old Wallace was poring over William Swainson's *Treatise on the Geography and Classification of Animals*, Darwin sketched out in thirty-five pages his theory of natural selection, a term he had invented to describe the outcome of the struggle for existence and the natural process equivalent to the artificial selection practiced on domestic animals and plants. Natural selection was an idea that had come to him in 1838, after reading Malthus's *Essay on the Principle of Population*, which would also play a seminal role in Wallace's thinking. In the summer of 1844, he expanded the sketch to 230 pages, which he copied out in more legible form. Hooker was the only man who had read this document, which was then put aside, with instructions to Darwin's wife, Emma, to publish it only if he died before he could complete a larger work. For the next fourteen years, he continued to conduct his research painstakingly and methodically, concocting "little" experiments to answer questions that sometimes seemed to touch tangentially on the subject of the origin of species—for example, testing his hypothesis that bees are necessary for both the self-fertilization and the crossing of kidney beans—and amassing a mountain of facts from his contacts with naturalists on nearly every continent, who provided him with specimens and data on request.[21] His object, he wrote to his cousin William Darwin Fox in March 1855, was "to view all facts that I can master . . . in Nat[ural] History, (as on geograph[ic] distribution, paleontology, classification, Hybridism, domestic animals & plants &c &c &c) to see how far they favour or are opposed to the notion that wild species are mutable or immutable: I mean with my utmost power to give all arguments & facts on both sides."[22]

As for the principle of divergence, the other critical element in evolutionary theory, scholars have suggested that Darwin already had alluded to the principle as early as 1837, for in one of his early notebooks he had used the analogy of a tree of life to describe the concept of gradual evolution (hence Darwin's reaction—"nothing very new"—when he read Wallace's "On the Law Which Has Regulated the Introduction of New Species").[23] But not until August 1857 did he use the term itself in a letter to Hooker. One month later, he sent a letter to the Harvard botanist Asa Gray outlining his theory of natural selection and the principle of divergence as it applied to the theory. "Each new variety or species when formed will generally take the place of, and so exterminate its less well-fitted parent," Darwin told Gray. "This I believe to be the origin of the classification or arrangement of all organic beings at all times. These always *seem* to branch and sub-branch like a tree from a common trunk."[24]

Both Wallace and Darwin were profoundly influenced by the same two men, Lyell and Malthus, and had access to a similar body of data—a fact that is no coincidence. But they approached the problem from opposite directions. Wallace discovered his principle of divergence before discovering natural selection, the mechanism behind that divergence. Darwin discovered natural selection first and joined it to his principle of divergence, the result of natural selection, more than a decade later. Wallace arrived at his conclusion from studies of organisms in their natural habitats; Darwin did so from studies of domesticated animals, barnacles, and plants. As the years progressed, Darwin's and Wallace's views would converge until the critical day in June 1858. Until that point, neither man fully grasped that each was about to collide with the other.

On June 18, 1858, Darwin made the following annotation in one of his journals: "interrupted by letter from A R Wallace." Clearly it was one of the lowest points in his life. He was so upset by what Wallace had sent him that the obsessive documentation of his daily activities and ruminations came to a sudden halt.[25] After reading Wallace's "On the Tendency of Varieties to Depart Indefinitely from the Original Type," he was faced with a serious dilemma—one that Lyell had predicted and that was now his to solve. His emotional turmoil is reflected in a letter he sent to Lyell that day:

> *My dear Lyell,*
> *[S]ome year or so ago you recommended me to read a paper by Wallace in the Annals, which had interested you, and as I was writing to him, I knew this would please him much, so I told him. He has to-day sent me the enclosed, and asked me to forward it to you. It seems to me well worth reading. Your words have come true with a vengeance—that I should be forestalled. You said this, when I explained to you very briefly my views on "Natural Selection" depending upon the struggle for existence. I never saw a more striking coincidence; if Wallace had my MS sketch written out in 1842 [sic] he could not have made a better short abstract! Even his terms*

now stand as heads of my chapters. Please return me the MS, which he does not say he wishes me to publish, but I shall, of course, at once write and offer to send to any journal. So all my originality will be smashed, though my book, if it will ever have value, will not be deteriorated; as all the labour consists in the application of the theory.[26]

Lyell did not answer at once. One week later, an apparently distraught Darwin wrote another letter:

My dear Lyell,
I am very sorry to trouble you, busy as you are, in so merely personal an affair; but if you will give me your deliberate opinion, you will do me as great a service as ever man did, for I have entire confidence in your judgment and honour. . . . There is nothing in Wallace's sketch which is not written out much fuller in my sketch, copied out in 1844, and read by Hooker some dozen years ago [sic]. About a year ago I sent a short sketch, of which I have a copy, of my views . . . to Asa Gray, so that I could most truly say and prove that I take nothing from Wallace. I should be extremely glad now to publish a sketch of my general views in about a dozen pages or so; but I cannot persuade myself that I can do so honourably. Wallace says nothing about publication, and I enclose his letter. But as I had not intended to publish any sketch, can I do so honourably because Wallace has sent me an outline of his doctrine? I would far rather burn my whole book, than that he or any other man should think that I had behaved in a paltry spirit. Do you not think his having sent me this sketch ties my hands? . . . If I could honourably publish I would state that I was induced now to publish a sketch . . . from Wallace having sent me an outline of his conclusions. We differ only [in] that I was led to my views from what artificial selection has done for domestic animals. . . . This is a trumpery affair to trouble you with, but you cannot tell me how much I should be obliged for your advice. By the way, would you object to send this and your answer to Hooker to be forwarded to me? for then I shall have the opinion of my two best and kindest friends.[27]

Darwin sent one final thought in a postscript to Lyell: "It seems hard on me that I should be thus compelled to lose my priority of many years' standing, but I cannot feel at all sure that this alters the justice of the case." Calling him a "first-rate Lord Chancellor," Darwin appealed to Lyell to resolve the problem.

Lyell followed Darwin's instructions by sending Wallace's paper and letters, along with Darwin's responses, to Hooker. Hooker, eight years younger than Darwin, was probably Darwin's closest scientific friend, and Darwin admired his keen intellect and blunt, constructive opinions. He could depend on Hooker's honesty, especially in the evaluation of his work. When Darwin had completed his manuscript on evolution in 1844, he suggested that Lyell edit his work; a decade later, he changed his mind, amending his will in order to assign the task of editing to Hooker.[28]

An emergency meeting of the Linnean Society—which, along with the Zoological and Entomological Societies, was among London's chief scientific conclaves—provided Hooker and Lyell with an opportunity to help Darwin out of his bind. The scientific societies generally disbanded for the summer holidays, not meeting again until October or November, but the death on June 10 of Robert Brown, the most eminent English botanist of the time, left a vacancy on the Linnean Society's board that had to be filled within three months, according to its bylaws. Out of respect to Brown, the regular June meeting had been postponed to July 1. On June 29, Hooker pressured Darwin to get some material—any material—together to be presented at the society in three days.[29]

In the meantime, not long after Darwin mailed his anguished deliberations, his infant son, Charles, developed scarlet fever and died; both his wife and his daughter were ill with diphtheria. Darwin responded to Hooker that evening:

> I have just read your letter and see you want papers at once. I am quite prostrated & can do nothing but I send Wallace & my abstract of abstract of letter to Asa Gray, which gives most imperfectly only the means of change & does not touch on reasons believing that species do change. I daresay all is too late. I hardly care about it.—But you are too generous to sacrifice so much time & kindness.—It is most generous, most kind. I send sketch of 1844 solely that you may see by your own handwriting that you did read it.—I really cannot bear to look at it. Do not waste much time. It is miserable in me to care at all about priority.[30]

Some thirty people, less than 10 percent of the society's membership, attended the July 1 meeting. Darwin, still grieving over the death of his son, remained at home. Lyell and Hooker railroaded the papers by Darwin and Wallace onto the agenda, displacing the papers of other men who had been scheduled to make presentations. [31] One of those was the botanist George Bentham, who felt "severe pain and disappointment" at having been preempted.[32] The secretary of the society, John Joseph (J. J.) Bennett, read an introductory letter from Hooker and Lyell explaining the Darwin–Wallace conjunction, which began:

> These gentleman, having, independently and unknown to one another, conceived the same very ingenious theory to account for the appearance and perpetuation of varieties and of specific forms on our planet, may both fairly claim the merit of being original thinkers in this important line of inquiry; but neither of them having published his views, though Mr. Darwin has for many years been repeatedly urged by us to do so, . . . we think it would best promote the interests of science that a selection from them should be laid before the Linnean Society.[33]

The papers were read—by George Busk, the society's undersecretary—in the order of their dates of composition: the first was Darwin's abbreviated abstract of his 230-page essay from 1844; the second was the "abstract of abstract" of his letter to Gray of September 5, 1857. The last offering was Wallace's "On the Tendency of Varieties to Depart Indefinitely from the Original Type." Hooker and Lyell reported that they had explained to Darwin "that we are not solely considering the relative claims of priority of himself and his friend, but the interests of science generally."

July 1, 1858, was the true birth of the modern theory of evolution, parented by both Darwin and Wallace. But the audience was somewhat baffled, if not overwhelmed, by the presentation. Hooker later said that Lyell's tacit approval of the papers as well as his own presence "as his lieutenant in the affair" must have "overawed" the fellows, who might otherwise have "flown out" against the doctrine. The papers generated little response and virtually no discussion. Bentham, whose article "On the Species and Genera of Plants, Considered with Reference to Their Practical Application to Systematic Botany" had illustrated facts and observations related to the "fixity" of species, was stunned and decided not to read his paper. Years later, he told Francis Darwin that he had realized that he would be forced, "however reluctantly," to give up his "long-cherished convictions."[34] At year's end, Thomas Bell, the president of the society, pronounced that 1858 was not "marked by any of those striking discoveries which at once revolutionize . . . the department of science on which they bear."[35] Not even Thomas Huxley fully grasped the significance of Wallace's contribution; he perceived the event through Darwinian eyes only. "Wallace's impetus seems to have [got?] Darwin going in earnest," Huxley told Hooker, "and I am rejoiced to hear we shall learn his views in full; at last I look forward to a great revolution being effected. Depend upon it in Natural History, as in everything else, when the English mind fully determines to work a thing out it will do it better than any other."[36]

But for Darwin, the joint presentation of the papers was a boon. Lyell finally had succeeded in bringing his views, imperfect as they were, into the light of day. Darwin was "much more than satisfied" at what had taken place, for he had thought that his letter to Gray would be read as an appendix to Wallace's paper, not second in order of presentation. He had begun to compose a letter to Wallace in which he relinquished all priority, but changed his mind when he learned of Lyell and Hooker's "extraordinary kindness" in the management of the affair.[37] The public support that Lyell and Hooker gave to his views convinced Darwin to accelerate the timetable for completing his book. He abandoned his original work, "Natural Selection," in favor of an abstracted version, which he set out to compose at an unprecedented pace, almost certainly imagining an equally eager Wallace nearing completion of his own species book. "It is really impossible to do justice to the subject," Darwin wrote to Lyell that July, "except by giving facts on which each conclusion is grounded & that will of course be absolutely impossible. . . . I look at this as so very important that I am almost glad of Wallace's paper for having led to this."[38]

Not all historians of science are satisfied with this account of a famous historical conjuncture. John Langdon Brooks has cast a less flattering light on the events of 1858. He believes that Darwin behaved in an underhanded manner to secure his place in history, outmaneuvering Wallace, who should be recognized as the first person to announce a complete theory of evolution. Brooks's investigation focuses on a sliver of time that Darwin specialists have elected to pass over. What really happened to Wallace's paper, he asks, after it was mailed from Ternate in March 1858? Brooks assiduously works out the elaborate journey of Wallace's essay to Down House and concludes that it could have arrived in England as early as May 18. (H. L. McKinney provides evidence that a letter to Henry Bates's brother, Frederick, dated March 2, 1858, but presumably mailed on the same day as the essay to Darwin, arrived in Leicester on June 3.)[39] Thus Darwin possessed Wallace's essay for at least two weeks—and perhaps for as long as a month—prior to his notation of its receipt in his private journal and the letter to Lyell. During this time period, Brooks boldly asserts, Darwin engaged in a bit of intellectual piracy, revising his notions about natural selection and divergence after reexamining Wallace's Sarawak Law paper, "On the Law Which Has Regulated the Introduction of New Species," and studying the new essay. These revisions, eventually embodied in the later chapters of *The Origin of Species*, resemble Wallace's conceptions more than his own previous sketchy ideas, which did not make as striking a connection among extinction, intermediate forms, and the natural system of classification.[40] That resemblance, he claims, is greater than Darwin scholars have acknowledged, and he spends a good portion of his book comparing sections of the *Origin* and Wallace's essays to substantiate his claim of intellectual theft. Without Wallace's two papers, according to Brooks, Darwin could not have completed the *Origin*. Darwin not only plagiarized Wallace but never gave him proper credit for his critical contributions in any edition of the *Origin* or in his autobiography.

Arnold Brackman, who shares Brooks's conviction, is even less sanguine about the outcome of the July 1, 1858, meeting of the Linnean Society. Brackman accuses Darwin, Hooker, and Lyell of conspiring to rob Wallace of his rightful claim to priority. It is a classic tale of class power, he says, one more example of an elite group of men trampling a lower-class rival.[41] Lost letters, a missing manuscript, pirated doctrines, behind-the-scenes maneuvering, and mea culpa letters to the offended party all lend themselves to assertions of foul play—Wallace and Darwin à la Mozart and Salieri.

But such views oversimplify the events of 1858 and presume that the participants possessed a greater knowledge of the future than they actually had. Despite their eloquent advocacy, Brooks and Brackman have failed to convince leading scholars that Darwin plagiarized Wallace, though many have criticized Darwin (in retrospect) for inadequately referencing Wallace's work in future editions of the *Origin*. In their eagerness to vilify Darwin, Brooks and Brackman barely acknowledge the breadth and intricacy of Darwin's research, which is amply documented in his private journals, notebooks, and vast correspondence—the last comprising several volumes that have been analyzed, annotated, and published by historians at Cambridge University. Even a cursory examination of these documents will impress all but the most hardened skeptics with the magnitude of Darwin's endeavors and reveal the hyperbole of Brooks's and Brackman's claims. As the editors of Darwin's correspondence justly note, "the impressive scope of Darwin's practical researches, demonstrating a desire to examine fully the consequences of his theories, provided the basis for the eventual presentation of natural selection as a theory firmly founded on observation and experiment."[42] Moreover, Brackman disparages the character of two honorable and highly respected scientists who sincerely believed that they were serving the interest of science in general and that of their friend in particular. There is little evidence that they bore any malice toward Wallace at the time. And although Wallace's class status affected the trajectory of his career, it played no role in this specific episode. Had Hooker and Lyell—reluctant rebels against ecclesiastical or political authority—foreseen that their actions would eventually unleash a firestorm of controversy, they might have been far less enthusiastic in their support of "an important line of [scientific] inquiry" than they appeared to be in July 1858.

Nevertheless, Darwin procrastinated in publicizing his theory, a lapse that nearly cost him enduring fame. Whatever the reason for this delay—failure of nerve, a passion for perfection, periodic debilitating illness—it was not until the unexpected appearance of Wallace's essay that the issue of priority suddenly reared its ugly head. Devastated by the communiqué from Wallace and distracted by a tragedy at home, he turned to his friends for help. The two men came through in a way more subtle than sinister; by presenting notes and papers chronologically, they implied that Wallace bolstered Darwin. Thus Wallace—from the perspective of history—was relegated to a secondary role in the greatest scientific revolution of modern times.

A more serious claim—and one that is harder to dismiss—is that Wallace and Darwin did not elaborate the same theory of natural selection, that their independent conceptions differed fundamentally. First, applying the terminology of the evolutionary biologist Ernst Mayr, Wallace's analysis appears to be a study of population ecology; that is, he seems to suggest that natural selection acted at the level of varieties and species, with the most unfit groups perishing. Darwin, though, believed that the struggle for existence occurred at the individual level, with the most fit winning the battle of life single-handedly and becoming better adapted to the surrounding conditions.[43] The confusion about Wallace's conception has arisen, in part, because of his terminology, which was expressed in a hastily written paper. At times, he used the terms "variety" and "variant individuals" interchangeably. Thus it is unclear if in 1858 he considered competition among individuals to be as important as competition among subpopulations in a species.[44] Second, according to Wallace scholar Malcolm Kottler, Wallace and Darwin did not mean the same thing when they described their principles of divergence. Wallace perceived divergence as mainly a linear process. At the beginning, for example, there is species *A*, which evolves into *a*1-*a*2-*a*3 . . . *a*10, species *a*10 being quite distinct from *A*. Darwin, however, recognized both linear divergence and branching divergence. In branching divergence, species *A* evolves into species *a*10 as well as species *m*10, *m*10 being as distinct from *A* as it is from *a*10. Darwin depicted this process in the *Origin* a year later in his famous schema of the probable effects of natural selection through divergence of character and extinction on the descendants of a common ancestor.[45]

But, as has already been noted, Wallace did acknowledge the concept of branching divergence, in both "On the Law Which Has Regulated the Introduction of New Species" and "Attempts at a Natural Arrangement of Birds." The problem is that in 1858 Wallace had not had time to elaborate fully his ideas on evolution: his publications were meant to foreshadow a future, better-documented work to be entitled "On the Law of Organic Change." Moreover, the point as to whether the differences between Wallace and Darwin are historically important is moot: in June 1858, the similarities and not the differences alarmed Darwin—and, knowing what transpired afterward, Darwin's perception is the crucial factor here.

And how did Wallace react to this improbable chain of events? If he was as devastated as Darwin, he never revealed his anguish, either in his journals or in letters to his family and friends. His response was measured and professional, consonant with the tone of the letters from two of England's best-known and respected scientists, Hooker and Darwin. Wallace wrote to Hooker from Ternate on October 6, 1858:

> Allow me in the first place sincerely to thank yourself and Sir Charles Lyell for your kind offices on this occasion, and to assure you of the gratification afforded me both by the course you have pursued, and the favourable opinions of my essay which you so kindly expressed. I cannot but consider myself a favoured party in this matter, because it has hitherto been too much the practice in cases of this sort to impute all merit to the first discoverer of a new fact or a new theory, and little or none to any other party who may, quite independently, have arrived at the same result a few years or a few hours later. . . . [I]t is evident that the time has now arrived when these and similar views will be promulgated and must be fairly discussed.

Wallace considered it a "most fortunate circumstance" that he had initiated a correspondence with Darwin on the subject of "varieties" and thus caused Darwin to move up the publication date of part of his researches. He had no doubt that Darwin's views were more complete than his own and secured for Darwin the right to claim priority: "It would have caused me much pain and regret had Mr. Darwin's excess of generosity led him to make public my paper unaccompanied by his own."[46] He also replied to Darwin in a letter that has also regrettably not been preserved but that was undoubtedly written in the same spirit.

On the same day that Wallace wrote to Darwin and Hooker, he sent a letter to his mother with the news, noting that he was "highly gratified" to have received communications from Darwin and Hooker and flattered that Lyell had thought highly enough of his paper to allow its immediate presentation at the Linnean Society. "This assures me the acquaintance of these eminent men on my return home," he told her.[47] To Stevens he exulted, "An essay on varieties which I sent to Mr. Darwin has been read to the Linnean Soc[iety] by Dr. Hooker and Sir C. Lyell on account of an extraordinary coincidence with some views of Mr. Darwin, being written but not yet published, and which were also read at the same meeting. If these are published I dare say that [the Society] will let you have a dozen copies for me. If so, send me three, and of the remainder send one to Bates, Spruce and any other of my friends who may be interested in the matter who do not attend the Linnean."[48] Darwin's and Wallace's views were indeed published—together, as they had been presented—in the August issue of the *Journal of the Proceedings of the Linnean Society of London.* "[If] you have any acquaintance who is a fellow of the Linnean Society, borrow the Journal of the Proceedings for August last," Wallace advised his friend George Silk, "and in the last article you will find some of my latest lucubrations and also some complimentary remarks thereon by Sir Charles Lyell and Dr. Hooker, which (as I know neither of them) I am a little proud of."[49]

These were not the reactions of a man concerned about priority. Wallace was clearly elated. To have commanded the attention and respect of three of England's most respected scientists was one of the greatest honors he could have hoped to attain. Only in retrospect, after fifty years of observing the vagaries of his and Darwin's evolutionary theory, did he express some annoyance at the outcome of the July 1 meeting of the Linnean Society. His essay had been hastily sketched, he later wrote, and he had had no opportunity to make revisions before it was printed. He later claimed that what no one but he had understood at the time—not even Darwin—was the idea of the constant variability of "every common species in every part and organ," a fact that could have been known only by someone who had examined, dissected, and prepared thousands of animals from different classes, orders, families, genera, and species. Virtually no one else had noted that "favorable variations" occurred frequently, constantly, in every generation, and in sufficient numbers for the principle of survival of the fittest to effect improvements in the species.[50]

But in 1858, the news was so exhilarating that Wallace emerged from a low point in his travels with renewed excitement in his work. If he had entertained thoughts about coming home earlier than planned, he now cast them aside. To Hooker, he commented that nothing could induce him to quit when his researches had reached "their most interesting point."[51] It was in this frame of mind that he set off for the island of Batchian to make further discoveries and strengthen his and Darwin's joint theory with more facts.

Wallace considered it to be the "greatest discovery" he had yet made, and he made it only three days after landing on Batchian. That morning, in virgin forest half a mile from his hut, he had overtaken Ali, who was in high spirits after a successful day of hunting. From Ali's belt hung a number of birds, one of which he pointed to with obvious pride. "Look, sir, what a curious bird," Ali said, holding up a specimen the size of a crow. What Ali showed him was puzzling, but after a few moments of study he realized that this was an undiscovered genus and species of bird of paradise, one differing in a remarkable way from every other known bird of the family Paradisaeidae. "The general plumage is very sober, being a pure ashy olive, with a purplish tinge on the back," Wallace noted with equal sobriety in his journal.

> The crown of the head is beautifully glossed with pale metallic violet, and the feathers of the front extend as much over the beak as in most of the family. The neck and breast are scaled with fine metallic green, and the feathers on the lower part are elongated on each side, so as to form a two-pointed gorget, which can be folded beneath the wings, or partially erected and spread out in the same way as the side plumes of most of the birds of paradise. The four long white plumes which give the bird its altogether unique character spring from little tubercles

> close to the upper edge of the shoulder or bend of the wing; they are narrow, gently curved, and equally webbed on both sides, of a pure creamy-white color. They are about six inches long, equalling the wing, and can be raised at right angles to it, or laid along the body at the pleasure of the bird. The bill is horn color, the legs yellow, and the iris pale olive.[52]

But to Samuel Stevens and the Zoological Society, Wallace could not suppress his excitement. "I believe I have already the finest and most wonderful bird in the island," he wrote. "I had a good mind to keep it a secret, but I cannot resist telling you. I have a new Bird of Paradise! of a new genus!! quite unlike anything yet known, very curious and very handsome!!! . . . Had I seen the bird in Ternate, I should never have believed it came from here, so far out of the hitherto supposed region of the Paradisaeidae."[53]

Wallace sent a crude sketch of the bird with its exact dimensions and his detailed description to Stevens, who passed them on to the ornithologist George Robert Gray. The provisional name given by Gray was *Paradisea wallacii*, subgenus *Semioptera*, in honor of Wallace's "indefatigable energy" in the advancement of ornithological and entomological knowledge by traveling to places rarely, if ever, visited by naturalists.[54] Gray called the bird "Wallace's standard-wing" because of the white epaulet-like shoulder feathers. A discussion ensued among ornithologists about the affinity of this bird with other birds of paradise, but eventually all agreed with Wallace. In the meantime, as the debate raged, any interested person could view Wallace's specimen in a square glass case in the bird gallery of the British Museum, along with the three other perfect specimens of birds of paradise—the great, lesser, and king—he had collected.[55]

Except for a two-week excursion to one of the neighboring islets, where he caught the rare Nicobar pigeon, Wallace spent nearly six months on Batchian, one of his more productive and satisfying periods in the archipelago. In mid-April 1859, he returned to Ternate by Dutch steamer and heard more about that other great discovery of his—the theory of natural selection. Another letter from Darwin, dated the previous January, awaited him. "Permit me to say how heartily I admire the spirit in which they [Wallace's letters to him and Hooker] are written," Darwin said. "Though I had absolutely nothing whatever to do in leading Lyell and Hooker to what they thought a fair course of action, yet I naturally could not but feel anxious to hear what your impression would be." Wallace had asked about Lyell's "frame of mind," but had to be satisfied with Darwin's tongue-in-cheek rendition. "He does not give in," Darwin wrote, "and speaks with horror often to me of what a thing it would be and what a job it would be for the next edition of the Principles if he were 'perverted.' But he is most candid and honest, and I think will end by being perverted."[56] Lyell, however, left Darwin to speak on his behalf. His failure to write, which was a great disappointment to Wallace, was a sign of his deep distress over a revolutionary doctrine that challenged his fundamental beliefs. Committed to the spirit of scientific inquiry, he seems to have viewed himself as a messenger rather than a spokesman for the Darwin–Wallace theory of evolution, which did not prevent some of his colleagues from misinterpreting his actions at the Linnean Society as a public endorsement of the theory.

Wallace's lost reply to Darwin's earlier letter is doubtless filled with numerous "lucubrations," but Darwin apparently had taken each of them into consideration. "You will see what I mean about the part I believe selection has played [in the development of domesticated animals]," Darwin said in response to a concern that Wallace must have raised about the differences between natural and artificial selection. Regarding Wallace's attention to the creation of birds' nests from the standpoint of evolution, Darwin replied, "I have done so [too], though almost exclusively under one point of view, viz to show that instincts vary, so that selection could work on and improve them." To Wallace's remarks about bees and beehives (Wallace apparently was uncertain that the construction of a beehive was an instinct, for bees had never been secluded in the larval stage and then loosed), Darwin replied, "I am delighted to hear that you have collected bees' combs. . . . This is an especial hobby of mine, and I think I can throw light on the subject." And he had made significant progress on his book: that January, he announced, he had only two more chapters to write. "My abstract will make a small volume of 400 or 500 pages," Darwin informed Wallace. "Whenever published, I will of course send you a copy."[57]

In the same letter, Darwin noted that "everyone whom I have seen has thought your paper very well written and interesting. It puts my extracts (written in 1839 [*sic*], now just twenty years ago!), which I must say in apology were never for an instant intended for publication, in the shade." Who these admirers were Darwin never said. A detractor, one Thomas Boyd, wrote a short and unfavorable critique for the *Zoologist*. Boyd recognized at once the novelty of the papers, which carried the argument about the variation of species "into ground scarcely touched before." Wallace's paper clearly impressed him more than Darwin's "sketch," which he barely acknowledged. Boyd, who was not an evolutionist, quibbled with Wallace's choice of the word "tendency," arguing that "a tendency is not a law of indefinite progress; a tendency to increase is not a law of indefinite increase, neither is a tendency to vary a law of indefinite variation." His most severe criticism was the absence of "facts, experiments and observations," precisely the charge that both Wallace and Darwin feared would be leveled against them with the premature publication of their theories. Nevertheless, he truly grasped-without-grasping Wallace's thesis. "Does he mean that by the tendency to vary we may explain all the differences that [obtain] between different varieties of the same genus, or between different genera of the same order?" Boyd asked. "Or, further still, that we may trace back all organic life, as we see it now, to some unknown root in the far-off geologic ages, some spore, or polype [*sic*], or vitalized cell, from which everything has since sprung? . . . [It] seems to me that painting such an ideal picture on the subject is like Science sitting down at the feet of Imagination."[58]

Meanwhile, despite the promising news on the scientific front, Wallace's family was alarmed by reports of his various illnesses. His brother-in-law, Thomas Sims, urged him to return home before his health was seriously undermined. Annoyed and unyielding, Wallace replied, "I feel that my work is here as well as my pleasure. Why should I not follow out my vocation? . . . I am engaged in a wider and more general study—that of the relations of animals to space and time, or, in other words, the geographical and geological distribution and its causes. . . . As to health and life, what are they compared with peace and happiness?" But there was another powerful reason for Wallace to remain in the archipelago—a financial one. He had not yet accumulated enough money to live on in England, and he felt that he would earn money more quickly by collecting than at any other occupation back home. There was nothing for him in England except surveying, and a job in surveying was not what he wanted. He had decided to devote his life to natural history. What did it matter if people called him an "enthusiast," Wallace told Sims. "Who ever did anything good or great who was not an enthusiast?" From his perspective, most men were enthusiasts at one thing only: "money-getting." As he told Sims, "It strikes me that the power or capability of a man in getting rich is in inverse proportion to his reflective powers and in direct proportion to his imprudence. It is perhaps good to be rich, but not to get rich, or to be always trying to get rich, and few men are less fitted to get rich, if they did try, than myself."[59]

The precarious state of his health did not prevent Wallace from pressing onward, obsessed by his mission to map zoogeographically the entire Malay Archipelago in a manner never yet attempted: through the lens of his theory of natural selection. On May 1, 1859, he left Ternate, circuiting the Moluccas on his way to Menado, on the northeastern tip of Celebes, before touching at Banda and Coupang, on the island of Timor. While waiting in Coupang for the next steamer to Celebes, Wallace wrote a paper that he characterized as an "imperfect outline" of a theory he hoped to "bring forward" on a future occasion. Called "On the Zoological Geography of the Malay Archipelago," it was a response to a recent article on the zoology of New Guinea by the ornithologist Philip Lutley Sclater.

Sclater had put forth the hypothesis that the straits of Macassar were a "determining line" that separated two zoologically distinct regions, the Asian and Australian, with the island of Lombok "debateable ground" between them, "as Mr. Wallace's investigations have shown." On one side were Borneo, Java, and Sumatra, allied "inseparably" with Asia; on the other, Amboyna, Gilolo, Timor, New Guinea, and "probably" Celebes, allied with Australia. After beginning on such a bold note, Sclater advanced no further, merely reviewing the literature published by French and Dutch naturalists and drawing up a list of birds from both sides of the dividing line without offering any explanation for the present-day distribution of animals in the archipelago.[60]

Wallace agreed with Sclater but went one step further, proposing that the division of the Malay Archipelago into two distinct regions "held good" in every branch of zoology. The purpose of his paper, however, was "to mark out the precise limits of each region, and to call attention to some inferences of great general importance as regards the study of the laws of organic distribution." Why, he asked, were islands that were in such close proximity and so similar in climate and geography populated by radically different faunal groups? Why were islands like New Guinea and Australia, which differed in climate and geography, populated by animals so closely related? Lyell's explanation for the geographic distribution of animals, based on ordained design, was not a viable solution. If Lyell were correct, one would expect a smooth transition of forms as one moved across the two zoological zones. "Were the Atlantic gradually to narrow till only a strait of twenty miles separated Africa from South America," Wallace hypothesized, "can we help believing that many birds and insects and some few mammals would soon be interchanged?" Yet such an interchange had barely occurred in the Malay

Archipelago, this despite the proximity of the two regions during a long period of geologic time. "Extreme zoological diversity" was the rule here, not the exception.

Celebes was the most anomalous island of all, Wallace wrote. Despite its central position within the archipelago, some of its species seemed to be related to species in Africa! A genus of "baboon-monkeys" (*Cynopithecus nigrescens*), the bizarre babirusa (or pig-deer, as it was sometimes called), and the cowlike ruminant known as the anoa indicated an African rather than an Australian or Asian origin. The Celebes roller (*Coracias temmincki*) was classified in Prince Charles-Lucien Bonaparte's *Conspectus Generum Avium* in the same genus as birds found in the savannas of eastern and southern Africa, Europe, and the Indian subcontinent. "I am aware of no other spot upon the earth," Wallace stated, "which contains a number of species, in several distinct classes of animals, the nearest allies to which do not exist in any countries which on every side surround it, but which are to be found only in another primary division of the globe, separated from them all by a vast expanse of ocean." The only explanation for such an anomaly was a "bold acceptance" of vast changes in the earth's surface. Facts suggested that Celebes was ancient—more so than its surrounding islands. At one time, he believed, there must have been a great continent, including Africa in the west and Celebes in the east, of which only fragments remained. The intervening land had become "submerged," leaving pockets of animals and plants that betrayed their past origins. Accidental transmission across the open seas accounted for only an estimated 20 percent of the fauna of Celebes; the rest, he believed, had arrived by a land route. Through similar inductive reasoning, one could develop hypotheses about the past history of other aberrant regions, such as the Galápagos, Madagascar, and New Zealand. "Geology can detect but a portion of the changes the surface of the earth has undergone," Wallace noted. "It can reveal the past history and mutations of what is now dry land; but the ocean tells us nothing of her bygone history. Zoology and Botany here come to the aid of their sister science."

Wallace postulated that the narrow strait between Lombok and Bali acted as an impassable barrier to the passage of most land animals, with the exception of the strongest flying birds. No significant connection in recent geologic epochs had existed between the Australian and Asian halves of the archipelago. Based on similar reasoning, he argued that the paucity of species and narrow range of genera in the Galápagos demonstrated that those islands had never had intimate connections with the South American mainland and that animals had arrived there by chance over widely spaced intervals of time.

But these observations explained only half the phenomena of speciation in the Malay Archipelago and other islands. The application of his Sarawak Law and the theory of natural selection to animal distribution could account for the remaining faunal and floral peculiarities. "The regular and unceasing extinction of species, and their replacement by allied forms, is now no hypothesis, but an established fact," Wallace boldly pronounced, as if every one of his fellow naturalists now believed in evolution. "It is the instances of identity of species in distant countries that presents the real difficulty."[61]

Wallace sent the article to Darwin, who had it read before the Linnean Society on November 3. "Your paper seems to me admirable in matter, style and reasoning," Darwin wrote. "And I thank you for allowing me to read it. Had I read it some months ago I should have profited by it for my forthcoming volume. . . . [Y]ou will see that my views are nearly the same with yours, and you may rely on it that not one word shall be altered owing to my having read your ideas. . . . I was aware that Celebes was very peculiar; but the relation to Africa is quite new to me and marvellous and almost passes belief."[62]

On one issue, however, the two men did not yet entirely agree: the means of colonization of oceanic islands. At the time, Wallace concurred with Hooker, Lyell, and most other naturalists about the existence of land bridges and now submerged continents at some point in the past to explain how animals had migrated over vast expanses of water from one region to another. Darwin had discovered other means, far simpler to prove than theories of lost continents and continental extensions, and would elaborate on them in his book.

Once Wallace reached Celebes, he confined his activities to the town of Menado and the surrounding villages from the end of June until September. In early August, he suffered a relapse of malaria and remained in town for two weeks, preparing his specimens while he recuperated. When he regained his strength, he proceeded to the seaside village at the farthest point on the peninsula and camped on a steep black-sand beach backed by a low, scrubby forest and framed on both sides by small rivers. His main object in coming to this wild and lonely place was to capture specimens of the maleo bird (*Megacephalon maleo*), one of the finest species in the family Megapodiidae, the mound builders. In August and September, when there was little or no rain, they descended "by scores and hundreds" to the beaches of certain bays far from human habitations. His illness had deprived him of witnessing the spectacle at its height, but pairs of birds (male and female) still were emerging, having traveled fifteen to twenty miles from the interior to lay and incubate their eggs in the hot black sand. Like other megapodes, the maleos made unusual nests—heaping up mounds of sand at the high-water mark of the beach by means of the vigorous backward scratching of their short, blunt claws.

Wallace had an opportunity not only to observe their behavior but also to examine the mounds closely. In the mass of sand, he found a number of holes about five feet wide and almost two feet deep. In these holes, the birds had buried seven or eight eggs, each one placed about seven inches apart. At first it seemed to Wallace that each egg must have been laid by a different bird, since on dissection of a pregnant female he discovered a solitary egg completely filling the abdominal cavity, squeezing aside the intestines, while the ovaries contained eight to ten eggs the size of small peas. Once the egg was laid and buried—by the assiduous activity of both male and female—the breeding pair left to feed on fruit in the forest. The natives asserted that the eggs were all from the same female and that the same pair returned every thirteen days to the same hole to lay another egg. Wallace was unable to confirm this cycle, but he judged it likely, given the size of the other eggs maturing in the ovaries.

The peculiar habits of the whole family of megapodes, which differed strikingly ("departed widely," to use Wallace's evolutionary terminology) from the behavior of all other birds, provided Wallace with much to speculate about. In a lengthy letter to Sclater, published as an article in the ornithology journal *Ibis*, he laid out the facts. First, because of the egg's enormous size, the female could lay only one at a time. Second, a considerable time would have to pass before laying another egg. Third, the total number of eggs that each bird produced every season was about eight, leaving an interval of three months between laying the first and last eggs. If the eggs were hatched in the typical way, he said, they would have to be laid on the ground, forcing the parents to watch each egg constantly to prevent attack by predatory lizards. Moreover, the birds would have to sit on each egg to ensure the successful development of the chick. Complete incubation would be impossible because of the difficulty of obtaining their highly specialized food in the forest. Therefore, Wallace concluded, "the *Megapodiidae must* behave as they do." A creationist would explain the animal's physiology as an adaptation to its instincts and habits, which explained nothing. From the evolutionary viewpoint, it was the other way around: to ensure the species' survival, the bird's incubating behavior was an adaptation to its physiology. "A little consideration of the *structure* of the species in question," he wrote, "and the peculiar physical *conditions* by which it is surrounded, would show [its behavior] to be the inevitable and logical result of such structure and conditions."[63]

Wallace returned to Menado on horseback, arriving just in time to catch the mail steamer to Amboyna. On the way, the boat stopped at Ternate, where he gathered his mail and sorted specimens. One of the letters was from Darwin, dated April 6, 1859. The letter confirmed that Darwin's book was ready for publication. "There is no preface, but a short Introduction, which must be read by everyone who reads my book," Darwin cautioned. Because his book was only an abstract, he would give no references but would allude to Wallace's two important theoretical publications, showing that "your explanation of your law is the same as that which I offer." But he was no longer vague about the book's contents and aims. "You are right, that I came to the conclusion that Selection was the principle of change from the study of domesticated productions; and then reading Malthus I saw at once how to apply this principle. Geographical distribution and geographical relations of extinct to recent inhabitants of South America first led me to the subject. Especially the case of the Galapagos Islands." Once again, he expressed admiration for Wallace's attitude toward the joint publication of their papers, and informed him of the letter not sent—the one in which he had stated that he would not publish before Wallace had published. Hooker and Lyell's subsequent handling of the matter, he said, had changed his mind.[64]

Wallace continued to Amboyna, where he spent the next month sorting and packing his collections before shipping them to London. He summarized his findings from Menado in an article written in October 1859 and published the following April in *Ibis*. One remarkable statement he made had nothing to do with his own observations. For the first time, he made it clear that the theory of natural selection was Darwin's theory, not his own. Regarding the maleo's unusual behavior, he said, "For a perfect solution of the problem we must, however, have recourse to Mr. Darwin's principle of 'natural selection,' and need not then despair of arriving at a complete and true 'theory of instinct.'"[65] It was at this point that he seems to have closed his species notebook. If he had any lingering hopes of completing his own major work on the subject, they came to an end with Darwin's announcement. Whatever internal conflict he may have had about the priority issue had been resolved. The position that he took on that occasion was the position he stuck to for the rest of his life: all credit to Darwin; none for himself.

Although Wallace had been forestalled by Darwin in one area of his research, he continued to puzzle over two unsolved problems: the geographic distribution of animals and the origin and diversity of humankind. He channeled his energies into the completion of his mission, zigzagging from island to island until he felt that he could do no more. In the meantime, his impressive collections arrived in London at regular intervals, keeping the taxonomists and anatomists busy cataloguing and classifying birds, mammals, and insects and reporting their findings to the respective scientific societies.

Wallace's activities astonished his colleagues, even though his own statements undercut his achievements. From Celebes alone, he sent back more than a hundred species of hymenoptera (bees, wasps, and ants)—all new to science. His birds were also admired.[66] While his collections were praised, his theories were ignored, as though the former were made by one person and the latter by someone else. His most loyal patron, the entomologist William Wilson Saunders, clearly demonstrated this attitude when he addressed the Linnean Society on the subject of Wallace's enormous collection of hymenoptera:

> [N]one has exceeded that whose contents are described in the present paper, in the beauty and variety of the species, as well as in the interest attached to their geographical distribution. . . . I would particularly call attention to the two forms of the worker of *Pheidole notablisi*. Though convinced that nothing is created in vain, and that every modification of form has its design, adapting it to the fulfillment of conditions necessary in the economy of the particular species, I feel quite unable even to conjecture the purpose of the enormously enlarged head of the worker major of that species.[67]

At the end of October 1859, Wallace left Ternate for the island of Ceram. "If you want fine birds," Wallace was repeatedly told by George Robert Gray, John Gould, and other ornithologists, "go to Ceram." Like a curse, those words would echo through his mind for the next two months as he trekked across the island but collected almost nothing. To Stevens, he complained that Ceram was "a *wretched* place for birds." By Christmas, he was half-starved after a stormy and rain-soaked journey by sea. He then sailed for

Amboyna, covered head to toe in boils that took nearly two months to clear up—"a not very pleasant memento of my first visit to Ceram."[68]

While recuperating in Amboyna, Wallace received a copy of Darwin's *Origin of Species*, which had been published on November 24, 1859. "If you are so inclined, I should very much like to hear your general impression of the book, as you have thought profoundly on the subject, and in so nearly the same channel with myself. . . . Remember, it is only an abstract, and very much condensed," Darwin had written on November 14, ten days before publication. "God knows what the public will think. No one has read it, except Lyell. . . . Hooker thinks him a complete convert, but he does not seem so in his letters to me. . . . I do not think your share in the theory will be overlooked by the real judges, such as Hooker, Lyell, Asa Gray, etc. . . . I think I told you before that Hooker is a complete convert. If I can convert Huxley I shall be content."[69]

Wallace read the book with "intense interest," but only after the fifth perusal, he later admitted to Thomas Sims, did he fully appreciate the strength of the entire work.[70] By his own account, the *Origin* awed him. He understood at once the magnitude of Darwin's achievement. "I know not how or to whom to express fully my admiration of Darwin's book," Wallace confessed to Bates, who recently had returned to Leicester from Brazil.

> To him it would seem flattery, to others self-praise; but I do honestly believe that with however much patience I had worked up and experimented on the subject, I could never have approached the completeness of his book—its vast accumulation of evidence, its overwhelming argument, and its admirable tone and spirit. I really feel thankful that it has not been left to me to give the theory to the public. Mr. Darwin has created a new science and a new philosophy, and I believe that never has such a complete illustration of a new branch of human knowledge been due to the labour and researches of a single man. Never have such vast masses of facts been combined into a system, and brought to bear upon the establishment of such a grand and new and simple philosophy![71]

Wallace told George Silk that the *Origin* was the greatest book since Isaac Newton's *Principia*, insisting that "his name should, in my opinion, stand above that of every philosopher of ancient or modern times."[72] He also outlined his initial impressions of the book to Darwin, who did not save Wallace's critique, but Wallace preserved Darwin's response: "Most persons would in your position have felt bitter envy and jealousy. How nobly free you seem to be of this common failing of mankind. But you speak far too modestly of yourself; you would, if you had had my leisure, have done the work just as well, perhaps better, than I have done it."[73]

It was not long before Wallace learned of the effect of *The Origin of Species* on the English and American scientific communities. He probably did not see Thomas Huxley's famous review in the *Times*, written anonymously but in an unmistakable style that combined erudition and lacerating wit. Immediately after reading the *Origin*, Huxley became a "convert," reportedly stating, "How extremely stupid not to have thought of that."[74] But Wallace followed the commentaries in the issues of the *Annals and Magazine of Natural History* of 1860, whose objections were predictable. According to one reviewer, the belief in the independent creation of species was so general an opinion among naturalists that it was almost an "axiom." Although independent creation could not be proved, he said, the theory of common descent was equally unprovable. He preferred to abide by the universally accepted theory as the one most in accordance with the facts.[75] Asa Gray, who reviewed the *Origin* in the August 1860 issue of the *Atlantic Monthly*, was one of the few men who gave ample credit to the contributions of Wallace's two seminal papers.[76] In the stacks of magazines in Wallace's Ternate cottage were reviews of the *Origin*, which he read with eagerness, like a parent who has given up his child for adoption but who nevertheless remains keenly interested in any news of its progress. He had many back issues of the *Athenaeum*, a magazine devoted to literature, science, and the fine arts that had a large circulation among the educated public. The editors were hostile to Darwin's book, disparaging it—and Darwin—at every opportunity. "He omits nothing and he fears nothing," wrote one critic. "The work deserves attention, and will, no doubt, meet with it. Scientific naturalists take up the author upon his own peculiar ground; and there will we imagine be a severe struggle for at least theoretical existence. . . . Having introduced the author and his work, we must leave them to the mercies of the Divinity Hall, the College, the Lecture Room and the Museum."[77]

Darwin's book and its implications were the highlight of the annual meeting of the British Association for the Advancement of Science, held at Oxford in early July 1860. Huxley once characterized the association as "a meeting of the savans [*sic*] of England and the Continent, under the presidency of some big-wig or other . . . for the purpose of exchanging information."[78] Perhaps Huxley's disparagement was a bit unfair, for the association's goals were lofty.

The objectives that its founding members had set down in 1831 were threefold: first, to give "stronger impulse" and more systematic direction to scientific inquiry; second, to promote the exchange of ideas among British scientists throughout the British Empire with one another and with foreign scientists; and, third, to incite the government and academic institutions to pay greater attention to science and to remove any impedance to its progress.[79] In the early nineteenth century, most scientists were wealthy dilettantes, not professionals; since its establishment, the British Association, through its annual meetings, helped to promote the professionalization of science and to generate the esteem with which scientists are regarded today. By midcentury, the British Association meetings had evolved into glittering, eagerly awaited affairs, and the 1860 meeting in Oxford would not disappoint, becoming the most famous in the association's history.

The association had the usual structure: a president, a vice president, a treasurer, and various councils. Every year, it convened in a different city in Great Britain—a boon to local shopkeepers and other businesspeople, as hundreds of scientists and their families descended on the city or town selected. Its gatherings were divided into six sections—Mathematics and Physics, Chemistry and Mineralogy, Geology and Geography, Natural History, Statistics, and Mechanical Sciences (engineering)—and each section had its own president as well as committees and subcommittees. Section D was the Natural History Section, covering zoology, botany, physiology, and ethnology.[80] It was at one of the meetings of this section that Thomas Huxley supposedly defeated the bishop of Oxford, Samuel Wilberforce, in a battle that earned him the sobriquet "Darwin's bulldog" and that ultimately would be oversimplified by Darwinian spin doctors into a battle between the forces of enlightenment and the forces of ignorance. The story has been told many times, but unfortunately no one recorded exactly what was said; the encounter was spontaneous and the speeches extemporaneous, and its telling suffers from the *Rashomon* effect. Such details that exist were reported in the *Athenaeum*, which Wallace read in Ternate, perhaps while savoring a cup of coffee on the veranda of his tropical pied-à-terre.

"The hospitality [in Oxford] has been limitless," a reporter wrote in the irreverent style favored by the magazine's editors. "The colleges, the private houses, have been full. The splendid and piquant New Museum has been open day and night. An unusual flutter of silk and muslin has warmed with a brighter glow the old caves of the Bodleian. Groups that Watteau would have loved to paint have been daily seen under the elms of the Broadwalk or in the shades of Magdalen. . . . Every morning has brought no less charming receptions."[81] The atmosphere was charged with excitement. People awaited the confrontation between the two volatile contestants, Huxley and Wilberforce, with an anticipation as frenzied as that at a major sports event. All Wallace would have been able to glean about the conflict was from the unsympathetic *Athenaeum* correspondent.

"The Bishop of Oxford," the reporter said, "came out strongly against a theory which holds it possible that men may be descended from an ape, in which protest he is sustained by Prof. Owen . . . and the most eminent naturalists assembled at Oxford. But others—conspicuous among these, Prof. Huxley—have expressed their willingness to accept, for themselves, as well as for their friends and enemies, all actual truths, even the last humiliating truth of a pedigree not registered in the Herald's College."[82]

Another account of the episode is provided by Joseph Hooker, who was present at the now immortalized duel of egos. Outraged by the bishop's remarks, in a letter to Darwin written on July 2 he perfectly captured the meeting's high drama and the passionate exchange between men who harbored murder in their hearts. Although Wilberforce had publicly stated that his mission was to "smash" Darwin in front of a popular assembly, he was not Huxley's main antagonist. Wilberforce was seen as a puppet of the eminent paleontologist and anatomist Richard Owen, who had once been Huxley's teacher and was now his greatest enemy. Tension between Huxley and Owen had been building for several years prior to the Oxford meeting. Neither wasted an opportunity to disparage the other. An article in the conservative scientific publication *Quarterly Review* that criticized *The Origin of Species* and that allegedly was written by Bishop Wilberforce bore Owen's unmistakable hand. Both Huxley and Hooker believed that Wilberforce was Owen's surrogate, and it was at Owen that Huxley lashed out. On Thursday, June 28, shortly before the famous confrontation, a paper by an early supporter of Darwin led to what Hooker called "a furious battle" between Huxley and Owen "over Darwin's absent body." The president of Section D, the Reverend John Stevens Henslow, asked Huxley to comment on a paper by Charles Giles Bridle Daubeny, a professor of botany at Oxford, entitled "On the Final Causes of the Sexuality of Plants, with Particular Reference to Mr. Darwin's Work on the *Origin of Species*," which strongly supported Darwin. But Huxley demurred, not wishing to debate a subject—the new theory of evolution—about which "sentiment . . . unduly interfere[d] with intellect." Owen picked up the gauntlet. As a fact to controvert Darwin, he pointed to the brain of the gorilla, which differed more from the brain of man, he said, than it did from that of the lower apes. Huxley, who already had investigated and written on this matter, profoundly disagreed. He demonstrated conclusively to his audience that Owen was wrong, that not even a "minor hippocampus" (as the *Athenaeum* later put it) distinguished the brain of a human being from the brain of a gorilla or a chimpanzee. Huxley's triumph humiliated Owen, who lusted for revenge. Darwin and his book became the kindling for a planned conflagration.[83]

Sensing a showdown, visitors packed the lecture hall in the Oxford University Museum of Natural History on Saturday morning, June 30. Henry Draper from New York was scheduled to read his paper "The Intellectual Development of Europe Considered with Reference to the Views of Mr. Darwin," but members of the association came not to hear Draper—whom Hooker called a "Yankee donkey" in his letter to Darwin—but to hear more about Darwin's theory and Wilberforce's rebuttal. The audience was so large that the crowd, numbering perhaps a thousand, had to move from the lecture hall to the more spacious West Room. Near the windows lighting the west side of this room were crowds of women, whose white handkerchiefs waved and fluttered in the suffocating air. Huxley dreaded the confrontation and almost did not attend, but at the last moment Robert Chambers, the still anonymous author of *Vestiges of the Natural History of Creation*, persuaded him not to "desert the cause." The meeting's chairman once again was Henslow, Darwin's mentor and a professor of mineralogy at Cambridge, whom Darwin considered "wise and judicious." Not himself an evolutionist, he nevertheless supported free thought and scientific investigation. Seated on Henslow's right was Wilberforce, with Hooker on his extreme left and Huxley in the center. Four other men, including Draper, shared the platform.

After Draper gave his hour-long presentation in a droning voice, according to Hooker, Henslow selected three men to make comments, but they were all "shouted down" for arguing on theological and not scientific grounds. There followed clamors for the bishop to speak. "Well, Sam Oxon got up and spouted for half an hour with inimitable spirit, ugliness and emptiness and unfairness," Hooker wrote to Darwin. "I saw he was coached up by Owen and knew nothing, and he said not a syllable but what was in the Reviews; he ridiculed you badly and Huxley savagely." Wilberforce—"Soapy Sam," as he was derisively called by Hooker—assured the audience that there was nothing to the idea of evolution. The permanence of species was a fact confirmed by every observation. All hybrid offspring were sterile; no experiments had ever confirmed the transformation of one animal into another. Rock pigeons were what rock pigeons had always been. He then turned to Huxley, Hooker reported, and with an insolent smile said, "I beg to know whether it is through your grandfather or your grandmother that you claim descent from a monkey."

The unabashed vulgarity of the remark, deemed ungentlemanly to Victorians, was a tactical mistake, for Huxley, a master of character assassination, was now given ample ammunition to mount an attack against his enemy. Letting Wilberforce finish his diatribe, he at first feigned a reluctance to descend to Wilberforce's level, but after a little encouragement he aimed his elocutionary arrow directly at his adversary and observed that he would rather be descended from an ape than from a man who plunged into scientific questions with which he had no real acquaintance in order to appeal to religious prejudice. According to Hooker, although Huxley's riposte was "admirable" and turned the tables on Wilberforce, Huxley's voice could not carry over so large an assembly. Moreover, Huxley did not allude to Wilberforce's weak points. As another observer would later remark, Huxley scored a victory over the bishop on the issue of good manners rather than good science. Nevertheless, it was a fierce battle, with much commotion among members of the audience; in the heat of the moment, a woman in the audience, one Lady Brewster, fainted.

Dissatisfied with Huxley's response yet cautious and mindful of his reputation, Hooker handed his name to Henslow as the next speaker. "I swore to myself that I would smite that Amalekite, Sam, hip and thigh if my heart jumped out of my mouth," he wrote to Darwin. "[T]here I was cocked up with Sam at my right elbow, and there and then I smashed him amid rounds of applause. I hit him in the wind at the first shot in ten words taken from his own ugly mouth." Hooker made several points, all of which were strictly limited to the botanical implications of Darwin's theory. First, that Wilberforce could never have read Darwin's book. Second, that he was "absolutely ignorant" of the fundamentals of botany. Existing species do not change form, Hooker said. All species varied, and it was on these variations that natural selection might act. He could not conceive how anyone who had read the book could make such a mistake. Darwin's hypothesis was the most powerful explanation of all the phenomena associated with classification, distribution, structure, and development of plants thus far postulated, and would lead to fruitful research. "Sam was shut up—had not one word to say in reply,' Hooker reported, "and the meeting was dissolved forthwith, leaving you master of the field after 4 hours' battle."[84]

Six years of Wallace's life had been spent in the Malay Archipelago. Now thirty-seven, he needed another six years or more to complete a zoological overview of the region, but two months of lost time recovering from the outbreak of boils convinced him that he could not remain in the archipelago indefinitely. If he was ever to return to England alive—and he had serious concerns that he might not—he had to accelerate his operations. Ali, his most trusted man, could not accompany him on every journey. He required a man with experience, one who could act as his surrogate in places he could not easily reach. With his heart and mind set on sailing home in the near future, he solicited the aid of his old assistant, Charles Allen.

Never one to burn bridges, Wallace had kept in contact with Charles since his departure in 1856. Charles had left Sarawak, having managed to survive several massacres by Chinese insurgents, and moved to Singapore to work on a plantation.[85] Wallace must have made him an offer he could not refuse—or perhaps he missed the itinerant life and the thrill of the hunt. By February 1860, he was in Ternate, ready to do Wallace's bidding. Now in his early twenties, he had matured into a good-looking young man over six feet tall, spoke fluent Malay, and was adept at traveling through the islands. Whether or not he had outgrown his irksome qualities did not concern Wallace any longer. Charles would be traveling independently most of the time and thus would not try his patience.

Wallace had not given up his quest for more birds of paradise. New Guinea was his ultimate destination, but he did not want to risk Charles's life in so dangerous a place. He outfitted him for Mysol, an island just south of the westernmost peninsula of New Guinea, where traders assured him that birds of paradise were more abundant and beautiful than anywhere else in the archipelago. Charles set off for Mysol at the end of the month. Wallace planned to go to the island of Waigiou, off the western tip of New Guinea, stopping one last time in Ceram and picking up Charles along the way. To expedite his journey, he had obtained a letter from the Dutch governor of the Moluccas requesting all village chiefs to supply him with boats and men.

The trip began auspiciously. At Wallace's first stop in Ceram, the local rajah assembled an impressive entourage for his next destination, where he was promised black and yellow lories and black cockatoos. He traveled in unaccustomed regal splendor, accompanied by four boats rowed by sixty men, with flags flying, drums beating, and spirited shouting and singing. The sea was smooth, the morning clear and bright, the scene exhilarating. When he disembarked, he was treated like a foreign dignitary by the local chief and other potentates, who dressed in their finest regalia in honor of his arrival, and was conducted to a house specially prepared for his use. But lories and cockatoos were not to grace his collections. Eastern Ceram was, like every other part of the island, a tropical paradise with a paucity of wildlife. The fanfare and the alluring tales of birds in profusion were all a ruse to bring a rich European, with his copious stores of spirits, to the village of a friend and ally. Annoyed, he left at once, sailing eastward along the treacherous southern coast, landing at another village, where he was stranded for a month and reduced to pleading in vain for a crew to take him elsewhere.

Eventually Wallace headed south toward the Ke Islands, which he had visited during his first journey to the Moluccas. On the island of Goram, he had a boat built for himself for the nominal price of £9; had he been less vigilant and not personally involved in its construction, he might have waited a year for its completion. After a month, he was the proud owner of a seaworthy prau outfitted with European accoutrements. Just before leaving, Wallace was treated to a preview of the dangers of New Guinea by six of its survivors, who described an attack by natives that left fourteen of their comrades murdered, including one of the rajah's sons. The village erupted in cries of lamentation and ululations at the terrible news, which put fear into the hearts of his own men. Only by means of threats and strong-arming was he able to assemble a crew willing to accompany him to Waigiou, a less bloodthirsty destination. They left Goram at the end of May 1860, but most of his crew slipped away in a dinghy one night. He raged in vain, vowing to punish the ringleader severely if caught, but they were never seen again.

With his remaining men, Wallace headed back to Ceram, where he was delayed for several days because of difficulty in recruiting another crew. A passing trader informed him that Charles was nearly out of food, pins, and other necessities. He was also ill and planned to return to Ceram if Wallace did not arrive soon. But Wallace never reached Mysol. The weather conspired with an ever-changing current, and he overshot the island several times, finally giving up in despair and sailing north. Finding three islands on his map, he helped his steersman navigate toward one of them, where he anchored and contemplated his next step. Unhappy with his tenuous anchorage, he sent his two best men ashore to cut some vines for rope to tow them to a safer spot. The boat tore loose, and as they drifted farther and farther from shore they called out to the two men, who panicked when they grasped what had happened and ran up and down the shore, too hysterical to notice the numerous logs strewn on the beach that might have made an excellent raft if tied together. Wallace realized that they had resigned themselves to their fate when he saw the smoke of a fire on the beach.

Wallace's own predicament was no better. With only a two-day supply of water on board and his two best men abandoned on a tropical isle, he guided his remaining crew to another island, where for three days they quenched their thirst with brackish water, reeking of rotten leaves, from a well they had dug near a stand of sago trees. With no hope of rescuing their companions, they sailed to Waigiou, which they reached—more by dint of their wits and luck than by knowledge—just in time to avert death by dehydration. Three more days of nerve-racking sailing brought them to the first human habitation they had seen. The headman, somehow apprised of their imminent arrival, came to their assistance with a present of coconuts and vegetables. Immediately after disembarking, Wallace hired three natives to search for his two men, but after ten days they returned because of bad weather. Wallace refused to give up. Bribing the search party with knives, handkerchiefs, tobacco, and other provisions, he convinced them to make one more attempt. This time they succeeded, and at the end of July he was rejoined by his two men, who were thin and weak but healthy and grateful for Wallace's persistence, having subsisted on roots, flower stalks, shellfish, and turtle eggs for a month.

The voyage was not all in vain. While he awaited the outcome of the rescue operation, Wallace obtained several good specimens of the red bird of paradise (*Paradisea rubra*), but he wanted more. He therefore made a short trip by outrigger canoe to a neighboring village, where for six weeks he rented a hut eight feet square by five feet high that was elevated on posts five feet above the ground. His spirits had improved considerably, and by early September he was able to write a playful letter to George Silk. It had been ten months since the date of his last letter from England, he said, but he did not miss the news. He did not care a "straw" for anything but his family and his personal affairs. He joked that while honorable members of the Royal Geographical Society were amusing themselves shooting partridges, he was on the other side of the world in a nameless village shooting—or trying to shoot—red birds of paradise. In his spare moments he read books, and his reading was eclectic: a history of prostitution by a Frenchman and Darwin's

Origin.[86] For the price of a handkerchief and string of beads per pair of birds, Wallace hired native bird catchers to collect as many birds of paradise as possible. Although these men were diligent, they could not—or would not—bring birds back in good condition. Carrying their booty long distances, they returned with birds invariably covered in ashes, pitch, or some other sticky substance that proved difficult to remove. A few of the specimens were rotting and stinking. Occasionally he was brought a live bird, which gave him a rare opportunity to study the habits of the species. He built a large cage to house them, but each died within a few days.

The technique of capturing a bird of paradise was an art practiced by no more than eight or ten men on the entire island of Waigiou. It involved the ingenious construction of lure and snare, which snagged the leg of a bird and prevented its escape. The little village was the headquarters of these Papuan Papagenos, but Wallace nearly starved to death in the process of adding to his collections. He was amply rewarded for his patience, however, and left at the end of September with twenty-five fine specimens of *P. rubra*, preserved in various stages of plumage—an achievement not only unique in the annals of ornithology but highly instructive for those unable to observe the bird in its natural habitat. But *P. rubra* was the only species he was able to collect; once again, he had been misled by his French predecessor, who had acquired seven species of bird of paradise at Waigiou, brought from the interior by native hunters.[87]

In early November, Wallace returned to Ternate, where Charles awaited him. Charles's collections were disappointing. He told Wallace that he had been forced to return to Ceram to replenish his supply of rice, which was delayed, and was able to return to Mysol for only two more weeks. Wallace spent the next two months organizing his collections and going through mail. He had sixteen thousand insects, birds, and shells to sort through, but none that he considered special or unusual. He told Stevens that Waigiou was "the very poorest island in the New Guinea zoological region." With demonic determination, he prepared Charles for another mission to New Guinea, this time to a place between Dorey and the island of Salwatty, south of Waigiou. For security, he sent along a lieutenant and two soldiers in the service of the sultan of Tidore. "If he does not succeed this time, I must give up the attempt in despair," he informed his readers in an article on the ornithology of Ceram and Waigiou published in *Ibis*.[88]

In early January 1861, Wallace left Ternate for Delli, in East Timor, where he remained until mid-May. In a letter to his brother-in-law, he described his plans to move into his mother's cottage (she apparently was no longer living with Fanny and Thomas) and asked for the dimensions of the rooms to see whether his collections could be accommodated.[89] But he was not ready to return to England at once and would remain for another year in the Malay Archipelago. From Timor, he sailed to the Bandas, and from there to Bouru, an island west of Ceram. At the end of June, he returned to Ternate for the last time. In the next two days, he packed his belongings and said his good-byes to the friends he had made there in the past two and a half years. He sailed to Menado and caught a steamer to Java, arriving on July 18 in Surabaya, in the eastern part of the island, where he remained until the end of October. Two days after his arrival, he already was enchanted, pronouncing Java to be "the garden of the East, and probably without exception the finest island in the world." In Java, Dutch colonialism was in full flower, exercised over a population that had experienced a long history of high civilization. Good roads, modern conveniences, the magnificent remains of ancient cities, and a luxuriant landscape appealed to him aesthetically and intellectually—particularly after nearly three years in the "wild and savage" Moluccas and New Guinea.[90]

For the next two weeks, he resided in a noisy hotel, preparing and sending off his collections, before embarking on a short journey into the interior. Along the way, he passed some of the greatest architectural monuments of the world—the Hindu temple complex at Prabanam and the Buddhist temple at Borobodur—whereupon he was "led to ponder on the strange law of progress, which looks so much like retrogression, and which in so many distant parts of the world has exterminated or driven out a highly artistic and constructive race, to make room for one which, as far as we can judge, is very far its inferior."[91]

Wallace gradually made his way to Batavia (Jakarta) before exploring the western portion of the island. He hiked to the top of a mountain ten thousand feet high in order to view the change from a tropical to a temperate flora and was surprised by the remarkable resemblance of the high-altitude plants to vegetation in faraway Europe. "The common weeds and plants of the top were very like English ones . . . all closely allied . . . but of distinct species," Wallace noted to Fanny Sims. "The fact of a vegetation so closely allied to that of Europe occurring on isolated mountain peaks, in an island south of the Equator, while all the lowlands for thousands of miles around are occupied by a flora of a totally different character, is very extraordinary." Some botanists, including Richard Spruce, believed that the closely allied but geographically separate species had been divinely created on two different occasions, but the phenomenon deserved a more scientific explanation. Darwin had proposed the following theory in the *Origin*: during the height of the glacial epoch, temperate forms of plants extended to the tropical zones, and then retreated up these southern mountains as well as northward to the plains and hills of Europe. Some northern species had even crossed the equator and reached the Antarctic. It hardly seemed credible to Wallace that the wide expanse of sea between Java and the European or Asian mainland would have permitted successful migrations of plants by the mechanism that Darwin proposed. For the time being, he remained puzzled.

The Javanese versions of buttercups, violets, wood thistle, and St. John's wort made him nostalgic for the English countryside. The greater beauty of the tropics was now a grand illusion to him, a dream mainly concocted by romantic travelers who wrote for an audience hemmed in by urban civilization and longing for a balm to soothe their daily miseries. "I still consider and will always maintain that our own meadows and woods and mountains are more beautiful," Wallace wrote to Fanny. "It is only the great leaves and the curious-looking plants, and the deep gloom of the forests and the mass of tangled vegetation that astonish and delight Europeans, and it is certainly grand and interesting and in a certain sense beautiful, but not the calm, sweet, warm beauty of our own fields, and there is none of the brightness of our own flowers."[92]

When Wallace returned to Jakarta, he met up with Charles, who, not surprisingly, had a complicated story to tell. After leaving Ternate in January, he had arrived without mishap in New Guinea. But when he explained his intention of obtaining birds of paradise to the local people, all sorts of objections were raised: it was three- or four-day journey through swamps and mountains; the people of the mountains were savages and cannibals who were sure to kill him; no man in the village would dare to go with him; and so forth. These were half-truths. In reality, Charles was eyed with suspicion. Birds of paradise were items of commerce monopolized by the coastal chiefs, who paid a tribute to the sultan of Tidore. Any direct trade with the people of the mountains would drive up the cost of a bird and lower profits. The chiefs blocked him at every turn. Eventually he was able to hire a prau to take him inland, but the people had somehow been warned of his arrival and refused to assist him. The Tidore lieutenant caused even more difficulties by quarreling with the locals. Charles had to intervene before blood was shed, becoming the protector of those who had come along to protect him.

As a white man, he did not feel threatened and used his position to assemble a crew, bribing them with knives, hatchets, and beads. This done, they trekked into the interior, where he remained without an interpreter for a month. Ultimately, he procured no new species of birds of paradise. "Thus ended my search after these beautiful birds," Wallace later wrote. "Five voyages to different parts of the district they inhabit, each occupying in its preparation and execution the larger part of a year, have produced me only five species out of the thirteen known to exist in New Guinea."[93]

On November 1, 1861, Wallace left by steamer for Sumatra. Dutch sovereignty made portions of Sumatra safe, and he was therefore free to travel wherever he liked within Dutch territory without fearing for his life. For a month he followed the numerous paths into the virgin and secondary forests or searched around streambeds, gathering small numbers of rare and interesting birds and butterflies. Relishing his remote location, Wallace told George Silk, "I am here in one of the places unknown to the Royal Geographical Society, situated in the very centre of East Sumatra. . . . It is the height of the wet season, and the rain pours down strong and steady. . . . Bad times for me, but I walk out regularly three or four hours every day, picking up what I can."[94] By January 1862, he was back in Singapore, making preparations for his final journey home. He left Ali his two double-barreled guns plus ammunition, his tools, and anything else he thought would be of use. He also commissioned a photograph of Ali, which he preserved for the rest of his life and published in his autobiography. The photo shows a young man in his early twenties, with a round, handsome face and dark, curly hair neatly combed back. His countenance is somber; his eyes, dark and intelligent. Wallace described him as "the best native servant I ever had, and the faithful companion of almost all my journeyings among the islands of the Far East." Now "quite rich," Ali returned to his wife in Ternate, after which nothing more is known about him.[95]

While Wallace was still in Singapore, his contacts notified him of two live birds of paradise, freshly brought by a trader from the eastern part of the archipelago. On inspection he found two healthy-looking juveniles of *Paradisea papuana*, not yet in full plumage, in a spacious cage five or six feet square. They were a noisy, active pair, squawking and flitting from perch to perch like jays. The trader asked an astounding £100 for the two birds, but Wallace took the risk and bought them. It paid off. He telegrammed the Zoological Society about his extravagant purchase and received an enthusiastic reply that the zoo would buy the birds at terms that Wallace found agreeable.[96]

The birds were a constant source of trouble and anxiety during the seven-week return journey, with Wallace worrying over them like a mother. Although they readily ate bananas and dried bread, he knew that they needed a more balanced and nutritious diet to survive the arduous journey through multiple climates. To his great relief, he discovered that they would eat cockroaches, which abounded on ships in the tropics, and every evening he went to the ship's storeroom to brush roaches into a biscuit tin. Another difficulty was keeping their cages clean. For all their beauty, birds of paradise were as filthy as the common crow. In Bombay, Wallace attracted a great number of curious visitors when he brought his birds out onto the hotel veranda. From Bombay, he sailed to Suez, and from there he traveled by rail to Alexandria. He spent a nervous night with his birds in the baggage compartment as the train crossed the cold, windswept desert. From Alexandria, they proceeded to Malta, where he bought smaller cages and remained for two weeks collecting cockroaches from a bake shop because they were difficult to find on board ship during the winter. From Malta, he went on to Marseilles, where he wired Philip Sclater at the Zoological Society to announce his imminent arrival in London with his prizes. On the way to Paris, railway officials refused to let him stay with the birds in the baggage compartment and insisted that they be treated as ordinary baggage, though it was now late March and the temperature hovered near freezing.

At long last, on March 31, 1862, he crossed the English Channel and landed at Folkestone. "I have great pleasure in announcing to you the prosperous termination of my journey and the safe arrival in England (I suppose for the first time) of the Birds of Paradise," Wallace telegrammed to Sclater. The birds had by now begun to develop their magnificent ornamental plumage. "Another year with a genial temperature, flying room, foliage, and abundance of food, and I hope they will be glorious."[97] When he alighted from the train in London, he gladly transferred his charges to the care of the zoo, where one bird survived until Christmas of 1863 and the other until the following March. Thus ended Wallace's travels to the Malay Archipelago.

CHAPTER 9

Beautiful Dreamer

As soon as Wallace stepped off the train in London, he took a cab to 5 Westbourne Grove Terrace, in Notting Hill, the home of Fanny and Thomas Sims, where Mary Anne Wallace also now lived. Eight years had passed since he had last seen his mother and sister. During those eight years, he had traveled more than 14,000 miles either on foot or by boat, amassing 125,660 specimens consisting of 310 mammals, 100 reptiles, 8,050 birds, 7,500 land shells, 13,100 butterflies and moths, 83,200 beetles, and 13,400 insects belonging to other orders—a quantity that promised to keep taxonomists and academic naturalists occupied for the foreseeable future.[1] From the northwestern coast of New Guinea and the surrounding islands alone, he had obtained more than 300 distinct species of birds not known to have existed there. Of these, 67 were new to science and 2 were immortalized with his name. Almost single-handedly, he had captured, prepared, described, catalogued, and shipped off his collections under circumstances as difficult to imagine in the 1860s as they are today. It was a stunning feat—the equivalent of scaling Mount Everest or trekking across the South Pole—that secured his position in the pantheon of the greatest explorer-naturalists of all time.

In 1862 Wallace was a famous man and one of English science's brightest stars. Thomas Huxley, who doled out compliments sparingly, remarked that "once in a generation, a Wallace may be found physically, mentally, and morally qualified to wander unscathed through the tropical wilds . . . to form magnificent collections as he wanders; and withal to think out sagaciously the conclusions suggested by his collections."[2] A year before his return to London, Wallace had been elected an honorary member of the British Ornithological Union. Only a few days after his arrival, the Zoological Society of London inducted him as a fellow in its exclusive ranks. His daring voyages and his indefatigable pursuit of scientific knowledge impressed entomologists, zoologists, ornithologists, ethnologists, and geologists alike, regardless of their position on evolutionary theory. Even the

Athenaeum, excusing his heretical views, hailed him—along with his colleagues Richard Spruce and Henry Bates—for extraordinary courage. As a reporter for the magazine noted:

> The great artery of South America has of late years been explored by three naturalists, of whom this country may well be proud, and whom science justly places in the foremost ranks of those who have extended her boundaries and enriched her with new and reliable observations. All three, dispensing with the pittance flung to explorers by Government, relied upon their own resources, and paid their way by sending home collections for sale to their agents in London. The fact that such an arrangement was successful speaks well for the industry of the travellers and the growing taste for the fruits of such labours at home. In tracking their steps over the immense territories explored, and in reviewing the bold enterprises which they accomplished, we ought not for a moment to lose sight of the fact that the whole was a private undertaking likely to collapse by a single false movement.[3]

In a large empty room on the top floor of the Sims household, Wallace installed all his journals and registries documenting eight years of work and travel in the Malay Archipelago. He also retrieved the specimens that he had entrusted for safe storage to his agent, Samuel Stevens, who had set aside three thousand bird skins and twenty thousand beetles and butterflies, in addition to land shells and a number of other insect orders. The remaining hundred thousand specimens had been sold off to museums or private collectors. "I found myself surrounded by a quantity of packing-cases and storeboxes," Wallace recalled in *My Life*, "the contents of many of which I had not seen for five or six years, and to the examination and study of which I looked forward with intense interest."[4] He engaged a carpenter to provide one side of the room with movable shelves and build a long table on which he could unpack and sort his specimens. He also bought a gross of cardboard boxes in three sizes in order to classify and preserve his bird skins, which he labeled with the generic or family name and arranged in order on the shelves, allowing him to find any species without difficulty whenever he needed. The unpacking and sorting took nearly a month.

Darwin was among the first to send a letter of greeting, less than a week after Wallace's return, extending an invitation to his house in the village of Downe, south of metropolitan London. He was eager to meet Wallace, who did not yet feel up to making the fifteen-mile journey from London, not even to meet Darwin. "As I am being doctored a little," he replied, "I do not think I shall be able to accept your kind invitation at present."[5] The polite refusal did not signify secret resentment toward Darwin about the priority issue. Wallace was simply physically exhausted. Stevens confided to Norton Shaw, the secretary of the Royal Geographical Society, that Wallace was in much better condition than might have been expected under the circumstances. (In the same letter, he reports the death of another naturalist from "jungle fever" in Siam, underscoring the inherent dangers to the traveling naturalist in the mid-nineteenth century.)[6] Like contemporary astronauts after weeks in space, Wallace needed time to recuperate, regain his strength, and readapt to his old environment. A photograph of him taken in Singapore in January 1862 on the eve of his departure for England shows a gaunt, bespectacled man in a wide-brimmed hat, with a bushy beard and loose-fitting clothes hanging on a skeletal frame.[7] Shortly after his return, he developed another outbreak of boils, a sign of a weakened immune system, which confined him to his apartment for ten days. As a gesture of goodwill and in a spirit of friendship, Wallace sent Darwin a wild honeycomb from Timor, promising to visit in the summer when his health had improved.

At the end of April, Bates came down from Leicester to visit Wallace. They went by cab to the Royal Botanic Gardens at Kew to meet Joseph Hooker, who had succeeded his father as director. Wallace's deference to Darwin had impressed Hooker, who was eager to be introduced to a scientist at once so illustrious and so lacking in vanity. But the introduction would have to wait. As a result of a miscommunication, Hooker left before they arrived. Wallace and Bates had to be content with a "delightful stroll" through the well-tended gardens, which gave them an opportunity to catch up on both scientific and personal matters during the past decade.[8]

Poor Bates! In Leicester, not long after returning from Brazil in 1859, he had met an illiterate young woman named Sarah Ann Mason. Only two months before he visited Wallace, on February 2, 1862, she had given birth to his child—a potential scandal since they were not yet married.[9] The birth complicated Bates's life both socially and economically. Although he had accumulated a substantial number of specimens during his travels, most were insects that earned far less than his friend Wallace's more exotic collection. He barely managed on an annual income of £123, part of which he was applying to a loan from the family business to fund his early years in the Amazon. Both his travels and his sexual indiscretion had taken their toll: Fanny's colorized portrait of him taken by Thomas Sims shows a man who looks much older than his thirty-seven years.[10]

Professionally, Bates's future seemed brighter. An important paper of his that had been presented at the November 21, 1861, meeting of the Linnean Society and published the following year elated both Darwin and Hooker. Innocuously called "Contributions to an Insect Fauna of the Amazon Valley," it would become a landmark in evolutionary science. The most substantive part of the paper dealt with mimicry among insects, especially butterflies—a phenomenon that Wallace had also observed but had not yet elaborated on. Its discovery boosted the evolutionary cause by providing the first concrete demonstration of the action of natural selection. Bates stated the problem and its solution succinctly: "Mimetic analogies . . . are resemblances in external appearance, shapes and colours between members of widely distinct families; an idea of what is meant may be formed by supposing a pigeon to exist with the general figure and plumage of a hawk." The facts led in only one direction once Darwin and Wallace had articulated a theory. "The explanation seems to be quite clear on the theory of natural selection," Bates concluded.[11]

Bates had the specimens to illustrate his theory. He laid them out before his colleagues at the meeting, and they were later reproduced for his article as colored plates from his own watercolors. Darwin, who was not present at the society meeting, was forcibly struck by the paper's power and jubilant about its validation of natural selection. "In my opinion," he told Bates, "it is one of the most remarkable & admirable papers I have ever read in my life." His only criticism was of the bland title. "I cannot but think that you ought to have called prominent attention in it to the mimetic resemblances. Your paper is too good to be largely appreciated by the mob of naturalists without souls; but, rely on it, that it will have *lasting* value. . . . You will find, I should think, that Wallace will fully appreciate it."[12] And, indeed, when Wallace inspected Bates's "mocking" butterflies one Sunday evening, Bates told Darwin that he "came to the conclusion that all Nature does not furnish so plain & striking a case of the origin of species & of new & complex adaptations to new conditions, by the simple process of variation & Natural Selection."[13]

While Bates was composing his paper, he was also working on his travel narrative, coached and coaxed by Darwin. Darwin had taken a paternal interest in Bates's professional affairs, grateful for his support of the theory of natural selection. He walked Bates step by step through the publication process, encouraging him to publish his findings in both the scientific and the popular press. Bates required considerable encouragement; writing was an occupation that did not come easily to him. "Some are born with a power of good writing, like Wallace," Darwin told him sympathetically. "Others like myself & Lyell have to labour very hard & long at every sentence."[14] Bates's *Naturalist on the River Amazons* [*sic*], published in March 1863, enjoyed critical success. Wallace warmly praised the book, thanking Bates for the "kind manner" in which he had mentioned his name and perhaps feeling some remorse for saying so little about Bates in *A Narrative of Travels on the Amazon and Rio Negro*, published in 1853, a few months before his departure for Singapore.[15] Darwin also extolled the book's merits, judging it superior to Wallace's account, which he had found "a little" disappointing. "Hardly facts enough," Darwin told Bates.[16]

In the meantime, Wallace set to work on his collections. He spent a considerable amount of time in the bird room of the British Museum, conferring with George Robert Gray, the ornithologist brother of John Edward Gray, the museum's director of the zoology department, who was describing and identifying many of Wallace's discoveries. Making up for years of social and professional isolation, Wallace immersed himself in London's scientific life and contributed his thoughts and opinions at meetings of the Zoological, Entomological, Royal Geographical, and Linnean Societies. He first appeared before the Zoological Society on May 27, 1862, to read a paper called "Narrative Search After Birds of Paradise," an apologia for what he considered to be the scanty number of species and specimens he obtained of this highly sought after group of birds. Into a few pages, he compressed the numerous journeys he had made in his search for the elusive Paradisaeidae. The audience doubtless was rapt, for Wallace wrote with dramatic flair in a style at times verging on the purple:

> A feather is itself a wonderful and a beautiful thing. A bird clothed with feathers is almost necessarily a beautiful creature. How much, then, must we wonder at and admire the modification of simple feathers into the rigid, polished, wavy ribbons which adorn *Paradisea rubra*, the mass of airy plumes upon *P. apoda*, the tufts and wires of *Seleucides alba*, or the golden buds borne upon airy stems that spring from the tail of *Cincinnurus regius*; while gems and polished metals can alone compare with the tints that adorn the breast of *Parotia sexsetacea* and *Astrapia nigra*, and the immensely developed shoulder-plumes of *Epimachus magnus*.[17]

Not until late July or early August did Wallace finally make his "long, toilsome" trek to Down House, as Darwin referred to the journey from central London. In June, flulike symptoms and recurrent fevers had nearly prevented him from presenting two more papers at the Zoological Society, one on new and rare birds from New Guinea and the other on three new species of *Pitta* from the Moluccas. Darwin had mailed explicit instructions to Wallace describing the easiest route to his home. From London Bridge Station, Wallace caught a train to Orpington, southeast of central London. From there, he boarded a stagecoach that passed through the village of Downe over jarring, unpaved country roads. His impending visit undoubtedly filled the Darwin household with great anticipation. Arriving was the man who had acted so nobly and honorably toward husband and father. Wallace was equally eager to meet the most famous—or notorious—man in English science, whose fame and notoriety he had helped to establish. If that thought crossed his mind on the way, it must have amused him. Had fate been otherwise, he might well be occupying Darwin's unenviable position and Darwin might be making the long, toilsome trek to his modest apartment in London.

The coach dropped Wallace and his bags in front of Darwin's white-stucco manor, which was situated on eighteen acres of meadows and groves of oak and ash trees. The house had not always been so appealing, having once been a wreck that Darwin renovated when he bought it twenty-five years earlier. The tranquility and beauty of its setting would have struck Wallace as ideal for the contemplation of the wonders of nature. Although he enjoyed the intellectual stimulation of London, he would soon tire of urban life and long for similar solitude in the English countryside. Joseph Paltrow, the family's longtime butler, welcomed Wallace into the house, and Darwin rushed forward to give his colleague a hearty handshake. Not far behind hovered Emma Darwin, her husband's constant companion, nurse, and confidante, and six of the children, spanning ages eleven to nineteen. Both men stood over six feet tall, but Darwin's habit of stooping shortened his stature. His craggy brow and thick eyebrows overshadowed his eyes, whose kindness softened an otherwise forbidding aspect. He was bald except for a tonsure of graying dark hair, and bushy muttonchops covered half his face. Wallace sported a healthy beard, which he had decided not to shave off after returning from the Malay Archipelago. By now, his full head of light brown hair, parted on the left side and swept off his brow, was prematurely graying. His penetrating blue eyes studied his revered colleague from behind wire-rimmed glasses. A world-weary air enveloped Darwin, who was fifty-three years old, but Wallace, at thirty-nine, still brimmed with youthful energy.

After tea and relaxing conversation, Darwin led Wallace into his study, where he began his interrogation, wishing to glean as many new facts as he could in a single visit. His workstation impressed Wallace—as it did many of his visitors—with its simplicity and orderliness. His dissecting bench, by the window, could be lowered to permit him to work while seated, since he tired after standing for too long a time. His dissecting tools and instruments lay scattered on the bench. A nearby table supported a tower of neatly labeled drawers and served as storage for the overflow of items that seemed useless to the casual visitor but that he saved because of his belief that "if you were to throw a thing away you were sure to want it directly."[18] There were shelves holding glasses, saucers, labels, biscuit tins for germinating seeds, and other necessities. Onto one shelf, he had piled books he had not yet read; onto another, he transferred those he had finished. Darwin always kept notes when he read a book and drew up an abstract from them, which he catalogued for future use. In every niche and corner of the room were bundles of miscellaneous manuscripts. Partitioned off from the rest of the room by a curtain was a well-used chaise to which he retired when fatigued or unwell. At its side, like a faithful pet, rested a chamber pot, which the butler made certain was emptied every day.

Darwin liked to show his visitors around the grounds of his small estate. Near the house, a swath had been cut out for a kitchen garden; there was an experimental plot and a greenhouse for his orchids. One of his favorite spots was the "Sand-walk," a narrow strip of land one and a half acres in extent, with a gravel path shaded by oaks and bounded by a hedge, over which were views of a quiet valley and rolling hillside. There, both alone or with family and visitors, he would take a midday constitutional. After a hearty lunch, Darwin retired to his study because of his ill health, allowing Wallace free range of his home and grounds.

The nature of Darwin's illness, never clearly defined, has sparked much speculation. Diagnoses have ranged from agoraphobia and panic attacks to Chagas' disease, a parasitic infection endemic to South America, with possible long-term effects on the heart and gastrointestinal tract. In March 1864, for example, Darwin wrote to Wallace from a spa that he had been suffering from fits of vomiting every day for two months, which had prevented him from doing anything. Such attacks, Wallace noticed, were brought on by "the least mental excitement."[19] From Darwin's numerous allusions to his infirmities in his letters, one might easily form the impression that he was neurotic. Just as he seemed on the verge of requiring the last rites, with friends and family plunged into despair, he would produce yet another work of genius. Only Proust in his cork-lined room suffered more than Darwin. But Darwin's friends and family took his complaints seriously. Emma Darwin thought of little else but her husband's comfort. According to his son Francis, Darwin's life was one of constant suffering, borne with a cheerful spirit. His father, he said, "never knew one day of the health of ordinary men, and . . . his life was one long struggle against the weariness and strain of sickness."[20] Wallace remarked in his autobiography that it was a "real wonder" that Darwin had accomplished so much despite his "almost constant and depressing" ill health. Had Darwin been less sickly, Wallace believed, he would have published a systematic series of volumes dealing with every subject outlined in *The Origin of Species*. During the last twenty years of his life, however, Darwin devoted his time to his observations and experiments with plants and worms, an occupation that he enjoyed and that constituted "relaxation" rather than genuinely productive work.[21] He had retreated from the world, observing it as through a telescope; thanks to good fortune and good birth, he could remove himself from the stink and chaos that others could not avoid. The pampered isolation he enjoyed at Down House allowed him to keep his hands busy and his mind clear.

This description of Wallace's visit with Darwin is partly imagined, pieced together from Francis Darwin's vivid recollections of the protocol his father followed with guests. Neither Wallace nor Darwin left an account of their first meeting, which is surprising given its historic importance, a fact that must have been apparent to both men even then. Darwin neglected to elaborate on the episode in his autobiography, written twenty years later, when Wallace's "impetus," as Huxley once called it, had faded from memory and Wallace had become an embarrassment to the scientific community. It is strange that Darwin devoted so much space to his friends Huxley, Hooker, and Lyell, as well as to less important people, and only a few lines to Wallace. He did note more generally that in later years he had "lost the power of becoming deeply attached to anyone" because of his illness, which prevented him from talking for more than an hour with anyone except his wife and children.[22] However, his substantial correspondence with Wallace belies this statement. The truth is that there was nothing more to say on the matter of Wallace: the record spoke for itself. Wallace puzzled Darwin from the moment of their first correspondence to the end of his life. And any hint of disparagement would have seemed like bad sportsmanship, raising the odious specter of the priority issue.

Wallace did not record or preserve his first impressions either, but probably for different reasons. To Wallace, Darwin's scientific judgment was inviolable, even if he later disagreed with some of Darwin's conclusions. He never wavered on the issue of priority, always minimizing his central role in the discovery of the origin of species. He never questioned Darwin's intellectual honesty. A description of his initial encounter with Darwin would have given undue emphasis to an event that he had elected to pass over. He was also too polite to cast an unfavorable light on the disputes that later arose between the two men, and his affection for Darwin, despite these disputes, seems genuine.

Darwin always warmly received Wallace at Down House. When he visited London, he would call on Wallace or lunch with him at the house of his older brother, Erasmus Darwin. But these visits were infrequent. The relationship between the two men developed principally through correspondence. Their letters are like love letters, especially in the 1860s, when each man held the other in high regard. Darwin often addressed his letters to "my dear Wallace," approaching his co-theorist with great delicacy. He worried about any misunderstanding or disagreement. During this pivotal period in the dissemination and acceptance of the theory of natural selection and common descent, Darwin and Wallace were the proud parents of a prodigy that they jointly nurtured in a hostile environment, striving to ensure its survival and maturation into a healthy and productive adulthood. But once the theory's success seemed certain, the natural divisions between the two men began to emerge. Class, education, and life experiences put a wedge between them. All they had in common was a passion for natural history. In a few years, it would become apparent that their harmonious coexistence had been an illusion, an accident of history. Gradually their fragile relationship, built on a mutual misunderstanding of each other's objectives, revealed its cracks and fell apart.

After his visit to Down House, Wallace left for Devonshire to continue his recuperation. "I . . . hope to lay in a stock of health to enable me to stick to work at my collections during the winter," he wrote to Darwin. "I begin to find that large collections involve a heavy amount of manual labour which is not very agreeable."[23] In Devonshire he spent a week with his old friend Sir James Brooke, the White Rajah of Sarawak, who now lived in comfortable and impecunious isolation on a small estate, Burrator, at the foot of

Dartmoor, amid the wild scenery that he enjoyed, and ruled his fledgling nation from afar. Together with Brooke's former private secretary, Spenser St. John, they strolled about the district or took a drive with Brooke's longtime confidante, Lady (later Baroness) Angela Burdett-Coutts, England's wealthiest woman, who supported and effectively dominated the ailing Brooke. Shortly after this visit, Wallace was invited by Alfred Newton to attend the meeting of the British Association for the Advancement of Science, which was held at Cambridge that year. Newton was an early convert to the new theory of evolution, having been impressed by Wallace's paper "On the Tendency of Varieties to Depart Indefinitely from the Original Type" and Darwin's abbreviated notes on natural selection, both published in the August 1858 issue of the *Journal of the Proceedings of the Linnean Society of London.* (Six years younger than Wallace, Newton had studied ornithology in Lapland, Iceland, the West Indies, and North America before settling at Cambridge, where in 1866 he would become Cambridge University's first professor of zoology and comparative anatomy.) The visit to Cambridge was Wallace's first, and he enjoyed the company of new scientific friends, mostly ornithologists. A recrudescence of "the rather painful dispute" between Richard Owen and Thomas Huxley, once mentor and student and now archenemies, on the differences between the brains of humans and apes briefly disturbed the air of collegiality.[24] But the meeting and its controversies apparently energized Wallace, who delved into his zoological work with renewed vigor.

In June 1863, he read one of his most important papers before the Royal Geographical Society. Entitled "On the Physical Geography of the Malay Archipelago," it secured for him a niche as the premier biogeographer of his generation. Wallace elevated biogeography from a mere descriptive field to the level of a true science—much as he and Darwin had done for biology and Sir Charles Lyell for geology—by applying the scientific method to the geographic distribution of animals. After examining its natural history, he deduced the past geologic history of the Malay Archipelago.

For several years, Wallace had been pondering the problems that the archipelago's geography and zoology posed—ever since the startling realization in 1856 that he had unexpectedly entered a new faunal world when he crossed from Bali to Lombok. Now he would receive accolades for a discovery that he could claim as his and his alone. The short, deep strait between the two islands acted as a faunal divide, partitioning Asia from Australia. As he performed the tedious task of cataloguing his specimens in his London apartment, he noted the relative ratios of "Asian" and "Australian" species among his collections from each island. In Timor, for example, he identified thirty-six species identical to those of Java and eleven that were closely allied, while thirteen species were identical to those of Australia and thirty-five were closely allied, indicating that species derived from Australia had inhabited Timor longer than species derived from Java. He was among the first biologists to apply such a statistical method to analyze the geographic distribution of animal species, though Alexander von Humboldt had invented a similar methodology for analyzing plant distribution fifty years earlier.

Although Wallace mentioned nothing about evolutionary theory to avoid offending Sir Roderick Murchison, the president of the Royal Geographical Society, who opposed both the theory and Lyell's uniformitarian doctrine, he prepared the minds of his audience for the inevitable conclusion: that evolution by natural selection and common descent, in conjunction with dynamic geologic forces, best explained the plant and animal distribution in the Malay Archipelago.[25] Several years later, Murchison publicly commented on Wallace's analysis following the publication of *The Malay Archipelago*:

> The result of Mr. Wallace's researches which chiefly interests us geographers is the establishment of a natural division between the eastern and western portions of the archipelago; a sketch of which, with the principal facts and reasonings leading to it, was given by him in a remarkable paper read before us soon after his return. So well has he elaborated his leading generalisation, and so thoroughly has he made it his own, that already writers are beginning to term the dividing channel between the two halves of the archipelago, "Wallace's line."[26]

As the evolutionary biologist Ernst Mayr has pointed out, further research in the Malay Archipelago did not uphold Wallace's original conclusions about the placement of his line of demarcation. Despite his extensive collections, Wallace significantly underestimated the true number of species present on each island. This lack of information, Mayr said, caused Wallace to single out what he considered to be typical representatives, so-called indicator species of an island's fauna. In reality, as Mayr demonstrated, the boundaries are more blurred than Wallace imagined, with a broad transition zone separating a continental flora and fauna from an island flora and fauna.[27] Wallace also did not envision masses of land shifting, as is now conceived in plate tectonic theory. Like virtually all naturalists and geologists of the era, he believed in the general permanence of oceans and continents, with connections or separations between islands or continents occurring through subsidence or elevation. Looking backward in time, the geography of the world had changed little to him except for the appearance and disappearance of land bridges or barriers, which made possible the migration and isolation of animals. But what is important today is Wallace's scientific methodology and statistical approach, to which all future biogeographers have been indebted. In honor of his seminal work, biogeographers christened the transition zone between the Asian and Australian faunas "Wallacea," an area bounded on the west by Lombok, on the south by the Lesser Sundas, on the east by the Moluccas, and on the north by Celebes and the Philippines.

In the three-year period between 1862 and early 1865, Wallace wrote twenty-eight papers—several on technical aspects of bird and insect taxonomy, the others on the geographic distribution of animals or natural selection—and also contributed observations on anthropological and ethnological subjects, all to Darwin's immense delight. "In this way the time passed rapidly," Wallace later wrote, "and I became so interested in my various occupations, and saw so many opportunities for useful and instructive papers on various groups of my birds and insects, that I came to the conclusion to devote myself for some years to this work, and to put off the writing of a book on my travels till I could embody in it all the more generally interesting results from the detailed study of certain portions of my collections."[28]

One paper that particularly delighted Darwin was Wallace's attack on the Reverend Samuel Haughton, a professor of geology at the University of Dublin. In May 1863, Haughton published in the *Annals and Magazine of Natural History* his paper "On the Form of the Cells Made by Various Wasps and by the Honey Bee; with an Appendix on the 'Origin of Species.'" Haughton ridiculed Darwin, using the construction of a beehive as a springboard for expressing his intense dislike of the theory of evolution by natural selection. In the *Origin*, Darwin had pointed out that the hive-making bees had solved a problem that awed mathematicians: how to make their cells of the proper shape to hold the greatest possible amount of honey with the least possible consumption of precious wax. The most skilled workman with the best tools would find it difficult to make the perfect cells of a beehive, yet bees accomplish this feat while crowded together in a dark hive. But the complex task of building a hive, Darwin said, could be shown to be due to just a few simple instincts. To support this idea, he looked for clues in existing families of bees. At one end of the spectrum are bumblebees, which use their old cocoons to hold honey and make separate and irregular rounded cells of wax. At the other end are the hive bees, which construct a double layer of six-sided cells—each one a hexagonal prism—the basal edges of which are beveled and joined to other cells to form an inverted, pyramidal comb. In between is the Mexican species *Melipona domestica*, which builds a regular wax comb of cylindrical cells for hatching the young and some large spherical cells for holding honey. The spherical cells of the *Melipona* are always made close together. When one cell rests on three other cells, there are three flat surfaces that unite to form a pyramid, a faint echo of the more elaborate pyramidal structure of the hive-making bees. In so constructing its cells, Darwin said, the *Melipona* saves both wax and labor. If one imagined that the *Melipona* made its spherical cells farther apart, equal in size, and arranged in a symmetrical double layer, the resulting structure would resemble the comb of the hive bee. Darwin concluded that if the instincts of the *Melipona* could be modified slightly, "this bee would make a structure as wonderfully perfect as that of the hive-bee. . . . By such modifications of instincts which in themselves are not very wonderful—hardly more wonderful than those which guide a bird to make its nest—I believe that the hive-bee has acquired, through natural selection, her inimitable architectural powers."[29]

Darwin then tested his theory with a series of elegant experiments on hive-making bees, showing that the construction of a hive was a "sort of balance" struck among many bees that all instinctively stood at the same relative distance from one another, working independently on their individual cell, building up or repairing the planes of intersections between these cells. He therefore concluded that the progenitor of the hive bee must have been an extinct species resembling the existing bumblebee. But this bee probably lived in large numbers that had to have a good supply of honey to survive as a community during the winter. Since honey is difficult to collect, those bees that economized by making their cells closer together and more geometric in shape would have had a greater chance of surviving:

> Thus . . . the most wonderful of all known instincts, that of the hive-bee, can be explained by natural selection having taken advantage of numerous, successive, slight modifications of simpler instincts . . . by slow degrees. . . . [T]hat individual swarm which . . . made the best cells with the least labour, and least waste of honey in the secretion of wax, having succeeded best, and having transmitted their newly-acquired

economical instincts to new swarms, which in their turn will have had the best chance of succeeding in the struggle for existence.[30]

Haughton found Darwin's explanation of the shape of the hive ludicrous. He believed that the properties of the cells were only a necessary consequence of their geometric form, which was the necessary consequence of mechanical conditions totally unrelated to the origin of the insects that made those cells. The shape of the beehive, he said, was caused by the "jostling and elbowing" of the bees in their crowded quarters. Because of such jostling and elbowing, they "cannot help making cells with the dihedral angles of 120 degrees of the rhombic dodecahedron." Economy of wax had nothing to do with the making of the cells. Other properties, such as maximum resistance to fluid pressure, "necessarily reside in the bee's cell because they are the inherent properties of the rhombic dodecahedron, which is the form affected by that cell." Adopting an ungentlemanly tone, he concluded, "How does it happen that a theory of the origin of species . . . is accepted by the multitudes of naturalists as if it were gospel? I believe it is because our naturalists, as a class, are untrained in the use of the logical faculties which they may be charitably supposed to possess in common with other men. No progress in natural science is possible as long as men will take their rude guesses at truth for facts, and substitute the fancies of their imagination for the sober rules of reasoning."[31]

Wallace was the first to come to Darwin's defense. He wrote that, having considered the problem of hive construction in the past, and having observed various species of bees and wasps during his extensive travels, he could easily dispense with Haughton's errors in the laws of physics and his comprehension of insect behavior. After doing so, he then accused Haughton of misreading *The Origin of Species* and reiterated the cardinal principles of natural selection and the ways that theory explained the behavior of hive-making insects. To Haughton's remarks about the "class" of naturalists, he replied that "this is the judgment of the Rev. S. Haughton on such men as Lyell, Hooker, John Lubbock, Huxley, and Asa Gray." A rereading of Haughton's paper would "enable one to judge how far Mr. Haughton himself possesses those 'logical faculties' which he is half inclined to deny to the mass of British naturalists."[32]

Darwin acknowledged Wallace's masterful coup de grâce. "In a letter received two or three weeks ago from Asa Gray," Darwin told Wallace, "[Gray] writes: 'I read lately with gusto Wallace's exposé of the Dublin man on Bee cells, etc.' . . . Further on Asa Gray seems to think much of your powers of reviewing, and I mention this as it assuredly is *laudatori a laudato*."[33]

But Wallace did not spend all his free time in defense of his and Darwin's theory. As in their youthful pre-Amazon days, Bates and Wallace set off on a few local expeditions in search of answers to difficult philosophical questions. They had both become preoccupied with the problem of the origin of life itself, a problem that the theory of natural selection did not seem to answer or even address. For guidance, they arranged to meet with the philosopher Herbert Spencer sometime in 1862 or 1863, after reading his *First Principles*, a book that Wallace felt went to the "root of everything."[34]

Their appointment was an audience, not a meeting. Bates and Wallace approached Spencer as a sage, and Spencer, for his part, thanked them for their practical exposition of evolutionary theory and expressed his hope that they would continue their fine work. A man inflated with self-importance, Spencer, at thirty-eight, was two years younger than Wallace. Another self-made intellectual, he had captivated many of the leading scientists of his generation with his voluminous writings on social and scientific subjects. Nearly a decade before Wallace and Darwin published their joint theory, he was an evolutionist. It was Spencer who coined the term "survival of the fittest" and who later applied evolutionary theory to nearly every sphere of social thought—but Lamarck, not Darwin or Wallace, inspired him. Spencer never fully accepted the Darwin–Wallace conception of evolution, upholding throughout his life his faith in the inheritance of acquired characteristics. *First Principles* was his attempt to reconcile science and religion, the great issue that divided the English scientific community in the latter half of the nineteenth century. "It is an incredible hypothesis that there are two orders of truth, in absolute and everlasting opposition," Spencer wrote. "It behooves each party to strive to understand the other, with the conviction that the other has something worthy to be understood; and with the conviction that when mutually recognised this something will be the basis of a complete reconciliation." Spencer touched on issues that Wallace himself had begun to grapple with, such as the Knowable versus the Unknowable. However one conceived of the origin of the universe, Spencer said, whether by some materialist principle or "divine" force, the idea of a First Cause was inescapable. Scientific reductionism and religious teleology inevitably converged, one ending where the other started, at a First Cause, which was "utterly unscrutable [*sic*]."[35]

Space, time, force, matter, and their interactions and origins engaged scientists, clergymen, and philosophers alike in the 1860s. Darwin, sensing a trap, steered clear of first causes. *The Origin of Species* focused on just that, and although the book was speculative, it speculated on facts. Only the knowable fell within the province of his science. As for the existence of an intelligent God or First Cause, "I cannot pretend to throw the least light on such abstruse problems," he wrote. "The mystery of the beginning of all things is insoluble by us." Once in the domain of the unknowable, the scientist entered the realm of subjective experience and no doubt risked his reputation and credibility. "I cannot see that such inward convictions and feelings are of any weight as evidence of what really exists," he concluded.[36] Wallace had no such fears. Having solved the riddle of the origin of species, he was not daunted by the double enigmas of the origin of life and the origin of the universe. Like a character in a Greek tragedy, he was touched with hubris—impelled to search for clues to the unanswerable. When Spencer pronounced the problem of the origin of the universe too fundamental for solution at that time, Wallace's hopes were temporarily dashed. Although the origin of life was probably knowable, Spencer said, man's ignorance of the essential constitution of matter and the various forces of nature left the problem of the origin of the universe unknowable. Scientists had to wait and work "contentedly" at minor problems. This was an answer that Wallace could not accept.[37] The solution to the origins of life and the universe became another obsession, the pursuit of which was to lead him on a lifelong journey through a metaphorical wilderness of headhunters and pirates intent on stripping the unwary explorer of all his earthly belongings. It would be a journey far more perilous than his earlier expeditions to the western and eastern tropics.

Of all issues in the new science of evolutionary biology, the application of the Darwin–Wallace hypothesis to the origin of human beings inflamed the most passions. Although Darwin had very little to say about human evolution in the *Origin*, few missed the implications of his doctrine. The war between Huxley and Owen was merely the most visible example of the polarization of the scientific community on this subject; throughout the 1860s, intermittent hostilities erupted everywhere between the opposing sides. The meetings of London's scientific societies became battlegrounds, where eloquent speakers challenged and vanquished their opponents with the masterful strokes of language, a weapon that mid-Victorian Englishmen wielded with great skill. Each side had its propagandists in the press, which significantly influenced the public mind on the issue of evolution. And Wallace made certain that he fought in the front ranks.

The question of human origin had long been the subject of speculation and debate among naturalists, philosophers, and theologians, but by the mid-1860s the biblical version could no longer be justified. The alternative explanation, however—humanity's kinship with "lower" animals, especially the great apes—was too repugnant even for most educated people to bear. Humans' special status, so clear and undisputed a decade earlier, had come under assault by a formidable foe bent on toppling the temples of the older order. But it was not just the theory of natural selection and its unpleasant inferences that forced the old order to look at the world in a different way. Another discovery shocked mid-Victorian sensibilities. In 1858 human artifacts, mainly flints and flakes, were found *in situ* with the remains of extinct animal species, like the mammoth and the "tichorine" rhinoceros.[38] Although prehistoric remains had been unearthed in Switzerland and Prussia earlier in the century, scientific men denied the implications of the evidence. However, the findings in Brixham Cave, four miles south of Torquay, England, were irrefutable: the flints and bones lay in stratified deposits of an age far greater than six thousand years. The discovery pushed human existence back long before the biblical account. Thus, beginning in the early 1860s, there was "a revolution of opinion" (in Lyell's words) about human history, both biological and historical, and its effects were profound.[39]

Among the major early innovators in the theory of human evolution were Thomas Huxley, Sir Charles Lyell, and Alfred Wallace. In his booklet of essays *Evidences as to Man's Place in Nature*, published in 1863, Huxley demonstrated on anatomic and behavioral grounds that human beings and the great apes had a common ancestry. In the same year, Lyell's *Geological Evidences of the Antiquity of Man* certified that human beings predated the biblical six thousand years, without advocating a belief in their evolution from other life-forms—a radical revision of his previous (only eight years earlier) belief in the human race's modernity. Wallace, building on the arguments of his two colleagues, applied the theory of natural selection to human origin and subsequent development in an anthropological essay. Together the three men paved the way for other investigators and, ultimately, Darwin's *Descent of Man, and Selection in Relation to Sex*, which was published in 1871.

Wallace's long-standing interest in comparative ethnology, as the study of humans was once called, can be traced to an essay that he wrote in the early 1840s, when he was still an apprentice land surveyor living and working in the Welsh town of Neath. Fascinated by a people that rejected the autocratic teachings of the established church in favor of the fire-and-brimstone oratory of the more democratic, dissenting religious sects, he approached the Welsh as an anthropologist, observing and analyzing the habits and characteristics of the poor farmers whose lands he surveyed. His essay, reproduced in its entirety in his autobiography, serves as a landmark in his intellectual development, a foreshadowing of his future scientific work.[40]

Even more than his experiences in Wales, phrenology inspired Wallace's views on human psychology. Long after its dismissal as a pseudoscience, and until the Freudian revolution, he remained a passionate defender of phrenology as "the only sound method of ascertaining the relations that exist between the development of the brain and mental faculties and powers."[41] In the 1830s and 1840s, phrenology, the science of "reading" the bumps on a person's head, excited many radical intellectuals, including Wallace. It was one of the more popular classes at the Mechanics' Institutes, and the young Wallace became an ardent student. Through the science of phrenology, its advocates predicted, true social reform and material improvement was possible; in the course of a generation, phrenology became a forerunner of modern sociology.

The Viennese physician Joseph Gall, who pioneered phrenology in the late eighteenth century, believed that he could describe a person's mental characteristics by means of an examination of the contours of the skull.[42] Gall made genuine breakthroughs, even if later research failed to corroborate most of his and his disciples' other scientific claims. After a long study of patients in an insane asylum and a painstaking comparison of the skulls (and casts of skulls) of hundreds of people whose mental characteristics were known, he postulated that the brain was the seat of the mind. From this deduction, he formulated two important concepts: first, that the brain was an aggregate of mental organs from which all faculties, tendencies, and feelings emanated; and, second, that these mental organs were topographically localizable. Both were radical ideas in the late eighteenth century, when anatomists shrugged off the brain as nothing more than haphazard coils of uncertain function and theologians condemned as heresy any suggestion that the body and mind were inseparable.

In the 1820s, George Combe, a Scottish barrister who became a phrenologist after observing the masterful dissections of the human brain by Gall's most prominent student, Johann Gaspar Spurzheim, felt that the principles of phrenology could be applied to social institutions. His book *The Constitution of Man Considered in Relation to External Objects*, published in 1828, was a best-seller among the literate members of the working classes. Combe asserted that sentiments and propensities, the two categories into which Spurzheim classified the higher and lower human mental traits, respectively, were out of balance in modern man. Imperial Britain marched in the vanguard of those who took the lower "propensities," not the higher "sentiments," as their guides. Only through a nonviolent revolution in the structure of England's social and political institutions, he believed, could a proper balance between the two opposing psychic forces be achieved—the attainment of harmony between human nature and the social, political, and physical environment. Thus phrenology and utilitarianism—the concept of the greatest good for the greatest number—were united. Combe's conclusion formed the centerpiece of Wallace's later writings about the flaws and failures of European civilization, a subject he would explore in great detail in the final decades of his life.

What had convinced Wallace of the truth of phrenology was his own phrenological analysis in 1846 or 1847, when he consulted two itinerant phrenologists who had come to Neath to lecture at the local Mechanics' Institute. On both occasions, he could afford only a brief examination, which entitled him to a partial sketch of his character summarized on a large preprinted sheet of paper containing the thirty-odd phrenological "organs" with their functions clearly stated. Opposite each organ, a number was placed to indicate its comparative size. The first examiner recorded three important characteristics:

> 1. "You will pay great attention to facts, but so soon as facts are presented you will begin to reason and theorize upon them. You will be constantly searching for causes."
>
> 2. "You will be a good calculator, will excel in mathematics, and will be very systematic in your arrangements."
>
> 3. "You possess a good deal of firmness in what you conceive to be right, but you want self-confidence."

The second examiner made five other observations:

> 4. "This gentleman should learn easily and remember well."
>
> 5. "He has some vanity but more ambition. He may occasionally ex-

hibit a want of self-confidence; but general opinion ascribes to him too much. In this, opinion is wrong. *He* knows that he has not enough."

6. "If *Wit* were larger he would be a good mathematician, but, without it, I do not put his mathematical abilities as first-rate."

7. "He has some love for music from his *Ideality*, but I do not find a good ear or sufficient *Time*."

8. "He is fond of argument and not easily convinced."

Wallace was sufficiently impressed by the accuracy and general agreement of the diagnoses, determined by two complete strangers on separate occasions, to save a written account of his examinations, which he reproduced in two of his books decades later for those who doubted phrenology's merits. Numbers 1 and 8, combined with the large organs of *Ideality* and *Wonder*, reflected his strong love of the beauties and mysteries of nature and furnished "the explanation of my whole scientific work and writings." Numbers 3 and 5 identified his lack of self-confidence. They explained his excessive shyness, a characteristic that caused him much suffering throughout his life and that was responsible for his making fewer intimate friends "than most men." Other estimates of his "innermost nature" he kept private but confirmed for him the value of phrenology.[43]

During his years in the Amazon, Wallace had examined the Brazilian peoples, assessing their mental powers and faculties with a phrenological eye. Here he was first exposed to human beings in their primal state, one "uncontaminated" by civilization. From his reading of Darwin's *Voyage of the Beagle*, he had expected to be as shocked and appalled as Darwin had been at the "wide difference" between "savage" and civilized man. Darwin believed that the people of Tierra del Fuego preferred uncouth savagery to the refinements of European culture. It was a conclusion he reached after a Fuegian who had been reared in London and who traveled with him on the *Beagle* unexpectedly shed his English accoutrements and manners and reverted to the barbarous customs of his naked and filthy kinsmen. "The difference between savage and civilized man is greater than between a wild and domesticated animal," Darwin wrote, "in as much as in man there is a greater power of improvement. . . . Viewing such men, one could hardly make oneself believe they are fellow-creatures, and inhabitants of the same world."[44]

By contrast, the aboriginal tribes of the upper Uaupés River, which snaked through regions where no white man before him had ventured, delighted Wallace. "I do not remember a single circumstance in my travels so striking and new," he recollected, "or that so well fulfilled all previous expectations, as my first view of the real inhabitants of the river." The surprise revelation of the experience, he said, was his surprise. He found the Indians as "original and self-sustaining" as the wild animals of the forests, living as freely as they had always lived before Europeans discovered America. It turned his notion of humanity on its head.[45]

In the Malay Archipelago, Wallace made more anthropological observations. In his species notebook, he questioned Owen's theory—presented at the February 9, 1855, meeting of the Royal Institution in London—that the climate produced the varieties of skin color in humans. As an example, Owen pointed out that the Jews of Syria were black, while those in northern European countries had fair hair, a light complexion, and blue eyes. But in Singapore, Wallace saw Armenian Jews who were equally fair, which seemed to contradict Owen's theory.[46] When his vessel, having set out from Macassar for Aru in December 1855, sailed into the harbor of the Ke Islands, he knew at once that Owen was wrong. "Between the Malayan tribes among whom I had for some years been living, and the Papuan races, whose country I had now entered," Wallace wrote in his journal, "we may fairly say that there is as much difference, both moral and physical, as between the red Indians of South America and the negroes of Guinea on the opposite side of the Atlantic."[47] Here was another classic Wallace's Line, where two long-established indigenous populations lived in close proximity to each other but differed not only in skin color, facial structure, and body type but in every manner and mode of expression. Climate was apparently irrelevant to the classification of the varieties of humankind.

But this was not all. In Borneo, Wallace had lived for a month among the Dyaks, famed as headhunters but with a moral sense more refined than that of the "higher" civilizations. Except for ritual head-hunting, crimes of violence were virtually unknown. The Dyaks also exercised temperance in food, drink, and sex. In his article "The Dyaks—By a Personal Acquaintance of Theirs," which appeared anonymously in the September 26, 1857, issue of *Chambers's Journal*, he described the ritual preparation of the hunted head. It was "the first accurate and objective account of this extraordinary ceremony," he believed, as well as of the grand celebration that followed, when friends and relatives of members of the hunting parties poured in from the surrounding areas to honor the head's capture. But the article, gleefully gruesome in its detail, aimed at something subtler. If one overlooked the macabre inspiration for the celebration, one could see that the demeanor of the visitors, their elaborate attire, endless chatter, good cheer, and tradition of providing copious quantities of food did not differ fundamentally from the behavior and customs of white Europeans on similarly momentous occasions, like weddings or funerals. The similarities, not the differences, provided the story's counterintuitive moral and hinted at the author's belief in a single human species.

After returning to London, Wallace continued his explorations of human nature. In late 1863 he began to attend the meetings of the Anthropological Society, founded the previous January and presided over by James Hunt, a thirty-year-old nonpracticing physician who hoped to elevate the study of man to a true science.[48] The rival Ethnological Society, bogged down (in Hunt's opinion) by biblical dogma, limited its scope to the history and science of nations and races. Hunt envisioned a discipline that united geology, anatomy, biology, physiology, and ethnology for the purpose of elucidating human origins, human development, and the investigation of the laws regulating the distribution of human populations. Above all, he wanted facts, not "speculation," as he liked to call theorizing. "We should endeavour to be careful not to fancy we aid the cause of science when we absurdly give our support to theories that no longer can be reconciled with established facts," he remarked in his opening address to the society in February 1863.[49]

Race and race relations provoked heated discussions at the first few meetings of the new society. It was the time of the American Civil War, and the British public had split over the emancipation issue. The prime minister, Lord Palmerston, and some of those in his cabinet, like William Gladstone, supported Jefferson Davis, the Confederacy, and the cause of white supremacy.[50] Hunt chaired presentations on such topics as "The Vitality of the Black Race," "Civilisation of the Negro," and "The Neanderthal Skull." The idea of classifying a European Caucasian, an African black, and an Australian Aborigine in the same species offended many in English society as much as the notion of human kinship with the great apes.

Reflecting prevailing social attitudes, scientists were divided into two groups on the race question: the "monogenists," who believed in a single origin of the human species, and the "polygenists," who believed in multiple origins. Traditional monogenists accepted the biblical account of origin from Adam and Eve, from whom all the peoples of the earth were ultimately descended. Within the two groups were subgroups: one could be a monogenist but not an evolutionist, a polygenist and an evolutionist, or some other permutation. Before Darwin, Huxley, and Wallace, the most prominent exponents of monogenism in England were two physicians: William Lawrence and James Cowles Prichard. As early as 1817, Lawrence had examined the mental, physical, and moral characteristics of the different varieties of the human species and had concluded that since the similarities outnumbered the differences, humans were a single species, not unlike the feral and domesticated varieties of cattle, sheep, horses, or dogs. But it was Prichard who influenced English thinkers more. Whereas Lawrence framed human characteristics naturalistically, which linked him at the time with postrevolutionary French materialist thought and caused him to be reviled as an atheist (thus nearly ruining his career as a physiologist), Prichard openly embraced revelation. For him, the biblical account was the starting point; he did not question the biblical timescale of human existence. Racial diversity had a historical root: the migration of peoples and diffusion of cultures linked separate nations to a single source. Nevertheless, Prichard discussed the physical differences among human groups within a scientific framework. Anatomy, physiology, physical geography, history, archaeology, and comparative philology—of which he made an impressive compendium in his magnum opus, *Researches into the Physical History of Mankind*—all pointed to a unity of humankind. And although he had difficulty reconciling his theory—subsumed under the appellation "ethnology"—with the increasing amount of data pouring in from all over the world (which to some suggested a polygenetic view), he held his monogenetic ground.[51] Before setting off for Brazil, Wallace had read Lawrence's lectures and Prichard's multivolume treatise on humans and eventually concurred with the single-origin theory, gathering supportive evidence in South America and the Malay Archipelago (though he, of course, rejected Prichard's Christian orthodoxy).[52]

Although his vision was lofty, and despite his profession of objectivity, Hunt often failed to suspend judgment. He dismissed the theory of natural selection as fantasy and believed that the theory of the unity of humankind was as much of an unproven hypothesis as the theory of the plurality of the human species (though he soon revealed himself to be an unrepentant polygenist of the worst type). His pompous, overbearing manner, with its class-conscious overtones, made him a number of enemies. Huxley detested him and pronounced him unfit for the presidency of the Anthropological Society. After a few meetings, he stopped attending. Wallace agreed with Huxley's assessment of Hunt's character but thought the society's meetings just as valuable as those of the rival group. "I cannot agree with you that 'there is not the slightest reason for its existence,'" he wrote to Huxley in February 1864. "It seems to me that its establishment is a good protest against the absurdity of making the *Ethnological* a *ladies'* society. Consequently many important and interesting subjects [like phallic worship and male and female circumcision] cannot possibly be discussed there—and as the Geographical is also a *ladies' Society* the *Anthropol.* is the *only place* where they can be discussed."[53] It was therefore before the Anthropological Society that Wallace presented one of his most important and provocative papers a year after Hunt had issued an unintended challenge. "It is a matter of uncertainty whether we shall ever be able to demonstrate by actual facts the *modus operandi* of Man's origin," Hunt had said in his opening address to the society in 1863, "but we may be able to ascertain the laws to which he owes his birth." It was his opinion that until the founding of the Anthropological Society, only "metaphysicians" had attempted to prove the logical necessity of the unity of mankind. "But is the origin of Man to be settled by the metaphysicians?" he asked.[54]

On March 1, 1864, Wallace read his essay "The Origin of Human Races and the Antiquity of Man Deduced from the Theory of 'Natural Selection.'" The subject matter of his essay was undoubtedly calculated to raise the hackles of the more conservative members of the Anthropological Society, especially Hunt.[55]

Wallace began by noting that most anthropologists no longer questioned the great antiquity of humans. It was certain, he said, that humans had inhabited the earth for at least a hundred thousand years; and it was even possible that humans had existed for a million years. Paleontologists had uncovered evidence that humans had coexisted with many now extinct animals, thus pushing the date of our origins farther back in time. But on other issues, anthropologists exhibited too much "dogmatism," especially on the question of the classification of humankind. Anthropologists had split into two groups: those who maintained that humans belonged to a single species, the differences between races being superficial and due solely to local and temporary variations; and those who believed that the genus *Homo* could be subdivided into multiple species, all created distinct and therefore incapable of changing. The latter group, the polygenists, frequently cited the historical record, which showed, for example, that blacks as depicted in the pictographs of ancient Egypt four thousand years ago were identical to blacks today. As a result, some subdivided the genus into 3, 5, 50, or 150 species. Wallace continued:

> This difference of opinion is somewhat remarkable when we consider that both parties are well acquainted with the subject; both use the same vast accumulation of facts; both reject those early traditions of mankind which profess to give an account of his origin; and both declare that they are seeking fearlessly after truth alone. I believe, however, it will be found to be the old story over again of the shield—about which the knights disputed; each party will persist in looking only at the portion of truth on his own side of the question, and at the error which is mingled with his opponent's doctrine.

He intended to reconcile the two opposing views by means of "Mr. Darwin's celebrated theory of 'Natural Selection,'" which would harmonize the conflicting theories of modern anthropologists.

After a brief recapitulation of the main points of his and Darwin's theory, Wallace stated his thesis: that the action of natural selection had been "checked" in humans. Once *Homo sapiens* had become a "social and sympathetic" creature who protected the weaker members of the tribe, the effects of natural selection shifted from the body to the brain. "Tribes in which such mental and moral qualities were predominant," Wallace concluded, "would therefore have an advantage in the struggle for existence over other tribes in which they were less developed, would live and maintain their numbers, while the others would decrease and finally succumb."

While other animals evolved better physical characteristics for pursuing prey or eluding capture, developing more versatile organs to cope with the vicissitudes of climate, man, "by the mere capacity of clothing himself, and making weapons and tools," took away from nature "that power of changing external form and structure which she exercises over all other animals." From that point on, the human mind was subjected to the same influences from which the human body had escaped. Every slight variation in mental and moral nature that guarded against adversity and enhanced mutual comfort and protection was preserved and therefore accumulated. The better and higher specimens of humanity increased and spread, while the lower and more brutal dwindled and eventually died out. Even the very lowest races of man were elevated far above the "brutes." The culmination of this evolution was the "wonderful intellect of the Germanic races." "And is it not the fact that in all ages, and in every quarter of the globe, the inhabitants of temperate have been superior to those of tropical countries?" Wallace asked. "All the great invasions and displacements of races have been from North to South."

Quoting Darwin, Wallace said that it was the same great law of "preservation of favoured races in the struggle for life" that led to the inevitable extinction of weaker and inferior peoples. Just as the weeds of Europe overran North America and Australia, extinguishing the less adaptable native varieties, so the morally and intellectually superior Europeans had overrun, outbred, and extinguished the less adaptable aboriginal peoples. Foreshadowing social Darwinism, Wallace united European imperialism with the law of natural selection in his essay on the human species.

Wallace imagined an epoch in the distant past when a single, homogeneous race existed, perhaps one without the capacity for speech, in some tropical region, at a time when the human body was still subject to the action of natural selection. Their superior survival skills permitted the members of the race to spread widely. As they were exposed to greater extremes of climate, new enemies, different types of food, and other environmental pressures, useful variations would have been selected and rendered permanent. These selected variations would have been accompanied, on the principle of correlation of growth (that is, the phenomenon of a useless characteristic varying simultaneously with a useful one), by corresponding external physical changes. Black skin, for example, might have been associated with protection from certain diseases. Hence the red, black, yellow, or "blushing white" skin; the straight, curly, or woolly hair; the scanty or abundant beard; the straight or oblique eyes; and the various forms of the skeleton evolved and became distinct characteristics. As humans mastered their environment and freed their bodies from the power of natural selection, Wallace hypothesized, natural selection would have acted on their brains. Speech developed, leading to further advances in mental faculties. The art of making weapons, the division of labor, the restraint of animal impulses, and all those qualities that we identify as human would have come under the sway of natural selection—"[and] we should thus have explained the wonderful persistence of mere physical characteristics, which is the stumbling-block of those who advocate the unity of mankind."

Humans, therefore, had a single origin, but their differentiation into races occurred at a very early period, when their mental faculties were "scarcely raised above the brute." There was no a priori reason, Wallace maintained, that the remains of humans might not be found at times earlier than supposed—even earlier than Lyell had postulated. "The absence of all such remains in the European beds of this age [the Tertiary] has little weight," Wallace suggested, "because as we go further back in time, it is natural to suppose that man's distribution over the surface of the earth was less universal than at present."

Wallace believed that the traces of our earliest progenitors would be found somewhere in the tropical belt, probably in eastern or southern Africa, in land not submerged beneath the ocean since Eocene or Miocene times. "It is there that we may trace back the gradually decreasing brain of former races, till we come to a time when the body also begins materially to differ," Wallace said. "Then we shall have reached the starting point of the human family. Before that period, he had not mind enough to preserve his body from change, and would, therefore, have been subject to the same comparatively rapid modifications of form as the other mammals."

It was this "mind" that gave humans clothing, weapons, the ability to farm, to domesticate animals, to build houses, to make a fire. The link between the human species and primordial organisms was seamless, even though *Homo sapiens* possessed unique qualities that differentiated him from other animals. "On this view of his special attributes," Wallace said, "we may admit that even those who claim for him a position as an order, a class, or a sub-kingdom, by himself have some reason on their side. . . . We can anticipate the time when the earth will produce only cultivated plants and domestic animals; when man's selection shall have supplanted 'natural selection.'"

Wallace concluded his essay in a utopian vein:

> While his external form will probably ever remain unchanged . . . his mental constitution may continue to advance and improve till the world is again inhabited by a single homogeneous race, no individual of which will be inferior to the noblest specimens of existing humanity. Each one will then work out his own happiness in relation to that of his fellows; perfect freedom of action will be maintained, since the well balanced moral faculties will never permit any one to transgress on the equal freedom of others; restrictive laws will not be wanted, for each man will be guided by the best laws; a thorough appreciation of the rights, and a perfect sympathy with the feelings, of all about him; compulsory government will have died away as unnecessary . . . and will be replaced by voluntary associations for all beneficial public purposes; the passions and animal propensities will be restrained within those limits which most conduce to happiness; and mankind will have at length discovered that it was only required of them to develop the

> capacities of their higher nature, in order to convert the earth, which had so long been the theatre of their unbridled passions, and the scene of unimaginable misery, into as bright a paradise as ever haunted the dreams of seer or poet.

Wallace's paper stimulated a lively discussion.[56] One member pronounced Wallace's analysis "a new era in anthropology." Most of the others, however, displayed a complete lack of understanding of the principle of natural selection as applied to both the human and other species. They saw no evidence, for example, that an exterminated race was less fit than the surviving one.

But the most acerbic remarks came from Hunt. Like everyone else, Hunt said, he was "charmed" with the paper. He was so charmed by the "elaborate promises" of natural selection that he was disappointed by the paper's conclusion, which struck him as nothing but speculation. Wallace's penchant for speculation irritated him, as did the entire Darwinian hypothesis, which was based not on facts but on assumptions. If Wallace was attempting to found his new theory about man's origins on known facts, then where were those facts? He strongly suspected that Wallace had no facts to bring forward. In reality, Hunt was too perturbed by certain aspects of Wallace's paper to be persuaded by any argument that Wallace could present in his defense. Hunt interpreted some of Wallace's remarks as a personal attack on his scientific judgment. "[He] brings . . . a charge against anthropologists—that they look to that portion of the truth that is on their side, and insist on looking at the errors on the other side," Hunt said. "I hardly think that such a statement is fair to anthropologists, ethnologists, and ethnographers; on the contrary, I believe there are many anthropologists living who are at least as capable of looking at the whole facts as any disciple of Darwin."

Hunt was also angered by what he perceived to be Wallace's defamation of the English national character. Based on the observations of the Romans, at one point Wallace claimed that the indigenous Britons had been savages when Julius Caesar conquered the British Isles. This was not a fact, Hunt said, but a tradition, based on the "barest" of historical evidence. Furthermore, he did not believe that there was evidence from any part of the world to prove that any "race" different from the present inhabitants had ever dwelled there. "Of course, if you go and take a Neanderthal skull as a type of race, although there is good evidence to believe it simply the skull of an idiot, you beg the whole question," Hunt said, concluding that "if the object of the paper is to assist in founding a science, that does not appear to have been carried out in the eloquent appeal which has been addressed to-night to the imagination."

Hunt demanded that Wallace withdraw his "beautiful dream" of humanity's future, which had nothing to do with his theory of humanity's origins or the theory of natural selection. Such a dream did not belong in a scientific paper, he said, because it could not be proved. "Then, I am sure, we shall all be very much indebted to him for coming before us this evening," he added sarcastically.

Unruffled, Wallace offered a spirited defense of his views, appealing to logic and common sense. To the objection that there was no evidence that when one race exterminated another the winner was somehow superior, he replied that the outcome was "not a question of intellect only, nor of bodily strength only. We cannot tell what causes may produce it. A hundred peculiarities, that we can hardly appreciate, may cause the one race to melt away . . . before the other. But still there is the plain fact that two races came into contact, and that one drives out the other. This is a proof that the one race is better fitted to live upon the world than the other." Contrary to Hunt's assertion, he did not promise to explain everything, Wallace said. He wanted to suggest a theory that would eliminate the difficulty of the "absolute contradiction" between two classes of ethnologists. There were plenty of facts—enough "to fill a volume." He did not have to remind anyone in the room that many of those facts he had collected himself, and that few men could claim more observations of humankind in its various states than he. His paper was necessarily a sketch, with brief allusions to those facts, a starting point for later work. As for the statement that archaeology did not show ancient races to be different from modern ones, Wallace could only throw up his hands in exasperation. "Well, that is a fact I quoted on my side," he exclaimed, "and his quoting it against me only shows that you can twist a fact as you like."

Wallace again clashed with Hunt at the next meeting, on March 15, when Wallace objected to a paper entitled "Notes on the Capabilities of the Negro for Civilisation." Although he agreed that the intellect of the black African was currently inferior to that of the white European, the black African was not the lowest grade of humanity. Blacks were energetic and intelligent, he believed, and with encouragement from a superior civilization they would rise higher than had ever been seen in the past. It was unfair to compare a group of people only recently freed from slavery with the Hindu or Chinese people, who belonged to the oldest civilizations on earth. Blacks were not lazier than anyone else—all people were lazy and required prodding to work. The European had never seen the black man under favorable circumstances, as he had in the Papuan regions of the Malay Archipelago, where he lived and worked with people who he believed were closely related to the tribal peoples of Africa. Hunt expressed outrage at Wallace's suggestion that with the help of Europeans blacks might rise to their level. There was an unbridgeable world of difference between the two races, Hunt said, and "it is the duty of anthropologists to oppose the opinion attempted to be established of the equality of the Negro and White Man."[57]

Although Wallace's paper was not well received at the Anthropological Society, others who read it were impressed. To Darwin, Hooker wrote, "I have just received Wallace's Anthropological paper and read ⅓ and am amazed at its excellence. It seems to me a very great move in advance and I am anxious to know what you think of it. It never struck me to account for the fixity of man as Wallace has done, and apparently with good reason. I am struck too with his negation of all credit or share in the Natural Selection Theory—which makes me think him a very high-minded man. I am burning to know your opinion of the paper."[58] Wallace's paper fascinated Darwin, who made numerous check marks in the margins as he followed the logic Wallace used to deduce his hypothesis. Although he was not sure if he agreed completely with Wallace's views, he had no doubt about its "remarkable genius."[59] "The latter part of the paper I can designate only as grand and most eloquently done," he wrote to Wallace.[60]

Darwin may not have doubted Wallace's originality and ingenious arguments, but he was developing doubts about the power of natural selection to explain every phenomenon in the biological world. He believed, for example, that he had accumulated enough facts to support his theory of sexual selection. It, not natural selection, explained the presence of those physical and behavioral attributes that seem to have nothing to do with the survival of the species in the struggle for existence but that have evolved instead from the competition for mates among members of the same species. Sexual selection was "the most powerful means" of evolving the various races of man, he told Wallace. He could show that the different races have a widely divergent standard of beauty. The most powerful of our ancestors, he said, would have had the pick of the most beautiful women—skin color having been one of their considerations—and these men would have left the most descendants. Darwin did agree with Wallace, however, that complexion might also correlate with "constitution." He had persuaded the director-general of the Medical Department of the British army to send a survey to the surgeons of all regiments in tropical countries to ascertain this point. His theory was that a "dusky" individual might escape "miasma" (malaria or smallpox).[61]

Wallace thanked Darwin for overestimating his "desultory efforts." "My great fault is haste," he admitted. "An idea strikes me, I think over it for a few days, and then write away with such illustrations as occur to me while going on. I therefore look at the subject almost solely from one point of view." But he disagreed with Darwin—and ultimately would never agree—on the relative role of sexual selection in evolution. Sexual selection would have had uncertain results in determining skin color, he said. In the most primitive tribes, polygamy was rare, and wives were "more or less a matter of purchase." Moreover, there was little difference in social condition in these tribes, and he had observed that no healthy and "undeformed" man, regardless of status, remained without wife and children. He doubted the truth of the "often-repeated assertion" that members of the English aristocracy were more beautiful than the middle classes. "Mere physical beauty—that is, a healthy and regular development of the body and features approaching to the *mean* or *type* of European man—I believe is quite as frequent in one class of society as the other."[62]

In the same letter, Darwin had also questioned Wallace's classification of humans. Wallace apologized for his lack of clarity. He had adopted Huxley's view, he said, which placed humans in one family, the Hominidae. If his theory—that while animals were undergoing modification of all of their body parts to a generic or even family degree of difference, humans were changing in the brain only—was correct, then humans as a species could be as old as other mammalian families. The origin of the family of man, he suggested, could date back to a period when the orders of other animals were originating.[63]

Lyell read Wallace's paper and complimented him on his achievement. However, he disagreed with his hypothesis that human beings may have existed as long ago as the Miocene. No human remains had been found in Miocene strata. Lyell was willing to admit only what geologic evidence had thus far shown: that humans coexisted with extinct mammals in the Pleistocene epoch, the epoch coinciding with the Stone Age (the "Recent era," as he called it). "If we reflect on the long series of events of the pleistocene and recent periods . . . it will be remarked that the time assigned to the first appearance of Man . . . is extremely modern in relation to the age of the existing fauna and flora, or even to the time when most of the living species of animals and plants attained their actual geographic distribution," Lyell had written in *Geological Evidences.*[64] Wallace's suggestion that humans may have originated in the Miocene, Lyell believed, was due to "a want of appreciation of the immensity of time at our disposal."[65]

Wallace replied to Lyell two days later:

> With regard to the probable antiquity of man, I will say a few words. First, you will see, I argue for the *possibility* rather than for the *necessity* of man having existed in Miocene times, and I still maintain this possibility, and even probability. The question of time cannot be judged of positively, but only comparatively. We cannot say a priori that ten millions or a thousand millions of years would be required for any given modification in man. We must judge only by analogy, and by a comparison with the rate of change of other highly organised animals.

Several extant genera of animals were known to have existed in the Miocene epoch, he pointed out, as did anthropoid apes related to the genus *Hylobates* (gibbons). If, as Huxley proposed, humans were classified as a distinct family, and if all agreed that the human family shared an ancestry with the anthropoid apes, then the origin of our family must date back to a period earlier than the Miocene. Second, the most important difference between the families of man and of anthropoid apes was the size of the brain and cranium. "While the Miocene Dryopithecus has been modified into the existing gorilla," Wallace said, "speechless and ape-brained man (but not yet *man*) has been developed into great-brained, speech-forming man." In "The Origin of Human Races," he had tried to show how humans had undergone a "more than generic change" in the size of their brains and crania, while their bodies had hardly changed from the anthropoid type. Since there was as yet no evidence to prove that we had evolved more rapidly than other mammals, he had to conclude that the family Hominidae had roots in a period more ancient than the one Lyell had conceived. "The immensity of *time*, measured in years, does not affect the argument," Wallace concluded.[66]

Lyell had written to Wallace in preparation for his upcoming September address in Bath, as president-elect of the British Association, in which he wished to introduce a paragraph or two about the division of the Malay Archipelago into two regions and the relation of this division to the races of man. Unexpectedly, however, he also raised the priority issue, noting that "the manner in which you have given Darwin the whole credit of the theory of Natural Selection is very handsome." Over the next few years, Wallace and Lyell carried on a lively correspondence on a number of issues—mainly geologic topics—but Lyell, who was twenty-six years older, also became a mentor of sorts for Wallace, almost a surrogate father. Lyell was one of the few men with whom he felt at ease, Wallace later wrote, because of his "refined and gentle manners, quiet humor, extensive knowledge of natural science and great liberality of thought." He was flattered that so eminent a man would "condescend" to discuss subjects on which he considered himself an "amateur."[67]

Lyell's intellectual waffling on the theory of evolution, especially in its application to human origins—which Wallace chose to overlook—disturbed Darwin, who told Hooker that he wished Wallace had written Lyell's chapters on humans in *The Geological Evidences of the Antiquity of Man*.[68] In that book, Lyell addressed the more profound philosophical questions engaging both men of science and men of religion. Like Spencer—but unlike Darwin—he sought a middle ground where the two opposing doctrines could be reconciled. He wondered if a materialist explanation of the origin of species could be combined with a supernatural explanation of the origin of the universe without invalidating either doctrine:

> The whole course of nature may be the material embodiment of a preconcerted arrangement, and if the succession of events be explained by transmutation [that is, the production of new species] the perpetual adaptation of the organic world to new conditions leaves the argument in favour of design, and therefore of a designer, as valid as ever. . . . It may be said that, so far from having a materialist tendency, the supposed introduction into the earth at successive geological periods of life,—sensation,—instinct,—the intelligence of higher mammalia bordering on reason,—and lastly the improvable reason of Man himself, presents us with a picture of the ever-increasing dominion of mind over matter.[69]

The friendship between Wallace and Lyell was to have a significant impact on Wallace's thinking, especially on his later views of humanity—to Darwin's great distress.

Also intrigued by Wallace's paper on the origin of the human races, Spencer invited Wallace to make contributions to a new magazine called the

Reader, which encompassed scientific, social, and political issues.[70] Over the next few months, Wallace expressed his opinions in this publication about human beings and human institutions with greater candor, perhaps at the instigation of Spencer, who encouraged him to publicize views that had been simmering in his mind but that he had not voiced since his return from the Malay Archipelago.

At the 1864 meeting of the British Association, Wallace made a mark with the presentation of a controversial paper, "On the Progress of Civilisation in Northern Celebes," in which he gave a glowing description of the regional capital, Minihasa, a town at the tip of the northernmost arm of the scorpion-shaped island of Celebes, where he had spent several weeks in 1859.[71] From lives as "complete savages" in bark clothing who practiced the "degrading" worship of demons and "garnished" their huts with trophy heads, the people had been transformed into the most industrious, peaceable, and civilized in the whole archipelago and now wore European-style clothing, worshiped the Christian God, and sat on chairs and at tables while they dined on a wholesome cuisine served on European china and eaten with European cutlery. All children went to school to learn Malay, which had replaced the babel of native dialects. Wallace had not forgotten the teachings of Robert Owen, the radical socialist thinker who had shaped his early character "more than I knew," or New Lanarck, Owen's idealized, experimental community. Indeed, he portrayed Minihasa, with slight exaggeration, as an East Indies version of New Lanarck, composed of model villages with beautifully cultivated fields, coffee plantations, landscaped gardens, and cottages "like those one sees upon the stage"—a place where the value of an honest day's work had been thoroughly inculcated and the general morality had undergone a vast improvement. In his opinion, the relation of a civilized to an uncivilized race was analogous to that of parent to child or teacher to pupil, necessitating a certain amount of paternal despotism. The Dutch practice of upholding and regulating the power of the native chiefs under the guidance of missionaries was far superior to the policy of "throwing open a country to the competition of a low class of European traders and cultivators which inevitably leads to the degradation of the natives, and a conflict of interests, inducing mutual animosity between two races." The consequent interference in personal freedoms was unfortunate, he admitted, but the restrictions were for the greater good and produced a vibrant, economically successful city.

Wallace then proceeded to attack the English system in India—the Indian mutiny of 1857 was still fresh in people's memories—where the British Raj had done nothing to improve the existence of the people either morally or intellectually. Wallace believed that unlike the British rule in India or other parts of the empire, the Dutch restrictions in the East Indies were temporary, a final stage on the path to civilizing a savage race. Nothing good came of free competition among different peoples, he said. Free competition between antagonistic races, as in New Zealand between the English and the Maoris, led not to the improvement of the indigenes but to their potential extinction. "Competition and free trade are excellent things of themselves and produce excellent results," Wallace said, "but we do not think of turning [out] our boys and girls at twelve years old to get their living and education by free competition in the world." In northern Celebes, the people became wealthy, reaped the benefits of the revenues without direct taxes, and were experiencing unprecedented population growth. He concluded, "I think that we should hesitate in applying the principles of free competition to the relations between ourselves and savage races if we ever expect them to advance in civilisation or even to maintain their existence upon the earth."

The premise of Wallace's paper offended some members of the audience, who were unprepared for such views from a naturalist. In the report from the British Association, his paper was pruned, with only the most inflammatory portions included and not their clearly reasoned humanitarian underpinnings. "The system which Mr. Wallace has lauded is simply the enforced coffee labour of Java . . . that treats the people as children," an indignant participant remarked in the discussion afterward. "But the people are not children; and something better than flogging them to make them work hard has been adopted at Singapore . . . where 40,000 to 50,000 have been converted to Christianity. We have done a great deal for the Maories [*sic*] and have treated them on terms of equality. We have civilised them from their abominable savagery." Wallace countered that his critic had "enunciated a doctrine with which he would find but few sympathizers in this room."[72]

Wallace was less enamored of the Dutch than he led people to believe. A few years earlier, he had admitted to his friend George Silk that in general he did not like the Dutch or the Dutch officials in the Malay Archipelago. Yet he could not "help bearing witness to the excellence of their government of native races, gentle yet firm, respecting their manners, customs, and prejudices, yet introducing everywhere European law, order, and industry."[73] He believed that all people required proper guidance to realize their full potential, to elevate them to a level of equality—a belief not shared by a large segment of the English population, who, like James Hunt, considered non-whites to be inferior varieties or even separate species of the genus *Homo*.

Wallace sharpened his attack on sacred English institutions in an article published in the June 17, 1865, issue of the *Reader*. In "How to Civilise Savages," he focused on the "barbarity" of the British settlers of Australia toward the Aborigines and elaborated on a theme from "On the Progress of Civilisation in Northern Celebes" that questioned the civilizing influence of Christianity, a system of belief that took root in Europe only after centuries of tears and bloodshed gradually molded it to the "mental constitutions" of its adherents. A paper read at the May 16, 1865, meeting of the Anthropological Society by the Reverend John William Colenso, the bishop of Natal, in South Africa, had influenced Wallace. Colenso had addressed the efforts of missionaries to proselytize the "savage peoples" of Africa, especially the Zulus. The paper "delighted" Wallace, for it proved that "the best effects" were produced when the missionaries sought only to do "good"—which, he believed, "would move the people more than aught besides."[74] "If the history of mankind teaches us one thing more clearly than another," Wallace said, "it is this—that true civilization and a true religion are alike the slow growth of ages, and both are inextricably connected with the struggles and development of the human mind." In his opinion, the missionary scheme had utterly failed among savages, for Christianity could not be maintained without "fresh relays" of priests sent from far distant lands. "Are these new Christians to be for ever kept in tutelage," Wallace asked, "and to be for ever taught the peculiar doctrines which have, perhaps, just become fashionable among us?" "Are they never to become men, and to form their own opinions, and develop their own minds, under national and local influences?" When a mission was successful, Wallace said, the success had less to do with religion than with the example of the civilized and educated people: their decency, cleanliness, and comfort, "their teaching of the arts and customs of civilisation, and the natural influence of superiority of race"—advantages that could be imparted without the inculcation of particular religious beliefs.

He concluded his paper not by attacking colonialism, an institution that he took for granted and did not oppose, but by suggesting a radical modification in the structure of the mission. If the savage races were to be improved, they would be improved by the power of example. Some of the funds from the missionary societies should be used to establish model communities, to show the people the benefits of the civilization that Europeans wished to introduce and serve as a visible illustration of the effects of Christianity on Christians.

In May 1865, Wallace published an impassioned retort in the *Reader* to the great English philosopher John Stuart Mill on the issue of public responsibility and the ballot. With parliamentary elections about to take place in July, the issue was timely. On moral and ethical grounds, Mill had expressed his opposition to the secret ballot in parliamentary elections, to which Wallace replied:

> It seems to me that, in the days of standing armies, of an elaborate Poor Law, of State interference in education, of the overwhelming influence of wealth and the priesthood, we have *not* arrived at that stage of general advancement and independence of thought and action in which we ought to give up so great and immediate a benefit to thousands as real freedom of voting, for the infinitesimal advantage to the national character which might be derived from the independent and open voting of the few who would feel it compatible with their duty to their families to struggle against unfair influence and unjust intimidation.[75]

From his redoubt in Downe, Darwin followed Wallace's activities with great interest. The *Reader*, a liberal forum sympathetic to the cause of evolutionists, was one of the journals he enjoyed at the time. Although he did not comment on Wallace's controversial political statements, he was concerned that Wallace was straying from his important scientific work. In a letter dated September 22, 1865, he asked Wallace whether he had made any progress on his travel journal. He was especially eager to see it, he said, after having read Wallace's early papers on the orangutan, with their foreshadowing of evolutionary theory, in the *Annals and Magazine of Natural History* in the summer of 1856.[76]

It was the first letter that Darwin had written to Wallace since early February. Throughout 1865, Darwin was plagued by illness, which prevented him from attending scientific meetings or communicating regularly with friends. In the same letter, he told Wallace that he had not yet had the strength to read Wallace's latest paper in the *Transactions of the Linnean Society* on the distribution of Malayan butterflies, "for though somewhat better I can as yet do hardly anything but lie on the sofa and be read aloud to." To distract him from his physical suffering, Emma Darwin read to him Edward Burnett Tylor's *Researches into the Early History of Mankind*, William

E. H. Lecky's *History of Rationalism in Europe*, and John Lubbock's *Prehistoric Times*—books that "much" interested him and would provide ideas for *The Descent of Man*. Darwin directed Wallace to the last chapter of Lubbock's book, which spoke of Wallace's "characteristic unselfishness" for ascribing the theory of natural selection solely to Darwin without sharing the credit, "a note about you in which I most cordially concur."

Characteristically, Wallace did not acknowledge Darwin's remark on the priority issue, but he offered brief, frank critiques of all three books. He found Tylor's study "disconnected and unsatisfactory from the absence of any definite result or any decided opinion on most of the matters treated." Although he liked Lecky's book, he considered it "rather tedious and obscure at times" and not original, since most of its content had been covered more thoroughly by another author. Lubbock's book was "very good," but the concluding chapter, the one that paid homage to him, was very weak. "Why are men of science so dreadfully afraid to say what they think and believe?" he complained.[77] It is not clear exactly what Wallace found "weak" or overly cautious about Lubbock's final chapter. Lubbock, who coined the terms "Paleolithic" and "Neolithic," agreed with Wallace's views of humanity's great antiquity—that is, as a family, not as a genus or species. He favored the opinion advanced by Wallace about the unity of the human race and agreed that "the Caucasian, the Negro, the Red Indian [were] derived by the operation of natural selection from one primitive type." He also agreed that the role of science was to make humans not only more comfortable but also more virtuous. Where Lubbock deviated from Wallace was in his allusion to religion. When people were convinced that "suffering is the inevitable consequence of sin, as surely as night follows day," Lubbock said, they would become wise and "leave off " sinning.[78] This remark, an obsequious concession to churchly views, may have irritated Wallace, who had rejected the doctrines of his church long before and held no religious convictions at the time. With Robert Owen and Thomas Paine, he asserted that organized religion was nothing but idolatry—a human invention set up, in Paine's words, "to terrify and enslave mankind, and monopolise power and profit."[79]

"It is very well for Wallace to wonder at scientific men being afraid of saying what they think," Hooker replied indignantly to Darwin, who apprised him of Wallace's sentiments. "He has all the freedom of motion *in vacuo* in one sense. Had he as many kind and good relations as I have, who would be grieved and pained to hear me say what I think, and had he children who would be placed in predicaments most detrimental to children's minds by such avowals on my part, he would not wonder so much."[80]

Wallace's comments were a sign of his restlessness, perhaps augmented by self-confidence or arrogance, as his reputation within the scientific community soared. Three years after returning from the Malay Archipelago, he no longer felt constrained to limit his investigations to zoology. Had he imposed such limitations on himself, he might have spent the rest of his life engaged in the analysis of his collections and nothing more, confining his philosophical musings to the privacy of his home. But Wallace branched out beyond zoology and taxonomy into the social, religious, and political spheres, choosing to share the results of these broader investigations with the scientific world—often with a candor that was unsettling and even offensive—and eventually shifting the debate about human origins to a different plane. The years 1864 and 1865 were pivotal in his intellectual development, a period in which he began to pursue a variety of interests with the zeal and thoroughness with which he had investigated all things scientific, including evolution, biogeography, physical geography, anthropology, and geologic time. It was the broader interests that would, in the end, cause him a great deal of trouble.

CHAPTER 10

A Turn Toward the Unknowable

ISHMAEL, MELVILLE'S NARRATOR in *Moby-Dick*, wryly observes that just because a man has seen the world does not mean that he will move with ease in the company of other people. Some of the greatest travelers, he says, possess the least assurance in the parlor. And twelve years of solo travel in the western and eastern tropics had not given Wallace a high social polish. He spoke plainly and expressed himself frankly, a trait that offended some of his peers, who preferred to navigate the thickets of Victorian etiquette with greater finesse.

In his interpersonal relationships, Wallace was equally heavy-handed. Having spent much of his life isolated from his fellow Englishmen, he felt most at ease alone. "I seldom have a visitor but I wish him away in an hour," he wrote to his friend George Silk in 1859 from the island of Batchian in the Malay Archipelago. "I find it very favourable to reflection." He hoped for "a happy future in England," where he could "live in solitude and seclusion, except from a few choice friends."[1] Instead, upon returning from the Far East he landed squarely in the midst of a scientific revolution. With his paper "On the Tendency of Varieties to Depart Indefinitely from the Original Type," he had launched the first volley. And now that the English scientific community verged on civil war, a life of seclusion and solitude was unthinkable—at least for one of that revolution's commanding officers. Although combative with his pen, by his own admission Wallace was extremely shy in his encounters with men and women he did not know. In fact, he was "so much disinclined to the society of uncongenial and commonplace people" (of all classes) that his natural reserve and coldness of manner amounted to rudeness.[2] It was impossible for him to make polite conversation with such individuals or even to acknowledge their trivial remarks. He believed he appeared gloomy, conceited, and arrogant to most people, when in reality he was only bored. This sort of behavior was not limited to people he disliked. The first time he lunched at the home of Sir Charles Lyell, in the summer of 1863, he did not make a favorable impression on Lady Lyell.

Lyell's new secretary, Arabella Buckley, the twenty-three-year-old daughter of a Brighton vicar who would become one of his few female friends, told him that his shy, awkward, and unpolished manner upset Lady Lyell. As always, Wallace had a theory for everything. Analyzing himself from a phrenological perspective, he concluded that his deficiency in the "language-facility" prevented him "from saying anything for the sake of politeness."[3] Lady Lyell was not the only woman upset by his gracelessness. His awkwardness sabotaged relationships with other women—particularly those whom he might marry. In the 1850s, this had not mattered. During his travels, marriage had been as far from his mind as the cultivation of aristocratic manners.

In his letters and early writings, he made few allusions to women, but his occasional comments were mainly ironic or unflattering. Only with rare exception did he find the native women attractive. In fact, he was bemused by the hyperbolic accounts of male travelers relating to native female charms. One of his favorite themes was the disparity between these fantasies and the disappointing reality of his own objective observations. "A couple of Indian belles," he mockingly observed in an 1854 article for the *Transactions of the Entomological Society*, "will often devote a spare half-hour to entomological researches in each others' glossy tresses, every capture being immediately transferred with much gusto to the mouth of the operator."[4] While visiting a rajah in Celebes a few years later, he took note of his host's daughters, who were gazing with great curiosity at their strange white visitor. As he recorded in one of his Malay journals,

> And here I might (if I followed the example of most travellers) launch out into a glowing description of the charms of these damsels, the elegant costumes they wore, and the gold and silver ornaments with which they were adorned. The jacket or body of purple gauze would figure well in such a description, allowing the heaving bosom to be seen beneath it, while "sparkling eye," and "jetty tresses," and "tiny feet" might be thrown in profusely. But alas! Regard for truth will not permit me to expatiate too admiringly on such topics. . . . The princesses were, it is true, sufficiently good-looking, yet neither their persons nor their garments had that appearance of freshness and cleanliness without which no other charms can be contemplated with pleasure.[5]

Despite this droll profession of sexual restraint in the face of temptation, which no doubt was intended to amuse his audience, there is no evidence that Wallace engaged in premarital sex, unlike Henry Bates. By all appearances, he exercised a rigorous self-discipline in sexuality, as in all other spheres of his life.

But now his wanderings had come to an end, and perhaps the prospect of a long life of solitude had lost its appeal—or, with his mother now too old to manage his domestic affairs, he may have decided that it was impractical to remain a bachelor. For the first time, he also enjoyed true financial security and could realistically contemplate marriage. Samuel Stevens's wise investments of the profits from the sale of his specimens ensured him an annual income of £300. His private collections, consisting of the best preserved and most interesting species from the Malay Archipelago, promised an additional annuity of £200.[6] It was not a huge income, but his net worth classified him as a gentleman, which meant that he could marry and live comfortably with a wife; to raise a family, he would have to find additional work.

In 1862, however, Wallace had no obvious candidates for marriage in mind. At the age of thirty-nine, he had to settle for someone less than his ideal. "I believe a good wife to be the greatest blessing a man can enjoy, and the only road to happiness," he told Silk in a letter from Singapore on the eve of his departure in January 1862, "but the qualifications I should look for are probably not such as would satisfy you. . . . I now look at intellectual companionship as quite a secondary matter, and should my good stars ever send me an affectionate, good-tempered and domestic wife, I shall not care one iota for accomplishments or even for education."[7]

An opportunity to marry arose unexpectedly in connection with Silk. An avid chess player, in the summer of 1862 Silk took Wallace to a private chess club in London, where they met a widower who lived with his son and two daughters. Silk and Wallace began to call on this man at his home. Eventually, Wallace visited alone for a game of chess, which was just an excuse. In the interval, he had become infatuated with one of the daughters, a twenty-eight-year-old woman whom he described as "very agreeable, though quiet, pleasant looking, well-educated and fond of art and literature." In his autobiography, he refers to this woman only as Miss L, probably to protect her privacy in the event that she or other relatives were still alive in 1905.[8] Consistent with his naïveté in matters relating to relationships with women, his approach to Miss L was anything but direct, in contrast to his approach to everything else. Rather than declare his feelings openly, he wrote her a letter a year after meeting her, asking for her hand in marriage. She expressed surprise at his proposal and stated that she had had no clue about his interest in her, despite several visits for tea, supper, and friendly conversations at her home. She refused him in a manner that he described as "kind" but that gave him some hope that she might reconsider. At the conclusion of her note, she begged him not to allow her refusal to marry him interfere with his friendship with her father. But the rejection left him in a state of emotional turmoil.

Wallace showed Miss L's letter to his mother and Fanny, who encouraged him to try a second time. But a year passed before he worked up the nerve to ask her once again to marry him. Fearing another refusal, he solicited her father, explaining his deep affection for the man's daughter. Fortunately, Mr. L gave his consent and arranged a private meeting between Wallace and his daughter. At this point she agreed to marriage, and the couple began meeting two or three times a week to work out plans for a wedding, which was finally scheduled for December 1864, the date having been put off several times because of her "ill-health and other causes." They sent out invitations, ordered wedding clothes, and set up a program for an unrecorded number of guests.[9]

In the spring of 1864, Wallace applied for a job at the Royal Geographical Society, a position that had become vacant after its assistant secretary died unexpectedly. His primary motive for seeking the position, he later explained, was to obtain "some permanent congenial employment" that would supplement his income and yet leave him time to study his collections.[10] Clearly, however, he was concerned about his ability to support a wife and raise and educate his children properly. In the mid-nineteenth century, a man of science born into the lower classes had few options for making a living in his area of expertise. In all of London, for example, there were only four or five positions for a zoologist or comparative anatomist. Thomas Huxley's antagonist Richard Owen, who was one of the most respected zoologists of his generation, earned only £300 in an academic position, which was less than the salary of the average bank clerk. A novelist, a poet, or an essayist could support himself by writing for a magazine or literary review, but no scientist could ever expect payment from a scientific journal at the time. "A man who chooses a life of science chooses not a life of poverty, but, so far as I can see, a life of *nothing*," Huxley remarked, "and the art of living upon nothing at all has yet to be discovered."[11] Richard Spruce told the botanist George Bentham that "poverty is such a positive crime in England, that to be there without either money or lucrative employment is a contingency not to be reflected on without dread."[12]

The vacancy at the society was an opportunity Wallace that could not afford to pass up. In the end, he did not get the job. In fact, the society never considered him a serious contender for the position. His alleged candidacy was omitted from the society's official records, an omission that Wallace either knew nothing about or refrained from mentioning. Reasons were never given. Six men were listed as candidates, and four were interviewed by the selection committee. One of the viable candidates was Bates, who eventually was offered the position.[13] Wallace seems not to have minded that Bates was hired. At the time, Bates was more in need, having virtually no income and a wife and child to support. Decades later, Wallace rationalized the society's decision on three counts: the position would bring Bates into central London and allow the two friends to spend more time together; Bates had a better head for business; and the demands of the job, including confinement in the unhealthy environment of the city, would have "greatly shortened" his life. Wallace did not brood over his rejection; in 1864 he was too busy with his descriptive work and writings to think much more about obtaining regular employment.[14]

Finding a steady job soon became a moot point, for sometime in September or October 1864, not long after the conclusion of the meeting of the British Association for the Advancement of Science, Wallace received devastating news. When he made his usual afternoon call on Miss L, a servant told him that she had gone away that morning and would write later. "Staggered," he returned home to await the letter, which came the following morning and not from her but from her father, who stated that his daughter wished to break off the engagement. "The blow was very severe," Wallace wrote forty years later, "and I have never in my life experienced such intensely painful emotion."[15]

Eventually Miss L wrote to him, giving her reasons for the decision, which had something to do with her suspicions about his interest in another woman, the widow of an army officer. "All this was to me the wildest delusion," Wallace later wrote. "The lady was . . . very pleasant and good-natured, and very gossipy, but as utterly remote in my mind from all ideas of marriage as would have been an aunt or a grandmother." He was "dreadfully" hurt. He replied with a "strong, bitter" letter to Miss L in an attempt to regain her confidence and reiterate his feelings for her, but his entreaties were ignored. The true reasons for her change of heart remain unknown, and he never again heard from Miss L or her father.[16]

Wallace told few acquaintances about his loss, and he continued to attend meetings of the various societies without any outward sign of perturbation. Among those he did tell was Darwin. "Though we have met so little," Wallace wrote to Darwin in late January 1865, "yet I look upon you as a friend, and as such hope you will pardon my boring you with my private affairs." For the past six months, he said, he had been doing "absolutely nothing" and feared that for the foreseeable future he would not be inclined to work. The reason was that he had suffered "one of those severe disappointments few men have to endure." After describing the details of Miss L's unexpected and ill-timed rejection, he asked Darwin to imagine "how this has upset me when I tell you that I never in my life before had met with a woman I could love, and in this case I firmly believe I was most truly loved in return."[17] Darwin replied sympathetically that the best way to banish painful thoughts was hard work—the usual recommendation he gave to a friend in emotional distress and the favorite remedy for his own troubles—but the advice apparently had little effect.[18] A few weeks later, Wallace's friend Alfred Newton, having succeeded Philip Sclater as the editor of *Ibis*, asked him to make contributions to that journal, but Wallace told Newton that he was still "too cut up" to work on his bird collections. Several more months passed before he felt himself able to comply with Newton's request.[19]

In mid-March, Wallace and his mother moved out of Fanny's home and into a small house at 9 St. Mark's Crescent, Regent's Park. He gave no reasons for the move, though his new accommodations were closer to the Zoological Society's library, which he frequently visited at the time for reference material. The decision may have been economic: by moving out, he freed up an apartment that Fanny and Thomas could now rent.

But something else was brewing in the Sims household. At some point—exactly when is unclear—Fanny had become an ardent spiritualist. It is unlikely, however, that her newfound passion influenced her brother's decision to leave, and there is no evidence of any serious disagreements between the two siblings. In fact, since his return from the Malay Archipelago, they had grown closer. Wallace later reported that he had always been Fanny's favorite brother, and if his letters to her are an indication of his feelings, he returned her affection. Despite any initial reservations he might have had about spiritualism, he respected his sister too much to dismiss her beliefs outright. And it is probable that she had been gently encouraging him to investigate the phenomena she had seen at public and private séances, which were becoming commonplace in mid-Victorian London. Not until late July—once the scientific societies had adjourned for the summer—did he agree to go to a séance to witness personally the mysterious rapping noises, levitations, automatic writing, and other extraordinary spectacles his sister had described to him with such enthusiasm.

Spiritualism as a sociological phenomenon had strange beginnings.[20] It originated in the United States in 1848, when a nine-year-old girl named Katie Fox from Hydesville, New York (near Rochester), was disturbed by mysterious knocking sounds that terrorized her household for months. A neighbor interpreted the rapping with the help of a crude alphabet inscribed on a tablet—the forerunner of the Ouija board—and determined that the knocks were a form of communication from a man who had been murdered and buried in the basement of the Fox home long before the Foxes had moved in. Investigators exhumed bones from the basement walls, corroborating the story told by the ghostly rapping. Reports of the exhumation created a sensation, with visitors besieging the Fox home out of curiosity about the young girl who could communicate with a dead man. Katie and her sister, who had developed similar powers, were treated as celebrities (or freaks) and paraded from town to town in western New York—a region that had been thoroughly evangelized in the mid-nineteenth century by revivalists, utopians, millenarians, and various otherworldly zealots—to display their unusual talents. On one occasion, they were subjected to a variety of moderately sadistic experiments designed to uncover fraud. A committee of ladies, it was reported, stripped the frightened girls, tied their dresses around their ankles, and forced them to stand on pillows to insulate rapping sounds they might make. Distinct raps from the walls and floor were still heard, and the committee declared the girls genuine mediums.

Others soon claimed the ability to communicate with the dead. The sudden and dramatic appearance of people—mostly, though not exclusively, young women—who called themselves mediums seized the collective imagination of a segment of the American public that had always been attracted to cults and messianic causes, coalescing rapidly into a movement that in short order spread to Great Britain and the Continent. The apparent confirmation of an afterlife provoked one United States senator to declare that the purpose of modern spiritualism was to bring humanity into universal harmony and prove the immortality of the soul. The voices and visions evoked at séances convinced many in the nineteenth century to study the phenomena scientifically. In a scientific age still in the shadows of a religious one, it was deemed the duty of science to prove or disprove the validity of the phenomena. Later science was expected to prove or disprove the sanity of those who believed in the reality of the phenomena.

In the United States, scientific societies composed of doctors, lawyers, and other professionals legitimized the spiritualist movement by taking it seriously. By 1870 there were 20 state associations, 105 societies, 207 lecturers, and 200 public mediums. At the movement's height, propagandists claimed 8 to 10 million members, a figure that was never verified. The number of adherents in Great Britain and continental Europe is also unknown, but in England spiritualism cast as wide a net as in its former colony. In France, the court of Napoleon III held spiritualist séances. Nearly every European country felt its influence. Several journals were devoted to the subject, their pages filled with testimonials from those who believed that they had scientific proof of communications with the dead. The most virulent opponents of spiritualism described the movement as an epidemic of delusion, a manifestation of mass hysteria. Others maintained that spiritualism exemplified the survival of "savage thought" among civilized people. Spiritualists were denounced as feeble-minded, gullible, or foolish.[21]

For years, Wallace had been an "utter skeptic" about the existence of "preter-human" or "super-human intelligences," never considering the wonders reported by spiritualists to be anything but fantasies or delusions. He states that he first heard about spiritualism while traveling in the Far East. Newspaper accounts described "the strange doings" of American and English spiritualists, some of which seemed to him "too wild and *outré*" to be believed. He could not have missed the disparaging commentary in the *Athenaeum*, which had as little fondness for spiritualism as it had for the Darwin–Wallace theory of evolution. He must have been shocked by the abuse it heaped on Robert Owen, his old socialist idol, who had become a spiritualist in the final years of his life. The *Athenaeum* called Owen a "sincere, deluded man" transformed over the course of sixty years from a young skeptic into an "old Spirit-rapper." Although he initially reacted to spiritualist claims as the "ravings of madmen," Wallace must have been intrigued, especially if a man of Owen's caliber had accepted them.[22] Another skeptic who became a spiritualist was John Elliotson, a physician who had been ousted from his position at the medical school of the University of London in 1839 for his experiments with mesmerism.

As a youth, Wallace had also dabbled in mesmerism. In 1770 the Viennese physician Anton Mesmer invented a form of therapy based on what he called "animal magnetism" that entailed the treatment of certain diseases with magnets and various forms of manipulation. He achieved a number of successes and won great celebrity. A special investigative commission that included Benjamin Franklin and the chemist Antoine Lavoisier quickly discredited mesmerism, but in the early nineteenth century a few daring physicians in Paris used Mesmer's techniques to elicit mind-reading, clairvoyance, and many other "apparently superhuman powers."[23] In 1829 Jules Cloquet, one of the most eminent surgeons in Paris, amputated a cancerous breast during mesmeric sleep, though the patient, who was partly awake, could converse. Cloquet performed many other painless operations, from simple tooth extraction to the more complex extirpation of the lower jaw. In the era before the discovery of anesthesia, surgeries were barbaric; the patient, screaming in agony, had to be tied down before any procedure could be completed. Such brutality had driven Darwin from the operating room at the University of Edinburgh's School of Medicine, where he had been studying to be a doctor, to the less gruesome quarters of Cambridge, which trained clerics.[24] Mesmerism seemed like a genuine breakthrough in the history of medicine, and in 1831 a commission composed of nine distinguished French physicians unanimously attested to the reality of some of the psychic phenomena that had manifested themselves during "magnetic sleep." In these experiments, the subject was not truly asleep but had been induced by the power of suggestion into an altered state of consciousness.

More than a decade passed before a few British surgeons seriously investigated mesmerism. One of the first scientific men to study mesmeric phenomena was Elliotson, an innovator who had introduced the stethoscope in English medical practice and championed the use of quinine for the treatment of malaria, both of which were initially rejected as quackery. Elliotson provoked controversy by advocating the use of mesmerism in the surgical suite. His medical associates objected on the grounds that the patient's reactions during the trance state opposed known physiological facts. For example, when a diseased leg was amputated, the sound leg should have reflexively contracted, but while the patient was hypnotized no such reaction occurred. They concluded that the patients were either naturally insensitive to pain or participating in a fraud. Despite the copious data that Elliotson and a few others supplied, orthodox physicians remained unconvinced. Surgically extirpating a jaw was one thing, but hypnotizing a mentally disturbed female who claimed to be clairvoyant so that she might predict the outcome of a proposed medical treatment was another. The uproar her predictions caused in the hospital ward brought Elliotson and his research methods under intense scrutiny. The *Lancet*, the conservative voice of medicine, denounced Elliotson's practice of animal magnetism as "immoral quackery" and a "worthless science" and warned that predatory men would abuse the procedure to make improper advances toward vulnerable young women. Although Elliotson himself was beyond reproach on this charge, some Russian aristocrats who came to London to attend his lectures were discovered to have such nefarious motives.[25]

Undeterred by the derogatory declarations of respected medical professionals, the twenty-one-year-old Wallace was fascinated by the hypnotic trance state. He took every opportunity to attend the lectures of Spencer Timothy Hall, the most eloquent spokesman for the new science of phreno-mesmerism, which combined the tenets of phrenology and the spectacle of hypnosis. Hall showed his students how to distinguish between the genuine mesmeric trance and any attempt at conscious imitation: in the mesmerized state, the eyeballs always rolled upward in their sockets. Wallace bought a small phrenological bust and soon became adept at hypnosis, which he practiced on two or three highly susceptible students at the school in Leicester where he was an instructor. After inducing a trance, he stood beside his subject and touched the mapped-out phrenological "organs" on the bust, which was on a table out of his subject's view. Within a few seconds, the student would change his attitude and expression to match the phrenological organ excited. "In most cases the effect was unmistakable," Wallace wrote years later, "and superior to that which the most finished actor could give to a character exhibiting the same passion or emotion."[26]

Wallace learned to produce paralysis of a limb or the whole body so that the boys would do something they could not do under normal circumstances. On the wrist of a boy's rigid outstretched arm, he would hang an ordinary chair while he wrote a short letter, stopping only when he feared causing serious injury. He gave another boy under hypnosis a glass of water to drink, telling him that it was a glass of wine. After drinking the "wine," the boy became intoxicated. He next told the boy that his shirt was on fire, and the boy tore it off in a fit of terror to save himself from being burned.

Through repeated experiments, Wallace convinced himself that the mesmeric effect was real and not caused by mere suggestion or the will of the hypnotist. What most mystified him was the apparent sympathy between himself and the hypnotized student. When he and his subject held hands, his subject sensed everything he did, including feelings, tastes, and smells. He created a human chain, with the hypnotized boy at one end and himself at the other. When someone pricked or pinched Wallace, the student flinched and complained, placing his hand precisely on the corresponding spot. If Wallace put a lump of sugar or salt in his mouth, the student would make sucking motions as though sucking the lump himself. For Wallace, such experiments confirmed Elliotson's (and Mesmer's) assertion that some sort of fluid or force was being conducted from the mesmerist through the chain to his subject.[27] Wallace was absolutely convinced of the reality of the phenomena, which he induced himself and whose effects he witnessed. From his personal experience, he understood that it was not necessary to resort to trickery to produce a trance state and the subsequent marvels. His demonstrations in front of competent observers relieved him "from the haunting idea of imposture which possesses most people who first see them." Later he and his brother Edward, another adept, would hypnotize the street urchins of Barra, demonstrating to their satisfaction that mesmerism operated even in people who knew nothing about the phenomenon. These experiments taught him two valuable lessons he was never to forget: keep an open mind and do not reject a belief or hypothesis on a priori grounds alone; and trust the evidence of one's own senses despite the objections of respected authorities.[28]

Spiritualism's vilification by most scientists made the investigation of spiritualist phenomena irresistible to the insatiably curious and intellectually mischievous Wallace. If it had been merely another religion, he probably never would have accepted an invitation to a séance. But spiritualism, like phrenology and mesmerism, had scientific pretensions and was frequently referred to as a "hypothesis." According to the spiritualist hypothesis, one allegedly could prove by induction that spirits from the world of the dead communicated with and were capable of influencing the living. Like a moth attracted to a flame, Wallace could not resist probing deeper. It was to prove a fatal attraction. His first séance on July 22, 1865, signaled an important turning point in a life full of momentous turning points.

Wallace later insisted that he had not been drawn to his investigations of the supernatural because of a "dread of annihilation" or an "inordinate longing for eternal existence." On three separate occasions during his travels he had narrowly escaped death, and all that he had felt on those occasions was a "gentle melancholy" at the prospect of entering "a sleep which might know no awakening." An agnostic from his youth, he believed that the great problem of "conscious existence" was beyond humanity's grasp, but he had retained a slender thread of belief in life after death; it was a notion he could not completely shake. He attended his first séance "unbiassed by hopes or fears" and with an "ingrained prejudice" against the very term "spirit."[29] Although curious, he did not expect anything of interest to result from his investigations. All this seems to be true. But it is probably no coincidence that he went to the séance in the summer of 1865, when he was still recovering from a failed love affair, too distraught even to take an interest in his collections. Suffering from a blow that struck to the core of his being, he needed something to distract and inspire him. An emotional crisis, not an intellectual or a spiritual one, drove him into the embracing arms of mediums.

The séance took place at the home of a friend—a skeptic, lawyer, and man of science whose identity Wallace never revealed. On that July afternoon, he sat in the living room with his friend and his friend's wife and two daughters, their hands placed palms down on a large wooden table as they waited for something to happen. The curtains were open, light flooded in, and there were no sounds except their expectant breathing. After several minutes the table began to move in curves, as though pivoting on one of its clawed feet, forcing them to shift their chairs to keep their places. Sometimes the movement reversed, and in this way the table zigzagged across the room. Soft but distinct tapping sounds came from somewhere. The table vibrated like an animal shivering in the cold, and Wallace felt the tremors from his hands up to his elbows. Over the course of two hours, these phenomena repeated themselves in various combinations.[30]

Several times that summer, Wallace returned to his friend's house. The strange movements of the table and the odd taps continued as before. The movements of the table may well have been produced by the conscious or unconscious muscular actions of one or another of the participants, but he felt certain that no one in the room could be producing the taps; most often, they sounded like long fingernails striking the underside of the table, but everyone's hands were always on top of the table. It occurred to him that a foot fitted with a sharp point could make similar sounds. To investigate this possibility, he asked each person to leave the table, one by one, certain that the tapping would cease if someone were consciously or unconsciously producing it. He was therefore surprised when, as the lone sitter, he heard two loud knocks—sounds that only he could have produced. But he was not producing them. If he was the victim of a deception, he thought, then everyone was deceiving him, a possibility that struck him as absurd. During these proceedings, Wallace took extensive notes. He approached the phenomena with the objectivity of a naturalist in the field, collecting data before formulating a theory. He could conclude only that there was "an unknown power developed from the bodies of a number of persons placed in connection by sitting around a table with all their hands upon it."[31]

On the recommendation of "a reliable acquaintance," he went to the home of a woman named Mary Marshall, who was the most famous public medium of the day. In September 1865, after the annual meeting of the British Association, he began a series of visits to Marshall, usually in the company of his sister and a friend, "Mr. R," who was a chemist and a skeptic. At the first séance, a small table rose a foot off the floor and remained suspended in midair for twenty seconds. Wallace and four other persons had their hands on top of the table, while Mr. R looked on from across the room, confirming that the table had risen and was freely suspended. At the second séance, a guitar that had been placed in one of the participants' hands slid to the floor, passed over Wallace's lap, stopped at Mr. R (who was now seated at the table), and, using one of his legs for leverage, slid onto the tabletop. On both occasions, the phenomena occurred in bright gaslight (which was unusual, for most séances were conducted in the dark), and Wallace and his friend had turned the table and chairs upside down beforehand and pronounced them ordinary pieces of furniture with no hidden mechanical devices or connections with the floor. After their examination, the medium allowed them to place the chairs anywhere they pleased. The phenomena, when they happened, seemed as real and astonishing to them as the motion of nails toward a magnet would have seemed to the uncivilized tribesmen Wallace had encountered in the Amazon or Borneo.[32]

Spiritualists referred to the movement of objects in space as physical phenomena. The second class of phenomena was known as mental phenomena. In spiritualist séances, mental phenomena usually took the form of names or words spelled out on a Ouija board, alleged communications from the world of the dead to that of the living. The medium would suddenly tell one of the participants that a communication was about to be made to him or her. The participant would go over the alphabet letter by letter with a pencil, pausing only when loud raps indicated the letters that formed the required name. Skeptics usually attributed the accurate spelling of a name to the astuteness of the medium, who, they assumed, closely observed the body language of a person as he or she moved a pencil or another object over the letters. Wallace therefore always took care to avoid favoring any of the letters on the board by moving the pencil over each letter "with steady regularity." On his first visit, "Para" was spelled out, and this was followed by the spelling of the names "Herbert Edward Wallace" and then "Henry Walter Bates." The only other person in the room who knew about Edward and Henry was Wallace's sister, Fanny, who had attended an earlier séance by herself at Marshall's house; but at no point did Wallace ever suspect his sister of having colluded in some way (intentionally or unintentionally) with the medium. Marshall was not informed of anyone else's name in advance; moreover, she would not have suspected that he and Fanny were related because their surnames differed.[33]

During the next visit, Wallace performed an experiment to trap the medium by inserting a sheet of paper with some private notations (to identify the paper as his, not someone else's) and a lead pencil under the central pedestal of the table. As a precaution, he thoroughly examined the furniture and floor, satisfying himself that there were no hidden devices or trap doors. Everyone's hands, including the medium's, rested on top of the table. Wallace soon heard taps, which, he was told, were communications meant for him. To his astonishment, he found the name "William" written in freehand on the piece of paper after he retrieved it. Once again, the recollection of a dead sibling haunted him. On yet another visit, after taking similar precautions against deception, one of Wallace's companions received a communication from his dead son. Perplexed, Wallace could not figure out how the medium had accomplished these feats. Although he had found no devices under the table or trap doors, it was impossible for him to imagine the medium slipping off her boots, seizing the pencil and paper with her toes, writing out a name she had to guess, and then putting her boots back on—all the while keeping her hands on the table.[34]

Wallace began to see spiritualism's connection with mesmerism, viewing the former as an outgrowth of the latter. The spirit circle resembled certain mesmeric experiments, when people gathered around a table and held hands, which enhanced the "fluid" or force thought to be responsible for mesmeric phenomena. Like all mediums, Marshall entered a trance state before communicating with the spirit world—one that differed little from that of hypnotized subjects. Just as he had witnessed in the boys he had hypnotized in Leicester in 1844, he observed that her eyes turned sharply upward, so that her pupils would not have been seen if he had raised her eyelids. He also noted the same change in countenance—an expression of semi-torpor and passivity—preceding the manifestations of phenomena. The mesmerized subject and the medium were like dreamers, with no recollection of what they had done only moments before and no recognition of the bizarre nature of the occurrences.[35]

The question confronting the scientific investigator of spiritualism was the same question that mesmerists half a century earlier had asked about their hypnotized subjects: What, exactly, was happening to the medium, or the hypnotized subject, when he or she lapsed into a trance? Was the medium a conduit from the invisible to the visible world or simply a fraud, an imposter? Scientists aligned themselves on one side or the other, the majority tending to suspect intentional deception or fraud. Despite scientific progress in the first half of the century, as late as 1867 James Hunt could still call the phrenological belief that the brain was the organ of the mind "a gigantic assumption."[36] To Wallace, it was an undeniable fact. If the brain was not the organ of the mind, he told his fellow anthropologists one evening, he might as well claim that "to assume that the eye is the organ of vision [is also] a gigantic assumption."[37] Phrenologists held that the brain coordinated mental functions, that peculiarities of the mind could be connected with distinct parts of the brain. But spiritualists were claiming something more radical: that "spirit" was the essential part of all sentient beings, while the body (including the brain) was merely the machinery and instrument that the spirit used to relate to matter and to other beings. Spirit was mind, and the brain and nerves were the magnetic battery by means of which the spirit communicated with the outer world. The spirit could quit the body in two ways: temporarily, by what spiritualists called traveling clairvoyance—that is, leaving the living body and visiting some distant place on earth or elsewhere in the universe—or permanently, upon death. Following death, a person's spirit survived in an ethereal body, gifted with new powers but mentally and morally identical to the man or woman he or she had been while alive. It was an extraordinary claim.

Meanwhile, in the autumn of 1865 Wallace was still wallowing in self-pity about his inability to find a wife, and he was progressing slowly with his scientific work. Darwin continued to press him to write up his travels, warning him that with each passing year the full import of Wallace's researches was in danger of slipping through his fingers. "I cannot bring myself to undertake them yet, and perhaps never shall," Wallace lamented, "unless I should be fortunate enough to get a wife who would incite me thereto and assist me therein—which is not likely."[38] But soon afterward his luck changed. Although he did not realize it at the time, his friend Richard Spruce already had introduced him to the woman he would marry.

Spruce had returned to London at the end of May 1864, after fifteen years of wandering through the Amazon and the Andes. For a year, he had rented an apartment in London near the Royal Botanic Gardens at Kew, whose collections he consulted, and he made short visits to his old friend Wallace when health permitted. His travels had taken an immense physical toll: he was a broken man, in worse physical shape than Darwin. Suffering from lung and heart ailments of an unspecified nature, as well as "rheumatism," he had barely survived the harsh climate of the Andean *altiplano*. In his final year in South America, he had eaten all his meals and written up his findings in his hammock. Because of the intermittent onset of stiffness in an arm or a leg, he had been forced to give up walking and horseback riding. Compounding his miseries was the loss of his life savings of £700: the money, deposited in a local firm in Ecuador, evaporated when two employees embezzled most of the funds. For his services to the British Crown and the British Raj—he had procured cinchona seeds for the establishment of cinchona plantations in India, which provided a source of the life-saving quinine used to treat malaria—Spruce received a small civil pension that was barely enough to live on.

Spruce had a friend named William Mitten, a specialist in mosses, who lived with his wife and four daughters in the village of Hurstpierpoint, eight miles north of Brighton. Spruce asked Wallace to accompany him to meet Mitten and his family in the summer of 1864. For Wallace, the visit proved a welcome break from London's grime and crowded streets. Mitten, a chemist and stationer by trade, had a passion for botany, especially mosses. Spruce had sent his collection of South American mosses to Mitten, who had behaved dishonorably by claiming some of Spruce's specimens as his own. The breach eventually was repaired, and the two continued to correspond and maintain a friendship, working amicably together at least through the end of the decade. Mitten and Wallace took an immediate liking to each other. But Wallace—who was engaged to Miss L at that time—did not show any immediate interest in Mitten's eldest daughter, Annie, who was eighteen years old when they were introduced.[39] A year later, not long after Wallace's letter to Darwin, Annie Mitten was somehow thrust forward as a matrimonial prospect despite the twenty-three-year difference in their ages. After a brief courtship, the two were married on April 5, 1866, in a quiet ceremony at the Hurstpierpoint Anglican church. It is unclear if the marriage was one of passion or convenience, but the two were to develop a long and loving relationship. Annie provided the peace, solace, and security that Wallace needed in order to pursue the work Darwin was encouraging him to do.[40]

As with other significant figures in Wallace's personal life, not much is known about Annie. Compared with two pages devoted to Miss L in his autobiography, Wallace wrote almost nothing about his wife. Except for a few allusions to Annie in his autobiography and correspondence, he protected her privacy. A handful of undated photographs of her is all that exists; they show a plain, unstylish woman but offer few clues to her personality. Annie shared her husband's love of the outdoors and frequently accompanied him and her father on excursions through the English countryside. Later in life she became an avid gardener, with a passion for primroses, which she cultivated and hybridized, suggesting a sophisticated understanding of the principles of botany. She also had a modest artistic talent; in middle age, she carved botanical scenes on panels of wood. Unostentatious, she remained in the background, apparently comfortable being out of the limelight. Annie also seems to have taken a passive role in Wallace's spiritualist investigations, attending his early experiments but then leaving him to investigate the phenomena with Fanny and others who were more enthusiastic.

Wallace stopped attending Mary Marshall's séances sometime in late 1865 or early 1866, and attempted to reproduce spiritualist phenomena in his own home. For months, he searched unsuccessfully among his friends for someone who had the power to elicit distinct taps, a phenomenon that no one, he believed, was capable of making on his or her own, either consciously or unconsciously. His clandestine scientific investigations might have ended there, unbeknownst to Darwin, Huxley, and his other evolutionist colleagues, which would have preserved his thus far unblemished scientific reputation. But in November 1866, Fanny discovered that her new tenant, Agnes Nichol, possessed the powers they had been looking for.

One Friday evening, Fanny and Nichol, a heavyset woman, arrived by coach at Wallace's apartment in Regent's Park to join him and four friends who met every Friday for a séance. Within a few minutes of sitting down and placing her hands on the table, Nichol fell into a trance and taps commenced. The taps seemed to come from various parts of the underside of the table and changed in tone and volume, one moment sounding like the tapping of a knitting needle or a sharp fingernail, the next like the blows of a fist, the drumming of fingers, or the squeaking noise of a damp finger rubbed hard on a wooden surface. All came with an amazing rapidity as one sound melded into another. Wallace asked everyone to kneel down under the table, but the taps persisted, making the table vibrate.

To satisfy himself that Nichol possessed genuine spiritualist powers, Wallace had secretly attached thin strips of paper beneath the claw feet of the table; the paper would be torn if anyone attempted to raise the table by wedging a shoe beneath one of the claws. In broad daylight, plainly visible to all and with everyone's hands pressing downward, the table rose a foot in the air. The paper strips, which he examined afterward, remained intact.[41] On another Friday evening, the company sat in the dark (he does not give a reason) around a table that had been centered under a glass chandelier. Wallace asked one assistant to hold both of Nichol's hands and a second assistant to strike a match at his signal in order to illuminate the room. In a few minutes, Nichol's chair slid away from her, forcing her to stand up as the assistant gripped her hands. "She is gone from me," he suddenly said as she glided noiselessly out of his grip. Then they heard a slight sound, as though someone were placing a wineglass on the table, followed by the faint rustling of clothes and a tinkling of the glass pendants of the chandelier. When Wallace ordered the match to be struck, all saw Nichol sitting quietly in her chair in the middle of the table, her head just touching the chandelier. The swift and almost noiseless translocation of a stout woman and her chair to the tabletop while five or six people sat nearby in the darkness defied logic.[42]

In mid-December, Fanny received a message through Nichol purporting to be from William. The spirit directed Fanny to "go into the dark at Alfred's this evening" to prove that he was "with" her. At the séance, instead of the usual raps, the room and the table shook violently. When nothing else happened, Wallace mentioned the message that Fanny had received and the group adjourned to the drawing room, which might be more conducive for producing the phenomena they expected. The need to move elsewhere did not trouble Wallace and his friends. In addition to a sympathetic audience, the medium required just the right atmospheric, physiological, and mental conditions or nothing extraordinary would happen. The weather outside had to be calm; the room, warm and dry; the chairs, comfortable. Thunderstorms, for example, ruined a good séance. Not everyone could communicate with the spirits in daylight or artificial light. Some mediums achieved success only in subdued light or complete darkness. Nichol, who was still a novice, was no doubt temperamental and easily distracted. Wallace and his fellow investigators were more than happy to accommodate her. After making certain that the doors and windows were firmly shut, they cleared away books from the drawing-room table, turned down the gaslights, and sat down, forming a circle by holding one another's hands. Only a weak, diffused light entered through the blinds; otherwise, they were enshrouded in total darkness. Raps soon began, instructing them to withdraw from the table, which they did. Wallace rose to turn up the lights to see how everyone was placed, but before he could do so Nichol cried out that something cold and wet had been thrown in her face; he held her hand to calm her as she trembled violently, and he was thus unable to light the lamp. In a few moments, several of the participants whispered that something was appearing on the table. Nichol thought she saw a hand, while others thought they saw flowers, which gradually became more distinct. Someone put his hand on the table and said, "There are flowers here!" The lamps were immediately lighted, and all were "thunderstruck" to see the table half covered with ferns and flowers—all fresh, cold, and damp, as though they had just been brought in from outside. True scientist that he was, Wallace omitted no detail, however insignificant. He counted fifteen chrysanthemums, six variegated anenomes, four tulips, five orange-berried solanums, six ferns of two sorts, and one *Auricula sinensis* with nine flowers—a total of thirty-seven blossoms, their dew evaporating in the heat of the room. There could be no other explanation: William had shown that his spirit was "with" them.[43]

After a year of experimentation, Wallace had immersed himself in the literature of the occult. He had read testimony after testimony of intelligent men who reported phenomena that defied all known laws of modern science—testimony that he could not easily dismiss, given the solid academic and scientific credentials of the investigators. But more than anything else, it was the discovery of the mediumistic powers of Nichol, who had come into his life from nowhere, that had a decisive effect on his thinking. Private experiences such as those produced in his own home, where deception seemed to him unlikely, convinced him that under special circumstances the impossible was possible. Years later, Wallace would write:

> My desire for knowledge and love of truth forced me to continue the inquiry. The facts became more and more assured, more and more varied, more and more removed from anything that modern science taught or modern philosophy speculated on. The facts beat me. They compelled me to accept them as facts long before I could accept the spiritual explanation for them; there was at that time "no place in my fabric of thought into which it could be fitted." By slow degrees a place

> was made; but it was made, not by any preconceived or theoretical opinions, but by the continuous actions of fact after fact, which could not be got rid of in any other way.[44]

Aware that his investigations would evoke scorn, Wallace felt certain that if any of his scientific friends could witness what he had seen in his own house, "under test conditions," they would be satisfied that the phenomena were genuine, even if they disagreed with the explanation suggested by spiritualists.[45]

The first scientific man of eminence Wallace invited to his Friday evening séances was William Benjamin Carpenter. The son of a Unitarian minister, he was a physician who had given up the practice of medicine to devote himself to a literary and scientific career. An expert microscopist, he studied the structure of the minute mollusks known as Foraminifera. He was also an expert physiologist and had written an important book on human physiology. His deep religious convictions prevented him from fully accepting the Darwin–Wallace theory of natural selection, but he had reviewed Darwin's *Origin of Species* favorably and given ample credit to Wallace's contributions. He was among the thirty or so men present at the historic presentation at the Linnean Society of Darwin's and Wallace's papers on natural selection. Respected for his scientific work and his leadership skills, Carpenter would eventually serve as president of the British Association.[46]

Wallace and Carpenter enjoyed a cordial relationship in the 1860s and spent a good deal of time together when they were neighbors in St. Mark's Crescent, Regent's Park. Wallace frequently called on Carpenter in the evenings, when he invariably would find him in his private study peering through his microscope at some obscure marine organism. Carpenter liked to hear himself talk and would "dilate" on a variety of subjects that Wallace did not find enlightening.[47] But their discussions veered into other areas that did interest him, such as psychic phenomena. As early as April 1864, the two had communicated about the reality of "thought-reading," or clairvoyance. Carpenter agreed with Wallace that "a force capable of acting at a distance, producing the phenomena of community of sensation [was] quite conceivable." But "after careful sifting" of the evidence, he was not yet convinced that such a phenomenon occurred.[48] As an experienced observer of mediums and séances, Carpenter joined Wallace and his friends one Friday evening. After hearing a few raps but nothing else, he never came again. A few years later, the relationship between the two men would sour over the issue of spiritualism. Until his death in 1885, Carpenter's attacks on Wallace and other scientific men who investigated spiritualist phenomena remained malicious and unrelenting.

The next person Wallace tried to interest was the physicist John Tyndall, who also attended séances in the past and told Wallace that he "really wish[ed] to see the things able to produce this conviction in a mind like yours, which I have always considered to be of so superior a quality." Wallace replied that he could guarantee nothing, but if Tyndall truly wanted to see the phenomena in all their manifestations, he strongly advised him to suspend judgment for the first two or three visits, passively observe the proceedings, and impose conditions later. Tyndall "forgot" or "purposely" ignored Wallace's advice, joking with Nichol and urging her to show him "something else" besides raps, which they both knew how to produce. The séance was a failure, and Tyndall never came again.[49]

In late 1866, not long after his early séances with Nichol, Wallace put together a booklet entitled *The Scientific Aspect of the Supernatural* and printed a hundred copies, twenty-five of which were to be set aside for friends. The books, which had been wrapped in paper into a single parcel, sat for four days on a table in Fanny's study before she had time to distribute them. One morning, she left the room for just a few minutes. When she returned, the parcel had been opened and the books were strewn about in every direction. Immediately suspecting the action of some spirit, she called her friend Agnes Nichol, but she already had guessed the meaning of the mystery. These books were to be distributed and not lay idle, the spirit rapped out after Nichol arrived. "One for my sister Frances, I have marked it," the spirit stated by way of the Ouija board. Thumbing quickly through the rest of the book, Fanny found red crayon marks on some of the pages (the spirit evidently had marked the book while it was still bundled up with the others). "If you could do this while the book was shut up," she said, "you could write my name in this book while it lays under my hand." After a few moments, she opened the book again and found her name, Frances Wallace, scrawled across the top of the first page. "Now dear spirit," she commanded, "write my marriage name." She closed the book and two minutes later found "Frances Sims" written below her maiden name, this time in a neater hand. The book and its strange tale (which Fanny recorded on two blank pages in the text) have been preserved.[50]

The booklet, a compendium of Wallace's research on psychic phenomena, targeted those skeptics who were convinced that the universe was governed by all-embracing, immutable laws uncovered by modern science. If a phenomenon opposed those laws, Wallace wrote, it had no place in scientific thought. If only one or two of the thousands of facts he could cite proved to be genuine, then the argument against impossibility "fell to the ground." As he noted, "The whole history of the progress of human knowledge shows us that the disputed prodigy of one age becomes the accepted natural phenomenon of the next, and that many apparent miracles have been due to laws of nature subsequently discovered." In the eighteenth century, the possibility of sending a telegram from one city to another three thousand miles away or taking a photograph in a fraction of a second would not have been accepted on any testimony except by the ignorant and superstitious who believed in miracles. The existence of intelligent spirits, perceived under certain circumstances, was at least a possibility, and should not be dismissed on a priori grounds alone. Although he could not yet personally attribute the spontaneous movements of objects in space to disembodied intelligences, he could not resist the testimony of a "vast number" of independent and reliable witnesses from the United States and Europe, many of whom were well-known literary and professional figures. "In short, the testimony has been so abundant . . . that either the facts must be admitted to be such as are reported, or the possibility of certifying facts by human testimony must be given up," Wallace concluded, quoting James Challis, Plumerian Professor of Astronomy at Cambridge University and director of the Cambridge Observatories.[51]

The anthropologist Edward Burnett Tylor, a staunch evolutionist, reacted with surprise to the booklet. Wallace had dropped no hint of his belief in spiritualism during a recent visit.[52] Thomas Huxley, another recipient of the booklet, was not disposed to issue a Commission of Lunacy against him, but he could not work up any interest in the subject. "I never cared for gossip in my life," Huxley told Wallace, "and disembodied gossip, such as these worthy ghosts supply their friends with, is not more interesting to me than any other."[53] Tyndall, however, was disappointed that Wallace described the experiences of others and not his own. "I see the usual keen powers of your mind displayed in the treatment of the question," Tyndall said. "That mental power may show itself whether the material be facts or fictions. It is not lack of logic that I see in your book, but a willingness that I deplore to accept data which are unworthy of your attention. This is frank—is it not?"[54]

James Hunt got wind of Wallace's investigations and published a scathing commentary in the *Anthropological Review*. "Mr. Alfred R Wallace, who is known to anthropologists by his advocacy of the unity of the origin of mankind on Darwinian principles," Hunt wrote, "has become, or at least is at this moment, a zealous Spiritualist, and has published his views. . . . This fact cannot fail to be of interest to those who are curious respecting the idiosyncrasies of men of science. . . . But probably Mr. Wallace's views have been communicated to him by some kind of departed spirit, perhaps that of the 'first man.'"[55]

Not everyone was unsympathetic. Robert Chambers, another recipient of the booklet, replied enthusiastically. The author of the notorious *Vestiges of the Natural History of Creation*, he had known "for years" that spiritualist phenomena were real and was gratified to know that Wallace, "of my friend Darwin's 'Origin of Species,'" was among the few men of science to admit the truth of the phenomena. Once it was fully accepted, he believed, spiritualism was poised to revolutionize thought on many important subjects.[56] The famous mathematician Augustus De Morgan was also supportive. "I doubt whether inquiry by men of science would lead to any result," De Morgan wrote to Wallace. "There is much reason to think that the state of mind of the inquirer has something—be it internal or external—to do with the power of the phenomena, and wish to justify it. I think it very possible that the phenomena may be withheld."[57]

There is no record of Wallace's having sent his booklet to Darwin, who could have learned something of Wallace's dabblings from Huxley or from the comments of Hunt in the *Anthropological Review*, a journal he read. At the time, however, he said nothing.

Despite reports of his further investigations with Agnes Nichol, Wallace's friends remained unimpressed. In 1868 George Lewes, companion of the author George Eliot, savagely attacked mediums and spiritualists in the *Pall Mall Gazette*. Lewes was among those whom Wallace had invited to his home to see Nichol, but Lewes was "too incredulous" to devote any time to the issue.[58] He proposed in his editorial that Tyndall sit with any medium and ask three questions for the spirits to answer correctly. Wallace, who by that time had fully embraced the spiritualist "hypothesis," was incensed by what he believed to be a simplistic—even irrelevant—approach to a complex series of phenomena, and he sent a lengthy letter of protest to the editor: "I admire and appreciate the philosophical writings of Mr. Lewes, of Herbert Spencer and of John Stuart Mill, but I find in the philosophy of Spiritualism something that surpasses them all,—something that helps to bridge over a chasm whose borders they cannot overpass,—something that throws a clearer light on human history and on human nature than they can give me." His letter was ignored and was never published, which prompted him to write angrily to Lewes claiming foul play. For the sake of Lewes's reputation, he said, he should write publicly on the subject of spiritualism only in journals that would permit a rebuttal.[59]

Wallace continued to promote his unorthodox views, quietly at first and then with increasing boldness—views that he understood could damage his scientific reputation and negate all that he had accomplished in the advancement of human knowledge. Over the next decade, he withstood the onslaught of criticism—which was considerable—with a fortitude that can only be called heroic. Whereas most men under similar circumstances would have withered in silent ignominy, Wallace actually flourished, like an organism peculiarly adapted to a hostile environment. In his quixotic quest for truth, he became inured to the opinion of others, confident that in time his convictions would be vindicated.

Alfred Russel Wallace, aged twenty-five (1848). (From *My Life*)

Alfred Russel Wallace, aged thirty (1853).
(Courtesy of the library of the American Museum of Natural History, New York)

Alfred Russel Wallace, aged forty-six (1869). (From *My Life*)

Alfred Russel Wallace, aged fifty-five (1878). (From *My Life*)

Alfred Russel Wallace, aged seventy-nine (1902). (From *My Life*)

Charles Darwin.
(Courtesy of the library of the American Museum of Natural History, New York)

Sir Charles Lyell, the great geologist and anti-evolutionist,
who paradoxically inspired the evolutionary ideas of Wallace and Darwin.
(Courtesy of the Wellcome Institute Library, London)

Joseph Dalton Hooker, a famous botanist and Darwin's closest scientific friend.
(Courtesy of the Wellcome Institute Library, London)

Thomas Henry Huxley, who was known to critics and friends alike as "Darwin's bulldog." (Courtesy of the Wellcome Institute Library, London)

Henry Walter Bates, who accompanied Wallace to Amazonia and whose discovery of mimicry in the insect world validated the theory of natural selection. (Courtesy of the Wellcome Institute Library, London)

Richard Spruce, a botanist and one of the greatest
but least known naturalists of the nineteenth century.
(Courtesy of the archives of the Royal Botanic Gardens, Kew)

Sir William Crookes, who was the discoverer of thallium,
inventor of the cathode-ray tube, and greatest scientific exponent of spiritualism.
(Courtesy of the Wellcome Institute Library, London)

William Benjamin Carpenter, an eminent physiologist and Wallace's great nemesis, who considered the spiritualist movement an "epidemic delusion." (Courtesy of the Wellcome Institute Library, London)

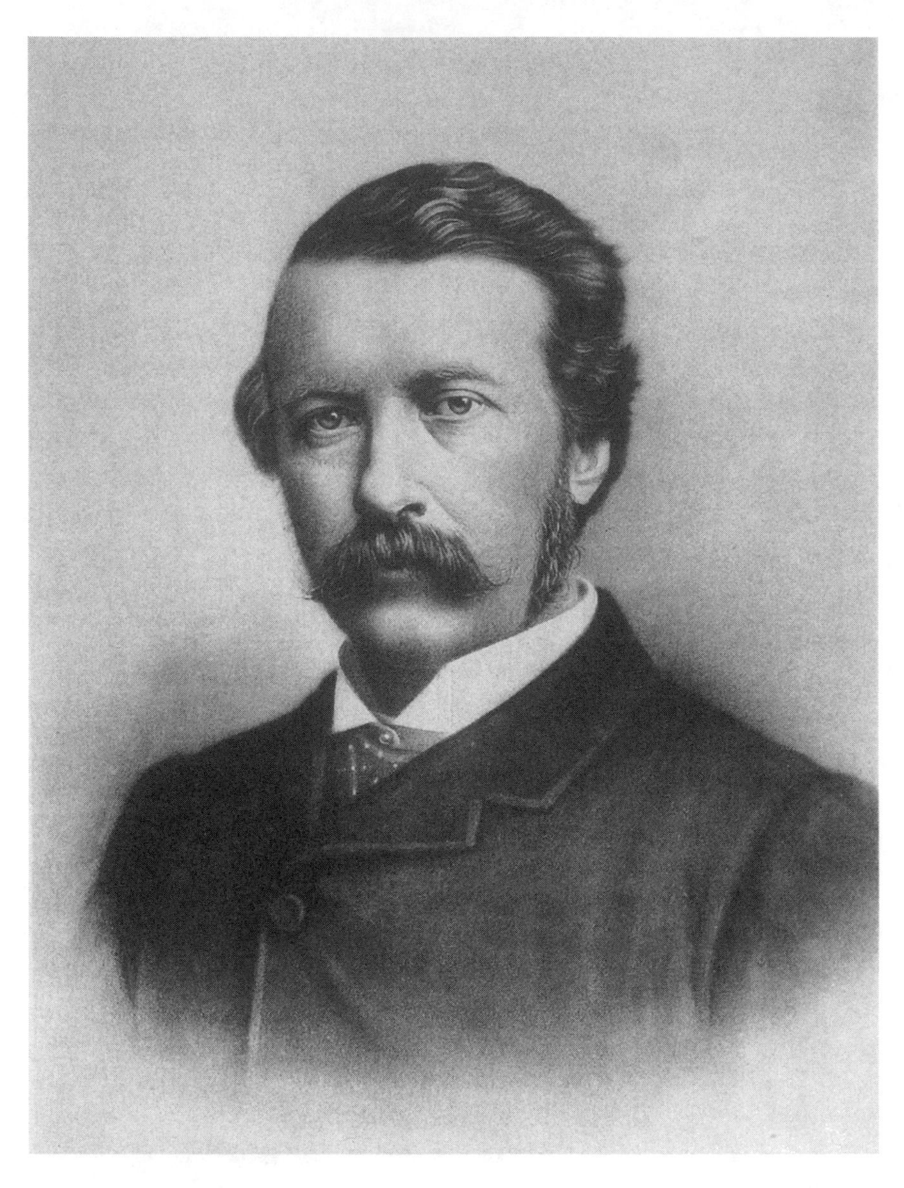

George John Romanes, who threatened to inherit Darwin's mantel, was
a vocal critic of Wallace, and yet, paradoxically, was a secret spiritualist sympathizer.
(Courtesy of the Wellcome Institute Library, London)

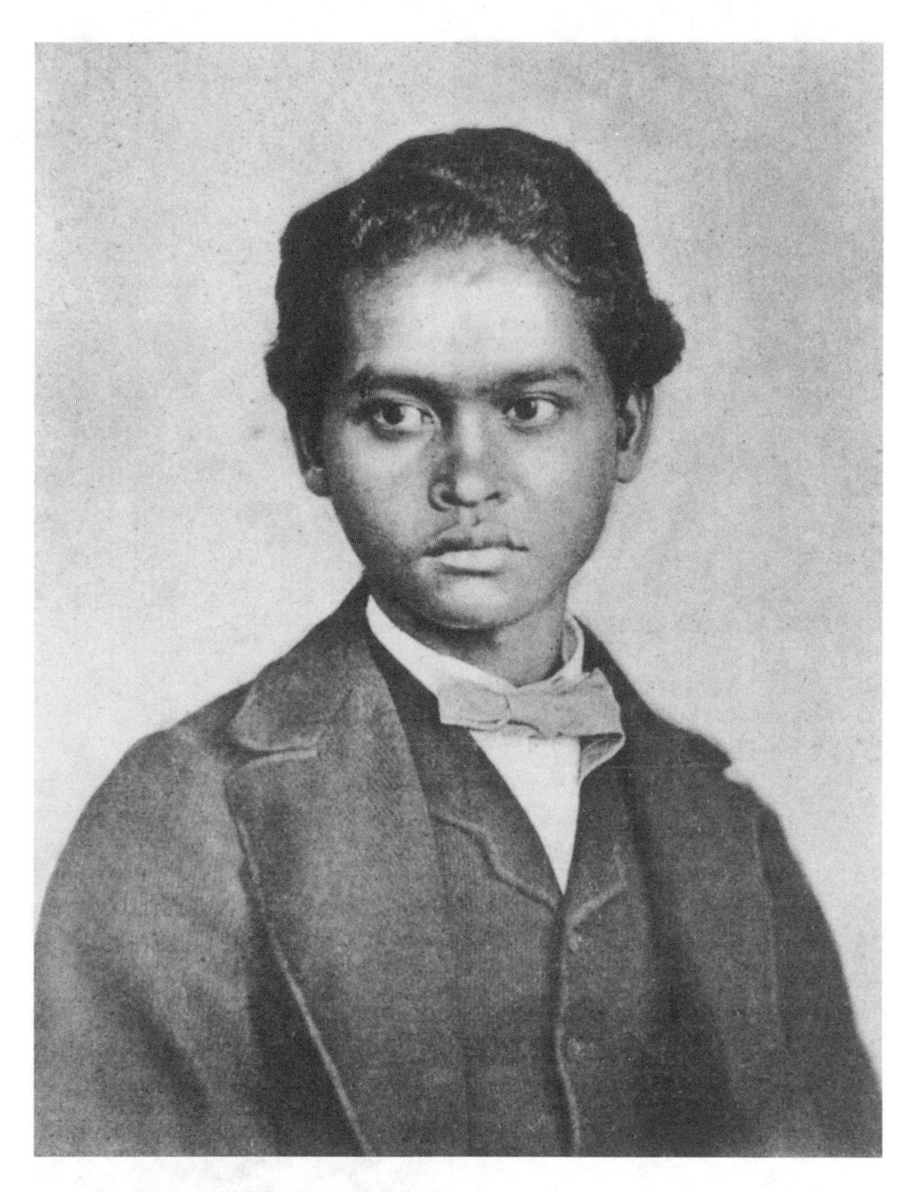

Ali, Wallace's devoted and accomplished assistant in the Malay Archipelago.
(From *My Life*)

Annie Wallace.
(Courtesy of Richard Wallace,
private collection)

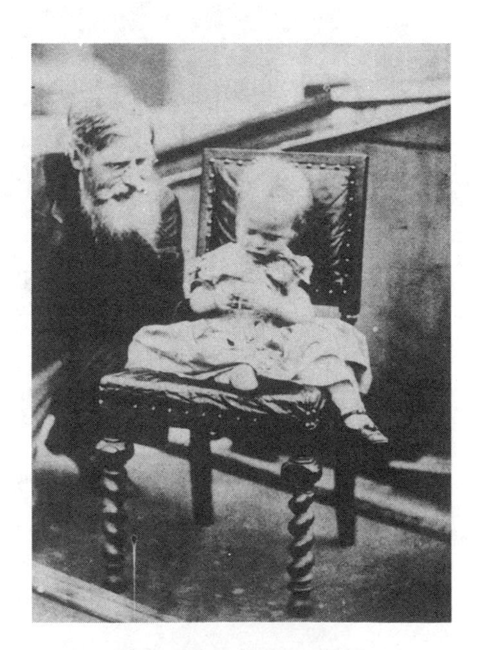

Wallace and his first child,
Herbert Spencer Wallace (ca. 1869–1870),
who died at age seven from
complications of scarlet fever.
(Courtesy of Richard Wallace,
private collection)

William Wallace, his third child. (Courtesy of Richard Wallace, private collection)

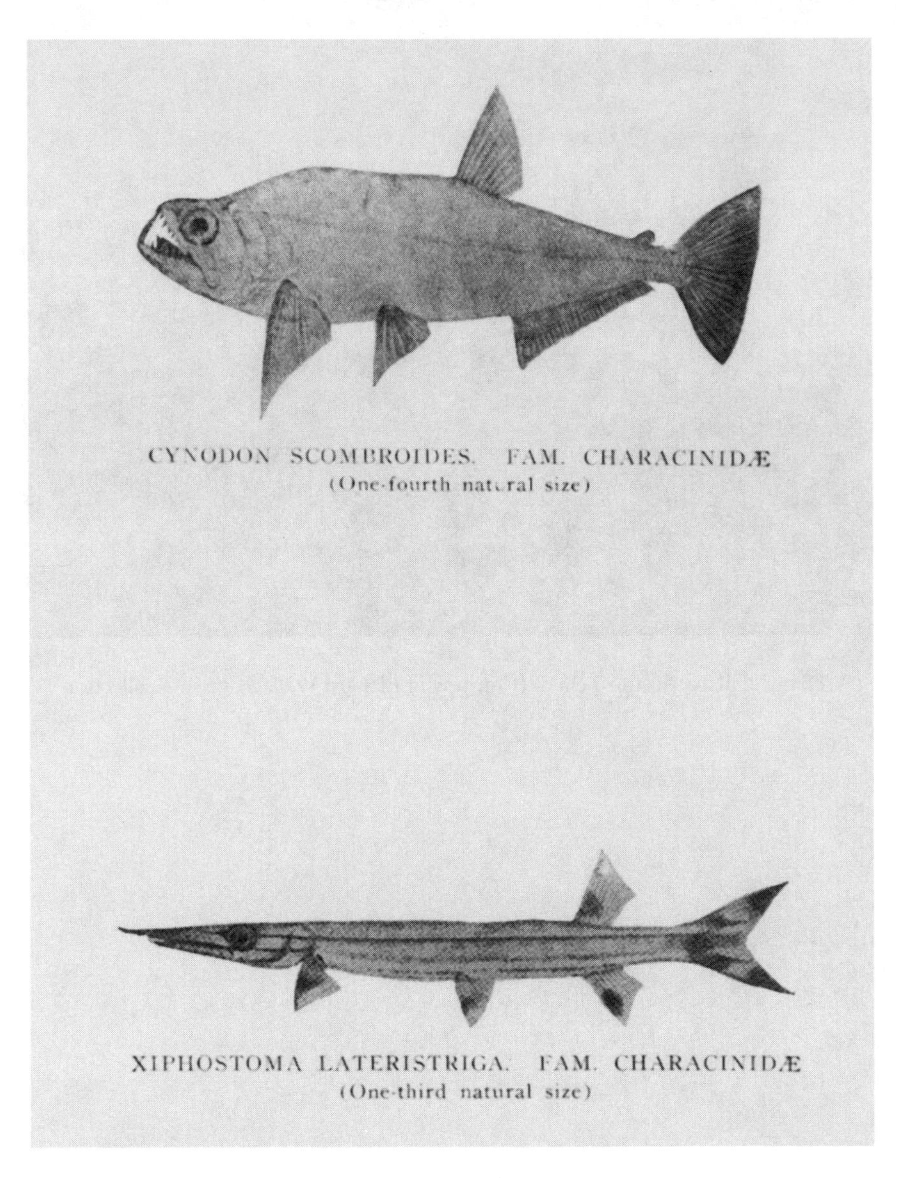

CYNODON SCOMBROIDES. FAM. CHARACINIDÆ
(One-fourth natural size)
XIPHOSTOMA LATERISTRIGA. FAM. CHARACINIDÆ
(One-third natural size)

Two sketches by Wallace of fish caught in the Rio Negro.
These two specimens, along with three years' worth of collections and journals, were lost at sea when the *Helen* caught fire and sank in August 1852. (From *My Life*)

Indian leaf butterfly (*Kallima paralekta*), the most outstanding example of protective resemblance in butterflies that Wallace had ever seen. (From *The Malay Archipelago* [New York: Harper, 1869]; book in author's collection)

Wallace's standard-wing (*Semioptera wallacei*), which Wallace considered his "greatest discovery yet"—even after formulating his theory of natural selection. (From *The Malay Archipelago* [New York: Harper, 1869]; book in author's collection)

Free Library, Neath. Wallace and his brother John were both the architects and the contractors for this building, originally the Neath Mechanics' Institute. (From *My Life*)

Old Orchard, Wallace's last house, in which he lived for twelve years, the longest of his residences anywhere. (From *My Life*)

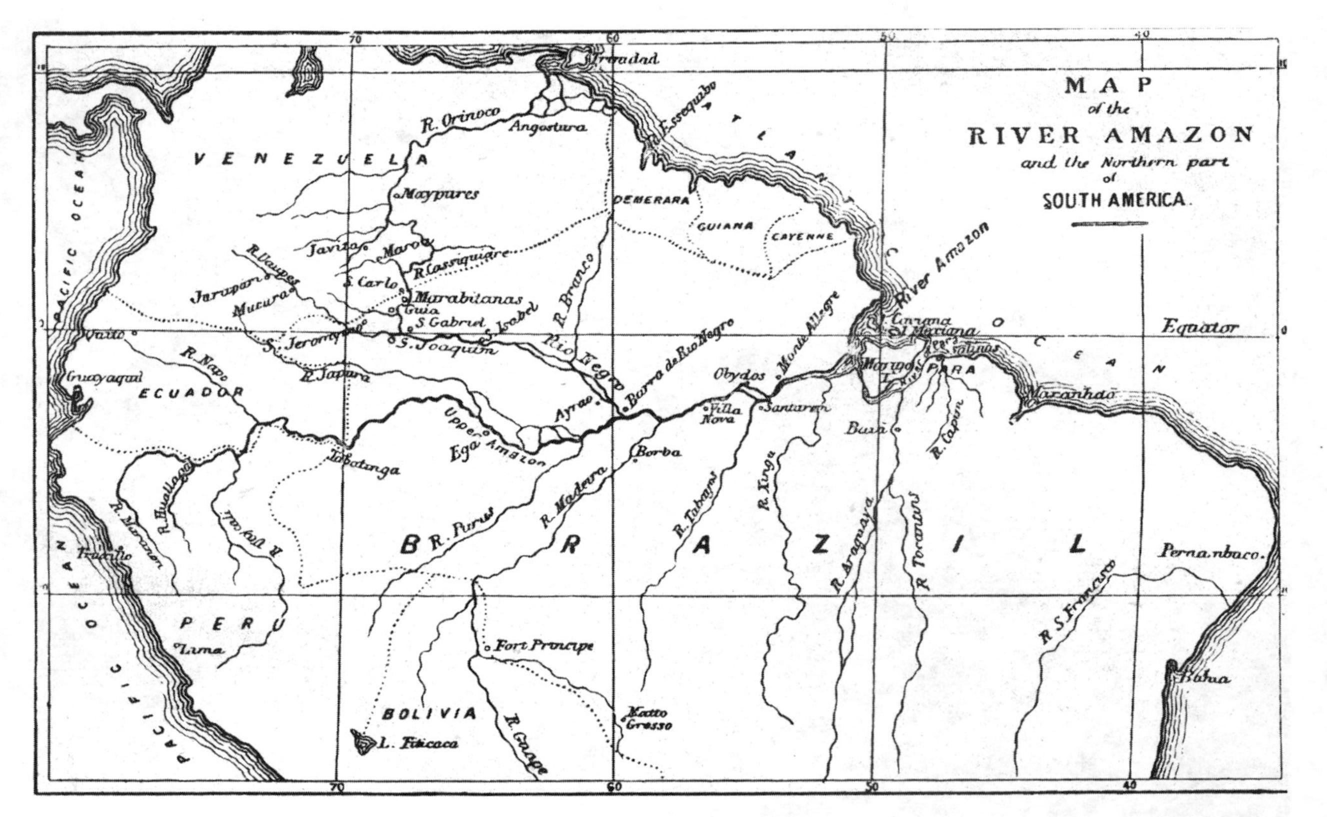

Map of the River Amazon and the northern part of South America. (From *A Narrative of Travels on the Amazon and Rio Negro*)

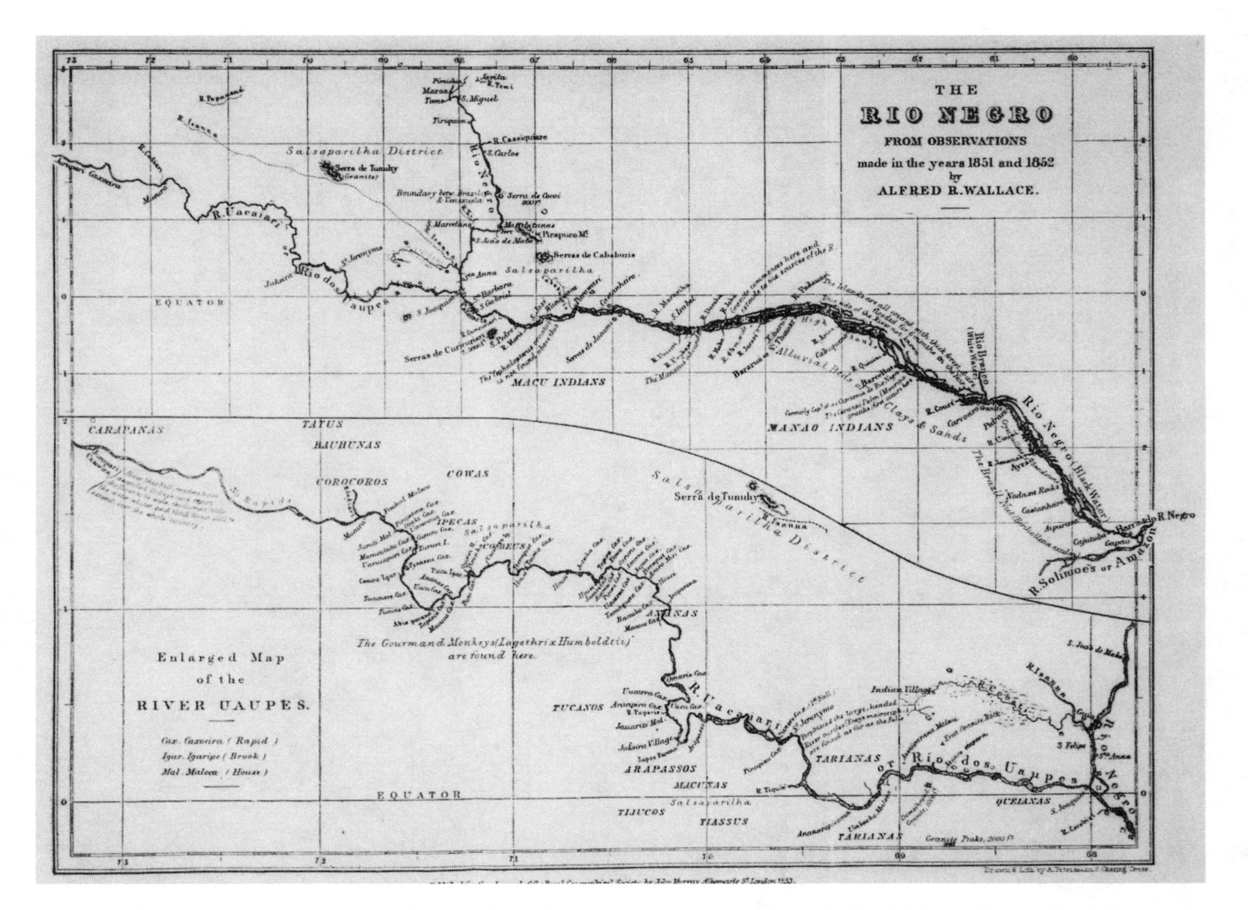

The Rio Negro from observations made in the years 1851 and 1852 by Alfred R. Wallace. (From *My Life*)

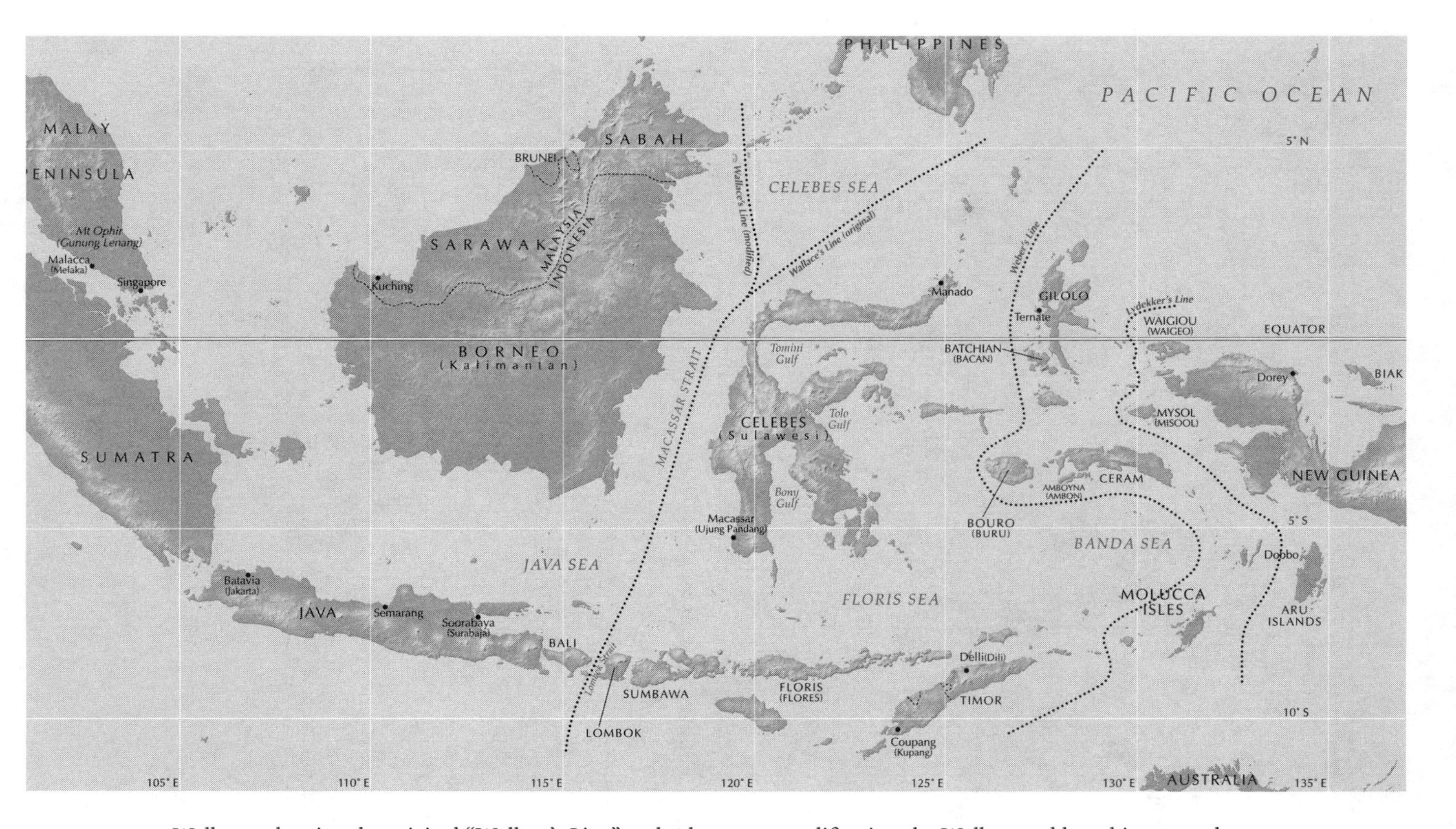

Wallacea, showing the original "Wallace's Line" and subsequent modifications by Wallace and later biogeographers. (Rendered by Dennis McClendon, Chicago CartoGraphics, 2002)

CHAPTER 11

The Olympian Heights and the Beginnings of the Fall

WALLACE'S INVESTIGATIONS of the spirit world did not interfere with his promotion of the principle of natural selection, which in the 1860s required an energetic defense against fierce detractors. His marriage to Annie brought him a good measure of personal satisfaction, and he resumed his biological work with renewed vigor. An article in the June 1866 issue of the *Reader* entitled "Materialism of the Present Day," by Paul Janet, particularly disturbed Wallace. Janet considered Darwin's weak point to be that Darwin did not see that "thought" and "direction" were essential to the action of natural selection.

"I have been so repeatedly struck by the utter inability of numbers of intelligent persons to see clearly, or at all, the self-acting and necessary effects of Natural Selection," Wallace wrote to Darwin in early July, "that I am led to conclude that the term itself, and your mode of illustrating it, however clear and beautiful to many of us, are yet not the best adapted to impress it on the general naturalist public." He thought the inadvertent image of an intelligent chooser created confusion. He preferred Herbert Spencer's phrase "survival of the fittest," which was not a metaphor, like the term "natural selection," but a clear definition of the mechanism for the origin of species. Wallace also asked Darwin to emphasize the importance of variation in the next edition of *The Origin of Species*. In the earlier editions, Darwin had given the impression that favorable variations were rare accidents. "I would ask [our opponents] to give any reason for supposing that any organ, etc. is ever *absolutely identical* at any *one time in all the individuals* of a species, and if not," Wallace argued, "then it is always varying, and there are always materials which, from the simple fact that the 'fittest survive,' will tend to the modifications of the race into harmony with changed conditions."[1] Although concurring with the spirit of Wallace's criticism, Darwin was not keen to make the recommended changes. The term "natural selection" had become so ingrained, he said, that it would be hard to give up. Moreover, Janet was nothing more than a metaphysician, with such an "acute" perception of the world that he often misunderstood the plain talk of "common folk."[2]

Only seven years after the publication of the *Origin*, evolutionists now occupied positions of power in the English scientific community. At the 1866 meeting of the British Association for the Advancement of Science, the Darwin–Wallace cause triumphed; the tide seemed to be turning. Thomas Huxley presided over Section D, now renamed the Biology Section. Wallace ("the traveler," as he was described by the *Quarterly Journal of Science*) headed the newly created Anthropology Department.[3] In his opening address to the group, he defined anthropology as "the science which contemplates man under all his varied aspects—as an animal and as a moral and intellectual being, in his relations to lower organisms, to his fellow man, and to the universe." Nothing pertaining to humans was unworthy of the anthropologist's attention. Anthropology's sole object was to discover man's true nature and, from the knowledge gained, to govern and improve uncivilized tribes as well as guide national and individual progress. Its agenda, therefore, straddled the scientific, social, political, and moral spheres.[4]

Once again, James Hunt was a thorn in the side of the evolutionists. In "On the Application of the Principle of Natural Selection to Anthropology: In Reply to Views Advocated by Some of Mr. Darwin's Disciples," the paper he read before the Anthropology Department, he attacked those disciples—Wallace and Huxley—for espousing views unwarranted "by logic or facts." Man was not a single species, he said; the Negro was a distinct species from the European. Between the Negro and the European, there was a far greater difference in intelligence than between the gorilla and the chimpanzee. No authorities besides Wallace and Huxley, he said, accepted the monogenetic view of humankind. Wallace's assertion that humans must once have been a homogeneous race was "startling." Why must humankind have been one race? No fact in history or archaeology supported this statement. "All we know is, that all science teaches man to be now much as he was when we first catch a glimpse of him at the dawn of history; and paleontology tells us that there were fossil apes," Hunt concluded. "Between these two facts all is darkness. . . . Happily, such teaching as [that of Wallace and Huxley] does not at present exert any great influence in this country. I must leave it for the audience to decide which are the false apostles and suffer from the effects of a 'heated imagination.'" Hunt ended by appealing to Darwin himself to come forward and declare whether or not the application of his theory led to unity of origin and would lead to a coming unity in the future.[5]

Darwin did not come forward, since he was not attending the meeting. About the evolution of humans he would remain silent for another five years. But Wallace did not need Darwin's bolstering to defend himself. Darwin's theory could be considered as nearly proved as any theory could be, he said. Even if Darwin did not apply his theory to *Homo sapiens*, there was no reason why it could not be so applied. It was logical to believe that the races of man, which resembled one another more than they differed, diverged from a common stock. It was also logical to believe that they were now gradually approaching one another again as the weaker were being exterminated.

Hunt replied that Wallace had misunderstood the thrust of his paper. He wanted people to keep their minds open on the question of the unity or plurality of the human species. In the present state of inadequate knowledge, he said, Darwin's theory did not have to be accepted. "We are not in a position to offer any opinion," Hunt insisted. "At present, we know absolutely nothing."[6] The matter remained unresolved, and neither Hunt nor Wallace budged from his respective position.

In February 1867, Darwin sent the following note to Wallace:

> On Monday evening I called on Bates and put a difficulty before him, which he could not answer, and, as on some former similar occasion, his first suggestion was, "You had better ask Wallace." My difficulty is, why are caterpillars sometimes so beautifully and artistically coloured? Seeing that many are coloured to escape dangers, I can hardly attribute their bright colour in other cases to mere physical conditions. . . . If anyone objected to male butterflies having been made beautiful by sexual selection, and asked why they should not have been made beautiful as well as their caterpillars, what would you answer?[7]

The purpose of coloration and ornamentation in the animal world preoccupied Wallace and Darwin during the 1860s. It was a sticking point for evolutionists that demanded an explanation. To Wallace, the theory of natural selection was proving to be comprehensive and powerful, capable of interpreting phenomena that most naturalists had otherwise construed as unaccountable quirks of nature, attributable only to the mysterious ways of God. Darwin looked for some additional mechanism. Unwieldy antlers, birdsong, the peacock's tail, and the size differences between males and females were among examples he cited as having evolved because of the allied phenomena of female choice and male competition for mates. On the principle of natural selection, brilliant colors also seemed equally inexplicable to him. He had therefore coined the term "sexual selection" to explain the existence of traits that violated the law of utility, the sine qua non of natural selection, because they did not improve the animal's (or the species's) chances of survival. Initially Wallace accepted Darwin's idea without reservation, but he grew skeptical about sexual selection as the decade progressed. He and Darwin would repeatedly lock horns over the relative importance of this form of selection, and their differing viewpoints signaled the beginning of a permanent schism.

But neither sexual selection nor natural selection seemed to account for the brilliant coloration of some caterpillars, which were asexual. An earlier theory—mechanistic and not creationist—had ascribed the origin of coloration in animals to external influences, such as climate, food, and soil type. However, anyone who bred domesticated species, like rabbits or pigeons, knew that the offspring varied widely in color; in such cases, the phenomenon clearly had nothing to do with any of those factors. In "Mimicry, and Other Protective Resemblances Among Animals," published in the *Westminster Review* in 1867, Wallace demonstrated with an overwhelming number of facts that for animals dependent on concealment for survival—either to escape detection by predators or to bring down prey themselves—protective coloration was the best strategy. Insects that resembled the bark of trees, lions crouching on the sand or among the rocks, tigers lurking in the grass, leopards waiting in the foliage, white rabbits hiding motionless in the arctic snows—all were examples of "the simple and effectual means by which animals are brought into harmony with the rest of nature." In the insect world, such adaptations to the environment were the most fully and strikingly developed. The most outstanding case of protective resemblance he had observed in butterflies was that of the swift-flying Malayan species *Kallima paralekta*, which had eluded him in the dry forests of Sumatra. He never observed this butterfly settle on a flower or a green leaf—and he always lost sight of it when it flew into foliage. Its defense was its resemblance to a dead leaf: the irregular outline of its folded and plain-colored wings resembled that of a shriveled leaf, while the tail of its hind wing was positioned to look like a stalk. Once Wallace realized this, he succeeded in capturing a number of specimens. Incessant variation, rapid multiplication, and survival of the best-adapted individuals within a species were the three pillars on which the principle of natural selection rested. Any slight advantage resulting from a change of color would increase the likelihood of success in times of stress. The phenomenon of protective coloration demonstrated the importance of studying the minutest details of anatomical structures and revealed the complex and delicate balance of nature. The meaning of the "strange travesty" of protective resemblance was not conscious "imposture and masquerade" but subservience to one great law: the preservation of the species.[8]

How, then, did Wallace answer Darwin's question? As he and Henry Bates had observed, many of the most spectacular and slowest-flying butterflies often had a peculiar odor and taste, which protected them from predation. So did white moths, according to the entomologist John Jenner Weir, who told Wallace one evening at a meeting of the Entomological Society that small birds in his aviary refused to capture and eat one of the most common white moths because it was unpalatable. "Now, as the *white* moth is as conspicuous in the *dusk* as a *coloured* caterpillar in the *daylight*," Wallace explained to Darwin, it stood to reason that the brilliant markings and vivid colors of certain caterpillars were a similar protective mechanism; that is, they served as a warning to potential predators.[9] A few days after receiving Wallace's letter, Darwin replied, "Bates was quite right; you are the man to apply to in a difficulty. I never heard anything more ingenious than your suggestion, and I hope you may be able to prove it true. That is a splendid fact about the white moths; it warms one's very blood to see a theory thus almost proved to be true."[10]

At the March 4 meeting of the Entomological Society, Wallace brought the subject of protective coloration to the attention of those present, requesting them to submit any evidence to support his "caterpillar-colouration" theory. He also wrote a letter to the popular outdoorsmen's newspaper the *Field, the Country Gentleman's Newspaper*, challenging readers to send notes or observations to him by the end of the summer. Only one person replied. But in 1869 Weir and another entomologist reported two years' worth of experiments and observations involving several of England's "gaily-coloured caterpillars" that upheld the theory. Weir found that the nocturnal-feeding "obscure-coloured" (green or brown) caterpillars were eaten by all species of insectivorous birds, whereas the brightly colored, spotted, or banded species always were rejected, as were the hairy and spiny caterpillars. Moreover, the last varieties are not protected by their hairs and spines, Weir said, but by their noxious taste. Weir's discoveries lent "immense support" to Wallace's view of the great importance of protection in determining coloration.[11] In his typical self-deprecating manner, Wallace later wrote, "It is, of course, only a wider application of the same fundamental fact by which Bates had already explained the purpose of 'mimicry' among insects, and it is a matter of surprise to me that neither Bates himself nor Darwin had seen the probability of the occurrence of inedibility in the larvae as well as in the perfect insects."[12]

In the meantime, important events were taking place in the Wallace household. On June 22, 1867, Annie gave birth to a baby boy, Herbert Spencer, named after Wallace's deceased younger brother and one of England's greatest living philosophers. Darwin—never a great admirer of Spencer—heartily congratulated him on the new member of his family but hoped that Wallace's son would "copy his father's style and not his namesake's."[13] After temporarily moving in with the Mittens in Hurstpierpoint, where Annie and the new infant were being looked after, Wallace started working in earnest on his account of his travels in the Malay Archipelago, driven to finish the manuscript more by economic necessity than by a desire to enhance his reputation. Hurstpierpoint, with its beautiful fields and clean air, was just the quiet and restful place he needed to write. To break up the monotony of his labors, he made several excursions into the countryside with William Mitten, a man ever in search of new ferns and mosses. By early February 1868, he had made considerable progress on the book. "I shall leave most of the Natural History generalisation, etc. for another work," he wrote to Darwin, "as if I incorporate all, I may wait for years."[14]

During Wallace's sabbatical from the London scientific scene, the two men carried on a lively debate on a variety of issues related to evolutionary theory. Their most intense discussions centered on the problem of sterility, hybridization, and species formation itself. To understand this interchange, it is important to realize that when Wallace and Darwin used the word "sterility" they were talking about "reproductive isolating mechanisms," a term that had not yet come into use. Reproductive isolating mechanisms are biological properties of individuals that prevent the interbreeding of populations that share a habitat.[15] "Disinclination to breed" was Darwin's term for explaining why members of two species did not mate in nature. Was such disinclination due to natural selection or to some other phenomenon? This was the question that Wallace and Darwin debated.

In *The Variation of Animals and Plants Under Domestication*, published in early 1868, Darwin stated that he had found no evidence to suggest that—

in animals at least—natural selection had anything to do with the phenomenon of sterility when two species were crossed. Although it was appealing to conclude that it would be advantageous for an incipient species to be "in some slight degree" sterile when crossed with the "parent" species or another variety of that species, there were great difficulties in trying to imagine the "many graduated steps" between "slightly lessened fertility" and absolute sterility. Moreover, he could not see what advantage there would be for two species to be mutually sterile—that is, either unable to produce hybrids at all or able to produce only barren hybrids, like mules. Thus the phenomenon of mutual sterility could not have occurred through the agency of natural selection. "We must infer," he concluded, "that [reproductive isolating mechanisms have] arisen incidentally during their [species'] slow formation in connection with other and unknown changes in their organisation."[16]

Wallace believed that "a slight degree of sterility" between two incipient species might be advantageous—not to the individuals that were sterile but to the species themselves—aiding rather than inhibiting natural selection. For example, if the hybrids of two incipient species, *A* and *B*, were less well adapted than *A* and *B* themselves to a "special sphere of existence," then *A* and *B* would eventually become separate species, outcompeting the hybrids, which would die out. If it could be proved that natural selection could produce such "sterility of hybrids," it might help to explain the phenomenon of species formation occurring without geographic isolation. Many critics had difficulty understanding how natural selection could cause a new species to arise unless some sort of physical barrier separated one part of a population from another. Darwin, however, stood firm in his belief that natural selection had nothing to do with sterility between species and insisted on looking for another mechanism. Over the next two months, a volley of letters passed back and forth between the two men, which did little to resolve the matter.[17] "Life is too short for so long a discussion," Darwin finally concluded on April 6. "We shall, I *greatly* fear, never agree." Wallace apologized for causing so much trouble with his ideas on sterility. "If you are not convinced, I have little doubt but that I am wrong; and I now think there is about an even chance that Natural Selection may or not be able to accumulate sterility," he said. "I will say no more [on the subject] but leave the problem as insoluble, only fearing that it will become a formidable weapon in the hands of the enemies of Natural Selection."[18]

The relative importance of protective coloration and sexual selection also came up again during these discussions about speciation. Darwin told

Wallace in February that he was hard at work on sexual selection, driven half mad by a number of side issues. "I am fearfully puzzled how far to extend your protective views with respect to the females in various classes," he wrote. "The more I work, the more important sexual selection apparently comes out." In mid-March, Darwin made a rare appearance in London to present a paper at the Linnean Society on the primrose and the cowslip, two species that frequently produced natural hybrids. His excursion to the city gave him an opportunity to collect facts on sexual selection, but the subject was growing more and more complex, he told Wallace, and in many ways more difficult and doubtful. "I have had grand success this morning tracing gradational steps by which the peacock tail has been developed: I quite feel as if I had seen a long line of its progenitors."[19]

Sexual selection was a theory that Wallace found difficult to accept. "I do not see how the constant *minute* variations, which are sufficient for Natural Selection to work with, could be *sexually* selected," he wrote to Darwin. "Sexual selection would seem to require a series of bold and abrupt variations. How can we imagine that an inch in the tail of a peacock, or a quarter of an inch in that of the bird of paradise, would be noticed and preferred by the female?" Darwin had greater confidence in the female bird's power of discrimination. An inch, a quarter of an inch, a fraction of an inch—the female would tend to be impressed by the male's "more gorgeous appearance," however that distinction was made. It was the gestalt, Darwin implied. "A girl sees a handsome man, and without observing whether his nose or whiskers are the tenth of an inch longer or shorter than in some other man, admires his appearance and says she will marry him." In a later letter, Darwin described the male as "the searcher," which, evolutionarily speaking, has "received and gained more eager passions" than the female. Thus the male would be unlikely to be attracted to a female for her beauty; his main concern would be to seize any female—or as many females as possible—with which to mate and produce offspring. The variations leading to beauty often must have occurred in the males alone, he said, and been transmitted to that sex alone, while it was somehow suppressed in females. This mechanism explained why males were generally more beautiful than females, which obviated the need to invoke the protective principle to explain the relative dullness of females. "Nothing would please me more than to find evidence of males selecting the more attractive females," he wrote. "I have for months been trying to persuade myself of this. . . . Perhaps I may get more evidence as I wade through twenty years' mass of notes."[20]

In "A Theory of Birds' Nests; Shewing the Relation of Certain Differences of Colour in Birds to Their Mode of Nidification," published in the March 1868 issue of the *Journal of Travel and Natural History*, Wallace agreed with Darwin that sexual selection probably determined the brilliant plumage and greater size of male pheasants and grouse, a result of the continual rivalry of males in terms of both strength and beauty.[21] But he could not believe that the theory explained the evolution of the spectacular coloration of some female birds, such as the toucan, parakeet, and macaw. Nor did the theory of sexual selection shed light on another problem: why some of the most colorful male birds mated with partners so dull and inconspicuous that the two were often taken to be separate species. This led him to propose a law that connected the colors of female birds with their nesting behavior. With very few exceptions, he believed, "when both sexes are of strikingly gay and conspicuous colours, the nest is . . . such as to conceal the nesting birds; while, whenever there is a striking contrast of colours, the male being gay and conspicuous, the female dull and obscure, the nest is open and the sitting bird exposed to view." He then listed two groups of birds to illustrate the truth of his hypothesis. The first group, in which both sexes were brightly colored, comprised twelve hundred species, or approximately one-seventh the number of all known birds, all of which built concealed or domed nests. The second group, in which the male was brightly colored but the female was dull or inconspicuous, comprised the majority of passerine (perching) birds, none of which, as far as he knew, built concealed nests. He excluded raptors (hawks and eagles) and wading birds from his analysis since, as he noted, female raptors and wading birds were rarely attacked. As for such anomalous birds as the Megapodiidae, the males and females were nearly identical in color, but the eggs were incubated in sand instead of by the sitting bird.

These facts, he concluded, firmly established the truth of his law. The exceptions were "too few and unimportant to weigh against such a mass of evidence." Females could be as brightly colored as males; only in those birds that built nests in the open had bright colors been "withheld or left undeveloped." This phenomenon could be explained, he wrote, by the action of both natural and sexual selection. The most brightly colored female birds could have acquired their coloration in only two possible ways: either the color was useful or correlated with some useful trait (natural selection) or it appealed to males (sexual selection). From the perspective of his and Darwin's evolutionary theory, any female bird that incubated her eggs in the open would have evolved plumage that rendered her relatively inconspicuous, since predators would have destroyed the more brightly colored females during incubation. Conversely, brightly colored females that incubated their eggs in concealed nests probably evolved their bright colors gradually, since predation would not have eliminated bright colors. He concluded his essay by stating that in the majority of cases, the greater or lesser need of protection determined the colors of female birds.

Darwin told Wallace that he was "deeply interested" in the birds' nest article. "I am delighted to see that we really differ very little. . . . You do not lay much or any stress on new characters spontaneously appearing in one sex (generally the male) and being transmitted exclusively, or more commonly only in excess, to that sex. I, on the other hand, formerly paid far too little attention to protection. I had only a glimpse of the truth." But he was not prepared to go as far as Wallace. He could not avoid thinking of the exceptions to the nesting rule, especially those "partial exceptions," when there is little difference between the sexes and yet the nests are built to conceal the sitting bird and the eggs. "I wish with all my heart I could go the whole length with you," he said. "You seem to think that such birds probably select the most beautiful females: I must feel some doubt . . . for I can find no evidence of it."[22]

In the summer of 1868, Wallace called a truce in the battle with Darwin and traveled to Switzerland with Annie, leaving Wallace junior, now a very active year-old toddler, with the Mitten grandparents and aunts.[23] He brought along papers and boards to dry and press the rare alpine plants that his father-in-law wanted, and boxes and pins for any new butterflies he might discover. But it was not only the opportunity to collect specimens or hike in the invigorating mountain air that drew him to the Alps. The geology of the region fascinated him. The most interesting geologic topic of the day was the effect and extent of glaciation, and Wallace's investigations of this phenomenon would influence his views of the forces that determined the past and present geographic distribution of animals.

Before 1840 no one suspected that glaciers had covered a significant portion of the temperate regions of the Northern Hemisphere in recent geologic times. A handful of Swiss geologists had claimed that so-called erratic blocks—massive chunks of granite or metamorphic rock—found throughout northern Europe on top of unrelated geologic formations and traced back to Scandinavia demonstrated the power of glaciation. Switzerland had the greatest concentration of these blocks in all of Europe, and a Swiss geologist named Jean de Charpentier was among the first to work out the laws of glacial motion, though it took him fourteen years to agree with E. T. Venetz, a Swiss mining engineer, who first postulated in 1821 that glaciers and not the Noachian Flood had deposited boulders in the Alpine valleys. Glacial action, Charpentier concluded after conducting surveys of the Rhône Valley, explained other ubiquitous features: moraines, the ridges composed of unstratified gravel and other debris left behind as the glaciers retreated; the enormous rounded and smoothed boulders known as *roches moutonnées*, with their vague resemblance to resting sheep; and the neatly gouged striae in bedrock. In 1836, a year after publishing a paper on the subject, Charpentier converted Louis Agassiz, a young Swiss naturalist who later immigrated to the United States to teach natural history at Harvard University. In July 1837, in an address before the Swiss Society of Natural History, Agassiz announced his theory of the Pleistocene Ice Age, during which, he claimed, ice had covered the world from the North Pole to the borders of the Mediterranean and Caspian Seas. The glaciers of the present were the remnants of this period of extensive glaciation, left behind as the massive floes of ice receded. Agassiz had arrived at this startling hypothesis after making surveys and seeing ubiquitous evidence of glaciation in northern Europe and Great Britain. In 1840 he published *Études sur les glaciers*, which presented his personal investigations, undertaken over three years, demonstrating that an ice age had existed in the recent past. On November 4 of that year he presented a paper, "On Glaciers and the Evidence of Their Having Once Existed in Scotland, Ireland and England," at the Geological Society of London. A few British geologists were won over, but most of them would continue to cling tenaciously to the biblical—or more catastrophic—interpretation.[24]

Wallace was an early convert to Agassiz's glacial theory, believing that it explained the facts better than any other hypothesis. Leaving behind Annie, who complained of altitude sickness but who was in fact pregnant with their second child, he proceeded up the wildflower-strewn slopes of Dent du Midi, but had to turn back before reaching the summit because of an imminent snowstorm. The next day, he and a guide walked down to the village of Aosta, to make an excursion on muleback to the summit of Becca de Nona, which took most of the day. He walked the last thousand feet alone, while his guide and mule waited below. Along the way, he stopped to examine the unique flora, including two species of the "exquisite Androsaces, the true gems of the primrose tribe,"[25] which at these altitudes endured intense sun, raging gales, and wide-ranging temperatures to win the battle of survival over less hardy species. From his top-of-the-world outpost, Wallace had a bird's-eye view of the glacially sculpted landscape below, etched out by forces that both Agassiz and the great Sir Charles Lyell had described so well. Lyell, whom Wallace engaged in extensive correspondence on the glacial issue, had adopted Agassiz's glacial theory ten years earlier but was unwilling to go as far as Wallace, who believed—correctly—that most of the lake basins of northern Europe and North America were of glacial origin.

Wallace ended his self-imposed exile that year by attending the meeting of the British Association in Norwich at the end of August. He was eager to hear the address of Joseph Hooker, now president of the association, who he hoped would "promulgate Darwinism."[26] In his speech, Hooker overcame his public reticence, apparently no longer afraid that his heretical views (which were not so heretical after all) would adversely affect the minds of his children or pain his many kind and good relations. Asking rhetorically what progress the Darwinian theory had made in the public's estimation ten years after the publication of *The Origin of Species,* he gave an answer that was more than the editors of the *Athenaeum* would ever concede. The book, he noted, had been translated into every European language and had passed through several editions; nearly every rising young naturalist had by now eagerly adopted its precepts. The scientists who publicly rejected the theory were few in number, most of them having been persuaded by Darwin's "true knight," Alfred Wallace, in the *Quarterly Journal of Science* and other publications. Above all, Hooker noted, Lyell, "that noble man," had bravely adopted the theory in his latest edition of *Principles of Geology.*[27]

Hooker was referring to Wallace's paper "Creation by Law," published in the October 1867 issue of the *Quarterly Journal of Science,* in which Wallace trounced the outmoded views of George Douglas Campbell, the eighth duke of Argyll, a vociferous opponent of the theory of natural selection. In his book *The Reign of Law,* the duke had espoused what Wallace called the "continual interference" hypothesis, which held that the universe, with all its laws intact, would be a sort of chaos—lacking variety, harmony, design, and beauty—were it not for the incessant interference and direct action of the mind and will of the Creator. Dismissing the idea of a self-regulating universe central to the Darwin–Wallace theory of evolution, Campbell denied that the myriad forms of life had an inherent power of adaptation to one another and to surrounding nature. "This adjustment," Wallace wrote, "necessarily leads to the greatest possible amount of variety and beauty and enjoyment, because it does depend on general laws, and not on a continual supervision and re-arrangement of details."[28]

The rest of the meeting failed to live up to his expectations. Wallace complained to Darwin that there were "no opponents left who know anything of natural history, so that there are none of the good discussions we used to have." In despair, he objected to an article by George Lewes advocating many distinct origins for some allied groups, "just as the anthropologists do who make the red man descend from the orang, the black man from the chimpanzee—or rather the Malay and orang one ancestor, the negro and chimpanzee another."[29]

After the meeting in Norwich, Darwin invited Wallace and his wife to Down House. He also invited at least two other naturalists, one of whom was John Jenner Weir. "If I could get several of you together," Darwin wrote, "it would be less dull for you, for of late I have found it impossible to talk with any human being for more than half an hour." Wallace accepted, though he had just hired a new servant and was uncertain if Annie could join them because "there is a baby at the mischievous age of a year and a quarter to be left in somebody's care." But in the end, Wallace managed to find a trustworthy caretaker for Bertie and arrived with Annie at Orpington Station at 5:44 P.M. in early September. The weekend was apparently a success, full of rousing discussions and heated arguments, but neither Wallace nor Darwin left any detailed recollections. "You will be pleased to hear that I am undergoing severe distress about protection and sexual selection," Darwin wrote to Wallace afterward. "This morning I oscillated with joy towards you; this evening I have swung back to the old position, out of which I fear I shall never get."[30]

Wallace had meanwhile determined to go on the offensive. "The more I think of your views as to the colours of females, the more difficulty I find in accepting them," he wrote shortly after his September visit. He laid out a complex argument, one that he hoped would demonstrate that females, which were more exposed than males to danger because of their reproductive habits or greater life expectancies, required "some extra protection."

"I am very much obliged for all your trouble in writing me your long letter, which I will keep by me and ponder over," Darwin replied. "To answer it would require at least 200 folio pages! . . . I grieve to differ from you, and it actually terrifies me, and makes me constantly distrust myself. I fear we shall never quite understand each other."

"Pray don't distress yourself on the subject," Wallace responded. "It will all come right in the end, and after all it is only an episode in your great work."[31]

This was an intellectual game for Wallace, who never lost sleep over any skirmish, whereas Darwin spent many a sleepless night agonizing over mere differences of opinion. To his friend Alfred Newton, Wallace confided, "To make sure of a discussion you ought to have *made a theory*, good or bad, and pretended to be *very sure about it*, —then you would have found lots of opponents and hence a *discussion*."[32] It was an impish tactic that might well backfire, and not everyone admired it as a way of carrying on an argument.

In November, Wallace received one of British science's highest honors, the Royal Medal, which the Royal Society awarded annually to two outstanding scientists, who usually were British. In the mid-nineteenth century, the Royal Society—founded in 1660 to promote "experimental philosophy" and improve "natural knowledge"—was Europe's most prestigious scientific organization, and its two Royal Medals were among its most coveted prizes.[33] Until 1825 the only medal a British scientist could hope to win was the Copley, "the highest public mark of approbation of the Royal Society," conferred on one scientist each year for a lifetime of contributions to science. But in that year, after significant pressure from influential scientists, the English Crown began formally rewarding "men of science" for original work in the physical and biological sciences. For years, members of the Royal Society had been complaining that the government had done too little to support scientific research. The French and Germans, who showered all sorts of honors, awards, and money on their most eminent scientific men, threatened to shove British science to the background. Sir Robert Peel, then home secretary, agreed. He spearheaded the creation of two gold medals, to be worth 20 guineas each and given annually to two deserving younger scientists, who would thus be encouraged to pursue their pioneering research. King George IV approved Peel's recommendation and increased the value of each medal to 50 guineas. Various schemes were devised to honor scientists in a number of different fields; in 1850 the scheme was simplified, with one medal designated for outstanding achievement in physics and the other for outstanding achievement in biology.[34]

At the June 20, 1867, meeting of the nominating committee of the Royal Society, the geologist Andrew Ramsay proposed Wallace as a recipient of the gold medal in biology. Huxley seconded the motion. If the Royal Society agreed that Wallace deserved the honor, it would be yet another coup for evolutionists; however, another man was chosen instead. Over the next several months, a flurry of correspondence passed among Darwin, Hooker, Huxley, Lyell, and others on the subject. Wallace seems to have known that his friends were pressing to have him awarded the medal. Lyell asked him to draw up a statement about his accomplishments, "[but] do it as if it was in the third person," because Lyell himself was "slow at composing such certificates."[35] On May 14, 1868, Wallace's name came up once again in the nominating committee.[36] Only Hooker had reservations about whether Wallace merited such an important award. "I hear that Wallace and [Fritz] Müller . . . are the most likely candidates for the Gold Medal for Biology and am puzzled a little to decide, but have so very high an opinion of Wallace that I incline to him," Hooker wrote to Darwin. "His work is so very good though less than one could have wished." Darwin, who had no such reservations, was pleased with Wallace's nomination. "Everything I have ever read of his gives me the highest idea of his extraordinary talents," Darwin replied to Hooker.[37]

Lyell had no reservations either. After having reread "On the Law Which Has Regulated the Introduction of New Species," which outlined Wallace's Sarawak Law, he had decided to quote certain passages in the next edition of *Principles,* "not in reference to your priority of publication but simply because there are some points laid down more clearly than I can find in the work of Darwin itself, in regard to the bearing of the geological and zoological evidence of species," he had written to Wallace a year earlier. Lyell had also asked Wallace to send him a list of the papers that he had published up to that point. On receiving it, Lyell said, "I was very glad to get your grand list of papers which ought to satisfy even those who vote species-describing to be the real work of a naturalist & all the profounder views of philosophy to be mere writing for the amusement of the public."[38] In the eleventh edition of *Principles,* published in 1872, Lyell would pay homage to Wallace for his contributions to evolutionary theory, allocating more space to him than Darwin had in the few cursory sentences in the introduction to the *Origin.*[39]

A month after the May meeting, Huxley officially nominated Wallace. William Benjamin Carpenter seconded the motion, and this time the committee gave their unanimous support. At the anniversary meeting of the Royal Society on November 30, 1868, which was held at the society's headquarters at Burlington House, near Piccadilly Circus, Wallace was formally presented with the Royal Medal for his "labours in practical and theoretical zoology." The members of the nominating committee considered his discovery of the dividing line between two great zoological divisions of the earth his most significant achievement. Huxley himself had designated the division "Wallace's Line," an appellation that still is used even though the line's position has been shifted many times as biogeographers have gathered more data from the Malay Archipelago.[40] "He has given an exceedingly ingenious and probable solution of the difficulties of the problem," the committee members stated, "while his method of discussing it may serve as a model to future workers in the same field." They did not criticize the principle of natural selection, as they blatantly had done by awarding Darwin the Copley Medal in 1865 for everything but his work in evolutionary biology. After only three years, the principle had won enough converts to permit an open reference to Wallace's radical work in this area of biological science. "Apart from its intrinsic merits," the committee members noted, "his 1858 paper will always possess an especial interest in the history of science, as having been the immediate cause of the publication of the 'Origin of Species.'" Nor did the committee forget the extraordinary circumstances under which Wallace had made the majority of his contributions, reminding everyone that "some of his most valuable papers are dated from places which some might consider so little favourable to study as Ternate and Sarawak." The president of the society, Sir Edward Sabine, a colonel in Her Majesty's Navy famed for his magnetic surveys of the globe, congratulated him on his selection: "Mr. Wallace, I have the pleasure of presenting to you this Medal in recognition of the great merit of your researches both in practical and theoretical Zoology, carried out in countries where such pursuits are necessarily attended with more than usual difficulties and dangers."[41]

About this honor, which some viewed as tantamount to a scientific knighthood, Wallace wrote very little. Forty years later, in his autobiography, he acknowledged Huxley's influence in granting the award.[42] Perhaps other events muted Wallace's reaction. Only two weeks earlier, on November 15, his mother had died of "senile decay" at the age of eighty. At the time of her death, she was residing with an artist and his family in the tiny village of Ifield, not far from Hurstpierpoint, doubtless to be closer to her grandson, Bertie. She was buried in the Ifield church cemetery rather than at the Wallace family gravesite in Hertford, probably because embalming was not yet a common procedure in England and transport to Hertford, some fifty miles north, would have meant undue delay.[43]

But Wallace was troubled for other reasons, his mother's death merely adding to his woes. In the late 1860s, he suffered several blows to his finances—and the 50-guinea Royal Medal did little to make up for the losses. Reflecting on this time of troubles in his autobiography, he admitted his incompetence in money matters and gave a frank account of the pitfalls into which he had been lured and that eventually had led to near ruin—a recapitulation of his father's decline and fall. Gradually he had divested himself of the proceeds from his Malayan travels, so carefully invested by the cautious Samuel Stevens in conservative securities; three acquaintances, acting in good faith, had persuaded him to make promising but highly speculative investments.

Such speculation was not unreasonable at the time. In the 1860s, England was enjoying a period of rapid economic growth and prosperity. Wallace bet some of his money on various British and foreign railways, "whose fluctuations in value I was quite unable to comprehend, and I began to find, when too late, that almost all my changes of investment brought me loss instead of profit." Another portion he handed over to men who bought up slate quarries in Wales. Without any knowledge of business or of the amount of capital investment required, he agreed to acquire shares and become a director of a corporation that existed on paper only. Machinery was purchased, the quarries were dug, and investors were sought, but after four or five years the operation foundered and he was suddenly a £1,000 poorer. He funneled yet another sum of money into English lead mines at the urging of his friend Frederick Geach, an English mining engineer he had met in Timor, who had examined the veins and vouched for their value. But competition came from an unexpected quarter—the territory of Nevada in the American West, where vast quantities of lead were being produced as a by-product of the extraction of silver and sold to Europe at bargain prices. The English export market would evaporate, along with his investment. By 1868 much of his money, "earned at the risk of health and life," was tied up in these reckless ventures, and he was strapped for cash.[44]

The lack of liquid investments, coupled with the expenses associated with raising one child and the imminent arrival of a second—his daughter, Violet, would be born on January 25, 1869—made the publication of his book of travels especially urgent. By the end of November 1868, he had nearly finished his manuscript on the Malay Archipelago and was actively negotiating with Alexander Macmillan of the Macmillan publishing house for the best deal possible. Macmillan's enthusiasm when he acquired the rights to the manuscript exemplifies the esteem in which Wallace was held at the time. "We have now in press a book of travels which we have reason to believe will be one of the most popular (and that not only for the moment, but for years to come) that has appeared in a long time," Macmillan wrote to his counterpart at Harper Brothers in New York, with whom he was negotiating the American rights for the book.

> It is the travels of Alfred Wallace in the Malay Archipelago. For charm of style and variety of interest, it seems to many of the best judges, who have seen it privately, that it cannot be surpassed. And it has beyond this a fine, unobtrusive, yet most effective current of far-stretching philosophy that, without making the book in the least degree dull for the general reader, yet gives it great value to the higher class. Mr. Wallace resided for about eight years in these regions, and saw them minutely and carefully in a way that no traveler ever saw them before. He has given such details of Natural History, manners and customs of the people, as could only be got after long experience, and has selected what to tell with such perfect judgment that the general result is a perfect work of art. . . . I enclose you an extract from the speech of the President of the British Association at the last meeting at Norwich, which will show you the estimation he is held in by the scientific world. . . . The work will be very fully illustrated, and by the best artists.[45]

Wallace consummated the deal with Macmillan on March 3, 1869. Macmillan advanced £100 and promised him a handsome proportion of the proceeds: 7 shillings, 6 pence per copy after the sale of the first thousand copies and 2 shillings per copy for the "cheap" edition. The cost of the book to booksellers was 24 shillings. Wallace was therefore contracted to receive one-third of Macmillan's profits for his book—a generous share at the time.[46]

Wallace dedicated his book, which he entitled *The Malay Archipelago, the Land of the Orang-utan and the Bird of Paradise; A Narrative of Travel, with Studies of Man and Nature*, to Darwin, although he warned Darwin that "it will be far too small and unpretending a work to be worthy of that honour."[47] On March 5, 1869, Darwin received a complimentary copy from the publisher. "What a capital book Wallace has published," Darwin wrote to Lyell.[48] For his part, Lyell told Wallace, "Nothing equal to it has come out since Darwin's 'Voyage of the Beagle.' I am not yet through the first volume, but my wife is deep in the second and much taken with it. It is so rare to be able to depend on the scientific knowledge and accuracy of those who have so much of the wonderful to relate."[49]

Wallace received congratulatory notices from a number of other eminent men. One came from the theologian and author Charles Kingsley. "Let me . . . compliment you on the modesty and generosity which you have shown, in dedicating your book to Darwin, and speaking of him and his work as you have done," Kingsley said. "Would that a like unselfish rivalry were more common—I do not say amongst scientific men, for they have it in great abundance, but—in the rest of the community." Wallace replied, "I cannot but feel flattered by your praise of my little book, which you . . . estimate far too highly. I compare myself to a guerrilla chief, very well for a skirmish or a flank movement, and even able to sketch out the plan of a campaign, but reckless of communications and careless about Commissariat; —while Darwin is the great general, who can maneuvre the largest army, and . . . lead on his formation to victory."[50]

With few exceptions, *The Malay Archipelago* was praised in every professional journal and in the popular press. The first printing of one thousand copies sold out, and the book remained in print into the first two decades of the twentieth century; clearly it was his most popular book. Joseph Conrad called *The Malay Archipelago* his "favorite bedside companion." Darwin and other scientist-explorers captured Conrad's imagination, but he was attracted especially to Wallace, whose generosity of spirit, immense courage in the face of suffering, and intense devotion to the elucidation of truth in the name of science radiated throughout the work. *The Malay Archipelago* inspired Conrad's *Lord Jim* and contributed background information for *Almayer's Folly*, *The Rescue*, and *An Outcast of the Islands*. Conrad himself had traveled through the East Indies between 1881 and 1889 but did not learn Malay and was unable to firmly grasp the Malay or Papuan character.[51] If Conrad ever corresponded with Wallace, no letter has survived and Wallace never mentioned ever having read Conrad. The darkness of Conrad's themes, which anticipated Freudian theory, would doubtless not have appealed to Wallace's more simplistic, phrenological views of humanity.

Throughout Wallace's life, perfect strangers wrote to tell him that they had read *The Malay Archipelago* over and over again and always took it with them on a journey. "This is the kind of thing I cannot understand," Wallace mused in his autobiography. "It is true, if I open it myself I can read a chapter with pleasure; but, then, to me it recalls incidents and feelings almost forgotten, and renews the delights of my wanderings in the wilderness and of my intense interest in the wonderful and beautiful forms of plant, bird, and insect life I was continually meeting with."[52] The book's publication and its critical reception fulfilled Wallace's youthful ambition to write a memoir that could stand with the classic travel narratives of Humboldt and Darwin. It was a remarkable milestone in an illustrious career in science. But it closed a period of unquestioned brilliance. Not until the final years of his life would he again enjoy accolades commensurate with his "extraordinary talents."

Although he had almost made his spiritualist beliefs public in 1868 in his (unpublished) letter to the editor of the *Pall Mall Gazette*, Wallace continued to hold back. He had been careful to keep spiritualism itself separate from the text of *The Malay Archipelago*, but his concluding note in that book reveals the moral tenets that informed the spiritualist movement and attracted him to it. In the note, he questioned the European notion of progress, which to him was more like retrogression. It was true that civilized people had surpassed savages in intellectual and material achievements, he said, but they had not advanced equally in the moral sphere:

> I have lived with communities of savages in South America and in the East, who have no laws or law courts but the public opinion of the village freely expressed. Each man scrupulously respects the rights of his fellow. . . . In such a community, all are nearly equal. There are none of those wide distinctions, of education and ignorance, wealth and poverty, master and servant, which are the product of our civilization. . . . But it is not too much to say, that the mass of our populations have not at all advanced beyond the savage code of morals, and have in many cases sunk below it.

Without according its less fortunate citizens a greater share of influence in legislation, commerce, and social organization, European civilization would never attain any real superiority even over the "better class" of savage. In Wallace's opinion, the European system of government remained in a state of barbarism.[53]

In the spring of 1869, shortly after the publication of *The Malay Archipelago*, Wallace put another tentative foot forward in the form of a review of Lyell's latest edition of *Principles*, saving his most controversial remarks for the end of the article. "I shall be intensely curious to read [it]," Darwin said with trepidation in late March, apparently having been warned of Wallace's heretical views. "I hope you have not murdered too completely your own and my child."[54] The "murder" took place the following month, in the April issue of the *Quarterly Review*. In his review, "Sir Charles Lyell on Geological Climates and the Origin of Species," Wallace said that the simple law of natural selection explained the development of the whole animal kingdom, from the lowest zoophytes up to the "animal" man. But natural selection could not explain the origin of consciousness. The laws of chemistry, biology, and physics determined how the body was constructed, grew, and reproduced, but they failed to explain how an assemblage of insentient atoms could give rise to consciousness. "We may go further, and maintain that there are certain purely physical characteristics"—he cited the brain, the hand, the organs of speech—"which are not explicable on the theory of variation and survival of the fittest." He raised several questions about the limits of natural selection. How could one explain the fact that the "lowest savages" possessed a brain identical in size and complexity to that of the "highest types" while their mental requirements were barely above the animals? Why was the hand equally perfect in all human beings no matter what their state of civilization? How did humans acquire their erect posture, their "delicate yet expressive" features, their beauty and symmetry of form? Of what use was the naked skin, which necessitated clothing?

"While admitting to the full extent the agency of the same great laws of organic development in the origin of the human race as in the origin of all organized beings," Wallace continued, "there yet seems to be evidence of a Power which has guided the action of those laws in definite directions and for special ends. And so far from this view being out of harmony with the teachings of science, it has a striking analogy with what is now taking place in the world, and is thus strictly uniformitarian in character." Just as humans guided and modified nature to suit their needs, why could there not be a "distinct intelligence" that guided the action of the laws of variation, multiplication, and survival for "nobler ends"?

Darwin congratulated Wallace on a review that he knew would gratify Lyell. Wallace's rendition of the theory of natural selection was especially masterful, and Darwin chided him once again for not alluding to his 1858 paper, "On the Tendency of Varieties to Depart Indefinitely from the Original Type," and taking proper credit for his own contributions. The article, he said, was an "immense triumph" for their cause, but had he not been warned, he would have thought that someone else had added the comments on humankind. "As you expected, I differ grievously from you, and I am very sorry for it," Darwin lamented. "I see no necessity for calling in an additional and proximate cause in regard to Man."[55] In the margins of the review, he had triply underlined "No, no!"[56]

Wallace had anticipated Darwin's negative reaction. He confessed that his views had been modified by consideration of a series of physical and mental phenomena that he had had every opportunity to test and that had demonstrated the existence of forces and influences not yet recognized by science. "This will, I know, seem to you like some mental hallucination," Wallace wrote. "I am in hopes that you will suspend judgment for a time till [I] exhibit some corroborative symptoms of insanity."[57]

Darwin feigned politeness and understanding, but he was profoundly distressed. He told Lyell about his disappointment with Wallace's new theory of human evolution, which he found "incredibly strange." Lyell replied that he did not oppose Wallace's idea about a Supreme Intelligence directing variation. He himself had alluded to such an idea in *The Geological Evidences of the Antiquity of Man*. "As I feel that progressive development or evolution cannot be entirely explained by natural selection, I rather hail Wallace's suggestion that there may be a Supreme Will and Power which may not abdicate its functions of interference, but may guide the forces and laws of Nature," Lyell said. But he had already written to Wallace to state that his arguments about the hand, the voice, the beauty and symmetry, the naked skin, and other attributes of human beings, implying a preparation for some other end, might easily be controverted.[58]

In the meantime, while completing his book on the Malay Archipelago and in anticipation of increasing expenses, Wallace had applied for the directorship of a proposed government-sponsored museum for the arts and natural history in the East End of London, to be called the Bethnal Green Museum. Nothing was yet settled, Lyell told Wallace, but Parliament had set aside money for such a museum. Since ethnology would form a "provisional part" of the new museum, he suggested that Wallace present his ethnological papers as part of his claims for the position. "You may send privately to me anything you take in the shape of a puff of your merits & qualifications," Lyell wrote to Wallace on January 24, 1868. "I will take care & keep it for myself. It will not fail to be a great help & do not shrink from doing it out of modesty."[59] Huxley created a *curriculum vitae* from Wallace's publications, which he attached to a written appeal endorsed by members of the Royal Society testifying to Wallace's credentials. Supporting his application were Richard Owen, William Benjamin Carpenter, Alfred Newton, Joseph Hooker, and John Tyndall, among others.[60]

But Wallace's timing was unfortunate. The July 1868 elections had ushered a Liberal majority into Parliament; on December 2, Benjamin Disraeli, the Tory prime minister, resigned and was replaced by his great rival William Gladstone. With a shift of power from the Tory to the Liberal Party, issues more important than the directorship of a minor museum preoccupied Parliament and the government ministries. Fearing that the exigencies of politics would brush aside his claims, an increasingly anxious and desperate Wallace pressed his friends, which prompted Lyell to speak to an influential member of the House of Lords. "I told him about your being kept in suspense: he has promised to do what he can which, in the present unsettled state of the whole affair, is not much," Lyell reported with some irritation. "I am sure he has no one else in view [for the position], and he has read the memorial in your favour."[61] Thus reassured, Wallace had no choice except to wait.

Without considering other options, he kept busy with a number of projects, including helping to launch a new scientific journal called *Nature*, whose first issue appeared on November 4, 1869. The journal was an outgrowth of the meetings of a group known as the X Club, formed by Huxley and Hooker in 1864. There were seven other members, and a tenth had been considered but never officially added—hence the name X—because no one could be found to fulfill the two requirements of membership: possessing "adequate" mental caliber and being on intimate terms with the existing members. (It is not known whether Wallace was ever considered for membership.)[62] The X Club, which began as an informal meeting of friends, came to wield considerable influence in the English scientific community and the history of science over the next thirty years. Five of its members were to receive the Royal Medal and three the Copley, six were to serve as president of the British Association, and three were elected president of the Royal Society. Their discussions of the pressing scientific issues of the day determined the agenda of British science. Guests were often welcomed at the boisterous two-hour evening forums, which were attended by many of the great names of English science and literature. They were held for a number of years at the St. George's Hotel, Albemarle Street, beginning at 6:00 P.M. on the first Thursday of each month (except the summer holidays in July, August, and September), before the more solemn 8:00 P.M. meeting of the Royal Society.

One of the X Club's priorities was to establish a first-rate scientific journal. In the view of its members and their followers, all the other journals had failed to achieve this goal. Moreover, none paid their contributors—articles were submitted gratis. The *Reader* had shown early promise as a mouthpiece for the "young guard" of science but folded by the end of the decade.[63] A young astronomer and solar physicist named Joseph Norman Lockyer, who had just discovered helium, was present at some of the X Club meetings and shared the members' desire to create a new journal. Having recently been hired by Alexander Macmillan as a science adviser, he appealed to the publishing house to oversee and subsidize the journal. Macmillan readily agreed and engaged Lockyer as editor. The decision to pay for articles was made in order to attract the contributions of leading scientists who were not independently wealthy and were therefore struggling to make ends meet.[64]

From its inception, *Nature* differed from any previous scientific journal. It was intended to be comprehensive, covering every aspect of science and providing something of interest for specialists and generalists alike. Intended for an international audience, the journal reported the proceedings of scientific meetings in other parts of the world as well as in London, though contributions from non-English scientists were accepted only if written in or translated into English. Macmillan and Lockyer maintained a high literary standard. Contributors did not shy away from controversy—in fact, the journal courted it. Battles, even outright wars, colored the pages of consecutive issues, playing themselves out over weeks and months. Wallace was often at the center of these skirmishes, as either victim or perpetrator.

Among his earliest contributions was an editorial on the subject of science reform in England, a priority of Gladstone's Liberal government. Mindful of the pending museum directorship, Wallace understood the importance of publicizing some of his views. At the time, many scientists were urging Parliament to apply tax revenues to science education both in the public (preparatory) schools—which were almost exclusively attended by the scions of the upper class—and in museums. Although Wallace believed as deeply in the importance of science education as did his colleagues, he objected to the use of public money to advance the interests of the privileged few. In the current inequitable state of English society, he said, such a direction of funds was not only "vicious in principle" but also a "positive wrong." The government had no moral mandate to use money raised by taxing all its citizens for a purpose that would not benefit the majority of Englishmen. "As it has no right to give class preferences in legislation," he reasoned, "so it has no right to give class preferences in the expenditure of public money." The expenditure of enormous sums of money on museums with vast collections of every obtainable zoological or botanical rarity, however obscure, and of value to only a limited class of people was equally unjust, in his opinion. The money earmarked for museums would be funneled into London institutions, while institutions and communities elsewhere in the kingdom would be denied their fair share. Without broad and fundamental reforms—that is, without the establishment of schools, museums, galleries, and gardens for the use and education of the public at large, independent of class, wealth, and place of residence—science did not merit a greater share of state patronage. "It will, perhaps, surprise some of your readers to find a naturalist advocating such doctrines as these," Wallace observed. "But though I love nature much I love justice more, and would not wish that any man should be compelled to contribute towards the support of an institution of no interest to the great mass of my countrymen, however interesting to myself."[65]

Readers were more than surprised. The response, in the form of a rebuttal published as the lead article in the very same issue of *Nature*, was hostile. "Is it really necessary to tell any educated man of the nineteenth century that science, art, literature, with one or two other matters, are simply civilisation," the anonymous author asked, "and that civilisation affects, not particular classes, but whole communities? Where, then, is the injustice of taxing all classes, in proportion to their means of commanding the results of science, for advantages which, if not so taxed, they would obviously gain at the cost of others?" The author had broader concerns in mind. At stake was England's hegemony, not only in the political and economic realm but also in the scientific realm. Centralization was critical to the nation's survival; administration at the local level had failed to maintain England in its position of primacy. Germany, with government intervention to a point "not as yet contemplated" by anyone in England, had produced a large number of first-rate scientists whose work was of the greatest practical and theoretical value. The ever-expanding German military–industrial complex posed a threat to the British Crown's colonies and commercial interests throughout the world. Men of science should guide the government in scientific affairs, he said, not vice versa.[66]

In his reply in the following week's issue, Wallace ignored the anxieties of those concerned with England's international primacy. In his view, the chief result of the cultivation of science was the elevation of the few to a higher mental and moral plane. The acquisition of "countless" physical, social, and intellectual benefits for the human race was a secondary result. No amount of money could guarantee the latter outcome; the history of science had demonstrated this. "[I]t is only those who in a noble spirit of self-sacrifice give up their time, their means, even their lives, in the eager and loving search after the hidden secrets of Nature, who are rewarded by those great discoveries from which spring a rich harvest of useful applications," Wallace said, clearly alluding to himself. Those individuals should receive their support from private sources—the patrons, the societies, and the classes who benefited most directly.[67]

In addition to his contributions to *Nature*, Wallace remained active in the various scientific societies. In February 1870, he was elected president of the Entomological Society of London, a position he held for two years. The presidency was a great honor, but consumed a considerable amount of his time and lacked remuneration. His finances uppermost in his mind at the time, he became entangled in an affair that would become "the most regrettable" incident in his life.[68] In January, an advertisement in the journal *Scientific Opinion* by a member of the Flat Earth Society had caught his eye. The sponsor, John Hampden, offered a prize of £500 to anyone who could prove that the earth was a sphere. Whoever took up the challenge would also be required to pledge an equal amount. The sponsor would "acknowledge that he has forfeited his deposit if his opponent can exhibit, to the satisfaction of any intelligent referee, a convex railway, river, canal, or lake," according to the advertisement. The offer was perhaps too good to be true, but because of his knowledge of the techniques of land surveying Wallace knew that he could easily win the bet, and the infusion of such a substantial amount of cash was something he could not afford to pass up. He consulted Lyell for advice. "It may stop these foolish people to have it plainly shown them," Lyell commented.[69]

Wallace accepted Hampden's bet after J. J. Walsh, editor of the *Field*, put up £500 on Wallace's behalf and agreed to act as Wallace's referee. Hampden's referee was William Carpenter, a journeyman printer who had written a book upholding the flat-earth theory, which neither Wallace nor Walsh had bothered to read. Wallace selected a lake in northern Wales for the experiment, but Hampden preferred another site—the Old Bedford Canal in the village of Norfolk, with a stretch of six straight miles between two bridges: the Welney and the Old Bedford. Since it was agreed that neither party had ever tried this experiment before at the site chosen, Wallace did not object. Walsh and Hampden each put up £500, which was deposited in a London bank for safekeeping, and Wallace signed an indemnity stating that he would repay Walsh after a victor was declared. The test was slated for early March 1870.

Wallace conducted the experiment flawlessly. He intended to demonstrate that when three objects were set in a straight line equidistant from the surface of the water and spaced three miles apart, the middle object, when viewed through a telescope, would appear to rise five feet or more above those at the two extremes, as calculated from the known dimensions of the earth. If, on the contrary, the earth was flat, the three objects would be seen to line up almost perfectly. Before undertaking the experiment, Wallace submitted a diagram to Hampden, Carpenter, and Walsh illustrating what they should see through the telescope in either situation. All three were satisfied.

The experiment was slated to take place on March 5. As it was foggy, and Walsh could not remain away from London for more than a day, a Mr. Coulcher, a friend of Walsh's who was a surgeon and an amateur astronomer, agreed to act as referee in Walsh's place. Carpenter accompanied Wallace during the placement of his markers to verify the measurements and watch for cheating. The following day, Carpenter, Coulcher, and Wallace went to the Welney bridge, looked through the telescope, and saw what Wallace had predicted. Coulcher sketched the results, and Carpenter signed the drawing as correct. Totally unexpectedly, Carpenter declared that what was seen proved nothing because the telescope was neither leveled nor had a crosshair.

Despite the absurdity of the complaint, Wallace borrowed a spirit level and a telescope with a crosshair from a local surveyor and again focused his telescope on the distant bridge and the central marker. Coulcher looked first, and then Carpenter looked once again and exclaimed, "Beautiful! Beautiful!" Hampden, who had come along this time but, rather inexplicably, declined to look through the telescope himself, asked whether "it was right." Carpenter replied that it was perfect—that it showed all three points in a straight line, as level as possible, and he even jumped for joy. Although baffled by Carpenter's odd behavior, Coulcher and Wallace said nothing, confident in their results. The four men next went by cab to the Old Bedford bridge, where Wallace set up his spirit level and telescope at the proper height to show the same curvature. Hampden again declined to look through the telescope, stating that he trusted Carpenter, who reacted just the way an inveterate flat-earth enthusiast would be expected to react: he refuted Wallace's interpretation of the results and declared Hampden the winner. Disgusted, Coulcher refused to discuss the question further. The two sides had arrived at a stalemate.[70]

Days passed before Hampden agreed to allow Wallace to appoint another umpire. Wallace chose Walsh, but Hampden agreed only because he had "unbounded faith" in Carpenter and believed that no honest man could think otherwise than that the earth was flat. As soon as Walsh looked at the reports of Carpenter and Coulcher, he declared Wallace the winner. He realized, however, that Hampden and Carpenter had acted dishonorably, for Carpenter already had tried the experiment on the same stretch of water, a violation of their initial agreement. Walsh had obtained and read the "little treatise" that Carpenter had written, published by the head of the Flat Earth Society under the pseudonym Parallax, which outlined his experiment on the Old Bedford Canal. Carpenter had, of course, arrived at results diametrically opposite those of Wallace's findings, thus upholding the flat-earth theory. "The good faith and perfect fairness of Mr. Carpenter were not, therefore, quite of the nature we then believed them to be," Walsh complained in the *Field* on March 26, "and we have no hesitation in affirming that he was a most improper person to be selected to act as referee in such a matter. Yet Mr. Carpenter and his master, 'Parallax,' both profess to be ardent in the cause of science; and that it has recoiled upon their heads can cause no regret to anyone who values the truth." Walsh affirmed that Wallace had proved what he set out to prove, and therefore decided in favor of Wallace. Hampden had forfeited his £500. "We have no doubt," Walsh said, "that all honourable men will agree in our decision."[71]

On April 2, Carpenter sent a convoluted and nonsensical rebuttal to Walsh about the decision to award the money to Wallace. Hampden also wrote a lengthy, crude diatribe in which he called the decision "monstrous" and accused Walsh of a "gross perversion" of the facts. He was indignant that he had been charged with acting unfairly, and he would denounce his own brother if he had departed "one jot" from the path of the strictest honor and rectitude! He demanded that any engineer in the kingdom prove that Wallace had succeeded. "Are you, or is any man in the possession of his senses, insane enough to call these apparent inclines a 'curve,' or a 'curvature'?" Hampden asked in a letter to the *Field*. "You . . . dare to call [this] a curvature, and assert that I have lost my money. No, Sir; at your peril touch it, and I will serve you both with a writ on the following day for conspiring to obtain money on false and fraudulent pretences."[72]

If Walsh and Wallace had called off the wager and returned the money to Hampden, the matter doubtless would have been dropped, but Walsh stuck by his decision in Wallace's favor and paid him £500, as stipulated in the contract. Carpenter wrote a letter to the *Field* in protest, and it was published the following week. Parallax himself sent a letter to the *Field* in response to a correspondent who upheld Wallace's claim. "If . . . other gentlemen of the same school of philosophy will only take the trouble to be correct before they speak or write," the president of the Flat Earth Society fulminated, "they will find that they have much to yield, and a great deal to learn."[73]

Wallace sank deeper into the quagmire by responding at the prodding of Walsh. "The 'men of common sense' to whom Mr. Carpenter so confidently appeals are very slow in coming forward," Wallace wrote in the *Field* on April 16 in response to the remarks made by Parallax. "Mr. Hampden, in his letter to me [published in the *Field* on April 2], continually appeals to 'public opinion' as being against the fairness of your verdict. It has, however, now clearly spoken through your widely-circulated columns, and, unless he can prove that letters on the other side have been refused insertion, he would do well, as a man of honour and of sense, to bow to its decision."[74]

Walsh's vote in favor of Wallace proved a mistake from several angles—legal, financial, professional, and psychological. Hampden, a paranoid schizophrenic with an Oxford education, suffered from delusions of persecution. To Wallace and Walsh, he was simply a harmless lunatic and crank—but a shrewd one. Both men played into his hands, fueling rather than dispelling his delusions, and he immediately began a campaign to impugn their reputations. Over the next several months, he sent Walsh letters, postcards, leaflets, and pamphlets calling him a liar, thief, and swindler. Then he turned his attentions to the Wallace household. "Madam—if your infernal thief of a husband is brought home some day on a hurdle, with every bone in his head smashed to a pulp, you will know the reason," Hampden wrote to Annie Wallace. "Do you tell him from me he is a lying, infernal thief, and as sure as his name is Wallace he never dies in his bed. You must be a miserable wretch to be obliged to live with a convicted felon. Do not think or let him think I have done with him."[75]

Wallace went at once to the police and pressed charges. Hampden was imprisoned for a week and threatened with further prosecution if he did not desist for three months. As soon as the three months had passed, however, he became more abusive, distributing tracts and writing to many small newspapers throughout England as well as to the presidents and secretaries of every society that Wallace belonged to. In January 1871, Wallace brought suit against Hampden for libel, but Hampden refused to plead, promising to declare bankruptcy if convicted. The jury ruled in Wallace's favor and awarded him £600 in damages. As promised, Hampden transferred all his assets to his son-in-law, filed for bankruptcy, and paid Wallace nothing, sticking him with all the legal expenses. Friends and strangers sent Wallace letters asking whether he knew what was being said about him. "I was grieved to see . . . that madman about the flat earth has been threatening your life. What an odious trouble this must have been to you," Darwin wrote sympathetically after reading in the *Daily News* a letter to the editor from Hampden in July 1871.[76] Not interested in prosecuting or imprisoning a man he recognized as certifiably crazy, Wallace reconciled himself to a torrent of written abuse that would doubtless last for the rest of Hampden's life.[77]

But the Hampden case continued to spin out of control. Hampden never tired of harassing him and the scientific societies to which he belonged. Wallace's lawyer advised that the best way to put a halt to Hampden's ravings was to return his letters unopened and to inform him that all future letters would be burned.[78] Nevertheless, Hampden's campaign against the alleged naturalist-scofflaw was surprisingly effective. To protect his reputation, Wallace was forced to draw up a public response outlining the facts of the case, which he published in the form of a pamphlet and distributed to everyone he could think of.[79] Once again, he prosecuted Hampden for libel. Ruling for a second time in Wallace's favor, the court placed Hampden on probation for one year and ordered him to publish an apology in the local newspapers. But it was a hollow victory, since Wallace shelled out more money for legal expenses.

Despite his public apology, Hampden renewed his vitriolic attacks at the end of the probationary period and was imprisoned for two months in 1873. A year after his release, he started up again. This time he was sentenced to one year in prison and two years' probation if he swore to keep the peace. Through the action of his flat-earth friends, he was liberated after six months and brought a countersuit against Walsh to recover his original deposit of £500, thereby doubling Wallace's misery. Both Wallace and Walsh had been unaware of a quirk in the law relating to wagers: although money lost through betting was not ordinarily recoverable, it was when a loser had claimed his money from the winner before it had been paid out by the stakeholder. According to this technicality, the money remained the loser's if he had made an immediate claim on it—and such a claim the court determined had indeed been made. Shortly after Walsh (the stakeholder) had decided in Wallace's favor, Hampden had written a note to Walsh asserting that the decision was unjust and ought to have been made in his favor. Hampden had saved Walsh's written denial of his request, which served as evidence that

Walsh had broken the law. Based on the facts of the case, Hampden won the lawsuit, and Walsh was commanded to repay Hampden. However, because Walsh already had disbursed the money to Wallace, and Wallace had signed an indemnity promising that if he lost he was obligated to repay Hampden's £500 deposit, Wallace had to reimburse Walsh. To make matters worse, the legal proceedings cost Wallace an additional £200. The £687 that Hampden owed Wallace for damages, if Hampden had not declared bankruptcy, should have neutralized Wallace's debt to Walsh. Unfortunately for Wallace, Hampden's bankruptcy prevented Wallace from enforcing his claim. "There were now legal difficulties on both sides," Wallace later reported. Hampden's bankruptcy was fraudulent and could be annulled, his lawyer said. But another lawsuit would have been expensive and the result uncertain. After negotiations among all the lawyers, a voluntary settlement was finally reached: Wallace owed £277 in legal fees, while Hampden owed Wallace £410—money that Wallace would never recover.[80] The costs of two lawsuits and four prosecutions for libel, the payments and costs of the settlement, and the expenses for the experiment itself—not to mention the public ridicule he was subjected to for having taken up the bet in the first place—thus exceeded the £500 he had received from Hampden. In 1876 the Hampden affair came to an ignominious end, resulting in financial loss and a permanent stain on Wallace's reputation.

CHAPTER 12

Wallace and The Descent of Man

IN MARCH 1870, before the outcome of the bet with John Hampden and the Flat Earthists, Wallace and his family moved to the town of Barking, on the eastern outskirts of London, in part to reduce expenses but also to prepare for his directorship of the nearby Bethnal Green Museum. The move proved to be the first in a series that would gradually cut him off from many of his scientific friends and associates. Darwin regretted Wallace's departure from central London. "I heartily congratulate you on your removal being over," he wrote after the Wallaces had settled into an old, run-down cottage that they named Holly House. "I much more heartily condole with myself at your having left London, for I shall thus miss my talks with you which I always greatly enjoy."[1] Although this did not happen often, Darwin's kind words must have touched Wallace. At this point in his life, Darwin rarely left his house, and he entertained few visitors except family members and his most intimate friends. He came to London only to spend a few weeks once a year with his brother Erasmus, paying visits to friends in central London when his health permitted. He did not make special excursions except to a spa to treat his puzzling maladies. A trip to Barking would have been too arduous an expedition for a man who was always ailing and needed constant attention. Wallace and Darwin maintained an active correspondence, but—with perhaps one exception—they never saw each other again.

Barking was a miserable locale, surrounded by miasmatic marshes and smoke-spewing factories—a fearsome place for someone concerned about his health—but Wallace thought of it as a way station and planned to leave as soon as he could afford to move to a healthier environment. In the meantime, he met several other unlucky inhabitants who satisfied most of his intellectual needs: one had a fine fossil collection from a nearby district; another shared an interest in spiritualism; and a third, a "utopianist," had invented an ingenious scheme to fertilize his farm with sewage.

A month before he moved in, Wallace had published his second major book with Macmillan. His move to Barking was not his only act of distancing from Darwin. *Contributions to the Theory of Natural Selection,* a collection of essays written over a fifteen-year period as well as a few others he had not yet published, marked his official departure from Darwin. The book included his two famous essays from 1855 and 1858—"On the Law Which Has Regulated the Introduction of New Species" and "On the Tendency of Varieties to Depart Indefinitely from the Original Type"—virtually unaltered; others were corrected, clarified, and enlarged. He presented his claims modestly:

> The present work will, I venture to think, prove, that I both saw at the time the value and scope of the law which I had discovered, and have since been able to apply it to some purpose in a few original lines of investigation. But here my claims cease. I have felt all my life, and I still feel, the most sincere satisfaction that Mr. Darwin had been at work long before me, and that it was not left for me to attempt to write "The Origin of Species." I have long since measured my own strength, and know well that it would be quite unequal to that task. . . . My own more limited powers have, it is true, enabled me now and then to seize on some conspicuous group of unappropriated facts, and to search out some generalisation which might bring them under the reign of known law; but they are not suited to that more scientific and more laborious process of elaborate induction, which in Mr. Darwin's hands has led to such brilliant results.[2]

His modesty and "more limited powers," however, did not prevent him from acting the gadfly. *Contributions* concluded with his two most provocative essays: "The Development of Human Races Under the Law of Natural Selection" and "The Limits of Natural Selection as Applied to Man."

The first was a reprint of Wallace's controversial 1864 presentation to the Anthropological Society: "The Origin of Human Races and the Antiquity of Man Deduced from the Theory of 'Natural Selection.'" He had planned to rewrite the essay but changed his mind, preferring to leave it as written except for changing the title and reworking "a few ill-considered passages" at the end. He removed the "dream" that James Hunt had ridiculed and replaced it with something many viewed as worse. The progress toward a homogeneous race was slow but steady, Wallace stated. Europeans had entered an abnormal period in history, when the inventions of science gave an illusion of moral and intellectual progress. The civilized societies of his day were too low intellectually and morally to know how best to make use of these inventions. Natural selection could not act in any way, because the mediocre, the lowest, the least moral, and the least intelligent were succeeding best and multiplying the fastest. Yet he had not given up hope. A high morality could still influence public opinion, he said. It was a sure sign that human beings were raised far above the animals and proof that there were other, higher existences from which our distinguishing qualities were derived and toward which we were "ever tending."

The second essay expanded on his review of Sir Charles Lyell's *Principles of Geology*. Wallace said that he would touch on a class of problems usually considered beyond the boundaries of science but that he predicted would someday fall within its domain. Above the law of natural selection operated some more general and fundamental law, one that required the intervention of higher intelligences. "It is . . . probable, that the true law lies too deep for us to discover it; but there seems to me," he wrote, "to be ample indications that such a law does exist, and is probably connected with the absolute origin of life and organization."

He also addressed the problem of the origin of consciousness by quoting John Tyndall, who, as president of the Mathematical and Physical Sciences Section of the 1868 meeting of the British Association for the Advancement of Science, had remarked that

> the passage from the physics of the brain to the corresponding facts of consciousness is unthinkable. Granted that a definite thought, and a definite molecular action in the brain occur simultaneously, we do not possess the intellectual organ, nor apparently any rudiment of the organ, which would enable us to pass by a process of reasoning from the one phenomenon to the other. . . . Were we intimately acquainted with the corresponding states of thought and feeling, we should be as far as ever from the solution of the problem, "How are the physical processes connected with the facts of consciousness?" The chasm between the two classes of phenomena would still remain intellectually impassable.

Tyndall was addressing the position of those scientific materialists who believed that "molecular groupings and molecular motions" explained everything. In fact, he said, they explained nothing; the problem of the connection between mind and body was as insoluble in the modern era as it had been in the prescientific ages.[3]

Wallace purposely chose Tyndall's comments to make a jab at Thomas Huxley, who had adopted a purely materialistic position by reducing the thinking process to molecular behavior. In his essay "On the Physical Basis of Life," Huxley had written, "Our thoughts are the expression of molecular changes in that matter of life which is the source of our vital phenomena. Consciousness is a function of nervous matter, when that nervous matter has attained a certain degree of organisation, just as we know the other actions to which the nervous system ministers, such as reflex action and the like, to be."[4] Huxley's theory, Wallace felt, was not only untestable but inconsistent with "accurate conceptions of molecular physics."

He then proceeded to give a discourse on the nature of matter, including a sketch of the most recent discoveries and speculations about the action of atoms. From atoms emanated attractive and repulsive forces, he said. By grouping atoms in symmetrical figures, all the general properties of matter could be explained. With more complex arrangements, the special chemical, electrical, and magnetic properties of various kinds of matter also could be elucidated. Each chemical element consisted of a molecule composed of simple atoms in greater or lesser numbers. Organic compounds were created by combining molecules. Combining organic compounds in ever greater complexity produced "organised beings." Wallace continued: "This view enables us to comprehend the *possibility*, of the phenomena of vegetative life being due to an almost infinite complexity of molecular combinations, subject to definite changes under the stimuli of heat, moisture, light, electricity, and probably some unknown forces." But this increasing complexity, even if carried out infinitely, could not have the slightest tendency to originate consciousness in such molecules or groups of molecules. "If a material element, or a combination of a thousand material elements in a molecule, are alike unconscious," he said, "it is impossible for us to believe, that the mere addition of one, two, or a thousand other material elements to form a more complex molecule, could in any way tend to produce a self-conscious existence." A more definite conception of matter was therefore required, one with clearly enunciated properties, explaining precisely how self-consciousness emanated from atoms. There was no escaping this dilemma; either all matter was conscious, or consciousness was something distinct from matter. If it was something distinct, then conscious beings were independent of what was termed "matter."

After accusing Huxley of using words "to which we can attach no clear conception," Wallace made statements equally abstruse. Matter was force and nothing but force, he said. Matter in the popular sense did not exist, but its reality was demonstrated whenever we touched something and felt a resistance or repulsive force. He identified two types of force: the first was "primary," which included gravitation, cohesion, heat, and electricity. The second was what he called will-force, which he defined as a power that directed the action of the forces stored up in the body. However minute the changes required in the cells or fibers of the brain to set in motion the nerve currents that excited the pent-up forces of certain muscles in the "animal machine," he said, some force was required to initiate those changes. That force was will-force. The origin of will-force could be traced not to something inside but to something outside humans—the will of higher intelligences or of one Supreme Intelligence. The theory claiming that matter, force, and consciousness were separate phenomena yet somehow interconnected was too complicated and contradictory and led to endless philosophical dilemmas. His theory was simpler. Matter as an entity distinct from force did not exist, and force was a product of Mind. But his theory that this Mind, or Supreme Intelligence, used the laws of organic development for a special end—that is, for humanity's spiritual development—did not negate the theory of natural selection. "I do not see that the law of 'natural selection' can be said to be disproved," Wallace wrote, "if it can be shown that man does not owe his entire physical and mental development to its unaided action, any more than it is disproved by the existence of the poodle or pouter pigeon, the production of which may have been equally beyond its undirected power."[5]

Darwin, who received his copy of *Contributions* in April, wrote to thank Wallace for the flattering remarks in the preface. "I hope it is a satisfaction to you to reflect—and very few things in my life have been more satisfactory to me—that we have never felt any jealousy towards each other, though in one sense rivals," Darwin said. "But I groan over Man—you write like a metamorphosed (in retrograde direction) naturalist, and you the author of the best paper that ever appeared in the Anthropological Review! Eheu! Eheu! Eheu!" After having reread that paper, he added, "I defy you to upset your own doctrine."[6]

Wallace's concluding essay also disturbed his old friend Henry Bates. "I have been having some conversations with the Editor of the 'Academy' about Mr. Wallace's last book & the appearance of backsliding from the Darwinian theory which it contains," Bates wrote to Darwin shortly after the publication of *Contributions*. "Other sincere friends of the pure truth have expressed a little surprise & bewilderment at the same phenomenon. The views of friend Wallace are so plausible & suit so well widespread prejudices that you no doubt think with me they might be controverted. But who is to criticise them? No one but yourself."[7]

But there were other critics. A Swiss naturalist from Geneva named Jean Louis René Antoine Édouard Claparède lambasted Wallace in the August 6 issue of the French journal *Revue des cours scientifiques* for what he felt were Wallace's inconsistencies. "While Mr. Wallace demands the intervention of a superior force to explain the foundation of the human races, and to guide man in the path of civilisation, he altogether denies the existence of such a force as assisting to produce the inferior races of animals and plants, which he attributes entirely to the operation of Natural Selection," Claparède wrote. He had sent a copy of his critique to Darwin and Wallace a month before its publication, prompting Wallace to write to Darwin that "his arguments in reply to my heresy seem to me of the weakest."[8] Darwin did not reply; in fact, he did not write again to Wallace until November. But he told Joseph Hooker, "I think well of Prof. Claparède's criticism and I think it would be well worth translating and publishing, partly because he is so capital a . . . naturalist, and chiefly because no sort of answer has yet appeared to Wallace. Bates thinks that Wallace's heterodox views have already done a great deal of mischief to the cause of evolution. . . . Wallace himself thinks Claparède's article very weak; but I conclude that he thinks so because Claparède has arrived at an unpleasant judgment."[9]

Claparède's arguments were summarized in the August 11 issue of *Nature*. Another review that appeared two months later in the same journal represented the prevailing opinion among scientific commentators. After paying homage to Wallace's important contributions to evolutionary theory, the reviewer remarked:

> To say that our brains were made by God, and our lungs by natural selection, is really to exclude the Creator from half His creation, and natural science from half of nature. All the phenomena we know are of necessity ultimately referable to the First Great Cause: the object of science is to discover their secondary causes; and if the theory of natural selection does not explain how the larynx or the brain of man were developed, then we must try to find another which will. To fall back for explanation upon the primary efficient cause of their existence and the design with which they were framed, is only to confuse two distinct branches of inquiry.[10]

Wallace had anticipated disapproval, though what the Darwinians perceived as an ever-widening gulf between themselves and Wallace he perceived as a mere wedge. In November 1870, he laid out his complete arguments against Claparède in *Nature*, but he was addressing all of his critics. None, he felt, had offered specific objections to the ideas set out in *Contributions*. They assumed that no characteristics of *Homo sapiens* differed in kind from those of other animals, whereas he had described substantial differences in kind that his critics could neither disprove nor deny. Claparède and others argued that natural selection must apply equally to humans and the rest of nature—or to neither. "But why must it do so?" Wallace asked. "I maintain . . . that man is descended from a lower animal form, but I adduce facts which go to prove that some other law or power than Natural Selection has specially modified him. If Darwin is not anti-Darwinian in admitting, as he does, the possibility that animals and plants may not have had a common ancestor, I may surely deny that I am anti-Darwinian when I show that there are certain phenomena in the case of man that cannot be wholly explained by the law of Natural Selection."[11]

Not all the reviews were damning. Although disagreeing with some of Wallace's conclusions, Anton Dohrn, a German naturalist, sympathized with his intellectual and philosophical dilemma. Since the question of how the motion of atoms and molecules could develop into thoughts was far from being solved, he said, one must concede Wallace's right to account for various psychic and organic phenomena by other principles. "We confess that Mr. Wallace's principles . . . admit of being methodically and consistently carried out. . . . If such principles do not directly help us onwards," Dohrn suggested, "they at least preserve us from onesidedness."[12]

Huxley would not let Wallace have the last word. He placed Wallace's objections at the center of current scientific debate, stating that *Contributions* was worthy of particular attention not only because of the competence of its author but also because of his willingness to raise important philosophical questions underlying all physical science. But after offering this faint praise, he attacked Wallace with his best weapon—his blistering wit. For example, in response to Wallace's remarks in his essay "The Limits of Natural Selection as Applied to Man" about the limited mental requirements of the "lowest savages," Huxley turned the tables on his friend, quoting a passage from another essay in *Contributions*, "On Instinct in Man and Animals," in which Wallace described the ability of savages to make long journeys requiring wide and accurate knowledge of the topography not only of their own district but of the surrounding regions. "In complexity and difficulty, I should say that the intellectual labour of a 'good hunter or warrior' considerably exceeds that of an ordinary Englishman," Huxley pointed out. "The Civil Service Examiners are held in great terror by young Englishmen; but even their ferocity never tempted them to require a candidate to possess such a knowledge of a parish as Mr. Wallace justly points out savages may possess of an area a hundred miles or more in diameter." The gauntlet that Wallace had thrown at his feet provoked Huxley. Such a challenge was impossible for someone of his prickly temperament to ignore. He had no use for Wallace's discussion of elementary physics. "With all due respect for Mr. Wallace," Huxley wrote, "it appears to me that his remarks are entirely beside the question. I really know nothing whatever, and never hope to know anything, of the steps [by] which the passage from molecular movement to states of consciousness is effected; and I entirely agree with the sense of that passage which [Wallace] quotes from Professor Tyndall, apparently imagining that it is in opposition to the view I hold."[13]

The verbal blitzkrieg especially delighted Hooker. Wallace, once Darwin's glorious knight, was now—in his mind at least—divested of his noble garments and forced to abdicate his honored position. "The tumbling over Wallace" was a great service to science, Hooker wrote to Darwin in a letter praising Huxley's attack on Wallace and Darwin's other critics.[14]

When Darwin finally published *The Descent of Man, and Selection in Relation to Sex* in January 1871, its message no longer outraged the world. The debate had been shaped and the way prepared for Darwin's grand entrance, which did not disappoint. Alone of all his fellow pioneers, Darwin and his work have survived down through the centuries. Other men may have been more courageous, but none could match him in sheer volume of material amassed and depth and breadth of analysis. The *Descent* towers above all other works on the human species published in the nineteenth century. Except for certain views inseparable from their era—women today would doubtless be repelled by Darwin's chauvinistic attitude toward their mental abilities, while human-rights activists might be appalled by his eugenic recommendations—the book still speaks eloquently to later generations of scientists and humanists. The *Descent* at long last addressed a heresy implicit in the pages of *The Origin of Species*: that the human species had evolved from "lower" animals. In Darwin's conception, inductive science compelled humans to forfeit their special status in the scheme of nature. As Sigmund Freud later remarked, "In the course of centuries the naïve self-love of men has had to submit to two major blows at the hands of science. The first was when they learnt that our earth was not the centre of the universe but only a tiny fragment of a cosmic system of scarcely imaginable vastness. . . . The second blow fell when biological research destroyed man's supposedly privileged place in creation and proved his descent from the animal kingdom and his ineradicable animal nature."[15]

Everything about the *Descent*, from its inception to its publication, underscored the great differences between the two men's personalities. Wallace courted controversy; Darwin detested it. Wallace was reckless in his publications and public statements; Darwin was slow, methodical, and restrained. The former disdained his critics; the latter had a sensitive—perhaps oversensitive—disposition. At an early point in his career, Darwin had taken Lyell's advice to heart. "I rejoice that I have avoided controversies," he wrote in his autobiography, "and this I owe to Lyell, who many years ago . . . strongly advised me never to get entangled in a controversy, as it rarely did any good and caused a miserable loss of time and temper."[16] If Lyell ever gave Wallace the same advice, it was ignored. Although Darwin's estimation of Wallace had slipped, it may be that he feared the power and persuasiveness of his rival's robust prose. "I have finished the 1st vol. and am half-way through proofs of 2nd vol. of my confounded book, which half kills me by fatigue, and which I fear will quite kill me in your good estimation," he wrote to Wallace in November 1870, a few months before the book's publication.[17] But Wallace still held Darwin in the highest esteem and looked on their differences as minor and not interfering with their general agreement on most scientific issues. Wallace remarked to Darwin that they agreed on nineteen points out of twenty, and in this he was partly correct.[18] Depending on one's perspective, however, that one point of difference was either negligible or immense. Where Wallace saw a negligible difference, Darwin did not, nor did Bates, Hooker, Huxley, or Darwin's other disciples. The one point subsumed the other nineteen.

The *Descent* was the obvious outgrowth of the *Origin*. His sole object, Darwin wrote in the introduction, was the consideration of three issues: (1) whether humans were descended from some preexisting form, (2) the manner of human development, and (3) the significance of the differences between the so-called races of *Homo sapiens*. He also elaborated on his theory of sexual selection, which he believed had played a crucial role in creating the distinctive differences between men and women and among the major human races. Throughout, he made more references to Wallace's views than to those of any other source.[19] This was not mere flattery, however. Darwin countered every issue that Wallace had addressed. For example, he saw no fundamental distinction between the mental powers of humans and those of the higher animals, and he was convinced that an inquiry into the origin of these mental powers was as futile as the attempt to solve the riddle of the origin of life: "These are problems for the distant future, if they are ever to be solved by man."[20] The belief in unseen or spiritual agencies was universal among the less civilized races, whereas the belief in monotheism was universal among the higher races. From the perspective of evolution, the progression from fetishism to polytheism to monotheism was natural and logical.[21] The implication was that Wallace's spiritualism was a step backward. But he adopted a distinctly Wallacean—or at least utopian—tone in his view of humanity's future. He believed that natural selection—"perhaps" with some help from the tendency to inherit acquired characteristics—and not some preternatural guiding force would shape the destiny of humankind. "Looking to future generations," he wrote, "there is no cause to fear that the social instincts will grow weaker, and we may expect that virtuous habit will grow stronger, becoming perhaps fixed by inheritance. In this case the struggle between our higher and lower impulses will be less severe, and virtue will be triumphant."[22]

Darwin refused to accept Wallace's latest views on human evolution and referred instead to "The Origin of Human Races and the Antiquity of Man Deduced from the Theory of 'Natural Selection,'" which in 1864 Wallace had read before the Anthropological Society, praising its boldness and cogency of argument regarding the action of natural selection on the intellectual and moral development of human beings. He predicted that the dispute between the monogenists and the polygenists would die "a silent and unobserved death." The weightiest argument against treating the races of humans as distinct species, he said, was that their supposed distinguishing characteristics graded into one another without any interbreeding having taken place. He doubted whether any character of a race was a constant or a universal trait. There were no anatomical differences between the races, and the tastes, dispositions, and habits of people of all races were remarkably congruent. "Now when naturalists observe a close agreement in allied natural forms," he wrote, "they use this fact as an argument that all are descended from a common progenitor who was thus endowed; and consequently that all should be classed under the same species."[23]

At times, Darwin's chapters on sexual selection read like one long argument against Alfred Russel Wallace. He seems to have felt that his greatest challenge was to overcome Wallace's objections. In most cases, Darwin admitted, the distinction between the effects of natural and sexual selection was unclear, yet the fact that males often possessed characteristics that were of no use and might, in fact, be injurious to them could be explained only by some agency other than natural selection, the doctrine of utility par excellence. Although the intense competition among males for females seemed to argue in favor of natural selection, "the power to charm the female has been in some few instances more important than the power to conquer other males in battle." The useless and injurious characters contributed to that power to charm.[24]

In the second part of his book, he exhaustively examined the secondary sexual characteristics of animals, from insects to the higher mammals. In his section on butterflies and moths, Darwin wrote, "Mr. Wallace has argued with much force in favour of his view that when sexes differ, the female has been specially modified for the sake of protection; and that this has been effected by one form of inheritance, namely, the transmission of characters to both sexes, having been changed through the agency of natural selection into the other form, namely, the transmission to one sex." Darwin confessed that he had been strongly inclined to accept this view at first, but after further study he found it improbable. His main difficulty with Wallace's theory was that he could not admit a dual selection process in one species, with its males rendered brilliantly colored by virtue of having outstripped their sexual rivals and its females dull colored by virtue of having escaped their natural predators. Various facts, he believed, supported the conclusion that the males, not the females, had been more modified. These facts were inextricably linked with the laws of inheritance, whose precise mechanism remained unknown, much to his consternation. An examination of closely allied species of butterflies, for example, showed that the females of almost all species in the same genus, and even in the same family, resembled one another more closely than the males, indicating that the males had undergone the greater modification.[25]

With regard to the origin of the extraordinary plumage of some male birds, Darwin said, it might seem unlikely that the female could appreciate the "fine shading and exquisite patterns," but this was indeed the case. "He who thinks that he can safely gauge the discrimination and taste of the lower animals, may deny that the female Argus pheasant can appreciate such refined beauty; but he will then be compelled to admit that the extraordinary attitudes assumed by the male during the act of courtship, by which the wonderful beauty of his plumage is fully developed, are purposeless; and this is a conclusion which I for one will never admit."[26]

From a general discussion of plumage, he moved to the specific discussion of ocelli, or eyespots, on the feathers of pheasants and peacocks, a phenomenon also present in some species of butterflies. Wallace had first called his attention to them, he said, showing him numerous gradations, from a simple, minute black spot to an "elegantly-shaded" ocellus in a series of specimens of a common species of English butterfly. The ocellus was therefore a highly variable feature. Darwin imagined a progenitor of the peacock, midway between the present species and an ordinary fowl with a short and spotted tail. Over time, the females would "unconsciously" select those males with the longest and most elegant erectile tail feathers, as well as those with the most dramatic ocelli, making the male peacock "the most splendid of living birds"—a clear case of a character arising from sexual selection.[27]

In total, Darwin devoted four chapters to sexual selection in birds. Two of those chapters addressed Wallace's theory of protective coloration in birds, which, Darwin said, rested principally, though not exclusively, on the nature of the nest. He acknowledged that to refute this theory required a "tedious discussion on a difficult point": whether the transmission of a character, at first inherited by both sexes, could be limited in its transmission, by natural selection, to one sex alone. He first looked for exceptions to Wallace's rules. The most serious objection was the existence of groups of birds in which the males were brilliantly colored and the females were "obscure," yet the nests were concealed. In England alone, he could find no close relationship between the colors of female birds and the nature of the nests they constructed. Of forty species he examined, twenty-eight built concealed nests despite being inconspicuous. The other exceptions were tropical species, like toucans and parrots, in which both the males and the females were brilliantly marked. In these birds, Wallace believed that as the males gradually acquired their brilliant colors through sexual selection, the females, which already enjoyed protection by virtue of concealed nests, inherited those brilliant colors, which were not eliminated through natural selection because the females were already protected. Darwin, on the contrary, believed that the female birds gradually modified their habits, building concealed nests instead of open ones, as they became more and more brilliant through inheritance. Darwin enumerated other examples that he felt supported his view.[28]

He next looked at differences or similarities in plumage between the sexes and between the juveniles and adults as proof that the males, not the females, had been modified. He defined six rules or classes of cases, the first two of which were the most important tests of his theory against Wallace's theory. In the first class, in which the adult males were more conspicuous or more beautiful than the adult females, the plumage of the juvenile offspring of both sexes more closely resembled that of the adult females. Any brightly colored or ornamented juvenile males doubtless would have been subject to danger and destroyed, whereas the duller colored and less ornamented juvenile males would have survived to adulthood. The most obvious conclusion, he said, was that the adult male alone had been modified by sexual selection: "No one probably will dispute that many gallinaceous birds that live on the open ground have acquired their present colours, at least in part, for the sake of protection. But can we believe that the very slight differences in tint and markings between, for instance, the female black and red grouse serves as a protection? . . . From what Mr. Wallace has observed of the habits of certain gallinaceous birds in the East he thinks that such slight differences are beneficial. For myself, I will only say that I am not convinced."[29]

In the second class, in which the adult females were more conspicuous than the adult males, the young of both sexes in their first plumage resemble the adult males: "Mr. Wallace, who first called attention to the singular relation which exists between the less bright colours of the males and their performing the duties of incubation, lays great stress on this point, as a crucial test that obscure colours have been acquired for the sake of protection during the period of nesting. A different view seems to me more probable." In most such species, he said, the females also were larger and more pugnacious than the males, and they drove away rival females in order to gain possession of the males: "On this view, all the facts become clear; for the males would probably be most charmed or excited by the females which were the most attractive to them by their brighter colours, other ornaments, or vocal powers." Sexual selection would operate at this level, gradually adding attractive qualities to the females while leaving the male offspring little modified.[30]

Darwin disputed every aspect of Wallace's theory of protective coloration when it clashed with his theory of sexual selection. The green of tropical birds might be protective, he said, but the crimsons, blues, and oranges were not.

Many species of woodpeckers were black or black and white, which would not be protective in a forest setting: "It is therefore probable that strongly-pronounced colours have been acquired by tree-haunting birds through sexual selection, but that green tints have had an advantage through natural selection over other colours for the sake of protection."[31] Darwin stuck by his strong belief that birds—especially females—chose their mates based on personal taste. Continuing up the evolutionary tree, he attributed the greater size of males and the vocalization, odor, and colors of mammals to one aspect of sexual selection: the battle between rival males for the attentions of the female. The conquering (that is, larger and more physically attractive) males left a larger number of offspring, and therefore these characteristics came to predominate.[32]

In the final third of the book, Darwin discussed sexual selection and human beings. He believed that many of man's unique features were attributable to sexual selection. The male beard, the naked skin, skin color, even mental differences between the sexes were the work of sexual selection. He speculated that during the dawn of humankind, aboriginal peoples lived in small groups, each male with as many wives as he could support and defend. The most powerful acquired the largest number of wives and therefore fathered the most children, to which they passed on their superior characteristics. The naked skin was a type of ornament, he said, the less hirsute being preferred over the more hirsute. According to this scenario, women lost their hair first, transmitting this characteristic to male and female offspring alike. Once human beings had lost most of their body hair, color differences were sexually selected. For some groups, a dark skin was a sign of great beauty; for others, it was a light skin. Darwin derided Wallace's latter-day theory of the origin of humanity's unique characteristics, footnoting remarks by the Reverend T. R. Stebbing made in a review of *Contributions* published in the *Transactions of the Devonshire Association for Science* in 1870. "It is surprising," Stebbing had said, "that he should picture to himself a superior intelligence plucking the hair from the backs of savage men . . . in order that the descendants of the poor shorn wretches might after many deaths from cold and damp in the course of many generations have been forced to raise themselves in the scale of civilisation through the practice of various arts, in the manner indicated by Mr. Wallace."[33]

Shortly after the publication of the *Descent*, Wallace let Darwin know that Stebbing's footnote offended him; Stebbing's remark was a caricature of everything he had ever written, he said. Darwin apologized and promised to omit it from future editions of his book.[34] But the damage was done, and Wallace sought to both clarify and defend his position. His review of *The Descent of Man* appeared in the March 15, 1871, issue of the *Academy*, a liberal scientific journal whose editors supported the Darwin–Wallace theory of evolution. He began by praising Darwin's overall achievement, referring to sexual selection as a "new branch of natural history, one of the most striking creations of Mr. Darwin's genius." Sexual selection was a potent force, he admitted, one that explained ornaments like crests, wattles, elaborate plumes, birdsong, and other physical or behavioral anomalies. But he felt that Darwin had overrated its influence. In his own view, natural selection was the first principle of evolutionary biology, whereas sexual selection was a distant second.

Some of their differences might have been resolved if the mechanism of inheritance had been elucidated. In the 1860s and 1870s, most naturalists believed in blending inheritance—that is, that traits were transmitted equally to both sexes from each parent. In blending inheritance, males and females would be expected to vary together. If natural selection acted on the male and the female simultaneously, how could sexual dimorphism be explained? Some other force had to be at work, Darwin said, and that force was sexual selection. But, Wallace asked, why were sexual differences beyond the power of natural selection? The pressures on each sex differed. Natural selection was perfectly capable of preserving traits favorable to either sex; it was not necessary to invoke another theory. The evidence for sexual selection in the lower animals remained relatively weak, he said. Male rivalry was an important factor, but female choice seemed improbable to him. When male butterflies swarmed around a female, what choice could she exert? The strongest won out. Throughout most orders of insects, there was no direct or indirect evidence of sexual selection. Coloration was determined by the degree of protection required. Asexual caterpillars, on which sexual selection could not operate, displayed many examples of ornamental appendages. "Now if all these beautiful and varied ornaments can be produced and rendered constant in each species, by some unknown cause quite independent of sexual selection," Wallace wondered, "why cannot the same cause produce colours and many of the ornaments of perfect insects, subjected as they are to so much greater variety of conditions than their larvae?" When traveling in the Malay Archipelago, he had noticed that species of widely varying genera of butterflies differed in exactly the same way as to color and form from allied species on other islands, suggesting some other law or laws of inheritance. Perhaps there were local modifying influences not yet discovered. In short, he believed that Darwin had not proved that the kind of sexual selection that depended on female choice existed in insects.

But the weakest portion of Darwin's book, in Wallace's opinion, was the discussion of sexual selection and humans. Wallace did not agree that the origin of the different races had anything to do with sexual selection. In his experience, men and women of each race admired the characteristics of their own race and abhorred wide differences, a fact that should prevent new races from forming and not cause them to arise. He did not accept Darwin's argument that polygamy contributed to sexual selection because he believed that infanticide, sexual promiscuity, and the pairing of all or most unmarried individuals would neutralize the effects of polygamy. Even if a more powerful man possessed many women and produced a large number of children, it had not been proved that women in monogamous relationships were less fertile. The "complete nudity" of humans had no parallel in the animal kingdom except in cases where natural selection had come into play. If male humans had hairless faces and chests and vividly colored buttocks, he would have found Darwin's analogy to other primates more forceful. He could not imagine a time before natural selection ceased to be a powerful force when there was a "universal and simultaneous fancy" among the most vigorous and hairiest men for the unnatural character of female hairlessness. This fancy would have had to persist for a long and unknown period of time before female hairlessness became a fixed and heritable trait. He concluded: "It is not easy to see how this severe struggle for existence and survival of the fittest would in any way aid sexual selection in abolishing the hairy covering."

What mattered most in human evolution, Wallace said, was the struggle among humans: family against family, tribe against tribe, individual against individual. Such a struggle would not have been possible unless the earth was heavily populated, so that conflict between different groups was inevitable. However, "[the] vast amount of the superiority of man to his nearest primate relatives is what is so difficult to account for," Wallace wrote. This vast superiority Darwin did not satisfactorily address. "It must be admitted that there are many difficulties in the detailed application of [Darwin's] views," he continued, "and it seems probable that these can only be overcome by giving more weight to those unknown laws whose existence he admits but to which he assigns an altogether subordinate part in determining the development of organic forms."[35] The laws that Wallace referred to were the laws that Darwin had described in the *Origin*: the "laws of growth," the "laws of development," the "laws of inheritance," the "laws of variation," the "laws of correlation," the "direct action of the environment," and the "laws of habit and instinct." It was these laws about which scientists were still ignorant and which Wallace ascribed to a controlling Mind or Supreme Intelligence.[36] Despite the apparent chasm between human beings and all other animals, however, Darwin repeatedly asserted that the differences between humans and the lower animals was one of degree and not of kind; he noted that even the "moral sense," the one mental attribute that philosophers pointed to as the most crucial distinction between humans and the lower animals, could be found in rudimentary form in other animals, like dogs and nonhuman primates.[37]

Darwin thanked Wallace for his "grand review" the day after the article appeared. He added that Lyell remarked that no one wrote better scientific reviews. But he was more candid with his daughter, Henrietta, to whom he wrote with irritation, "I see I had no influence on him, and his Review has had hardly any on me." If Wallace's intransigence disappointed Darwin, it did not disappoint Darwin's older brother, Erasmus. "It seems to me there is a good deal to answer in Wallace's review if possible," Erasmus observed to Henrietta. "I think the way he carries on the controversy is perfectly beautiful, and in future histories of science the Wallace–Darwin episode will form one of the few bright points among rival claimants."[38]

The theory of sexual selection has had a rocky course. Wallace may have been Darwin's first significant critic, but he was not his last; their impasse survived them. For half a century, the problem of sexual selection and female choice continued to be a stumbling block for evolutionary biologists. Good taste (Darwin) and good sense (Wallace) seemed irreconcilable. But since the birth of modern evolutionary theory (Modern Darwinism) in the 1930s—when it was shown that natural selection, combined with Mendelian genetics, could explain better than either theory alone the variation and geographic distribution, as well as the origin, of life on earth—natural selection and sexual selection are no longer conceived of as mutually incompatible. "For modern Darwinism," writes the philosopher Helena Cronin, "nothing remains of the traditional idea that the intraspecific and social nature of sexual selection sets it apart from natural selection. . . . It is now routine to regard relations between organisms, particularly members of the same species, as highly significant selective pressures. . . . Indeed, it is now standardly recognised that social competition among members of the same species, not merely for mates but for any resources, can be a powerful force for a co-evolutionary spiral." She adds that "the modern conception of sexual selection is not about discriminating animals but about discriminating genes."[39] For reasons that neither Wallace nor Darwin could have envisioned, both men were theoretically right. Nevertheless, many mysteries of mating persist and the debate among Darwinists, one that Wallace initiated in the 1860s, continues into the new millennium.

As for that other conflict—the origin of humans' apparently unique intellectual and physical characteristics—today's evolutionary biologists are satisfied with a biological explanation. To them, even mathematical abilities and musical gifts—just to name two apparently nonadaptive (at least to Wallace) traits—are not beyond the powers of natural selection. However, the widespread persistence of a belief in intelligent design (whether Creationist or not) into the twenty-first century suggests that the conflict between Wallace and Darwin on this issue is far from over.

CHAPTER 13

The Descent of Wallace

IN FEBRUARY 1871, Wallace acquired four acres of land near the village of Grays, twenty miles south of London and perfectly suited for gardening and a little farming. He had fallen in love with the commanding view of rolling hills stretching to the Thames and the picturesque old chalk quarry surrounded by majestic elms and other grand trees. For almost a year, he had been negotiating for the property. "I am . . . being dreadfully ridden upon by a horrid old-man-of-the-sea," he complained to Darwin, "who has agreed to let me have a piece of land I have set my heart on, and which I have been trying to get of him . . . but who will not answer letters, will not sign an agreement and keeps me week after week in anxiety." But in the end, he got what he wanted. After signing a ninety-nine-year lease at a rate of £28 a year to be paid to James Theobald (the old-man-of-the-sea) or his heirs, he asked Darwin to rejoice with him in his acquisition of the "chalk pit."[1]

As with every project, Wallace threw himself full force into his new diversion. He hired an architect to design his house, supervised the digging of a hundred-foot well, and constructed two artificial ponds. With his friend Frederick Geach, the mining engineer who had suggested the ill-fated investment in lead mines, he combed the engineering shops of London for a windmill, which they would install themselves. He and a laborer laid out all the plumbing, with branches and taps for watering various parts of the garden. Over the next year, he prepared his garden and built its walls, dug a winding road into the property, carted away gravel to make the land more cultivatable, and planted a number of rare and delicate shrubs and flowers, some culled from Darwin's garden and those of other friends. The Dell, as he called his property, was ideally situated: a short walk to the Grays train station and a quick ride of thirty-five to forty minutes into London, yet far enough in the country to provide him with peace and solitude, the perfect spot for contemplation and the indulgence of his passion for gardening.

The approach to The Dell was along a serpentine drive that ended at a four-room entrance lodge. The house itself was three stories high and was outfitted with all the conveniences of the day. A porch surrounded the ground floor. The spacious interior consisted of a drawing room with a bay window, dining room, library, lavatory, and large kitchen with ample storage and work space. The dining room opened onto a conservatory and greenhouse. There were two water closets, one for the servants and one for the family. Two stone staircases led to the first floor, which was subdivided into four bedrooms, each with a fireplace and bathroom. The second floor had four airy bedrooms, all warmed by stoves. Every lavatory and sink was supplied with hot and cold water. In March 1872, at the height of his fame, Wallace moved into his house with his young wife and three children (the third, William, was born on December 31, 1871). He was proud of his new residence; having grown up in humble surroundings, he now lived like a squire in a house every bit as imposing as Darwin's.[2]

All was not perfection, however. Halfway through construction, the contractor absconded with Wallace's money, failing to pay his men or Wallace's suppliers. Bills for bricks, timber, stone, and ironwork appeared in an endless stream and required payment. Wallace accepted all this in stride since the position at the Bethnal Green Museum, with the assurance of a steady income, was to be offered any day. But month after month had passed with no word from the government, and Darwin was concerned for him. "I have looked out in the papers for any notice about the curatorship of the new Museum," he informed Wallace, "but have seen nothing. If anything is decided in your favour, I beg you to inform me."[3]

The blow came at the end of August. Sir Charles Lyell learned that Bethnal Green would serve as a repository for art rather than natural history specimens and would be managed from London. A curator was no longer needed. He broke the news gently to Wallace, who was devastated after his great exercise in patience. While awaiting the government's decision, he had been kept financially solvent only by means of makeshift employment. Through Henry Bates, he had obtained positions in 1870 as an assistant examiner for the Departments of Physical Geography and Geology at the Indian Civil Engineering College and for the Royal Geographical Society. Eventually he took another job grading examination papers for a government agency whose examiner-in-chief was David Thomas Amsted, an eminent professor of geology at King's College in London. The job would last until 1897. For three weeks each summer, he was paid £50 or £60 to grade a thousand or so papers. The work was tedious, but the money he earned for the brief commitment of time eased his anxieties.

His losses forced him to sell his collection of exotic birds—one of the best in the world—to the British Museum. They also compelled him to write. Between 1872 and 1876, Wallace published more than twenty-five papers on scientific, spiritual, and social subjects, for which he received modest payments.[4] Concerned about his deteriorating financial situation, his friends came to his aid. Lyell hired him at £5 an hour to assist in the revision of a new edition of *The Geological Evidences of the Antiquity of Man*, as well as three or four editions of his other books. Darwin offered him an opportunity to edit *The Descent of Man*, but Bates expressed reservations. "You will perceive that he is at present unaware of the scope & nature of the revision required, & I should think they ought to be very exactly defined if he is employed," Bates counseled. "Otherwise he would be likely to query the reasoning."[5] The admonition was unnecessary. Wallace was well aware of the opinion that Bates and most of his other peers held regarding his "heresies." Quelling their concerns, he assured Darwin that he would not offer any criticism. He had no idea how long editing the manuscript would take. As for his fee, he thought £7 an hour fair; the £5 that Lyell paid him was not quite enough for the class of work required, he told Darwin in strict confidence. Wallace's economic problems were humiliating. Even if couched euphemistically as editing, the acceptance of financial help from friends was not easy. His letters hinted at the awkwardness. You can see the upper lip stiffening as he wrote to Darwin: "No doubt you will get it done for very much less by any literary man accustomed to regular literary work and nothing else, and perhaps better done, so do not in the least scruple in saying you decide on employing the gentleman you had in view if you prefer it." Darwin decided to have his son George edit the *Descent* instead.[6]

Wallace meanwhile lost himself in gardening, an interest spurred by Augustus Mongredien. A member of the Corn Exchange, Mongredien was a savant with an excellent knowledge of classical Latin and Greek and several modern European languages; he also had written a book on trees and shrubs suitable for English gardens. Wallace contacted him and was invited to his nursery, which occupied several hundred acres of land on the Bagshot sands, located on the border of Surrey and Berkshire, southeast of London. Wallace and Annie spent several weekends at Heatherside, Mongredien's estate in Bagshot, to which Mongredien frequently invited a number of English intellectuals, and it was at these gatherings that Wallace got to know Richard Owen, Thomas Huxley's great nemesis. They already were acquainted; while traveling in the Malay Archipelago, Wallace had corresponded with Owen about his orangutan specimens from Borneo, and after returning to England he occasionally had conversed with him at the British Museum or at scientific meetings. Wallace liked Owen despite the calumnies that Huxley had heaped on him, finding his pompous manner charming rather than annoying, and he respected Owen's immense fund of knowledge. When their visits coincided, the two took long walks on Mongredien's beautiful grounds, breathing in the scent of the magnificent conifers and discussing gardening and botany. Not long after meeting Wallace, Mongredien turned his nursery into a joint-stock company and persuaded Wallace to become one of its directors. It is not clear if Wallace parted with any money, but if he did it was a mistake: the company was dissolved in 1873 or 1874, and Mongredien and his business associates lost their entire investment.[7]

While living in Grays, Wallace found it more difficult to attend evening meetings in London than he had anticipated; access to the city was not as simple as he had been led to believe. Nevertheless, he went to London as often as he could, spending the night with Fanny and Thomas Sims at their home on Westbourne Grove Terrace. He also entertained numerous visitors at The Dell, including several eminent foreigners, among whom were Russian and German spiritualists.[8] After stepping down from the presidency of the Entomological Society in 1872, he set to work on a general review of the geographic distribution of animals, a project that Alfred Newton and Philip Sclater had suggested to him a couple of years earlier. Previous presentations were unsatisfactory, Wallace told Newton, because of inconsistencies and a failure to interpret the facts of distribution in light of the new thinking on the formation of species.[9] Darwin also had encouraged him to write a full account of the subject. As the acknowledged master of zoogeography, Wallace could be expected to produce a work of immense value to every naturalist. The task daunted him, however; the subject was vast in extent, the amount of details he needed to consider overwhelming, and his knowledge of the different classes of animals inadequate. Since he intended to examine animal distribution in both time and space, knowledge of the fossil record was absolutely essential but complicated by the absence of material on many important animal groups. Moreover, the classification system itself had fallen into a hopeless state of confusion. After some preliminary research, he gave up for the time being, feeling that he could not yet do the subject justice.[10]

Beginning in 1872, Wallace's reputation in the scientific community began to slip precipitously. A two-part article he wrote for *Nature* in support of Henry Charlton Bastian—a respected physician at the University College

Hospital of London who had resurrected the theory of spontaneous generation, which he disguised as "archebiosis"—raised a number of eyebrows. In his book *The Beginnings of Life*, Bastian attempted to prove that all the lower forms of life—bacteria, amoebae, and paramecia, for example—were spontaneously generated from inorganic matter due to the influence of unknown laws. The book's apparent solution to one of the problems of evolutionary theory—the enormous amount of time required for evolution of even the lowest life forms—appealed to Wallace.

Only a few years before the publication of Bastian's book, the eminent physicist William Thomson (later Lord Kelvin) had asserted that the earth had not existed long enough to allow for evolution by natural selection. Thomson towered over other figures in nineteenth-century mathematics and physics, and most scientists took his every opinion seriously. Thomson's estimation of the age of the earth, based in part on his calculations of the rate at which the earth must have cooled down to its current temperature, was a mere 100 million years, a potentially fatal objection to evolutionary theory. So formidable were his credentials and so forceful were his arguments that he dominated the debate about the earth's age for nearly a half century. Even if his calculations and assumptions have been proved wrong—the discovery of radioactive decay in the early twentieth century, which provided a measurable timeline for the history of the earth, finally subverted his theory—his signal contribution to an understanding of the earth's history was his application of his second law of thermodynamics to geology. This theory holds that in any closed system, the total amount of energy remains constant, but that every physical process (an open system) results in an irretrievable loss of energy in the form of heat. Thus the universe, including the sun and the earth, was running down through the dissipation of energy. Thomson negated Lyell's pivotal idea (which he oversimplified) of the earth as a great perpetual-motion machine by imposing limits on the earth's existence. The age of the earth was calculable, not indefinite or beyond human comprehension, as Lyell insisted. Moreover, he convinced his peers that Lyell's conception of the conservation of energy, implying endless cycles ("no vestige of a beginning, no prospect of an end," in the memorable phrase of Lyell's predecessor, James Hutton), was erroneous and violated the laws of physics. Thomson mainly objected to Darwin's application of Lyellian uniformitarianism to the development of life because it denied the existence of design and order in the universe, one of the foundations of his cosmology.[11] His objections, eloquently argued in journals and scientific meetings and couched in seemingly invincible mathematics, sent a tremor through the community of proponents of the Darwin–Wallace theory and prompted Darwin to refer to Thomson as "an odious spectre" whenever anyone mentioned his name.[12]

Although on empirical grounds Darwin doubted Thomson's estimate of the age of the earth, Wallace (like most of his contemporaries) accepted Thomson's argument and sought creative ways to fit the theory of natural selection into Thomson's new geologic paradigm. If Bastian was right, Wallace wrote in July 1872 (with Thomson in mind), then once the earth had arrived at the proper conditions for supporting organic life, the spontaneous generation of bacteria and amoebae would have saved an immense amount of time. These primordial microorganisms would have served as the substrates on which natural selection could act to produce higher organisms. Instead of a single tree of life, he suggested, there might have been a vast number of trees, the majority becoming extinct and all existing life stemming from the survivors.[13] Wallace recommended Bastian's book to Darwin, who found the general argument for archebiosis "reasonably strong," though he wished that Bastian had arrived at his conclusions inductively rather than deductively. "I know not why," Darwin said, "but I never feel convinced by deduction." Moreover, he could not accept archebiosis as applied to more complex forms of single-cell organisms, insisting that "my mind can no more digest such statements, whether true or false, than my stomach can digest a lump of lead." And he had a number of other reservations about Bastian, including his competence as a scientist. Bastian had made a fool of himself by mistaking a fragment of sphagnum moss that had contaminated his abiotic solution for a "vegetable growth" spontaneously generated. As a result, Huxley had rejected every one of Bastian's other observations in his usual contemptuous manner, a judgment that Wallace found illogical.[14]

Anton Dohrn, his erstwhile champion, spoke for other naturalists when he sent the following assessment to Darwin: "Poor Wallace completely drifts away and now most unfortunately associates himself with such men as Bastian! His two articles in *Nature* are the worst thing he ever did in his life—and it becomes really difficult for friends to speak with respect for him."[15] But Wallace cared little what Dohrn and others thought about him. He did not believe strongly enough in Bastian to mount a campaign on his behalf, but Thomson's charge remained a formidable obstacle. Either the earth was more ancient than Thomson estimated or a force other than natural selection was boosting the rate of evolution. No evolutionist offered a convincing rebuttal, and the matter was quickly swept under the evolutionary carpet, concealed from troublesome critics.

Wallace was not an easy man to ignore, even if his colleagues could not speak respectfully about him. His theories, correct or errant, stimulated thought. Without apology, he moved on to the next issue that attracted his attention, so-called animal instinct. Inspired by a report from Darwin on his belief in the heritability of behaviors, Wallace wrote to the editor of *Nature* in February 1873, offering a theory to explain how a dog shut up in a basket in a coach during a journey with its owner could find its way back home if it was somehow lost. The majority of naturalists believed that an animal's ability to return home was an instinct. Wallace disagreed. The animal reasoned by its sense of smell, he said. In the anxious state of confinement, the animal would take note of successive odors to orient it, accumulating a series of impressions as distinct and prominent as visual images. The detection of these odors would lead the animal back home, no matter how many turns and crossroads were encountered. Wallace's theory had ramifications. For example, it might explain how certain species of turtles found their way back to tiny specks of land in the midst of a great ocean.[16] His casual hypothesis caused a minor war in the London scientific community. For four months, a steady volley of letters appeared in *Nature*, with one side slinging arrows at the other. Charles Kingsley and Charles Darwin wrote in support of Wallace, but a good number of others were unconvinced.[17]

While he remained active in scientific debates, Wallace also continued to pursue his spiritualist investigations. Among the friends with whom he investigated spiritualist phenomena in the early 1870s was Lyell's personal secretary, Arabella Buckley, whose own mother was a spiritualist. Although her mother was convinced that she could communicate with a deceased daughter, Buckley remained skeptical. Respected by Darwin, Lyell, Huxley, and Wallace, she was an intelligent woman whose aspirations appeared to transcend her secretarial career. Wallace especially admired her review of *The Descent of Man*, written for *Macmillan's Magazine*, a popular publication with liberal leanings. Entitled "Darwinism and Religion" and signed only A. B. (no doubt because she was venturing onto ground not usually trod by women), the review asserted that Darwin and his theory of human evolution did not "blot out of existence" all communion with God and all hope of immortality. In fact, such beliefs and hopes were not inconsistent with either the theory of evolution or that of creationism. By approaching the subject of morals and intellect "from the side of natural history, [Darwin's theory] gives us the means of testing metaphysical arguments by the touchstone of physical facts."[18]

In 1870 Wallace invited Buckley to a lecture by Emma Hardinge Britten, a leading spiritualist. He wanted to show her that not all spiritualists were prejudiced and unreasonable, and that some were imbued with a true scientific understanding. He hoped that she would appreciate spiritualism's moral teachings, which were "elevated and beautiful and calculated to do good." His belief in the future progress of humankind was founded on observable phenomena that could be examined in the present, not on phenomena alleged to have occurred thousands of years ago, when records were imperfect, as in the case of the biblical accounts. He advised her not to judge spiritualism by public séances alone.[19] Buckley delayed her scientific studies of spiritualism until 1873, at which point Wallace recommended that she keep notes of her experiences. He educated her about the subtleties and complexities of spirit communications. For a time she believed that she had mediumistic powers, which excited Wallace. "It will be splendid if you really become a good medium for some first-rate unmistakable manifestations that even Huxley will acknowledge are worth seeing," he told her.[20] But in the end, she remained conflicted. On one side were the believers, her mother and Wallace; on the other side were the skeptics, Huxley, Darwin, and Lyell. Was she wasting her time pursuing a "will-o-the-wisp," she asked Wallace. "It seems to me as if Spiritualism from my experience leaves one always in the state of the man who 'never is, but always to be, blessed.'"[21]

Another fellow investigator was the physicist Sir William Crookes. If Wallace earned praise from spiritualists as spiritualism's great philosopher, Crookes earned praise as spiritualism's great experimental scientist. The two men had met in London on some unknown occasion and by the early 1870s had collaborated on a number of investigations that soon established them as the most prominent scientific exponents of the spiritualist movement. Crookes supplied the experimental proof to bolster Wallace's convictions. What they saw together gave Wallace the lifelong strength to stand by his beliefs in the face of mounting ridicule and charges of credulity.

Crookes was born in London in 1832, the son and first child of a tailor. Like Wallace, he was self-educated, though at the age of fifteen he entered the Royal College of Chemistry in Hanover Square, London, where he received some formal training in organic and inorganic chemistry. In 1854 he became superintendent of the meteorological department at the Radcliffe Observatory in Oxford, and the following year he was appointed a lecturer in chemistry at the Chester Training College. His discovery of the element thallium in 1861 brought fame early in his career, leading to his rapid elevation into the upper echelons of British science; he was elected a fellow of the Royal Society in 1863. Acknowledged as a mechanical genius, he invented the cathode-ray tube, known as "Crookes's radiometer," and constructed apparatuses that measured the weather and solar eclipses. He experimented with the new art of photography in an attempt to record bullet trajectories. But Crookes lacked charisma and never attracted a flock of admirers or protégés. "In his presence one did not feel the worshipful enthusiasm which some of the great men of science have aroused," one biographer noted. Wallace recollected a cold and reserved man who was secretive and unintentionally mysterious.[22] Nevertheless, he felt an intellectual kinship with him. Intrigued by spiritualist phenomena, Crookes eventually believed as strongly as Wallace did in their reality, if not the interpretation of their source. In July 1869, he attended the séances of Mary Marshall, possibly in the company of Wallace. He also befriended Daniel Dunglas Home, the most famous medium of the day.

Of Scottish origin and raised in the United States, Home had returned to Great Britain in 1855 and quickly established his reputation in the parlors of the aristocracy. Tall, thin, and sickly, he charmed his audiences with his pleasant demeanor, somewhat effeminate and delicate features, and impeccable dress. The phenomena associated with his séances were reputed to be among the most remarkable ever recorded. Despite the numerous examinations he was subjected to, no one ever uncovered fraud or deception. More than anything else, it was his honesty, which he took great pains to demonstrate, that led to a large number of spiritualist conversions, especially among the upper classes. Although he often conducted his séances free of charge, he never refused the lavish gifts that his devoted followers showered on him.[23]

Home's séances began in the typical fashion. Participants sat around a table and held hands as he fell into a trancelike state in order to communicate with the spirits. Furniture shook, swayed, and levitated; raps sounded; spirit arms and hands appeared; and breezes wafted through the room. He was most noted for his accordion, which produced its own music while it dangled from his left hand as he played the piano with his right hand. What distinguished his séances from the others, however, was his attitude toward light. Whereas the majority of séances were conducted in the dark, Home permitted some natural and artificial illumination. One of Home's most sensational séances occurred in December 1868 before a distinguished audience that included Robert Chambers, author of *Vestiges of the Natural History of Creation*, and Home's close friend, the young Lord Adare, Windham Quin, later the fourth earl of Dunraven—a flamboyant figure who would become a war correspondent during the Franco-Prussian War and hunt with Buffalo Bill in Texas. According to an account published by Adare shortly after the séance, Home was seen to levitate while in a trance, float out of the sitting room and into the bedroom, and then out the window at a height of seventy feet above the street before returning to the sitting room. Two other witnesses corroborated Adare's account. Another of Home's impressive feats was the fire test, also corroborated by several independent observers. While in a trance, he grabbed a glowing coal with his bare hands from the hottest part of a fire and carried it around the room. In front of gasping spectators, he placed the coal on the head of one of the participants. The man's white hair glowed like silver threads over the red coal. Home drew the strands of hair into a pyramid as the coal, still red-hot, showed underneath. When the coal was removed, neither the man's scalp nor his hair was blistered or singed. It seemed to Wallace that Home had somehow conveyed to the man the same resistance to burning.[24]

Wallace and others urged Crookes to undertake a scientific inquiry of spiritualist phenomena. Crookes agreed, selecting Home as his primary subject because he believed that Home had the most developed psychic power of nine or ten other persons he screened. He also chose Home in order to refute the charge that mediums were mainly "tricky" women who used their charms to sway even the most objective scientist. After beginning a secret investigation, Crookes was exposed by a correspondent of the *Athenaeum* in 1870 and was forced to announce his intentions prematurely. "I consider it the duty of scientific men who have learnt exact modes of working, to examine phenomena which attract the attention of the public, in order to confirm their genuineness, or to explain if possible the delusions of the honest and to expose the tricks of the deceivers," Crookes wrote in the July issue of the *Quarterly Journal of Science*, adding, "But I think it a pity that any public announcement of a man's investigation should be made until he has shown himself willing to speak out." He told his readers that he had as yet no explanation for phenomena whose reality he did not doubt but that could not be explained by any known physical law. However, he had seen nothing to convince him of the truth of the spiritualist theory. In such an inquiry, the proof of a spiritual (that is, supernatural) cause must be absolute and no other explanation plausible. "It must be so strikingly and convincingly true," he wrote, "that we cannot, dare not deny it." He proposed a scientific approach to observing and interpreting the phenomena associated with a medium at a séance. First, the investigator had to be certain of the facts; second, conditions for obtaining those facts had to be controlled; third, the laws behind the facts had to be elucidated. He faulted most spiritualist evidence as either fraudulent or obtained under conditions that were not controlled or reproducible. "The pseudo-scientific spiritualist professes to know everything: no calculations trouble his serenity, no hard experiments, no long laborious readings. . . . He talks glibly of all sciences and arts, overwhelming the inquirer with terms like 'electro-biologize,' 'psychologize,' 'animal magnetism,' etc. Popular science such as this is little able to guide discovery rushing onwards to an unknown future." If, say, a spiritualist claimed to have witnessed manifestations of a force equivalent to thousands of "foot-pounds," the scientist, believing firmly in the conservation of force, should be able to ask for such an exhibition in his laboratory, where he could weigh, measure, and submit that force to "proper tests."[25]

Crookes's announcement of his intent to investigate spiritualist phenomena in a scientific manner was universally lauded by his colleagues. Wallace later summarized this praise in an essay on spiritualism. One man stated that he was "gratified to learn that the matter is now receiving the attention of cool and clear-headed men of recognised position in science." Another wrote that "no one could doubt Mr. Crookes's ability to conduct the investigation with rigid philosophical impartiality." A third remarked that "if men like Mr. Crookes grapple with the subject, taking nothing for granted until it is proved, we shall soon know how much to believe."[26]

Wallace was present at one of the séances with Home and Crookes, which took place at the house of a woman who lived in the West End. As the participants sat in a spirit circle around a table in a well-lighted room, Home lapsed into a trance and communicated with the spirits. All heard raps and watched the table levitate—nothing unusual so far. But then Home brought out his famous accordion. Wallace witnessed "that most wonderful phenomenon," an accordion playing music by itself. Home, who was holding the accordion with one hand, invited him to examine it as it played, and as closely as he pleased. Wallace had to squat under the table to observe the keyboard, which dangled out of view. He determined that no one was touching it, yet at one point he saw a shadowy, ill-defined spirit hand, coming from he knew not where, stroke the keys. "This is too vast a phenomenon for any skeptic to assimilate," he wrote to Arabella Buckley shortly after that séance.[27] Crookes apparently had not devised any experiment to test this feat; he chose instead to test Home's ability to make the table levitate.

Wallace watched as Crookes suspended the table from a spring balance; Home was able to make the table light or heavy at Crookes's request, and an indicator registered the shift in weight. To make sure that the apparatus was working properly and that none of the participants was unconsciously (or deliberately) influencing the results, Home was asked to make the table light by the power of his will when everyone's hands rested on top and heavy when everyone's hands were underneath. The difference in weight was approximately forty pounds, Wallace said. Crookes also asked Wallace to place a candle under the table while it rose into the air to show that Home's feet and body had nothing to do with the act of levitation.[28]

But Crookes was not satisfied with this simple demonstration of Home's alleged powers. He invited Home to the laboratory he had set up in his own house to test Home under conditions that Crookes could control. Without assistance from Home or any other medium, Crookes had constructed elaborate mechanical apparatuses that were designed to detect a force while preventing the subject of the experiment from physically affecting the measurement. Using these instruments, Crookes was able to discover and measure an indefinable "psychic force" emanating from Home. In a paper that he presented to the Royal Society for possible publication, he described his methods and instruments in great detail and included diagrams. He also invited the two secretaries of the Royal Society, the physicist George Gabriel Stokes and the physiologist William Sharpey, to witness his experiments with Home and examine his experimental apparatuses, cautioning them that Home might not be able to produce his phenomena immediately and that they might have to return two or three times. Sharpey politely declined the invitation; Stokes said that he would "endeavour to call" at Crookes's house, but in the end he made excuses for not being able to come and instead pointed out possible sources of error without personally examining Crookes's instruments.

Undeterred, Crookes published the results of his laboratory experiments in the *Quarterly Journal of Science*, of which he was the chief editor, affirming that they confirmed "*beyond doubt* the conclusions at which I arrived [earlier], namely, the existence of a force associated in some manner not yet explained, with the human organisation, by which force, increased weight is capable of being imparted to solid bodies without physical contact."[29] A cardinal law of physics, he knew, was that no force could manifest itself without the corresponding expenditure of another form of force. Home had been emotionally and physically drained after each of these laboratory experiments and usually would collapse to the floor in a near faint, pale and speechless. Home's state of prostration convinced Crookes that a drain of what he called a vital force or nervous energy accompanied the development of the "psychic force." These terms were vague, he realized, but he was trying to keep the phenomena within the range of experiment and argument.

Wallace was grateful for Crookes's work and felt personally vindicated, even if most of the scientific world remained incredulous. "Spiritualists are greatly indebted to you for furnishing them, in your self-registered experiments, with a final answer to [our opponents' objections]," he told Crookes in October 1871 after reading his article.[30] However, other scientists reacted with predictable hostility. Crookes had been forced to publish his findings in his own journal because the review committee of the British Association for the Advancement of Science had refused to allow him to present his paper at the association's annual meeting (nor was he ever given the opportunity to present it at the Royal Society). The excuse that Stokes, as head of the committee, made to Crookes was that he had received the paper late and had had only forty-five minutes to decide whether to include it. "I am not prepared to give my adhesion without a thorough sifting by more individuals than one," Stokes told Crookes. "I don't see much use discussing the thing in the sections, crowded as we already are: but if a small number of persons in whom the public would feel confidence choose to volunteer to act as members of a committee for investigating the subject, I don't see any objection to appointing such a committee. I have heard too much of the tricks of Spiritualists to make me willing to give my time to such a committee myself." Another delegate expressed the association's general view: he saw no reason to inquire into a matter that was unworthy of further study or investigation. Crookes's painstaking work was summarily rejected, and a few of his colleagues asked him to "get better witnesses before he [could] be believed."[31]

As for Wallace, his participation on a commission sponsored by the London Dialectical Society did not enhance his reputation. The origins of this society, "with its high-sounding name" (as a reporter from the *Times* later derisively noted), are obscure.[32] In January 1869, the society council had appointed a commission "to investigate the phenomena alleged to be spiritual manifestations, and to report thereon." Eventually thirty-three people—mostly professional men—agreed to sit on its six subcommittees. (Huxley had been invited to sit on the commission but declined, responding as he had to Wallace a few years before, though less politely: "The only good that I can see in the demonstration of the truth of 'Spiritualism' is to furnish an additional argument against suicide. Better live as a crossing-sweeper than die and be made to talk twaddle by a 'medium' hired at a guinea a séance.") For nearly three years, Wallace and his co-investigators sifted through an immense amount of evidence and correspondence before issuing their report in 1872.[33]

After fifteen meetings to assess the evidence presented by the six subcommittees, the commission concluded, as part of its four-hundred-page report, that the subject of spiritualism deserved serious attention and careful investigation. The witnesses of the phenomena possessed high character and great intelligence, the report stated. One subcommittee had met forty times for the purpose of experimentation. Initially the great majority of that subcommittee's members had been skeptical, firmly believing the phenomena to be the result of imposture, delusion, or the involuntary muscular action of the participants. However, by the end of the investigation most had been "slowly and reluctantly" convinced "by irresistible evidence" that what they had observed were indeed "veritable facts" and not delusions. "The motion . . . produced in solid bodies without material contact by some hitherto unrecognised force operating within an undefined distance from the human organism, and beyond the range of muscular action" was an undeniable fact and should be subjected to further scientific examination to ascertain its true source, nature, and power. None of the subcommittees were prepared to attribute the phenomena to a supernatural origin.[34]

On December 26, 1872, the *Times* published a caustic critique of the Dialectical Society's report, calling it "a farrago of impotent conclusions, garnished by a mass of the most monstrous rubbish it has ever been our misfortune to sit in judgment upon." Dismayed that Britain's "savants" were too busy giving themselves up to such matters as the descent of man from monkeys and the fertilization of the world by means of mossy stones falling through space (an alternative theory to archebiosis), the *Times* reporter had attended four séances, two of which were conducted by Home, who, to his great surprise, agreed to every condition that the reporter set. Throughout the séance, Home's hands and feet were bound. Despite all attempts to discover imposture, the reporter could find none. A table floated eight inches above the floor; the table was made light or heavy at the reporter's whim; an accordion produced a melody, though Home held the instrument in his outstretched right hand while his left rested on the table, or the reporter himself held the accordion; as he held the accordion with one hand, a flower was placed in the other by a finger and thumb from under the table; when he dropped the flower, it was immediately "given back" to him. He struck a match and searched under the table, even making a "sub-mahogany expedition" on hands and knees with a spirit lamp. "There was nothing during the whole evening except the phenomena themselves to suggest imposture," he concluded, though with less assurance than he had evinced at the beginning of his article. He did not deny the fact that the table had risen from the ground or that a flower had been placed in his hand by a disembodied hand, but he had no explanation for how these things had happened. "The nature of the phenomena and of human nature are such as to force us to suspect imposture and legerdemain until we can satisfy ourselves of the true causes, whatever these may be," he wrote, clearly confounded by his experiences.[35]

Public enthusiasm for spiritualism peaked in England in the 1870s. Wallace's old friend the physiologist William Benjamin Carpenter, whom he had hoped to recruit in defense of the movement, referred to spiritualism as an "epidemic delusion," not unlike the waves of mass hysteria that afflicted communities from time to time. The claims of spiritualists, Carpenter said, defied common sense and arose from the very natural desire to communicate with loved ones after death. All the manifestations at séances could be explained by the unconscious muscular actions of the participants. Spiritualists were reacting to the intense materialism of the times, which had weakened or even destroyed that belief in a higher future state so intrinsic to human nature. He attributed the acceptance of such nonsensical doctrines by even the most brilliant minds to a defect in early education, when learning rather than wisdom was imparted. This defect led to a childish persistence of the inability to distinguish between reality and illusion—a defect that could have been corrected by a more assiduous study of science, which revealed the true laws of nature.[36]

Darwin had never taken spiritualism very seriously, but he found himself surrounded by friends and members of his family who were greatly interested in the manifestations witnessed at séances. His brother Erasmus investigated spiritualist phenomena. Hensleigh Wedgwood, his first cousin and brother-in-law, a distinguished philologist, was a confirmed spiritualist. His cousin Francis Galton, though not a card-carrying spiritualist, had seen and been impressed by Home and other mediums. Even his former family physician was a spiritualist! On one occasion, Darwin's son George hired a medium to conduct a séance for the family at Erasmus's home. The large party included George Lewes and George Eliot. Henrietta Litchfield, Darwin's daughter, later wrote that Lewes caused trouble by making jokes throughout the evening and preferred to sit silently in the darkness rather than play the game "fairly." Wedgwood and Galton also were present. The medium, Charles Williams, caused various objects—a flute, a candlestick, "fiery points"—to jump about in the drawing room in a manner that astounded everyone. The séance occurred in subdued light, and Galton and Wedgwood held on to Williams the whole time. Darwin had decided to attend but, overwhelmed by the room's warmth and the claustrophobic atmosphere, left before the miraculous "jugglery" occurred. "How the man could possibly do what was done, defies my understanding," he wrote to Joseph Hooker. "The Lord have mercy on us all, if we have to believe in such rubbish. Francis Galton . . . says, it was a good séance."[37]

Intrigued but suspicious, George Darwin arranged a second séance a few days later at Wedgwood's house. Fearing the physical and emotional strain on his health, Darwin decided not to participate, but at his urging Huxley agreed to attend incognito—though whether Huxley could ever appear anywhere in London incognito is unclear. Darwin trusted the judgment of his son, who was a physicist, and Huxley "more than any two people in the world." It was George and Huxley's task to uncover evidence of imposture if any occurred.[38] Huxley reported his experience to Darwin in a letter written shortly after the séance. Five people were present, including Charles Williams, the medium. Wedgwood sat opposite Williams. George was on Williams's right and Huxley on his left, firmly interlinking their hands and feet with those of Williams. A fifth, unidentified man sat between Huxley and Wedgwood. In the middle of the table were a guitar, an accordion, two paper horns, a Japanese fan, a matchbox, and a candle in a candlestick holder. Nothing happened in the first half hour, so the participants (including the two confederates, Huxley and George) agreed to darken the room further by closing the shutters and drawing the curtains. Even so, three rays of light shone from the hallway outside and struck the wall in such a way that Huxley felt he could detect any movement by Williams if the medium attempted to obliterate them. Huxley focused on rays *A* and *B* (as he called them), which must have been increasingly difficult as time passed. He soon felt a muscular twitching coming from Williams's body, not unlike the convulsions of a galvanized frog. The twitching ceased, and rays *A* and *B* were by turns obliterated and restored; Huxley surmised that Williams somehow had bent over the table—a hunch that was confirmed by the noise of the guitar and accordion shifting positions. To Huxley's great "disgust," however, George remarked that he saw two spots of light and that something had "occulted" one of them. "I blessed him for spoiling my game," Huxley told Darwin.

Williams then complained that the light from the door was distracting him and asked that the door be shut. While a candle was lighted to perform this task, Huxley quickly glanced at the table and noted a longish thread of catgut attached to the end of the guitar, leading to Williams's place at the table. When someone asked why nothing was happening, Williams rearranged the sitters. Huxley remained where he was, while Wedgwood took George's place. "Good medium," Huxley thought. "Now we shall see something." The candle was snuffed out, and Williams soon began to twitch again. The participants heard a dragging sound, and an empty armchair from a corner of the room moved up against Wedgwood's legs and suddenly was lifted onto the table, shoving aside the guitar. For a brief moment, Huxley and Wedgwood had lost contact with Williams, who cried out, "Don't leave go of my hands." The separation was so brief that Wedgwood was convinced of its unimportance, but Huxley concluded that Williams was a cheat and an impostor and that he had somehow distracted Wedgwood while moving the chair against Wedgwood's leg and "coolly" lifting it onto the table.[39]

Huxley's report, which coincides with George's, delighted Darwin. George Darwin felt that he had been given a lesson in the worthlessness of the evidence of the senses; in the future, he would not doubt his ability to be fooled. But Galton did not cease his investigations. He asked Darwin to keep an open mind and encouraged him to investigate the subject further in his capacity as a scientist. Darwin demurred, citing his wife's opinion that it would be impossible for him to take on any other work in his present physical condition, but he urged Galton to continue: "Do not give up yourself. Can you not get men more known for physical science to join you?" Darwin expressed great interest in Crookes's laboratory apparatus, which would settle "once and for all" the question of whether any power emanated from certain individuals. In his opinion, "This would be a grand discovery."[40] Informed about the séance at Wedgwood's house, Home offered to conduct a séance for Darwin at no charge—a "remarkably liberal offer" that Darwin turned down. It is unclear whether his encouragement of his cousin was sincere or merely courteous. In an earlier letter to Galton, he had not minced words, noting that "it is rather dreadful to think what we may have to believe."[41]

In early 1874, only two years after giving up in despair, Wallace felt that the state of knowledge had advanced enough for him to reconsider writing a general review of the geographic distribution of animals from the perspective of evolutionary theory. Data were accumulating at a logarithmic rate. What had changed his mind was the publication of several important catalogues and treatises on systematics. Moreover, scientists on board the HMS *Challenger* were mapping the depths of the oceans as they circumnavigated the globe on a science mission, providing invaluable information to the scientific world. (In fact, Wallace had served on the advisory board to help define the *Challenger*'s mission.) He therefore resumed a project that would become a labor of love, for the masterwork he ultimately produced would be of value only to the specialist. He struggled in a way that he had never had to before when writing on a scientific subject. Normally he was capable of completing a lengthy essay in a matter of days, with few additions or corrections; the delay in the publication of *The Malay Archipelago*, complex as the work is, was due to procrastination, not writer's block. *The Geographical Distribution of Animals* differed from every previous literary project. It became a great monster to be wrestled to the ground. He worked fitfully, and there were endless distractions at The Dell, including his three children, visitors, straggling plants, leaky pipes, and ghost pictures.

Ghost pictures—or spirit photographs, as aficionados called them—had been circulating among spiritualists for a few years. They purported to show forms produced by invisible intelligences, clothed in material that could make them recognizable to people on earth. Such figures made their appearances on a photographic plate only in the presence of a medium, who might also have seen and described the figure as the photographer was preparing to photograph the sitter or sitters. Wallace acknowledged that a large number of these photographs were fraudulent, but he was intimately acquainted with two photographers—his brother-in-law, Thomas Sims, and his physicist friend William Crookes—both of whom, Wallace believed, knew how to set up conditions to prevent tampering. With a medium present, Crookes took a series of photographs of a female spirit-form. The medium had spent a week with Crookes and his family, sharing a room with one of Crookes's servants, and was under constant surveillance. It would have been impossible for a second outsider to enter a house full of people without ever being detected, Wallace reasoned, yet the same spirit-form never failed to appear during a séance. The spirit's features were like the medium's, as though the two were sisters, but a close comparison of the photographs, which Wallace examined, showed that the spirit-woman and the medium were not the same, for the spirit-woman was a head taller. Crookes even managed to obtain a photo of himself standing beside the two women, demonstrating the respective differences in their heights.[42]

Wallace also knew two amateur photographers, Mr. and Mrs. Guppy. At this point in her career, Mrs. Guppy—the former Agnes Nichol—had progressed far in her mediumistic powers, and the materialization of spirit-forms had become a regular feature of her séances. The Guppys generated some of the first spirit photographs. On March 14, 1874, Agnes Guppy accompanied Wallace to the studio of a professional photographer named Hudson, who had made portraits of the Guppys that revealed a white-robed figure with "Oriental" features in the background. Wallace expected that if he were to obtain a spirit photo of his own as he posed for the photographer, the materialized figure would be that of his oldest brother, William, in whose name Guppy often had received messages. He watched as the film was developed and detected no opportunity for trickery. Much to his surprise, two of the three photographs contained not William but a female figure, holding flowers; this figure was the first to appear as the developing fluid was poured on, while his own portrait formed afterward. One of the figures bore some resemblance to his mother. He showed the proofs to Fanny and sent another set to John in California. John had an old photograph of their mother and confirmed the startling likeness. John had no reason to agree with Wallace, since neither he nor his wife knew anything about spiritualism; in fact, "they were both prejudiced against it." Convinced of the reality of these photographs, Wallace wrote, "I see no escape from the conclusion that some spiritual being, acquainted with my mother's various aspects during life, produced these recognisable impressions on the plate. That she herself still lives and produced these figures may not be proved; but it is a more simple and natural explanation to think that she did so, than to suppose that we are surrounded by beings who carry out an elaborate series of impostures for no other apparent purpose than to dupe us into a belief in a continued existence after death."[43]

In the meantime, Bertie, Wallace's older son, had become severely ill with scarlet fever. Confined to bed by his doctor because the slightest activity exhausted him, he was moved to Hurstpierpoint, where his grandparents and aunts could nurse him while Annie attended to their other two young children, Violet and Will. As Bertie's condition worsened, Wallace felt frustrated with his son's medical care. His suggestions, based on an idiosyncratic view of disease causation, were dismissed or ignored. "Our orthodox medical men are profoundly ignorant of the subtle influences of the human body in health and disease and can do nothing in many cases which Nature could if assisted by proper conditions," Wallace complained to the always sympathetic Arabella Buckley. "We who know what strange and subtle influences are around us can believe this."[44]

Wallace considered taking Bertie to an unorthodox specialist in a nearby village only five miles away, but he feared that Bertie would not survive the journey. After four weeks without any sign of improvement in his son's condition, Wallace, now desperate, gave up on traditional medicine and turned to an unspecified form of unconventional therapy. To his surprise and relief, Bertie seemed to rally. Buoyed by the belief that his son would continue to improve, he left Hurstpierpoint by coach on April 23 and spent the night with Fanny and Thomas in London. But when he arrived in Grays on the following evening, a letter awaited him. During the night, there had been a sudden turn for the worse. Bertie's grandparents, who did not share their son-in-law's spiritualist convictions, called in their Anglican minister to baptize Bertie and give him the last rites, which would ensure a proper Christian burial. (Wallace, having abandoned his faith in organized religion years earlier, had elected not to baptize any of his children.) Bertie died at 8:00 A.M. on April 24, shortly before his seventh birthday. He was buried in the cemetery of the Hurstpierpoint Anglican church.[45] Except for a brief and almost bloodless account of the facts of Bertie's illness sent that day to Buckley, Wallace recorded nothing more about his feelings. His anguish can only be imagined, the faintest glimmer of which can be detected as displaced anger toward the medical profession. Had the unconventional treatment been started two weeks earlier, he believed, Bertie's life might have been saved. In the future, he would rarely talk about Bertie. Whenever the subject arose, Buckley later noted, tears came to his eyes and he would quickly move on to some other topic. In his autobiography, written thirty years after Bertie's death, he avoided mentioning his son's name, as though he had had only two children and not three.[46]

During what was certainly the darkest moment of his life, Wallace composed an essay entitled "A Defence of Modern Spiritualism," which was published in the May and June issues of the *Fortnightly Review*. The editor of the magazine, John Morley, was an acknowledged agnostic and freethinker interested in psychic phenomena. He had commissioned the article, which was written with great "diffidence," Wallace confessed, but with an "imperative sense of duty." He was convinced that spiritualism, though treated with ridicule and contempt, embodied truths of the greatest import to human progress. He knew that his essay would try the patience of many of his readers, but he begged them to set aside preconceived notions and follow his argument, which he promised to support with a "vast mass" of facts. In part, he hoped to address some of the virulent attacks on spiritualism by his new enemy, William Benjamin Carpenter. It was Carpenter's firm conviction that spiritual communications came from within, not from without, and he discredited and impugned the judgment of the investigators, including Crookes. Wallace's article is laced with anger and bitterness toward Carpenter and those other colleagues who openly disparaged the reputations of serious scientists.[47] After outlining the history of the spiritualist movement, he gave evidence of "incontrovertible" facts from the experiences of the prominent men who accepted the truth of the phenomena they had personally witnessed and investigated. Addressing John Tyndall's criticism of his initial defense of spiritualism, he now provided numerous examples from his own experience. And then he leveled with his critics, asking:

> Now what do our leaders of public opinion say when a scientific man of proved ability again observes a large portion of the more extraordinary phenomena, in his own house, under test conditions, and affirms their objective reality; and this not after a hasty examination, but after four years of research? Men "with heavy scientific appendages to their names" refuse to examine them when invited; the eminent in the society of which he is a fellow refuses to record them; and the press cries out that it wants better witnesses than Mr. Crookes, and that such facts want "confirmation" before they can be believed.

Wallace stated that the phenomena in their entirety did not require further confirmation. They were proved as facts, as in other sciences. The only way to disprove their truth was through new facts and accurate deductions from those facts. Spiritualism explained a whole array of phenomena in human history, from the oracles of antiquity to the miracles of the Old and New Testaments and even the "so-called superstitions" of savages, providing convincing proof of ethereal beings and their power to act on matter. This discovery revolutionized philosophy, Wallace believed, by demonstrating mind without brain, intelligence disconnected from the material body. Once the body disintegrated in death, the person's existence could continue. The dead still lived and guided humankind, but without the proper conditions they could not make their presence known. "[Spiritualism] thus furnishes us that proof of a future life which so many crave," he wrote. "This is the answer to those who ask, 'What use is it?'" For men like Herbert Spencer, who believed that certain domains remained by their very nature unknowable, he had evidence to the contrary.[48]

Wallace concluded his essay by uniting the philosophies of spiritualism and phrenology. His prose soared, like a homily fired off by an articulate and impassioned preacher. Wallace predicted that those who indulged the lower faculties of passion, selfishness, and acquisitiveness at the expense of the higher faculties of affection and other intellectual powers exposed themselves to a life of misery and an arduous, uphill struggle for a higher existence after death. The spiritualist was deterred from crime by the knowledge that its consequences would cause him ages of remorse. The bad passions that such behavior encouraged would be a perpetual torment "in a state of being in which mental emotions [cannot] be laid aside or forgotten amid the fierce struggles and sensual pleasures of a physical existence." Phrenospiritualism was a science of human nature, Wallace said, not the vestiges of a primitive tendency to believe in the supernatural. As an experimental science, spiritualism laid "the only sure" foundation for a true philosophy and a pure religion. Spiritualism united all creeds and all humankind in a universal brotherhood, abolishing forever the ceaseless discord and incalculable evil perpetrated by revealed religion.[49]

Wallace had crossed the Rubicon, and there was no turning back. By publishing "A Defence of Modern Spiritualism," he had declared war on the scientific materialism of his age. His peers reacted predictably to this beautiful dream of a new science of human nature. Buckley reported that Lyell was not convinced a bit by it, nor were Darwin and Huxley.[50] Publicly, however, his opponents remained silent. Darwin, who also had suffered the loss of a beloved child and could empathize with Wallace, finally sent a note in mid-December, asking Wallace to meet him for lunch in London. It would be "a great pleasure" to see him, Darwin wrote, but he was forced to cancel at the last minute because of illness.[51] It is not certain if they ever met.

On February 22, 1875, Sir Charles Lyell died at the age of seventy-eight. His remains eventually were interred at Westminster Abbey, the result of a petition circulated among members of the Royal Society. It was a fitting tribute to one of Britain's greatest scientists. Darwin told Buckley that almost everything he had done in science he owed to the study of Lyell's great works. Wallace could have said the same. For some reason, Wallace did not attend the funeral, but years later he remembered Lyell with deep fondness, looking back on their friendship "with unalloyed satisfaction as one of the most instructive and enjoyable episodes in my life-experience."[52]

Between June 1, 1874, and Lyell's death, Wallace published only two articles, remaining preoccupied with his book on geographic distribution. Hard work, as Darwin once advised, was the best way to banish sad thoughts. As he gathered more data, the work grew at an alarming rate; tables and chapter headings were revised not once but many times. Wallace had amassed so much material from so many sources of such variable quality that he often lost his way. His habit of committing facts and arguments to memory now worked against him. Had he created a file system, as Darwin did, he might have struggled less. Instead, he scanned the margins and flyleaves of books in his library for personal annotations on whatever subject he happened to be thinking about at the time. Living outside London was an additional inconvenience. He often had less than a day to search for information in London's libraries and museums, which led to hasty verification of references, careless documentation, and failure to properly acknowledge sources. His work suffered from this carelessness, but he was less interested in the minutiae, which were easily amended, than the larger picture, preferring to paint with broad strokes and leave the details for others. Above all, he strove for clarity and simplicity. But Wallace lacked Darwin's patience, organizational skills, and vast network of international correspondents. He confessed to Darwin that his "horrid book" was driving him mad with drudgery.[53]

Gradually Wallace was able to divide the earth into six zoogeographic divisions, a number still accepted in principle. His divisions followed those originally proposed by the ornithologist Philip Sclater, who in 1857 had established six primary zoological regions based on the distribution of the chief genera and families of birds. But Wallace broadened his analysis to other classes of animals. Although he admitted that a perfect division of the earth into different zones was impossible, he believed that a broadly marked, easily remembered set of regions was a convenient, natural, and intelligible guide to the geographic distribution of animals, even if somewhat contrived. Ideally, he said, the possession of "peculiar" families or genera was the main character of a primary zoological region. But the absence of certain families or genera was of equal importance, for even if no physical barrier prevented their entrance in recent times, some physical barrier must have existed at some point in the past. A distinct and well-balanced set of organisms in the present era thus had created a "living" barrier that served to keep out intruders.

The Palearctic region included Europe, "temperate" Asia, and North Africa from the Mediterranean coast south to the Atlas Mountains of Morocco. The Ethiopian region included Africa south of the Atlas Mountains, Madagascar, the Mascarene Islands (Mauritius, Réunion, and Rodriguez), and southern Arabia. The Indian region included India south of the Himalayas, southern China, plus Borneo and Java. The Australian region included Celebes and Lombok, east to Australia and the islands of the Pacific. The Nearctic region included Greenland, North America, and northern Mexico. The sixth and last region was the Neotropical, which included southern Mexico, Central America, South America, and the islands of the Caribbean. Wallace then drew up tables of the families and genera found in each region and subregion. He immersed himself in paleontology, possibly having obtained advice from Richard Owen during their strolls at Augustus Mongredien's estate. By reading the "best authors," he wrote in the preface of his work, he was able to sketch the relation of extinct mammalia to existing groups. From the outset, he had decided not to include *Homo sapiens* in his analysis, his reason being the universal distribution of the human species. He also concentrated on genera and families rather than on species or varieties—a practical rather than a philosophical decision.[54]

Wallace divided his two-volume book into three parts. Part 1, in which he enunciated the general principles on which zoogeography was based, consisted of a series of short chapters on the mechanisms underlying the proliferation and dispersal of animals, the effects of physical geography and geologic forces on distribution, the reasons for dividing the world into six zoological regions, and the issues that the new science raised about the system of classification. In part 2, he examined each region and subregion, focusing on the peculiar characteristics of groups and their relationships with groups in other regions and elaborating on the dynamic relationship between changes in the earth's surface and the movements of animals through time and space. For example, he asserted (correctly) that the North and South American continents had been united and separated multiple times, with forced intermingling of two different faunal regions. The appearance in the north of so-called Neotropical forms—that is, animals from South America—was a comparatively recent event, when climatic conditions resulting from the union of North and South America were favorable to their immigration. As a rule, successive glacial epochs forced the more dominant northern forms southward. In part 3, he surveyed all known families of the major animal classes and provided descriptions of the distribution of each genus. He designed his maps to show how the depths of oceans, the elevations in the earth's surface, and the character of regional vegetation posed physical barriers that delimited the distribution of organisms.

Wallace intended his section on zoological geography (part 2) for traveling naturalists, who would be interested in the numbers and kinds of animals they were likely to encounter as they moved from region to region. Specialists would find the section on geographic zoology (part 3) equally practical, with its descriptions, lists, and distributions of each known family and genus of the higher animals. It was undoubtedly the text that Wallace had longed to have at his side on his travels to South America and the Far East. He had created the book for the field naturalist, not the laboratory scientist, calculating that it would appeal to both the collector and the philosophical biologist.[55] He acknowledged his text's deficiencies to his friend Alfred Newton, imagining that in a matter of only a few years certain sections would become obsolete as more data poured in from all corners of the earth. He also knew that his designated regions would be criticized as either too static or too wide-ranging:

> The most radical question is, what are our zoological regions to be? Or rather, are we to have any at all, or are we to have *ornithological* regions, *reptilian* regions [and so forth], and if these, why not *coleopterous* and *lepidopterous* regions? . . . But I hold, and I think you will agree with me, that they would be confusing and almost are not even *zoological* regions but *biological* regions—for I believe that plants and animals *must* be radically affected by the series of geographical regions which our "regions" are intended to summarize.[56]

How best to illustrate his work bewildered him. Three hundred maps alone would have been required to acknowledge each family of animals. The solution came from Newton, who suggested a single map showing regions and subregions, which could be referred to by a series of numbers. Wallace also worked closely with his artists, instructing them to create typical scenes of the fauna of a particular region. His plates were of the highest quality, as they had been for *The Malay Archipelago*.

Wallace outlined the book for Macmillan in February 1875. His agent, G. L. Craik, proposed to publish two thousand copies of the first edition. Wallace hoped to retain the generous terms of his agreement for *The Malay Archipelago*. For the second edition, he expected a "considerably higher" percentage of the profits, since a large amount of revision would be required. The contract did not meet all his expectations, but it was satisfactory. He was advanced £200 for the first thousand copies and a royalty of 20 percent of the selling price for all two thousand. For the next edition, the arrangement remained the same unless he undertook a major revision.[57]

The book went to press in early 1876, but not without travail: there was an inordinate delay at the printer, accompanied by complaints about the difficulty of the work, about which Wallace became irate. Nine months earlier, he had reviewed the requirements with the printer in the presence of Alexander Macmillan himself; one month earlier, he had made the necessary corrections. "Your excuse for Messr. Clay is a very bad one," he wrote to his agent. "Either . . . Mr. Clay was, in August last, ignorant of his business and gave an estimate he could not possibly carry out,—or he willfully deceived us in order that we might not go elsewhere."[58] Once the book left the printer, other troubles arose. For example, Wallace was annoyed by a lack of symmetry in the cover design. The title, *The Geographical Distribution of Animals*, was arranged in a circle surrounding an ornate rendition of his initials, ARW. The word "Distribution" bothered him because it was not centered properly over his first initial; by centering on the "r," the word was split into five and seven letters, whereas by centering on the "i" the symmetry would have been maintained. He disliked the arrangement of the cover page. Among other requests, he asked for the phrase "Volume I" to be brought down to the center to give a proper division and balance to the page. He preferred a dark green color for the book's binding, which would show off the gold lettering to far greater advantage; he was thinking of the book's appearance on a library shelf. There was also some quibbling about the final length of the book. His agent had expected a total of 700 pages for the two volumes, but the final number, including blank pages, had ballooned to 934.[59]

When it was published in the spring of 1876, *The Geographical Distribution of Animals* was received with high praise. Critics immediately recognized the striking originality of the book's premise. The reviewer for *Nature* regretted Wallace's too brief introductory outline of its general argument: "It must be borne in mind that the very idea of the existence of any regular laws of distribution is a novelty to most people—even . . . to many who call themselves naturalists." Despite disagreement over some minor matters, he was astonished by Wallace's achievement:

> The mass of details to be gone through in bringing together the most prominent facts connected with the mammals, birds, reptiles, amphibians, fishes, butterflies, beetles, and land-shells of every different part of the world's surface, is a task that the boldest naturalist might well stand aghast at. The selection of such second-hand information and its reduction into a uniform shape, is of itself a task of appalling magnitude, and we can only congratulate Mr. Wallace on having had the strength and leisure to accomplish such a Herculean labour.[60]

Writing in the *Athenaeum*, Alfred Newton lauded the book as one of the two chief events of zoology during the past year.[61] At the 1881 meeting of the British Association, Joseph Hooker would refer to Wallace's book as one of the two most important general works on distribution that had appeared since the association's founding.[62] Wallace himself considered this study to be his most important scientific work. In the preface, he stated that his book bore a similar relation to the eleventh and twelfth chapters of *The Origin of Species* as Darwin's *Variation of Animals and Plants Under Domestication* bore to the first chapter of the *Origin*, being an expansion and elaboration of themes only briefly touched on. "Should it be judged worthy of such rank," he humbly said, "my long, and often wearisome labours, will be repaid."[63]

Darwin concurred with the reviewer from *Nature* and congratulated Wallace for his "dispatchment" of the imaginary continent theory, which had postulated now submerged vast continents to explain the presence of certain anomalous animals in different parts of the world.[64] Wallace had not ruled out continental extensions, like the subsidence and reemergence of a land connection between North and South America, but the existence of such continents as Lemuria and other geologic lost Atlantises—an idea he had long ago rejected—he put to rest forever. No facts could be found to support such a theory, he said, and none were necessary to explain present-day animal distribution.

For the next eighty years, *The Geographical Distribution of Animals* would serve as the definitive text on the subject of biogeography. But as knowledge of the distribution of animals increased, some biogeographers became dissatisfied with the rigidity of the zoogeographic classification system. Instead of thinking of fixed regions, they said, it was better to think of a more fluid process. This so-called faunal approach—the one widely accepted today—rejects the mainly descriptive, regional, and nondynamic picture of geographic distribution that some of Wallace's followers espoused. Contemporary biogeographers prefer to study Neotropical fauna, not a Neotropical region—a broader conception more clearly rooted in evolutionary theory. Moreover, the theory of plate tectonics has replaced the idea that the earth's continents and oceans have remained essentially unchanged throughout geologic time, except for the bridges and breaches formed by elevation or subsidence.[65] Wallace himself was less certain about the correctness of his approach than were some of his successors, as his letters to Newton suggest. However, the complex interaction of geologic forces and animal distribution, refracted through the lens of evolution and natural selection, remains the centerpiece of his analysis and is still valid. As he traveled through the Malay Archipelago two decades earlier—observing and collecting a bewildering array of insects, birds, and mammals and puzzling out the patterns and underlying causes of their distribution—a mental map had formed in his mind, one encompassing not only space but time. It was this multidimensional model that altered biologists' conception of geographic distribution and laid the foundations for the work of future biogeographers.

The publication of Wallace's book coincided with the unpleasant judgment passed down against him in his suit against John Hampden. Reduced to pleading for financial help from Alexander Macmillan, on March 28, 1876, Wallace asked for his contract to be restructured to give him a lump-sum payment upon publication. This, he said, would save him from having to sell his property at a loss. He also requested an advance of £500 at 5 percent interest to be paid from the royalties of the German and American editions. Macmillan advanced Wallace the money on his proposed terms, trusting to the book's merits and marketability among a select audience.[66] Just in the nick of time, Wallace had been saved from the shame of declaring bankruptcy.

CHAPTER 14

The War on Spiritualism

BY JUNE 1876, even the indefatigable Wallace felt beaten down. The Dell had lost its magic. He told friends that he was sick of the climate, which destroyed the delicate plants in his garden, and frustrated by his isolation from London, which prevented him from attending evening scientific meetings and social gatherings with friends.[1] But there were other concerns, both financial and personal. He was most alarmed by the health of five-year-old Will, who was wasting away from an unexplained illness, a frightening replay of Bertie's fatal condition. And he had received a communication from his dead brother William—in the form of automatic writing by way of Agnes Guppy—warning him to leave Grays as soon as possible.[2] The advance from Macmillan allowed him to hang on to his property just long enough to sell it at a reasonable price. He then moved his family to Dorking, forty miles to the west and only a short train ride to London. The immediate improvement in Will's health convinced him of the astuteness of his decision.

Not long before the move, Wallace had been unanimously elected president of Section D, the Biology Section, of the British Association for the Advancement of Science, which was scheduled to meet in Glasgow in early September. Michael Foster, a physiologist at Cambridge University and one of the general officers of the association, gave him the news in March. The president's duties, he said, required that he also take charge of Section D's Anthropology Department, while the two vice presidents headed the other two (Zoology and Botany). "This year it seemed right that the Department of Anthropology should have the honour of having the President—and the council felt it could not be in better hands," Foster said.[3] It was the highest honor the organization had awarded him since he began attending its meetings in 1862. Wallace assented. The next step was likely to be the presidency of the association itself. But the 1876 meeting would be his last. In that year, he reached both his apogee and his nadir.

The meeting opened auspiciously for Wallace. His address to the section, given on September 6, was eloquent and thought-provoking. After paying obligatory homage to Darwin's greatness and reaffirming his belief in the general truths of the doctrine of natural selection, he reiterated his conviction that Darwin's theory of sexual selection as the chief cause of coloration in birds and insects was insufficient to explain all the facts. He was currently studying the evolution of species on islands, he said, and had discovered striking examples of phenomena that could not be attributed to sexual selection. Large size and pale color characterized many unrelated species of butterflies endemic to islands, for example, while their allies on the mainland varied more in size and color. These were facts for which only natural selection could account. Because of the smaller number of predators and the less intense pressure from inter- and intraspecific competition, animals with traits that might be disadvantageous to mainland species flourished on islands.[4]

Wallace extended the theory of natural selection to insular floras. The peculiarity of these floras, he suggested, was due to the scarcity of flower-frequenting insects. On the most isolated oceanic islands, plants with showy flowers were rare. The unusually high proportion of ferns was due not to an equable climate and perennial moisture but to the relative absence of competing "phaenogamous" (flowering) plants, which, in turn, was a consequence of the scarcity of pollinating insects. On islands where flowering plants outnumbered ferns and "flower-haunting" insects were rare, the flowers were mainly small, green, and inconspicuous. Only those plants not requiring insect pollination won the struggle for existence and proliferated. Wallace felt that such facts were not appreciated because of the inadequacy of museums, which failed to group floras and faunas by location. If science hoped to make any progress in the understanding of geographic distribution, museum displays and collecting cabinets had to be arranged geographically.

Wallace then touched on the subject of humankind's antiquity and origin. He was treading on dangerous ground and knew it, but since he was also chairman of the section's Anthropology Department that year, he could not avoid the topic. Against his better judgment, he had given in to the secretary of the association, who had urged him to discuss recent discoveries in anthropology and ethnology.[5] He reminded his audience that only twenty years earlier, scientists had scoffed at the idea that human beings had coexisted with now extinct animals. No one doubted any longer that the human species was derived from the lower animals, but some believed that the human mind "and even some of [humankind's] physical characteristics" arose from the action of other forces. "We need hardly be surprised [at the] tendency among men of science to pass from one extreme to the other; from a profession (so few years ago) of total ignorance as to the mode of origin of all living things, to a claim to almost complete knowledge of the whole progress of the universe, from the speck of living protoplasm up to the highest development of the human intellect. Yet this is really what we have seen in the last sixteen years." Wallace believed that opposition was the best incentive to progress; even good theories should not be allowed free rein.

One such theory was the linearity of human progress from a primitive to a civilized state. There was much evidence to the contrary. In Java, for example, the magnificent ruins of Borobodur and Prabanam loomed over the hovels of the current inhabitants, suggesting degeneration, not advancement: "The course of [our] development has been far less direct and simple than has hitherto been supposed; . . . instead of resembling a single tide with its advancing and receding ripples, it must rather be compared to the progress from neap to spring tides, both the rise and the depression being comparatively greater as the waters of true civilisation slowly advance towards the highest level they can reach."

Wallace's address, carefully constructed to skirt controversial issues, won praise, but his luck changed later that week. As chairman of the Anthropology Department, he presided over the reading of a paper that had aroused considerable debate among council members before its presentation. A single vote had prevented the paper's rejection, an outcome that angered many at the Glasgow meeting.

The stirring days of the wars between Thomas Huxley and Richard Owen had passed. The press, which had been clamoring for years for something more newsworthy from the British Association, was calling the 1876 meeting one of the dullest on record. The *Times* reminded men of science of their duty to engage the interest of the public and spend less time on esoteric matters. But a paper on clairvoyance was not what anyone had in mind.[6] Suddenly all eyes turned to the Anthropology Department and William Fletcher Barrett, a thirty-two-year-old physicist at the Royal College of Science in Dublin and a former student of John Tyndall. Barrett was also a hypnotist. His paper, "On Some Phenomena Associated with Abnormal Conditions of the Mind," was the most exciting event of the week. More than a thousand people filled the lecture hall, with two hundred more filling the aisles.

Barrett's paper addressed two subjects, mesmerism and spiritualism. He first described the case of a girl he had hypnotized whose most unusual aspect was "the degree of exaltation of [her] perceptive powers." She displayed the usual responses to pressure on different parts of her skull, altering her behavior and expression depending on the place touched, as predicted by phrenologists. What was most remarkable, however, was her uncanny sensitivity to his voice. If he whispered her name, she responded at once. As he moved farther and farther away, even out of sight and earshot, she continued to respond, though more weakly as the distance increased. When he tasted or smelled something or touched something hot or cold, even while he was standing behind her and at some distance, she reacted as he did. No one else was able to produce these effects on her; the rapport between the two was extraordinary. Barrett disputed William Benjamin Carpenter's interpretation of this uncanny affinity. Carpenter denied the existence of any special rapport between hypnotist and subject. According to Carpenter's theory of "unconscious cerebration," Barrett said, the sensitive girl already would have been convinced of the hypnotist's ability to exert a special influence over her before entering the trance state, just as people follow the commands and actions of a charismatic leader. "I do not think that the whole mystery of this so-called 'rapport' can be disposed of quite so easily," Barrett asserted.[7]

To demonstrate his point, Barrett described a few of his other experiments with this girl. After having induced the trance state in her, he picked a playing card at random from a drawer, glanced at it, and concealed it inside a book. Bringing the closed book to her, he asked her to describe what was inside. She said that she saw something with five red spots. The card was the five of diamonds. When he inserted a card in the book without having looked at it, she could not describe it. He next asked her to travel, by means of her imagination, to Regent Street in London and name the shops. Although she never had been to London, she described one shop that he was thinking of at that moment. Her astonishing ability convinced him of the existence of clairvoyance, or thought reading, which he likened to the conduction of electricity through a material across space. The nervous activity that constituted thought could be excited in the clairvoyant by corresponding nervous activity in another individual, which was transmitted across space by some sort of radiant energy unperceived by any of the senses. He believed that the facts he was presenting justified further inquiry: "All I wish to urge is, that it is not wise to push forward a natural feeling of incredulity on this matter, as a barrier to a possible extension of knowledge."

Barrett then turned to the topic of spiritualist phenomena, which he divided into two classes. One class manifested itself in either subdued light or total darkness and was witnessed by only the "favored few." The well-known case of Daniel Dunglas Home floating from one room to another or handling hot coals with impunity belonged to this class. Eminent men whose sincerity and honesty could not be doubted either saw or thought they saw these phenomena; in his opinion, the latter was closer to the truth. Relaxed by the serene ambience, their senses "enfeebled" by near-darkness, they would yield to the emphatic suggestions of the medium. To test this hypothesis, he once hypnotized a young man and drew his attention to a pair of shoes on a table. He told the young man that he was standing in the shoes and was now going to rise up and float around the room, and then he pointed to the successive stages of his imaginary flight. Upon awakening, the subject was convinced that Barrett had floated around the room.

The other class of phenomena occurred in broad daylight and was witnessed by the skeptical and credulous alike. The usual explanation was "clever conjuring," which, though often true, could not explain all the facts. "I am well aware that I shall arouse feelings of incredulity, if not contempt, on the part of those who have never investigated the phenomena," he predicted. He related the case of a young girl—the daughter of a gentleman of good social position—who was troubled by episodes of knocking. The knocking, which occurred when the child was in a "passive" condition, displayed some "intelligence"—say, by keeping time with a tune or spelling out words. Even in the full glare of the sun, when every precaution to prevent deception had been taken, the knocking was heard in different parts of the room, beyond the reach of the child, whose feet and hands he watched closely. The knocks were not limited to the house but occurred on the lawn as well. After dozens of tests, and with skeptical friends attesting to their reality, he was compelled to conclude that the sounds were real and not produced by any visible agency. "Is it not possible that there may be some foundation for the stories of occasional supernatural eruptions into the present visible universe?" Barrett asked. "I do not here refer to the great mysteries of religion, but to the numberless cases which float in families wherever you go; many, no doubt, are sheer hallucination, but still a formidable residue exists that cannot be explained."

Barrett rejected Carpenter's highly publicized opinion that these matters, which defied common sense, should be regarded as "a diluted insanity," an opinion he found astounding. "Common sense," Barrett protested, "is set up as the tribunal before which every fact must pass muster before it can be accepted." He concluded by appealing to the true scientific spirit, which did not repudiate ideas on a priori grounds alone, and he recommended a patient, impartial, and systematic inquiry by a specially appointed scientific committee.

After Barrett had finished his presentation, Wallace addressed the audience: "Ladies and gentlemen, I think you must all agree with me that I have hardly heard in this room a paper which, whatever you may think of its subject matter, has been treated in a more careful, and a more truly scientific manner, than this paper of Professor Barrett's." Here he was interrupted by applause. "We may hardly wonder at this when we consider that he is a thoroughly trained man of science," he continued. "As this subject is one which will naturally excite a considerable amount of discussion, and perhaps some feeling, I hope you will support me in my endeavours to restrict the discussion of it within certain limited bounds. I think it absolutely necessary, as this paper is a record of facts solely, that we should not allow persons to speak who know nothing whatever of the facts." More applause. "I will call upon those gentlemen who, I believe, are prepared to add something to our knowledge on this interesting though mysterious subject."

Sir William Crookes was the first to speak. Crookes said that he had come to the meeting to listen, not make comments, but Barrett's paper, which he praised, brought several things to mind. He objected to Barrett's characterization of the state of the scientific men who had witnessed phenomena at Home's séances:

> A physicist shows an experiment before an audience of physicists. Persons might say, "Your audience think they see it; you have very wonderful . . . mesmeric powers; you throw a kind of glamour over your audience; nothing takes place; but you make them think they see them." That is exactly the kind of reasoning which I have heard and read applied to certain very extraordinary phenomena which I had been investigating for some years, and which I am glad to see Mr. Barrett is just nibbling at the edges of.

As a scientist, he said, he never had investigated phenomena except under his own test conditions: at his own house, among his own select friends and spectators, and with apparatuses he had designed. He always had let the device test the phenomena, avoiding the evidence of his senses as much as possible. "But when it is necessary to trust to my own senses, I must entirely dissent from Mr. Barrett, when he says a trained physical inquirer is no match for a professional conjuror. I maintain a physical inquirer is more than a match."

The physicist John William Strutt (Lord Rayleigh), who likewise expressed his original intention to remain on the sidelines, instead came forward to congratulate Barrett for his moral courage. Strutt was thirty-four years old at the time and would later discover the element argon, for which he would win the Nobel Prize. He had been attracted to the subject of spiritualism after reading Crookes's articles on the subject in the *Quarterly Journal of Science*. "I have seen enough to convince me that those are wrong who wish to prevent investigation by casting ridicule on those who may feel inclined to engage in it," he said. For most of his life, he was intrigued by psychic phenomena and would not hesitate to associate his weighty name with psychic research, but he was less convinced than Wallace and Crookes that such phenomena had a supernatural origin.

Charles Ottley Groom-Napier, author of *The Book of Nature and the Book of Man*, next gave a long-winded account of his own spiritualist powers, including examples of his clairvoyant talents, which provoked grumbles from the audience. Some grew impatient with him and laughed at his descriptions of his experiences. "Give us facts!" someone shouted. Undeterred, Groom-Napier continued to give personal testimony until he was interrupted by more laughter, shouting, and ridicule and then finally told to shut up. Humiliated, he returned to his seat.

A Reverend Thompson then rose to tell the audience that he had been waiting for thirty-five years for a good scientific demonstration of spiritualism and had yet to see one. Spiritualism, he said, was not a science like other sciences, and discussions of it did not belong at the British Association meetings. This time Wallace interrupted. "Please do not give us generalisations," he said, "give us some fact or statement." His request was seconded, but to no avail. After reading and hearing about these matters, Thompson continued, he had not found anyone among his circle of friends and acquaintances who was convinced of the truth of the phenomena. Wallace once again interrupted him, warning, "I must stop you if you go on in that way. I have told you we cannot have mere opinions and statements of disbelief. We must have facts, and if you cannot give us any, I shall rule that you must sit down."

"I am a fact," Thompson declared, evoking laughter from the audience. "Why should those gentlemen who hold opinions on the opposite side be afraid to have their opinions overhauled?" He tried to go on, but was derided. "Will the President kindly keep order for me?" he pleaded. Wallace asked for silence and allowed Thompson to sum up his argument. When Thompson had finished, Wallace said, "I think in order to limit this discussion within reasonable time, it ought to be made a rule that the speakers should reply to the paper, and not to the accidental remarks of other speakers. If they do so we shall have other gentlemen replying to them, and others to them, till we get miles away from the subject with which we commenced. I therefore beg, as chairman of this meeting, to rule that we will have no answers by one speaker to another unless the point refers directly to a matter brought forward by the reader of the paper."

"Hear, hear!" someone cheered.

Thompson now raised further objections to the findings of séances. Why confine séances to a few friends, he wondered. Let them be tested publicly. To this suggestion, some members of the audience objected, and the room erupted in a cacophony of conversation, accompanied by the stamping of feet. "I am glad you approve of my sentiments," Thompson shouted sarcastically. Wallace demanded that he sit down.

"You won't hear my case because it is too strong," Thompson complained, getting in one last word before taking his seat.

Carpenter, who had been uncharacteristically silent up to this point, asked Wallace for permission to speak. The audience quieted as he rose from his chair. He began by raising objections to Barrett's quotations of his work, which, he said, had been taken out of context. "For instance, when I say of phenomena, that [they are] 'impossible to believe,' I do not mean at all to assert that we know everything. . . . I do not think anyone has a right to say I have ever been unwilling to receive new evidence on any fact." He admitted that some individuals seemed to have the curious power of thought reading, which he had not yet investigated. "I believe it is one of the special gifts of a detective policeman [who] has a greater power of what is called 'seeing through a millstone' than other people." The detective perceives certain signals—tone of voice, facial expressions, and gestures—not consciously but intuitively. While not denying the possibility of direct communication between one nervous system and another, Carpenter held that body language was a more probable explanation. "[W]e are not at all unwilling to investigate, if only the things are presented to us in a way which enable us to investigate." This was followed by applause.

Over the years, Carpenter continued, he had attended many séances and been told over and over again that his disbelief interfered with the manifestations. Lately, however, he had gone to séances held by Henry Slade, a celebrated American medium who was visiting London, and had witnessed astonishing things. He proposed that Slade come to his house and conduct a séance. "If Dr. Slade can do in my house, with my chairs, with my tables, with my slates, without any previous preparation, what he showed me in his own room, I will then honestly say this is a case for thorough investigation." The audience enthusiastically applauded his challenge.

A Reverend Dr. McIlwaine, who spoke after Carpenter, impeached the credibility of Barrett's witnesses, one of whom, he said, was a relative of his and not fit to give evidence. In his own experience, the majority of mediums were young girls, who were not proper persons to base great superstructures on, a remark that drew both laughter and hisses from the audience. But McIlwaine went further than Wallace could tolerate when he referred to what he called the great facts of the Bible. "I cannot allow theology to be introduced into this discussion," Wallace interrupted. McIlwaine dropped the biblical references and began an account of a séance held in Belfast in 1856. Wallace asked whether he had been present. McIlwaine said that he had not been present but could give evidence from secondhand sources.

"We cannot have it," Wallace declared.

"I will give you the name of the gentleman," McIlwaine replied.

"That won't prove that he is trustworthy," Wallace said. "You have accused one of your own [relatives] of being untrustworthy; how do we know that these people are trustworthy?"

McIlwaine backed off again and instead offered an American professor, whose name he could not remember, who had visited his town and produced a number of phenomena.

"You must not allude to people without giving the name," Wallace said.

The professor was a charlatan, McIlwaine continued, ignoring Wallace's objection. The man had caused all sorts of mischief before his true identity was discovered. McIlwaine concluded that spiritualist phenomena were due to three things: the excitability of women, the credulity of the observers, and imposture. But before Wallace could again intervene, Barrett defended himself by repudiating McIlwaine's credibility. McIlwaine's relative was not the person to whom he had alluded in his paper. That person was an Englishman not even remotely connected to McIlwaine and someone he could not have known. As for the mesmeric phenomena, they had occurred in the house of McIlwaine's brother-in-law, a man with whom McIlwaine had not communicated for many years and whose credibility McIlwaine had impugned because of a difference of religious opinion.

Hyde Clark, the chairman of the Psychology Committee of the Anthropological Institute, interjected that he did not believe that Barrett's paper contained any new facts. It was an old story, he said, told time and time again. The phenomena had to be presented in such a way that "men of sense" should be able to reproduce them themselves. Carpenter agreed, noting that scientists had accepted Crookes's statements about his various apparatuses because they could be reproduced by anyone. If Crookes could reproduce spiritualist phenomena in the same manner, those phenomena would receive the same attention.

It was Wallace's turn to speak. He had never seen a perfect case of clairvoyance himself, he said, but he believed in its probability:

> There are certain phenomena you cannot bring before you. They must be sought for, and a case very much in point is that of the meteorolites [*sic*], the fall of which was for many centuries disbelieved by scientific men, and it was only after a considerable number had been actually recorded that they were accepted as a fact. According to the general system of unbelief, we ought to disbelieve [in them] even now, because all the scientific men cannot prognosticate when a meteorolite is going to fall, and we cannot go and see them fall.

Barrett rose once again. He congratulated the association for allowing the presentation and full discussion of phenomena previously shunned at such meetings and praised the audience for its polite reception of a controversial topic. Although he was not personally prepared to accept the spiritualist explanation of the phenomena, he lauded men like Wallace and Crookes, who had maintained objectivity in their investigations. He challenged Carpenter to account for *every* phenomenon as fraudulent, not just a great many. He called on all his fellow scientists who had witnessed these phenomena to have the courage to describe their experiences in public and accept criticism if it was due. He started to propose a resolution, but shouts from those opposing him blocked him, while others cried out for order. "I am out of order in asking the section to pass the resolution," Barrett shouted, raising his voice for the first time, "but I think I am not out of order in asking that the resolution be referred to a committee of the section." Barrett proposed that a committee of scientific men be appointed to inquire into the phenomena of mesmerism and spiritualism to remove any uncertainty that currently prevailed. Ignoring the suggestion, Carpenter returned to Barrett's earlier objection that he had characterized the belief in spiritualism as "a diluted insanity." He had used that phrase, Carpenter said, about beliefs that were on "extremely" inadequate grounds, not about every belief.

"I am only too glad to be corrected on this point by Dr. Carpenter," Barrett said, "for it is further confirmation of the facts in my paper with regard to the gradual change of opinion."

Carpenter asked Barrett what those facts of spiritualism were: "If, for instance, I visit Dr. Slade, and the chair jumps up and then falls back again, is that a fact of spiritualism? How do I know that it is a fact of spiritualism till I have carefully investigated that chair, till I have found whether there are concealed springs and mechanism to make that chair jump up?" In refutation of clairvoyance, he told a story about his friend Sir James Simpson, who placed a £500 note in a sealed box and offered it as a prize to any clairvoyant who could tell him the note's serial number. No one had ever succeeded.

Carpenter's remarks angered Wallace, who had himself visited Slade only a couple of weeks before the British Association meeting and had been convinced of his authenticity. Wallace wanted to make a "slight" correction to Carpenter's clarification of his phrase "diluted insanity." Carpenter had applied that phrase to Crookes, and everyone knew it, he said. It was not right to put a new slant on it. As for the monetary note, one negative was insufficient to disprove established facts. Why should anyone believe that there was a note in the box in the first place, he asked. "How do we know that this was not a piece of blank paper? We are asked to take these things on hearsay." If Carpenter disbelieved Barrett's facts, why should Wallace believe Carpenter's?

"I will not stand here and allow the character of Sir James Simpson to be called in question," Carpenter exclaimed.

"I call no man's character in question," Wallace retorted.

But his disparagement of Simpson offended others besides Carpenter. "You said, 'How did you know that it was a genuine note?'" Carpenter replied heatedly, encouraged by cheers of support. Wallace protested that those were Carpenter's recollections, not documentary evidence, at which point Barrett interrupted, fearing that the discussion threatened to degenerate into a shouting match or worse. His congratulations to the group for their gentlemanly behavior had been premature. People were leaving in disgust. "Don't let us break up in anger," Barrett pleaded. "There has at least been sufficient evidence given to show that full inquiry ought to be made." But the discussion came to an abrupt end, and those in the audience who remained clamored for the next paper. Barrett's recommendation to establish an ad-hoc committee to study the validity of psychic phenomena was tabled indefinitely and would never be brought up again at a meeting of the British Association.

The issue of spiritualism and science, simmering beneath the surface for five years, had finally arrived at its Armageddon. Sides were taken, and the battle commenced. The first sortie came from Edwin Ray Lankester, a twenty-nine-year-old fellow of Exeter College, Oxford, and a professor of zoology at University College, London. Lankester spoke for the scientific establishment when he wrote a scathing letter to the *Times* accusing Wallace of having degraded the British Association's annual meeting.[8] Wallace replied indignantly to Lankester and attempted to show that the charge was "wholly without foundation." The secretary of Section D had brought Barrett's paper before the subcommittee of the Anthropology Department, Wallace said, but after a full discussion the subcommittee had vetoed its presentation. The paper next passed to the section committee, which decided to let the paper be read, though the decision was not unanimous. "Professor Lankester is evidently ignorant of the fact that the reading of this paper was decided after a vote taken in two Committees. . . . As to [his] opinion as to what branches of inquiry are to be tabooed as 'degrading,' we have, on the other side, the practical evidence of such men as Lord Rayleigh, Mr. Crookes, Dr. Carpenter, and Colonel Lane-Fox—none of them inferior in scientific eminence to Professor Lankester, yet all taking part in the discussion, and all maintaining that discussion and inquiry were necessary." Moreover, the paper received the attention of the president of the British Association, Thomas Andrews, and of a crowded audience that seemed eager to hear more on the subject.[9]

But a member of the Anthropology subcommittee, a man named J. Park Harrison, thought that Wallace had unfairly tried to place blame on the committee and subcommittee of Section D, when in fact it was Wallace's own underhandedness that had been responsible. In a rebuttal to Wallace in the *Times*, Harrison wrote that the subcommittee had voted by a slim majority to refer the paper "in the usual way"—that is, to the committee of Section D itself. Wallace, however, had spoken against this proposal, urging that such a move would practically shelve the paper since the majority of the committee's members undoubtedly would oppose any discussion of spiritualism at the association. Nevertheless, the subcommittee voted to pass along the paper. Despite a request to have the paper forwarded in time for a general meeting of the committee, the paper reached the secretary's office after most of the members had left Glasgow. The scientific merit of the paper was therefore discussed before a much smaller meeting. Once again, Wallace had pleaded on the paper's behalf, and by a majority of one—his own vote—he had won.[10]

Augustus Henry Lane-Fox (later General Pitt-Rivers), president of the Anthropological Institute of London and a member of the Anthropology subcommittee, came to Wallace's defense. Contrary to his statement, Harrison had been in a minority "in wishing to taboo the subject." Lane-Fox continued, "So far from meriting or receiving obloquy from men of science on this account, I believe that the Department of Anthropology as a practical science will only do itself honour by boldly grappling with the errors of our times." How was science to play a leading role in preventing the public from "falling into the absurdities of past ages [with respect to] necromancy" if men professing to be anthropologists were afflicted with "a superstitious terror" of this subject and were "content to limit their investigations exclusively to old mounds, old scratches, or the relative position of people's toes?" He agreed that science had a perfect right to veto the "intrusion of the empirics" on ground that it already had reclaimed and mapped out on its own, but who said that science had claims on the domain of the will and consciousness? "It is absurd," he noted, "to speak of all previous experience in a field of inquiry where there is no experience."[11]

Lane-Fox's impassioned plea notwithstanding, Lankester and his supporters were bent on further mischief. He and a former Oxonian, Horatio Donkin, an assistant physician at the Westminster Hospital, plotted to expose Henry Slade as a fraud in order to humiliate Wallace, Crookes, and other scientists who had put their faith in what the Lankester camp considered nothing more than charlatanry and self-delusion. Slade, a spiritualist from Michigan, had stopped off in London in July 1876 while en route to Russia at the invitation of a committee from the Imperial University of Saint Petersburg that wanted to test his powers. During the previous fifteen years, Slade had impressed numerous skeptics in America by producing written messages on blank slates, which had been sealed and placed in full view of investigators. Slade charged a nominal fee for his séances, operated in broad daylight, and performed other feats, including having an accordion play a tune in the manner of Home. He dared the foremost scientific figures of the day to attend his séances and investigate him.

Lankester first visited Slade on September 11 at a house on Upper Bedford Place. Slade led him into a well-lighted room furnished with a four-legged table and invited him to take a seat at the table. Slade sat catty-corner to him, took his hand, and asked if he had ever attended a séance; Lankester replied that he had been to two of them. As he lapsed into a trance state, Slade suddenly began to shiver, as though exposed to cold, and told Lankester that the influence was very strong that morning. There were three taps, which shook the table, and Slade said that these phenomena had been occurring for twelve years, ever since the death of his wife, Allie. It was Allie's spirit that wrote to him and produced the phenomena, he said. He showed Lankester a clean double-sided slate, free of writing, with a pencil resting on top of it. He then put the slate beneath the table and held it flat against the underside with the fingers of his right hand, his thumb braced on the table and the pencil "supposedly" between the table and the slate; his left hand was plainly in view. The slate was so firmly pressed against the table, Lankester noted, that no hand or finger could get at the piece of pencil. But he immediately heard a scribbling noise. The spirit, Slade said, was at work. The slate was removed, and Lankester saw a message that read: "I am here to help you . . . Allie."

Slade repeated this process several times, sometimes leaving the slate on his knee for a few moments. Meanwhile, Lankester kept a cool eye on the medium's every gesture, ignoring what he considered to be intentional distractions—raps, gentle kicks, and movements of the table—and detecting subtle movements of Slade's right arm while his wrist and fingers were under the table supposedly holding the slate. Each time, the message was longer and better written, though not always in the same flowing hand. Sometimes Slade placed the slate on top of Lankester's head, at which point the spirit-writing would allegedly ensue. Throughout the remainder of the séance, other messages appeared, including one from Lankester's "Uncle John," though he had no such uncle. Lankester feigned an "ardent belief" in the mysterious nature of what he saw. However, he was utterly convinced that Slade wrote the messages while the slate rested briefly on his knee. He arrived at this "hypothesis," he said, by noting the delay that invariably occurred between his being shown the slate, with its two clean sides, and Slade's placing it beneath the table or on top of Lankester's head. During this delay, Slade made various excuses, bit the piece of pencil, or cleared his throat—all meant, Lankester believed, to keep him off guard. At one point, a hand touched his leg two or three times, startling him—no doubt a delay tactic, Lankester thought. "Did you feel anything?" Slade asked. "Yes, a spirit touched my leg," Lankester replied. Finally, Slade said that there would be no more messages that day. He then joined his hands with Lankester's and leaned forward, and his chair rose perpendicularly off the floor, hovered for a moment, and dropped back down. The table also suddenly levitated. Lankester expressed amazement at the feat. Slade promised that if he returned, he would see more remarkable things. Lankester thanked Slade for the séance and agreed to return in a few days. But it was his intention to put his hypothesis to the test by seizing the slate from Slade's hand just before the spirit was about to begin writing on it.[12]

Lankester returned with his friend Donkin on Friday, September 16. As a witness to the event, Donkin would help him prove to the world that Slade was writing the messages himself. At the critical moment, he would grab the supposedly blank slate. If Lankester found writing already there, he said, he would conclude that only those who had not lost their reason would remain unconvinced.

During this second séance, Donkin observed the same to-and-fro movements of Slade's arm, with some contraction of the flexor tendons of the forearm as though in the action of writing. The message that appeared was barely intelligible; Donkin believed that Slade had produced it by writing with a minute piece of pencil placed under the nail of his right middle finger. The next message, which came after "a considerable interval" while Slade was clearing his throat and making short remarks, was more legible and written in a "straightforward, undisguised hand." After the slate was once again cleaned, Slade told Lankester that the spirit probably would write more distinctly for Lankester and he therefore would allow him to brace the slate against the underside of the table. Donkin watched Slade go through the same maneuvers, apparently writing while the slate was out of sight on his knee. And then the critical moment came. As Slade was about to give the slate, still supposedly free of its message, to Lankester, Lankester seized it and opened it in front of Donkin, revealing the writing. "You have already written on the slate." Lankester said. "I have watched you doing it each time."[13]

But the cool, scientific account of this "critical moment" that Lankester and Donkin gave to the readers of the *Times* was in reality less polite. Lankester later admitted that when he grabbed the slate and saw the writing, he smiled triumphantly and pronounced Slade a "scoundrel," accusing him of having written the message himself. Slade said nothing but slumped back in his chair, apparently exhausted and bewildered as he emerged from his trance state. Lankester and Donkin rushed out of the room, slate in hand as evidence, creating a commotion in the waiting area, where they accosted Slade's assistant, Geoffrey Simmonds, and called Slade a "damned liar." Simmonds astonished the two scientists by reacting with delight: their accusation, he said, was the best advertisement that Slade could have hoped for, and would draw at least two hundred people to see whether they could detect a swindle. Lankester and Donkin threatened to go to the press, which only pleased Simmonds more. They then stormed out and wrote up an account of their experience, which they posted to the *Times*.[14]

Lankester and Donkin were not Slade's only visitors from the British Association. Barrett wrote to the *Times* that he, too, had been suspicious of Slade's methods at first and had witnessed the same movements of the tendons of the wrist, the coughing, and the fidgeting. But instead of forcibly interrupting Slade and discovering writing when none was supposed to be present, he performed a different experiment. He had taken the precaution of bringing his own blank slate, which he placed face down on the table and anchored with his elbow, the pencil underneath the slate but not touching the writing surface. He then grasped one of Slade's hands, while the tips of the fingers of Slade's other hand barely touched the slate. As he focused his attention on Slade's hands, neither of which moved perceptibly, Barrett was astonished to hear scratching from the underside of the slate. When he turned the slate over, he found it covered with writing. A few days later, he reproduced these results during a second séance. The only time Slade failed to elicit writing occurred when the slate was sealed in a box—he had refused to make an attempt.[15] In a letter to the *Times* on September 23, Slade defended his refusal to submit to Barrett's last experiment. He had no objection to people bringing their own ordinary slates, single or folded, but he objected to locks, boxes, or seals: "I claim to be as honest and earnest in this matter as those who call upon me for the purpose of investigation. To my mind it would be as reasonable to sever the wire and then ask the operator to send your message as it is to violate the conditions which experience has taught me are essential in these experiments in order to obtain successful results."[16]

Annoyed that Lankester and others had impugned his reputation and integrity in the popular press, Wallace sat again with Slade on October 7 and 16. On the first day, not much happened. The only message that appeared was "Can't now," written faintly in pencil, the pencil dust still present, indicating that the message had just been written. For the second session, he brought his own blank slate from home, a double slate that was folded shut and that only he had handled. Slade slipped into a trance and remained in a constant state of agitation, moving about uncontrollably. As he took the slate from Wallace's hands, he accidentally struck Wallace in the face and apologized for his loss of control, but through it all Wallace never lost sight of the slate. Although Slade could have inserted a finger between the two faces, he could not have written anything. But when Wallace opened the slate, a message was there, written in a clear hand with the "i" dotted and the "t" crossed. The message read: "Is this proof, I hope so."[17]

In the meantime, Lankester, now obsessed with rooting out spiritualism from the hallowed halls of modern science, had hired a lawyer and brought charges against Slade and his assistant for conspiracy and fraud under the Vagrant Act, which had been passed by Parliament early in the century to protect the "weak-minded" from the depredations of fortune-tellers. As his co-plaintiffs he named five other men, including William Carpenter, who objected that his name had been added without his permission and refused to sign a petition against Slade that had been circulated among the members of the British Association. Slade was represented by two attorneys; one of them, Charles Carlton Massey, had become a confirmed spiritualist after reading Wallace's passionate "Defence of Modern Spiritualism" and had witnessed forty or fifty messages and other phenomena at Slade's séances, in both England and America.

Slade and Simmonds were brought before a judge on October 2, 1876, in a courtroom packed with spectators.[18] Throngs of people clamored to get inside. By the time the judge pounded the gavel to initiate the proceedings, they had managed to do so, with the street nearly empty except for broughams awaiting the departure of several aristocratic female observers. As everyone knew, the trial had nothing to do with Slade and everything to do with the powerful egos of London's men of science. Lankester was the prosecution's star witness and had the implicit support of the British Association. As the association's unofficial representative, he understood that it was his duty to destroy spiritualism and its credulous advocates once and for all. The pressure on Lankester was intense, and his victory was far from certain. The proceedings were followed closely by the press, which reported every detail. Wallace stayed away at first, no doubt hoping that the affair would die a natural death.

The judge, a man named Flowers, was not an impartial observer. At the outset, he stated that he had no sympathy for the defendants. This nonprofessional attitude contributed to the circuslike atmosphere of his courtroom, with the trial spinning out of control on numerous occasions. He seemed one step behind everyone else, recognizing his mistakes only after he was no longer able to do anything about them. His first error was to allow John Nevil Maskelyne to appear as a witness for the prosecution. When Maskelyne stepped up to the witness stand, the spectators went wild. Maskelyne, one of the most famous magicians of his day, specialized in exposing the "tricks" of mediums. Every night he entertained crowds at the Egyptian Hall, a haven for antispiritualists. Once Maskelyne took the stand, the trial became a farce, according to one *Times* reporter. Flowers was completely helpless. The last thing he wanted, he said, was a magic show, but a magic show is what the audience was given. Maskelyne proceeded to describe or demonstrate every technique a medium might resort to in the art of slate-writing, though he had never attended any of Slade's séances. Convincing or not, his performance was at least entertaining.

The trial dragged on for weeks. At first, the spectators had seemed to favor Lankester. But by the time one of Slade's lawyers gave an impassioned speech about the freedom to promulgate new ideas without ridicule, the composition of the observers in the packed courtroom had shifted, with the antispiritualists making up a distinct minority and the women outnumbering the men. Many were distressed by the likely outcome. "From Galileo downwards the pioneers of every new movement which clashed with the prejudices of the day have been subject to persecution," one of Slade's lawyer thundered. "What is laughed at today might be very differently regarded tomorrow." Wallace now sat among the defendants' sympathizers and watched the proceedings for a while before taking the witness stand for the defense. When called, he boldly described his own investigations of the subject of spiritualism over the preceding eleven years. His account was terse and unemotional, in contrast to the virulence and fervor of his opponents. But his courage in risking his reputation on behalf of liberty and justice did nothing to sway the judge. On November 1, the "Great Spiritualist Case at Bow Street," as the trial was called, came to an end, with Flowers upholding the prosecution. Although the Vagrancy Act antedated spiritualism, he believed that it was broad enough to cover the "subtle tricks and devices" used by Slade. While admitting that he had been influenced by his recollections of the mischief perpetrated by other professed mediums, he nevertheless inflicted the maximum penalty: three months' imprisonment with hard labor.

A reporter for the *Times* concurred with the magistrate's judgment. Men of science, he said, were not the proper investigators of "affairs of this kind," as a professional magician had proved. He was pleased with spiritualism's "comeuppance," though he did not expect the trial to have much effect in putting down spiritualism. "As long as there are weak and silly people in the world," he wrote, "so long will there be rogues ready to take advantage of them."

Not everyone in the press shared this reporter's view. The *Spectator*, whose literary editor and co-proprietor, Richard Holt Hutton, had an abiding interest in psychic phenomena, complained that the sentence imposed on Slade was a severe application of a law intended for something different. The defendant was a foreigner, with no prior criminal record, and had committed acts that were not illegal in his own country. The commentator (possibly Hutton himself) noted with uncommon sagacity:

> The plain truth of the matter is that the belief in art-magic and the disposition to inquire into phenomena apparently ultra-natural revive periodically, whenever accepted faiths are shaking, or accepted physical knowledge is enormously and suddenly increased. A society suddenly amazed by a new learning, as Europe was in the Renaissance, or by a new set of religious and philosophic ideas . . . loses its old landmarks, thinks anything possible, and either believes marvels or, which is the much more frequent phenomenon, sees no reason why it should not investigate marvels. Its sense of the limitations of power is temporarily obscured.

Darwin, who had followed the proceedings closely from Down House, wrote to Lankester to congratulate him on his success and offered to contribute £10 to the costs incurred for the prosecution of the case.[19] Emma Darwin seemed to be speaking for the whole family when she wrote to their son Leonard, who was stationed in Malta at the time, "I think that the sentence was too severe, at least as to hard labour, viz. three months' imprisonment. If people are so credulous some allowance ought to be made for the rogues."[20]

Slade appealed his conviction in late January 1877. Wallace appeared once again to lend his support for the spiritualist cause. The public, however, seemed to have lost interest in the case, and the courtroom was less crowded. This time, Slade and Simmonds (or their wealthy supporters) had hired a shrewder attorney. Drawing the attention of the presiding magistrates to the precise wording of the Vagrancy Act, this attorney pointed out that Judge Flowers had purposely omitted the words "by palmistry or otherwise," as stated in the original act. After reading Flowers's judgment verbatim to the appellate judges, the attorney said, "To my astonishment—and I think it would excite the astonishment of every member of this Bench—the conviction does not declare or charge the appellants with anything of the kind." The actual material words, "by palmistry or otherwise," under which Slade could be convicted by the Vagrancy Act were "by some sleight of hand" omitted from the conviction. Thus the charge against Slade could not be supported "for five minutes." Speaking lawyer to lawyer, he said, what he pointed out had to be regarded as a fatal objection to the conviction. After deliberating in their chambers for an hour, the judges returned with their verdict. The words "by palmistry or otherwise," they said, were of vital importance to the case. "The reasons for this omission and for framing the conviction in its present form are not far to seek. If the particular description 'by palmistry' were applicable to the case it was unnecessary to avoid it; and if the fact had been such as to bring the case within the meaning of the Act, preceded by the description, it would be sufficient to quote the language of the enactment." They were therefore compelled to overturn Flowers's decision.[21] "The fault in this case may not be in the law," the *Times* huffed. "We do not presume to decide; but that the case should have broken down on such grounds is certainly discreditable to the administration of the law."[22]

Just as the eventual acquittal of Slade failed to vindicate spiritualism, the publication of Wallace's masterwork on the geographic distribution of animals failed to rescue his tainted scientific reputation. His central role in the brouhaha at the British Association led to the adoption of more stringent rules about the type of papers that could be read at its scientific meetings.[23] Wallace was never again asked to preside over a section. It may be that he never would have wanted to; in his memoir, he states that he had "pretty well exhausted" the interests of the association, preferring to take his autumn holiday with his wife and children elsewhere.[24] Thirty years after the fact, he could afford to affect a nonchalant attitude, but in 1876 he may not have been so indifferent. With a single exception, he never attended another meeting of the association. In 1881 he made a brief appearance, but only because the meeting was held in York, near the home of his friend Richard Spruce, who lived fourteen miles northeast in Coneysthorpe, and provided a good excuse for visiting him.

The rank and file of the British Association were equally annoyed with Wallace. The *Lancet*, the voice of English medicine, which had praised *The Geographical Distribution of Animals*, expressed the general bewilderment of his scientific colleagues at what appeared to be Wallace's bizarre attitude toward science and the supernatural, wondering how "the feebleness which begets folly may coexist in a mind with so much general strength and earnestness."

> Side by side with acumen we find obtusity [*sic*]; a vigorous and exacting judgment compounded with impressibility amounting to a credulous abnegation of intelligence; a tame submission to any mystery claiming to be supernatural, which dethrones the faculty of reason, and in its place sets up that tyrant and traitor puppet-sense, susceptibility, by some men called "faith," by others "docility," by a third-class "truth-seeking," but which is either pretentious ignorance aping the worship of the unknown, or an over-strained and dazed mind ready to take refuge from the harass of doubt in any delusion—offered from without. On this wise, and this alone, can we explain the unquestioned fact that many intelligent and strong intellects exhibit the extraordinary spectacle of a belief in "spiritualism."[25]

The immediate consequence of Wallace's outspokenness was the forfeiture of the respect of a considerable number of people in the scientific community. One friendship that ended abruptly that fall was with Henry Bates. The relationship had been in its death throes anyway. For several years, the two men had been drifting apart, both personally and professionally, as Bates lost patience with Wallace's heterodoxies. The split with Darwin over the question of the origin of the human mind had been almost intolerable, but the way in which Wallace had tried to ram spiritualism down the throats of his fellow scientists was the final straw for Bates. The two never talked to each other again. Wallace left no record documenting this final break. Although he had few serious and permanent disagreements with any man, he severed all relations when the boundaries of Victorian social intercourse were breached. Bates had crossed that line, but what he said or did is unknown. Only two months after the Glasgow meeting of the British Association, Wallace accepted an invitation to lecture before the Royal Geographical Society on zoogeography, but the formal tone of his letter to its secretary, addressed not to Bates but to "Dear Sir," underscored the degree of disaffection between the two men.[26] Fifteen years later, when he was asked to write Bates's obituary notice for *Nature*, Wallace at first refused, stating that he knew absolutely nothing about Bates's life during its last two decades, and only after some arm-twisting did he consent. What he produced lacked the affection and vividness of the obituary notice that he had written for Spruce, a man he had seen much less often but whom he always regarded as a brother.[27]

Wallace ceased proselytizing and maintained a dignified isolation from his scientific colleagues. When he wrote to Darwin in January 1877, following the acquittal of Slade, he seemed almost lighthearted: "I . . . am very idle and feel inclined to do nothing but stroll about this beautiful country and read all kinds of miscellaneous literature."[28] Darwin knew better. Wallace was too restless to be idle. Despite mounting financial woes, he had no prospects for future work.. The Great Depression of Trade, which began in 1874, led to the evaporation of most of his remaining investments. He was also saddled with another lawsuit. The contractor who had absconded with his money and defaulted on his obligations to his suppliers had the audacity to demand £800 or £900 in back pay for not having been allowed to complete the house. Wallace refused to reimburse him. The contractor sued, forcing Wallace to retain a lawyer. Depositions, affidavits, replies, and objections sailed back and forth between the opposing sides for two years. Ultimately the suit was dismissed as groundless, but the grief cost him £100 in legal fees. "That was my experience of English law," Wallace complained years later, "which leaves the honest man in the power of the dishonest one, mulcts the former in heavy expenses, and is thus the very antithesis of justice."[29]

In the meantime, his former friend Carpenter had positioned himself as spiritualism's Antichrist. Carpenter's opposition to spiritualists had reached fever pitch, and he was relentless in his persecution of Wallace, Crookes, and other eminent scientists who openly espoused spiritualist beliefs—this despite the fact that he himself dabbled in spiritualism well into the late 1870s. While the Slade case was in progress, Carpenter invited Tyndall and Huxley to a séance at his house. Tyndall and Huxley did not wish to get mixed up with Carpenter in an inquiry of this kind, but they reluctantly accepted. "I would not trust the Virgin Mary herself if she professed to see a medium," Huxley wrote to Tyndall. "But we shall see what we shall see."[30] What they saw they did not say, yet the trio continued to attend séances. It may appear strange that all three men persisted in their investigation of spiritualist phenomena while simultaneously expressing contempt for the "deluded," but séances were oddly compelling and grandly entertaining for believers and nonbelievers alike. Carpenter, Huxley, and Tyndall—though they had refused Wallace's and Crookes's invitations—surreptitiously went elsewhere, as though fearing that they might be forced to admit the reality of phenomena produced under rigid test conditions by two serious and reliable scientific men.

Despite his own indefatigable investigations, Carpenter pursued Wallace and Crookes as zealously as Javert hunted Jean Valjean, publishing articles and letters in a number of journals and provoking Wallace to note in a letter to the *Athenaeum* "the extreme inconvenience of Dr. Carpenter's erratic mode of carrying on a discussion."[31] At the same time that he was arranging to attend a séance, Carpenter complained in the *Spectator* that "men like Mr. Crookes . . . and Professor Barrett seem to me to resemble Baron Reichenbach . . . and other Physicists, twenty-five years back, in their ignorance of the nature of their instruments of research; putting as much faith in tricky girls or women, as they do in their thermometers or electroscopes."[32] Barrett wryly observed Carpenter's curious attitude toward spiritualism:

> Here I yield to none in my admiration for the perseverance with which Dr. Carpenter has for a quarter of a century endeavoured to arrive at the bottom of this matter. Hardly any professional medium escapes him. At one time it is Mr. Foster, then Mr. Home, now Dr. Slade, and again . . . Mr. Kane. In fact, were it any one else but Dr. Carpenter, whose philosophy places him above the contagious influences of epidemic diseases, we should suspect such an ardent investigator to be wanting in that intellectual fortitude which comes from a steady reliance on "unconscious cerebration."[33]

Crookes, for his part, wondered why, if all the phenomena ascribed to spiritualism were imposture, Carpenter was wasting his valuable time interviewing and sitting with mediums: "Does he regard the subject as his own special preserve, and may his demonstrations against other explorers in this domain of mystery be looked upon as the conduct of a gamekeeper towards a suspected poacher?"[34]

Such remarks enraged Carpenter. A prolific and eloquent writer, he often resorted to ungentlemanly tactics to humiliate anyone who dared to challenge or controvert him. Wallace was attacked after publishing a critical review in *Fraser's Magazine* of Carpenter's *Mesmerism, Spiritualism, &c Historically & Scientifically Considered*, a book illustrating Carpenter's theory that these various "isms" could be explained by mental action, not mysterious external agencies and forces. Wallace had upheld him "as an example of what prepossession and blind scepticism can do for a man . . . how it makes a scientific man unscientific, a wise man foolish, and an honest man unjust."[35] Carpenter took Wallace's review personally, accusing him of assailing his "honesty and good faith" and deprecating his scientific character by stating that he manufactured evidence to support his views. He threatened to seek the intervention of such mutual friends as Huxley and Joseph Hooker, though what these men would, or could, have done to Wallace is uncertain.[36]

Wallace referred Carpenter to the exact words he had written in his review, pointing out misunderstandings of the grammatical constructions. For example, he "never for a moment" intended to suggest that Carpenter had manufactured evidence in the case in question. "The words 'his disposition to manufacture evidence' plainly refer to and are supplementary of 'the untrustworthiness of the authority,'" Wallace wrote privately to Carpenter. "On the previous page I have referred to this 'authority' as 'not an unbiassed witness' and have asked for independent proof that 'the chemical was not applied to the flowers after the séance to keep them fresh.' That is clearly the 'manufacturing testimony' to which I subsequently refer as possible, and which can by no possibility be taken to implicate you." Wallace also had described Carpenter as an expounder of popular and educational science; but Carpenter sensed sarcasm, believing that Wallace had meant "mere" expounder. Wallace denied that he had ever "depreciated" Carpenter's character in such a manner. However, he agreed to apologize if Carpenter could prove any genuine misstatements—noting that this was more than Carpenter had been willing to do when he made deliberate misstatements about Crookes.[37]

Unmoved by Wallace's clarification of his position, Carpenter hit below the belt. Focusing on one of Wallace's assertions, Carpenter cited the belief in a flat earth as "a parallel case," equating Wallace's tactics with those of the infamous John Hampden. In a letter to the editor of *Nature*, Carpenter wrote:

> It must be perfectly obvious to any one who is capable of reasoning logically, that nothing which I said of Mr. Wallace . . . can be twisted into the implication that he is either "a fool or a knave." John Hampden is

> continually saying this of Mr. Wallace and everybody who upholds the rotundity of the earth. And I mildly suggested whether . . . Mr. Wallace is not assuming an attitude in some degree similar, that is, setting himself up as the one wise and honest man that everybody else is either stupidly or wilfully blind to the evidence he presented.[38]

Carpenter called Crookes and Wallace a "tower of strength" to the various orders of tricksters in the world, contending that the two men followed methods that were "thoroughly un-scientific" and that they accepted on implicit faith statements that a good scientist would reject as untrustworthy. Forty years of theoretical studies had given him an "unusual power" to deal with the subject of spiritualism. It was his mission to prevent the spread of a "noxious mental epidemic" in England, and he had received many notes of thanks for curing those "bitten" by the malady. He was a martyr for the cause. Although not thick-skinned, he was content to "brave all" in his crusade against a dangerous disease imported from the United States.[39]

Comparing himself and Wallace with the small body of earnest students of alchemy who detected "germs of truth amidst the ravings and juggleries of the gold makers" and laid the foundations of modern chemistry, Crookes countered that the Carpenter of that period would have denounced those pioneering students as "scientific advocates of the system of alchemy," feeling that it was his duty to humanity to undermine their reputations. Crookes added: "This is an act of disinterested kindness which recalls to me the exquisite truth of Dean Swift's remark, 'No enemy can match a friend.'"[40] Undeterred, Carpenter carried his battle to the pages of the *Athenaeum*, in which he published a short article, "The Curiosities of Credulity," in December 1877. Wallace, like Crookes, was growing weary. He could only hope, he replied in a subsequent issue, that a new audience might see through Carpenter's obfuscations.[41] To Barrett, he confided that he had advised friends not to waste their time on "Dr. C." "Nothing would tend to lower Dr. C. in public estimation on this subject more than his being forced to acknowledge that what he has for more than thirty years declared to be purely subjective is after all an objective phenomenon."[42]

The epistolary war eventually ended, and Wallace and Crookes moved underground, attending the meetings of the British National Association of Spiritualists, where they found peers who were more sympathetic. This association had been founded in 1874 in order to consolidate the British spiritualist movement and organize its expansion. It was also an outgrowth of the Psychological Society and sought to promote the study of a new field of science. Its membership included a large number of well-educated and respectable men and women, mainly from London and the surrounding principalities. At one of these meetings, the famous explorer Richard Burton read a paper on his mystical experiences in the Middle East. Burton had advanced the spiritualist cause by coming to the defense of a pair of mediums known as the Davenport brothers when they were "unjustly" attacked, but he kept a respectful distance and did not wholly embrace spiritualism. After the paper's presentation, Wallace was asked to make some remarks. "It appears to me . . . that he believes everything we do," Wallace said, "only he puts a different interpretation upon it." He wished that Burton had elaborated on his own experiences, not those of the Eastern authorities he cited in his paper. Crookes agreed and asked Burton for an account of objective phenomena, not the subjective phenomena of peering into crystals or searching for meaning in a drop of ink. Burton demurred, but he praised the courage of men like Wallace and Crookes. "I [see] no stronger raison d'être for a Spiritualist Society than that of giving greater boldness to men in expressing their belief," Burton said, "whether true or false, especially when their beliefs are unpopular," adding, "I very much doubt whether the 'new truths' are so valuable as the new fact of encouraging men to tell the truth about old things."[43]

In their public defense of spiritualism, Wallace and Crookes obviously had touched a raw nerve, and everything they did afterward became suspect. By the late 1870s, Wallace occupied an anomalous, Janus-like position. In 1858 he had spearheaded a scientific revolution. Twenty years later, he was in the vanguard of a counterrevolution—a paradox to everyone but himself.

CHAPTER 15

Phoenix from the Ashes

IN 1878 WALLACE PUBLISHED with Macmillan a scientific ode to the living world: *"Tropical Nature" and Other Essays.* It anticipates a concern for the environment that would not fully emerge until the twentieth century. No one in English science at the time—and certainly no one of Wallace's stature—showed much interest in the ecological consequences of human encroachment on the natural world. *Tropical Nature* is the spiritual forerunner of Rachel Carson's *Silent Spring,* which was written some eighty years later, long after *Tropical Nature* had vanished into obscurity and a prophet from an earlier generation had been forgotten. Its main thrust, however, was to provide a detailed response to Darwin's theory of the origin and purpose of coloration in the higher animals. If Wallace's message about the destruction of the environment was lost, it was because he distracted his critics with other issues—in particular, his endless battle with Darwin over the importance of sexual selection.

Wallace began his study with a description of the distinctive climatic, floral, and faunal characteristics of the tropics. He called attention to the wanton clearing of virgin forests. Plant life and the climate were intimately related. In tropical and even in temperate countries, the rains were periodic and often torrential; when forests were cleared, the heavy rainfall quickly stripped off the topsoil, destroying in a few years the fertility that had been built up over centuries. Without plants, every particle of moisture in the subsoil evaporated in the intense heat of the sun. This process of desiccation interacted with the climate to produce long droughts relieved only by sudden and violent storms. Further cultivation thus was impossible. He noted that wide tracts of fertile land in southern Europe had been devastated in just this way. "Knowingly to produce such disastrous results would be a far more serious offence than any destruction of property which human labour has produced and can replace," Wallace wrote. "Yet we ignorantly allow such extensive clearings for coffee cultivation in India and Ceylon, as to cause the destruction of much fertile soil which generations cannot replace, and which will surely, if not checked in time, lead to the deterioration of the climate and the permanent impoverishment of the country."[1]

He then described in greater detail this fragile world threatened by humanity's ill-considered actions. The abundance of animal life in the tropics had nothing to do with the sun's light and heat, but with the uniformity and permanence of terrestrial conditions. Successive glacial periods had devastated temperate zones, whereas the equatorial regions remained relatively stable. The vicissitudes of climate imposed limits on variation, weeding out any form or color injurious to an animal, but such disadvantages were not experienced in the tropics, where the struggle for existence was always less severe. In the tropics, food was plentiful, shelter was easily obtained, geologic change was uncommon, and natural selection kept the "teeming mass of organisms" in harmonious balance. "The equatorial regions are then, as regards their past and present life history, a more ancient world than that represented by the temperate zones, a world in which the laws which have governed the progressive development of life have operated with comparatively little check for countless ages, and have resulted in those infinitely varied and beautiful forms . . . which delight and astonish us in the animal productions of all tropical countries."[2]

The second part of his book consisted of an elaboration of his theory of coloration. Color, he insisted, had a purpose in nature; it was its absence, not its presence, that required explanation. Like the abundance of wildlife, color had nothing to do with the direct action of sunlight; contrary to the popular notion, brilliantly colored birds, insects, and flowers were no more numerous in tropical than in temperate regions.[3] Wallace defined four classes of colors in animals: protective (mimicry being a subclass), warning, sexual, and typical (that is, species-specific). Every day, new cases of protective coloring were being discovered—though one could "hardly tell by the mere inspection of an animal whether its colours are protective or not."[4] While dwelling on the nature of color, Wallace digressed to discuss the wave theory of light, the structure of the retina, and the retina's ability to perceive colors. Colors resulted from the underlying molecular structure of an object as well as chemical changes. Heat alone could bring about a color change. Pigmental colors varied according to their position in the integument: epidermal colors were deep and rich and tended not to fade after death; hypodermal colors were lighter, more vivid, and faded quickly. Like any trait, color varied and occasionally appeared where it had been absent. Natural selection eliminated tints that were injurious to a species and preserved and intensified those that were useful.[5]

Darwin, Wallace said, attributed almost all colors in birds and insects to sexual selection. The difference in coloration between the sexes of a species resulted from the transmission of color variations either to one sex only or to both sexes by some unknown laws. Wallace continued to believe that the primary cause of such sexual dimorphism was the need for protection. Bright colors were somehow suppressed in many female birds to make them inconspicuous while nesting. The "greater vigor" and "higher vitality" of the male led to his more vivid coloration. As evidence of the influence of vigor and vitality, he cited the dull coat of ill health and the glossy coat of well-being. The deepest intensity of color occurred in the breeding season, when "vitality was at a maximum." The most vigorous and energetic male was more successful in siring offspring; hence natural selection would preserve or intensify color if color depended on or correlated with vigor. He reasoned that if variations in color could be traced to chemical or structural differences in the integument, then abnormal developments of hair, horns, scales, feathers, and other useless ornaments could just as easily be traced to the effects of increased vigor on different parts of the integument.[6]

Wallace disagreed with Darwin about the function of recognition colors. Recognition colors helped the sexes distinguish themselves from closely allied species and also guided the young to their parents. The white tail of the rabbit, which Darwin found useful only to the sportsman, was to Wallace a signal to the young to escape danger. The zebra's stripes, which Darwin regarded as useless on the open African plains, alerted a straggler to the location of the herd or camouflaged the animal when foraging or resting in the bush. "Until the habits of the zebra have been observed with special reference to these points," Wallace wrote, "it is surely somewhat hasty to declare that stripes 'cannot afford any protection.'"[7]

While admitting that females exerted a choice, Wallace insisted that it had not been proved that color determined their choice; when a male displayed to a female, it might well be his persistence and energy rather than his beauty that won the day. Wallace could even explain the unwieldy headgear on some insects as arising from natural selection: "The long, pointed or forked horns [of beetles], often divergent, or movable with the head, would render it very difficult for . . . birds to swallow such insects and would therefore be an efficient protection."[8] But it was the study of Darwin's case of the Argus pheasant that had caused Wallace to lose faith in the importance of sexual selection in birds and insects. He found it "absolutely incredible" that the long series of gradations—from simple daubs of color to the perfect ball-in-socket pattern in the sumptuous tail feathers of the males of the present-day species—could be traced to the preferences of tens of thousands of female birds over thousands of generations. Although he had no solution, he suggested that the answer might be found in unknown laws of physics or chemistry. He did, however, offer an explanation for the elaborate crests and other erectile feathers on male birds. These ornaments served to frighten away enemies rather than attract females, since they were displayed when the bird was angry or preparing for combat. The most pugnacious and defiant male would survive and pass his traits to his offspring. If sexual selection played any role in the natural world, it was that the most aggressive and vigorous males, which also happened to be the most ornamented, exerted the choice. "Natural selection, and what may be termed male selection," Wallace wrote, "will tend to give them the advantage in the struggle for existence; and thus the fullest plumage and the finest colors will be transmitted, and tend to advance in each succeeding generation."[9] Natural selection, an "admitted *vera causa*," explained it all.

Edward Percival Wright, in a review for *Nature*, complimented Wallace for writing "a most interesting volume on the peculiarities of tropical life." Wallace's theory of color, which Wright called a "molecular" theory, was for him the most important part of the book. Since it opposed Darwin's theory of sexual selection, *Tropical Nature* could not fail to attract the attention of anyone interested in the subject. "Doubtless this theory will give rise to much controversy," Wright concluded. "And in the course of this, no doubt, many important facts will be elucidated."[10] Darwin, however, would not give an inch. He found the terms "vigor," "activity," and "vitality" as incredible as Wallace had found his ideas on sexual selection. "I could say a good deal in opposition to you," Darwin responded, "but my arguments would have no weight in your eyes, and I do not intend to write for the public anything on this or any other difficult subject."[11] Not long afterward, a thirty-year-old philosophical biologist named Charles Grant Blairfindie Allen (better known as Grant Allen) sent Darwin and Wallace a copy of his own book, *The Colour Sense*, which adopted sexual selection as a *vera causa* of evolutionary change. Wallace praised the book's overall quality but disagreed with its conclusions. Darwin, of course, was pleased. "I have no fear about [sexual selection's] ultimate fate," he told Allen. "Wallace's explanation of, for instance, the display of a Peacock seems to me mere empty words. . . . For years I have quite doubted his scientific judgment, though admiring greatly his ingenuity and originality."[12]

For the third time in two years, the Wallaces moved, this time to the village of Croydon, fifteen miles north of Dorking and closer to London. The reason Wallace gave for the move was that Croydon had a more pleasant climate and better schools, but in reality he was reliving the gypsy life of his childhood, when his family moved from place to place to reduce expenses. Having nearly drained his savings, he remained solvent with earnings from his writings for a variety of publications. But unless he secured a steady source of income, he would continue to totter toward bankruptcy.[13] Despairing over his finances, he solicited help from Alexander Macmillan, who had rescued him in the past. "I am exceedingly anxious to obtain some regular literary employment that will bring me in a fixed income however moderate, having had a series of losses and misfortunes that render it very hard for me to get on," Wallace confessed to Macmillan in July 1878. He proposed an assistant editorship for *Nature* if Joseph Norman Lockyer could use the help and Macmillan could afford the extra salary.[14]

Before Macmillan had a chance to reply, Wallace learned of an opening for a superintendent at Epping Forest, the last remaining wilderness area in the vicinity of London. Convinced that he was the best candidate for the job, he withdrew his request to Macmillan and directed his energy to procuring a position that he believed was more suited to his talents. The man he appealed to was Joseph Hooker, to whom he sent a petition for signature. "You will understand how unpleasant it is for me to have to lay before you so egotistical [a thing] as the accompanying paper," Wallace wrote, "but you are no doubt well aware that it must be pretty strongly worded if it is to have any effect."[15] Hooker complied, and Wallace circulated the petition among other friends and acquaintances. In September he contacted Darwin, whose fame he felt would certainly improve his chances of obtaining the job.[16] "I return the paper signed," Darwin replied, "and most heartily wish that you may be successful, not only for your own sake, but for that of Natural Science, as you would then have more time for new researches."[17]

Wallace attacked his new project in his overly zealous way, mapping out a vision of Epping Forest that differed from that of the politicians, who could not have cared less about its improvement. The members of the government's search committee saw the forest as the site of excursions, picnics, and dinners for politicians at public expense. Hearing rumors that he planned to turn Epping Forest into a "Kew Garden," Wallace dashed off an article for the *Fortnightly Review* to clarify his position and outline his plan to restore the forest to its original pristine state.[18] One hot summer day—too hot for the black overcoat he usually wore when he went outdoors—he visited the forest with Annie, Will, and Violet, promising all sorts of delights if they moved nearby. He tried to interest nine-year-old Violet and seven-year-old Will in the uncommon plants he discovered, but they were too hot and uncomfortable to pay attention. After checking the ordnance map he invariably brought along on his hikes, he led them to a stream and delighted the children by showing them how to use a long piece of black rubber tubing as a straw to quench their thirst, while Annie, who worried about the quality of the water, looked on in some distress.[19]

As usual, the selection committee dragged its feet in making a decision. A year passed before Wallace was asked to attend a subcommittee interview as one of twelve candidates for the position. In the end, he was rejected in favor of a landscape gardener.[20] Once again, his hopes had been dashed. Darwin wrote to express his condolences.

Wallace now considered applying for some post at a college of science in Birmingham, perhaps as registrar, curator, or librarian—anything.[21] But he seems to have given up before even applying. Aged fifty-five, too old and too controversial, with no marketable skills or university degree, he could not find regular employment. England in 1879 appeared to have no use for a famous explorer-naturalist-spiritualist.

No one outside the immediate family comprehended the true nature of Wallace's financial woes except Arabella Buckley. Without consulting Wallace, she turned to Darwin, suggesting that he and Hooker, being men of influence, find Wallace a modest post in which his expertise in natural history could be utilized and which would provide him with financial security.[22] Moved by Buckley's request, Darwin promised to do his best. It had already occurred to him to seek a government pension for Wallace; he told Buckley that he would write to Hooker for his opinion, which he did.[23] Hooker replied at once, unleashing a torrent of words that took Darwin by surprise:

> Wallace has lost caste considerably, not only by his adhesion to Spiritualism, but by the fact of his having deliberately and against the whole voice of the committee of his section of the British Association, brought

> about a discussion on Spiritualism at one of its sectional meetings. That he is said to have done in an underhanded manner, and I well remember the indignation it gave rise to in the B. A. Council. . . . Then there is the matter of his taking up the Lunatics' bet about the sphericity of the earth, and pocketing the money.

In good conscience, therefore, Hooker could not solicit his friends to sign a petition on behalf of a man who had sullied the scientific profession. And even if he did support such a petition, he thought that it was the duty of scientific men to let those in the government know that Wallace was a leading spiritualist. If they did not, he warned, they would find themselves in an embarrassing position once the government learned the truth. Although he claimed to have no special "animus" toward Wallace, his letter was full of hostility. He was not moved by Wallace's alleged desperation, noting that "after all Wallace's claim is not that he is in need, so much as that he can't find employment."[24] Darwin replied that spiritualism and the lunatics' bet had never crossed his mind. He had thought only of Wallace's distress and his service to natural history. He thanked Hooker for his insights and gave Buckley the bad news. Buckley was disappointed but philosophical. She had always felt, she told Darwin, that Wallace's "want of earthly caution" would return to haunt him.[25]

While others argued his merits, Wallace was hard at work on another book on his favorite subject of geographic distribution. The main impetus for writing it was his desire to improve his scientific position, which was "continually" misunderstood. He also eyed the literary marketplace, informing Darwin in a letter dated January 1880 that he hoped to write something more accessible, popular—and profitable.[26] Wallace had not communicated directly with Darwin for more than a year, though they lived closer to each other than ever, Croydon being less than ten miles from the village of Downe. Between September 1878 and January 1880, they had not exchanged a single letter. Toward the end of his life, when he reflected on this period, Wallace attributed the silence to the fact that Darwin was working on subjects that he knew nothing about, implying that Darwin contacted people only when he needed data.[27] But Wallace made no overtures himself. The two men were worlds apart on certain issues. Wallace, a pariah, was a man best left alone to brood in isolation. Buckley would not have had to reveal the contents of Hooker's letter to apprise him of his status in scientific circles.

An article by Wallace entitled "The Origin of Species and Genera," published in the January 1880 issue of the periodical *Nineteenth Century*, prompted Darwin to warm up the cold peace. "You must allow me to express my lively admiration of your paper. . . . You certainly are a master in the difficult art of clear exposition," Darwin wrote. He agreed with "almost everything . . . excepting the last short sentence."[28] Wallace had concluded his article with the following observation:

> I have . . . attempted to show that the causes which have produced the separate species of one *genus*, of one *family*, or perhaps of one *order* from a common ancestor, are not necessarily the same as those which have produced the separate *orders, classes,* and *subkingdoms* from more remote common ancestors. That all have been alike produced by "descent with modification" from a few primitive types the whole body of evidence clearly indicates; but while individual variation with natural selection is proved to be adequate for the production of the former, we have no proof and hardly any evidence that it is adequate to initiate those important divergences of type which characterise the latter.

In his letter Darwin did not elaborate on their differing views. Twenty years later, Darwin's son Francis asked Wallace if he truly held the opinion expressed in the final paragraph of his essay. Wallace did indeed hold this opinion, he told Francis, but the stress must be placed on the word "proof." He said that Darwin, too, had maintained that there was no proof that the several great classes or kingdoms were descended from a common ancestor. Wallace, on the contrary, maintained that they were. Whether one could attribute to the action of natural selection the evolution of the different classes and kingdoms from a few primordial types was another question. "I do not say [it was] not sufficient," Wallace told Francis. "I merely urge that there is a difference between proof and probability."[29]

Wallace replied to Darwin's letter with a touch of sarcasm: "It is a great pleasure to receive a letter from you sometimes, especially when we do not differ very much." He moved on to the more congenial subject of his upcoming book, briefly outlining its contents. This book would be his last, he said, as he preferred to find some "easy" occupation for his declining years, one not too confining or involving much desk work. "You see I had some reason for writing to you; but do not trouble to write again unless you have something to communicate," he advised. Darwin would not bother him again for another ten months.[30]

In early fall 1880, Wallace published—again with Macmillan—*Island Life*, a sequel to *The Geographical Distribution of Animals*. In his newest book, he applied scientific principles to the special case of islands. He had selected islands for study for two reasons: first, they were restricted areas with well-defined boundaries whose geographic and biological limits conveniently coincided; and, second, they were laboratories for the evolution of species. He wrote his book for a general audience that could more easily comprehend the laws governing island biogeography than the more complex laws regulating the larger and less well defined faunal boundaries of continental landmasses.[31] Including geologic, paleontological, astronomical, meteorological, and other physical phenomena in the purview of evolutionary theory, he suggested how organisms were dispersed, underwent modification, and came to occupy their present island habitats. Presenting copious data, he demonstrated that the fragmentary groups and isolated forms found on islands were the relics of once widespread species, which had been preserved in a few localities where the physical conditions were conducive to their survival and the intensity of inter- and intraspecific competition had diminished significantly in comparison with the competition they faced on continents. It had taken him three years to complete the book. To fill in gaps in his knowledge, he had read extensively and corresponded with experts in the field. He even quoted his father-in-law, William Mitten, who alerted him to the curious affinity of Andean and British mosses.[32] Although he had a clear plan in mind, he depended on inspiration to answer some of the special questions about plant and animal distribution. For example, he tackled the great problem of climate, hitting on what he considered a "true solution" of its cause only midway through. "Like most of my other theories," Wallace confessed to his friend Raphael Meldola, a chemist with an interest in protective coloration, "it came upon me while writing, for when I began my book I had no notion of how to treat it."[33] *Island Life* delighted Darwin. It was almost as if the old Wallace had returned to the fold. "I have now read your book, and it has deeply interested me," he wrote to Wallace in November. "It is quite excellent, and seems to me the best book which you have ever published; but this may be merely because I have read it last." *Tropical Nature* may have wearied him, but *Island Life* did not. He sent seven pages of notes along with his letter, many critical, citing facts collected and kept in his files. Among other elements in the book that were "rather too speculative for my old noddle," he disagreed with Wallace's argument about the dispersal of seeds and spores from one mountain range to another or across other great distances, arguing that "I do not believe that there is at present any evidence of their being thus carried more than a few miles."[34]

Wallace rarely yielded on any issue, and he did not yield on this one. He had new evidence that seeds could be carried great distances. He cited an article in *Nature* in which a botanist from Kew had described a number of alpine plants of Madagascar that were identical to several species flourishing on the mountains of Abyssinia, Cameroon, and other African regions. "[These] alpine plants could hardly have migrated over tropical forest lands," Wallace argued, "while it is very probable that if they had been isolated at so remote a period, exposed to such distinct climatal and organic environments as in Madagascar and Abyssinia, they would have in both places retained their specific characters unchanged. The presumption is, therefore, that they are comparatively recent migrants, and if so must have passed across the sea from mountain to mountain."[35] Darwin replied that Madagascar had once extended far south during the glacial period, which allowed it to be "peopled" with African continental species through the agency of birds and sea currents. "How lamentable it is," he wrote, "that two men should take such widely different views, with the same facts before them; but this seems to be almost regularly our case, and much do I regret it."[36]

Wallace's book garnered a slew of admiring reviews and letters. Archibald Geikie, one of the foremost geologists of his day, wrote in his review that Wallace "deserves the thanks alike of geologists and of biologists for a treatise, the appearance of which marks another epoch in the history of the doctrine of Evolution."[37] Hooker was also impressed. He told Wallace that *Island Life* was a first-rate book that brushed aside "more cobwebs" than any other book on the subject. To Darwin, he wrote (alluding to Darwin's recent book *The Power of Movement in Plants*), "I am only two-thirds through Wallace and it is splendid. . . . [T]hat such a man should be a Spiritualist is more wonderful than all the movements of all the plants."[38]

Island Life rekindled Darwin's efforts to get Wallace a civil pension. Changing tactics, he bypassed Hooker and enlisted Thomas Huxley's services instead. A compassionate man who had forgiven Wallace for his peccadillos long ago, Huxley accepted his mission "like a true knight."[39] He asked Darwin to draw up a full and condensed statement of Wallace's claims. Once he had that in hand, he would approach Hooker himself and two or three others. Darwin then wrote to Buckley, requesting her to supply him with any hints or advice about Wallace's present circumstances, adding, "I do *most earnestly* hope that we may succeed."[40]

At this point, Buckley revealed to Darwin that Wallace was now fully aware of the scheme to help him. Wallace hesitated at first, not considering himself worthy of such an honor, but when she reminded him that two other prominent scientists, James Joule and Michael Faraday, had received civil pensions, he told her: "I confess it would be a *very great relief* to me and if such men as Darwin and Huxley think I may accept, [I] suppose I may." Buckley composed a terse account of Wallace's economic misfortunes. He was dependent primarily on his writings, she said, which were not very lucrative, "in an inverse ratio to their true value." She listed three major accomplishments: his extensive collections from the Malay Archipelago, his contributions to the theory of natural selection, and "above all" his application of the theory to the geographic distribution of animals. (Darwin added a fourth: the mechanisms of coloration in animals.) Buckley suggested contacting George Douglas Campbell, the eighth duke of Argyll, who served in William Gladstone's cabinet. The duke, an amateur geologist of some note, was not one of Darwin's favorite people, in part because of his strong public resistance to the theory of natural selection and other "unphilosophical" views. These reservations aside, Darwin wondered about the propriety of asking a cabinet minister to sign a petition to the prime minister. It was decided that Campbell would act as intermediary.[41]

One morning, Darwin composed a long and full statement about Wallace, polishing and elaborating on Buckley's notes, and dispatched it to Huxley for emendations, remarking that "I hardly ever wished for anything more than I do for the success of our efforts." Huxley was able to coax Hooker into cooperating by dispelling one of his "grave" objections: the bet with John Hampden. He insisted that Wallace had not pocketed the money but had given it away to charity. It seems that Huxley confabulated this story to mollify Hooker, since Wallace had had no money to give away except what he paid to his lawyers. The other objection, the belief in spiritualism, was no longer sufficient in Huxley's (or Darwin's) mind to diminish Wallace's "claim on the gratitude" of his country. Since Wallace already had made his heretical views public, Hooker suggested that they were free to inform the prime minister of the heresy while emphasizing that in the view of his fellow scientists it did not invalidate his right to a pension.[42] Darwin was pleased with Hooker's change of heart and congratulated Huxley on his wonderful management of the affair. "I cannot see that there is the least necessity to call any Minister's attention to Spiritualism," Darwin concluded, "or to repeat (what you said) to Gladstone—that Spiritualism is not worse than the prevailing superstitions of this country!"[43]

For two months, the "Wallace affair" obsessed Darwin, leaving him little time and energy for anything else, according to his daughter Henrietta.[44] When he finally heard from the duke, who reported that he had written to Gladstone to express his "high" approval for the pension, Darwin was jubilant. The price he had to pay, however, was a social call. "The Duke wants to come to Down," Darwin complained to his son Francis. "The Lord have mercy on me—but I shall write and offer to call on him when next in London."[45]

The memorandum composed by Darwin and Huxley was passed around the scientific community. For a month, it languished in Henry Bates's hands without further action, until an anxious Darwin contacted him. Greatly disappointed at the delay, he feared that the prime minister would soon be too busy with the new session of Parliament to deal with this private concern.[46] That day, Bates forwarded it to Huxley, and on January 5, 1881, it was presented to Gladstone, who approved it within two days, announcing that he had done so in a personal letter to Darwin. Darwin exclaimed to his wife, "Hurrah, hurrah! . . . Was it not extraordinarily kind in Mr. Gladstone to write himself? . . . The Duke of Argyll's private note to Mr. Gladstone seems to have done good service."[47]

Darwin reported the full details to Wallace on the day of Gladstone's announcement. He wished to spread the credit among many other people, especially Huxley and Buckley, and he quoted Gladstone's note: "I lose no time in apprising you that although the Fund is moderate and at present poor, I shall recommend Mr. Wallace for a pension of 200 pounds a year." In his usual meticulous manner, Darwin promised to keep the note on file in case Gladstone's government fell and the next government reneged on the agreement. "I hope that it will give you some satisfaction," he informed Wallace, "to see that not only every scientific man to whom I applied, but that also our Government appreciated your lifelong scientific labour." Wallace was grateful, responding that news of the pension was a "very joyous surprise," coming as it did on the day before his fifty-eighth birthday. "As I am assured both by Miss Buckley and by Prof. Huxley that it is to you that I owe in the first place this great kindness, and that you have also taken an immense amount of trouble to bring it to so successful issue, I must again return you my best thanks, and assure you that there is no one living to whose kindness in such a matter I could feel myself indebted with so much pleasure and satisfaction."[48]

It has been suggested that Darwin's efforts to secure Wallace a pension were an act of expiation for his behavior in what one commentator has characterized as "the sordid conspiracy and cover-up surrounding his obsession with priority."[49] If so, no one appears to have harbored this perception at the time. Hooker, one of the alleged conspirators, evidently felt no guilt or remorse. Nor does anything in Darwin's extensive correspondence support this view. What seems more likely is that Darwin was glad to do a favor for a man who had sacrificed his self-interest on Darwin's behalf in the name of science. Obtaining a lifetime pension for Wallace was small repayment for his extraordinary generosity. If Darwin felt any guilt, it was for his privileged position and cloistered life as a near-invalid. At the conclusion of the manuscript of his *Autobiography*, he added—but did not publish—the following remark: "I feel not remorse from having committed any great sin, but have often and often regretted that I have not done more direct good to my fellow creatures."[50]

Although the annual sum of £200 was not considerable, its regularity alleviated much of Wallace's anxiety and stress. Supplemented by the income from his writings, the pension allowed Wallace to hope that he could live in modest comfort for the rest of his life and continue to devote himself to those causes calculated to do "direct good to [his] fellow creatures."

Not long after obtaining the civil pension, Wallace built a small house in the village of Godalming, more than thirty miles southwest of central London, and the family moved once again in the early summer of 1881. He had been partly induced to move there, he later wrote, to be near his old friend Charles Hayward, who had owned the bookshop in Neath and had introduced the young Wallace to the scientific literature. Wallace and Hayward had remained in touch over the years. Hayward had moved to Godalming after the death of his wife and lived with a nephew, an architect whose children were about the same age as Violet and Will. At the age of seventy-two, Hayward had taken up the new hobby of painting in watercolors, creating his own pigments from natural products and producing what Wallace judged to be "a number of bold and effective landscapes." Perhaps more important to Wallace than its charming rural scenery was the fact that Godalming had an excellent school for his two children. Nutwood Cottage, as he called his new home, was located on half an acre of land ideally suited for a good working garden. He added a small greenhouse and cultivated more than a thousand species of plants, most of which he acquired free from Kew Gardens. But if his peers hoped that Wallace would henceforth concentrate on scientific endeavors, they were to be disappointed. He now applied himself to more pressing issues. While Darwin had been preoccupied with securing Wallace's future, Wallace had already embarked on a project that had nothing to do with either his science or his personal finances: the land question.

Wallace's concern for social injustice had intensified with age. As a young land surveyor, he had not been goaded into action by the inequitable distribution of land that he witnessed firsthand, though he sympathized with the plight of the landless majority. When he returned from the Amazon, he read Herbert Spencer's *Social Statics*, published in 1851, which, among other issues, addressed the subject of political and social reform in relation to the use of the land. This book inspired him to mobilize some of his friends to discuss the implications of Spencer's analysis, but no one was interested in gathering for the weekly meetings he proposed. Fifteen years later, in 1869, his indictment of contemporary English society in the epilogue of *The Malay Archipelago* attracted the attention of John Stuart Mill, who asked Wallace to serve as a member of the general committee of his proposed Land Tenure Reform Association, an invitation Wallace accepted. Unfortunately, it dissolved shortly after Mill's death in 1873.[51]

Although Wallace intermittently ruminated about the land problem for the remainder of the decade, between 1871 and 1879 he wrote only a handful of articles on social and political issues. Nevertheless, as the decade progressed he became embittered with English justice and its labyrinthine legal system, in part because of his own unfortunate experiences with it. His profound ignorance of the conduct of domestic and international business led him to the brink of financial ruin, providing him with a firsthand example of the vulnerability of ordinary citizens to the predations of ruthless speculators, and his ostracism by the orthodox scientific community exemplified the intolerance of the majority toward minority views. By 1879, he doubtless felt besieged on all sides. After a period of isolation, contemplation, and research, he projected his frustrations outward. In 1879 he set out to attack those institutions and policies responsible for the miseries of the common man—and his own. Thus began the next phase of his life: a foray into the world of politics, not as a politician but as a philosopher acting behind the scenes to effect a change in social policy. His energies—directed first toward evolutionary biology and then toward spiritualism—now found a practical application in the political sphere.

Wallace's political debut was marked by an examination of the question of free trade. It was a subject that had piqued his interest a number of years earlier during his brief friendship with Augustus Mongredien, who had written a book on free trade. What now provoked Wallace was a recent book entitled *Free Trade and Protection,* by Henry Fawcett, a professor of political economy at Cambridge, whose arguments astonished and offended him. Wallace believed that England's trade policy was antiquated because it was based on fifty-year-old assumptions and tailored to a world then dominated by English manufacturers. For more than a generation, England had thrown open its doors to the world, importing food and other goods from all nations without imposing import taxes. By the 1870s, other countries were producing similar commodities and challenging England's hegemony in international trade. Many of these countries taxed British imports while subsidizing and exporting their own goods, which resulted in an imbalance of trade in favor of the foreign competitor at the expense of English manufacturers and the English working classes. These competitors enjoyed a virtual monopoly in their own countries, profiting on goods sold at home and underselling the surplus to the open markets of England. English manufacturers watched their profits plummet, while English workers were forced into unemployment. Fawcett advocated imposing tariffs in order to take revenge and make the people of these nations suffer too. He singled out France and the United States, whose policies inflicted loss and inconvenience on the English peoples. Such an argument, Wallace said, was unworthy of a man of Fawcett's high reputation. In an article published in the April 1879 issue of *Nineteenth Century* (later collected in a book of his essays), Wallace wrote: "The desire of our manufacturers and workmen to enjoy the legitimate benefits of free trade and to be guarded against the injury admitted to be done to them by the arbitrary and uncertain departures from its principle by other nations is a very different thing from 'retaliation' or a revengeful wish to make others suffer." Wallace proposed the simpler, more equitable solution of reciprocity—that is, the imposition of the same import duty on articles from protectionist countries as was imposed on identical exports from Britain, thereby restoring the balance of trade: "[The] great thing is, that we shall obtain stability. Our capitalists and workmen will alike feel that foreign protectionist governments can no longer play upon our industries as they please, for their own benefit."[52] Wallace's theme was justice for the working class, the backbone of English manufacturing.

The reply to his article was swift and caustic. The respondent, Robert Lowe, the first viscount Sherbrooke, who had been Chancellor of the Exchequer under William Gladstone from 1868 to 1874, was famed for his irascibility; he lambasted Wallace as a political naïf who had stepped too far beyond the boundaries of his scientific work. Wallace's proposal was protectionism in sheep's clothing and contained nothing commendable. He reserved his most conclusive objection to the very last, "an objection which appears to me so complete and absolute that I am convinced, had it occurred to Mr. Wallace, the article we are considering would never have been written." Wallace's retaliatory measures, he said, were in effect an admission of weakness where there was none. Only the weak imposed protective duties; a superior commodity did not require protection. Lowe concluded that "protective duties may be a sword in the hand of the weak; they can never be a weapon of offence in the hands of the strong."

Lowe's "very forcible, not to say violent and contemptuous article" did not intimidate Wallace, who was unimpressed by his aristocratic assailant's "gutter tactic, . . . usually exercised in the House of Commons," of flatly denying facts that a previous speaker had presented as undisputed. "Amused and disgusted," he responded in the July issue of *Nineteenth Century*. "The fact that such protected goods are imported into this country, and do compete successfully with our own, must surely be known to Mr. Lowe," Wallace wrote in "A Few Words in Reply to Mr. Lowe," "and I am afraid the most charitable view we can take is, that this article was written with some of that want of consideration which he so confidently alleges against myself." Lowe's privileged status, Wallace implied, made him insensitive to the true consequences of a regressive policy that had been established in a different era under different circumstances, when the cheapness of English goods guaranteed English prosperity and when no other nation, it was believed, could or would compete successfully with England. While a large segment of the English population in the 1870s suffered as a result of the importation of cheap goods, the wealthier minority benefited because they did not lose their jobs as a result of an imbalance in free trade and were free to buy whatever they wanted. "Mr. Lowe's arguments and sarcasms may pass for what they are worth," Wallace concluded.[53]

The eruption of unrest in Ireland in 1879 and 1880 over the tyranny of Protestant landlords inspired Wallace to articulate his latest views on the land question. Wallace objected to a proposed government buyout of the landlords and the conversion of their tenants into "peasant-proprietors" who would repay the mortgage of the land over thirty-five years. This policy, Wallace believed, would only create another privileged class, while the rest of the Irish remained disinherited from their native soil.[54] In a November 1880 article in the *Contemporary Review,* he proposed that the land gradually revert to the state, from which tenants could then rent it. To make the transition as palatable as possible, he said, the state could assess the value of the land in two ways: according to its inherent value as determined by "nature and society," and its value as a result of improvements made on it by the owner and his predecessors.[55] The proposal was audacious, especially given Darwin's attempt to obtain for Wallace a civil pension from the Gladstone government. Like most members of the upper classes, Gladstone believed not only that landlords were a part of the natural rural order but also that no solution to the land question was possible without their active participation. Gladstone had initiated his own version of land reform in 1870 during his first administration—a half measure intended to mollify both proponents and opponents.[56]

Although Wallace's opinion apparently did not influence Gladstone's decision to award him the civil pension, his article created a stir among proponents of land reform. Spearheaded by a political activist named A.C. Swinton, several meetings were held to discuss a definite program, which led to the formation of the Land Nationalisation Society in early 1881, with headquarters in London at 57 Charing Cross.[57] Wallace was elected its first president, a position he would hold until the end of his life. It is unclear if he was compensated for his work, but he felt so strongly about this cause that he doubtless would have accepted the post without remuneration. Underscoring his commitment to the issue of social equity, he later took as his motto *Fiat justitia, ruat coelum* (If Justice be done, the heavens will fall; that is, the impossible will happen).[58] He and his small circle of land reformers quickly issued a mission statement identifying three goals: (1) to affirm that the state holds land in trust for each generation, (2) to restore to all people their natural right to use and enjoy their native land, and (3) to obtain revenue from the land for the benefit of the nation as a whole. The society would supply lecturers to anyone willing to support their cause. Because the founding members were impecunious, they made a special appeal to all those who realized that "the gigantic and perilous pauperism and demoralisation of the nation are mainly due to the existing land system."[59] One person Wallace solicited for support was Herbert Spencer, who wrote back to express his general sympathy for the society but declined to join. He recommended first directing the public's attention to "the abstract inequity of the present condition of things" and to a law that already existed and tacitly denied absolute private ownership, leaving the state the right and power to repossess the land for a number of reasons after duly compensating the owner. He feared that the government would object to a premature proposal of a specific scheme without wider public support.[60]

Although he was disappointed by Spencer's refusal to join, Wallace accepted his advice and composed a handbook to introduce the land question to a larger audience. As usual, he read everything he could find on the issue, including voluminous reports on agriculture, Irish famines, and other relevant material collected by the government. His investigations only increased his disgust with the social injustice at the heart of the land monopoly in Great Britain. Land, he felt, was as much a basic right as food, clothing, and shelter. It was also every person's natural heritage, like air and water. Yet in the late nineteenth century, a handful of the population held the majority of land in Great Britain, wielding enormous power over the lives of everyone else. Wallace uncovered numerous facts, many of which were unfamiliar to the general public. For example, 536 peers owned 15 million acres, or 20 percent of the British Isles, and extracted £18 million a year in rent "without lifting a finger." Taxes from tenants amounted to £44 million a year, while the peers, through their influence on legislation, contributed less than 1 percent of their general share of revenue from the lands they owned. Land monopoly necessitated the importation of enormous quantities of food from foreign countries—commodities that easily could have been produced at home if England's landowners played fair. Instead, despotic landlords taxed tenant farmers to such a degree that farming was neither safe nor profitable. There was not one aspect of English life, Wallace believed, that the monopolization of land did not touch. Every social, economic, or political problem could be traced back to the land question—including the depression of trade, widespread poverty, starvation wages, unemployment, homelessness, urban overcrowding, and even drunkenness, immorality, and crime. It was a great web ensnaring everything. Without land reform, Wallace felt, England could not hope to become a just society.[61]

Not long after the formation of the Land Nationalisation Society, Wallace read American economist Henry George's *Progress and Poverty*, which he called "the most startling and original book" of the past twenty years.[62]

One chapter in particular caught his attention. George had subverted the Malthusian doctrine that population growth would outstrip the food supply and lead to a reduction in wages and greater poverty. The injustice of society, George asserted, not the limits of nature, caused destitution and misery. Under the right social conditions population growth ought to increase the collective power of people to provide for one another, not diminish it. Furthermore, overproduction, as other social theorists claimed, had nothing to do with the phenomenon of the impoverishment of the working classes. The fundamental cause, he said, was that people were shut off from the land, which was the main source of wealth in any society. And the only way to redress this wrong was to return the land to the people, who would rent it from the state. He also proposed a single tax on that rent to simplify the burdensome tax system that plagued both the old European societies and the newer American one: all taxes on labor, industry, and trade that impeded their free development should be abolished. *Progress and Poverty* stirred the collective imagination of the English public and became a best-seller in 1882.[63] A hero in the Wallace mold, George had left school at the age of thirteen and voyaged around the world, working as a cabin boy for three years before disembarking in San Francisco and entering the printing business. As he watched a small number of men amass great fortunes while many others became paupers, he felt great compassion for the working people. He emphasized the paradox of American society: that increased wealth led to greater, not lesser, poverty. Wallace was pleased by the remarkable similarity in their views, having identified the same root problem. George's message harmonized with Wallace's own utopian vision of a better world.

Wallace tried to whip up support for the Land Nationalisation Society among his friends. Arabella Buckley was one of his sounding boards; Darwin, another. Wallace recommended George's book to Darwin, noting the elaborate discussion of Malthus, "to [whom] you and I have acknowledged ourselves indebted." Darwin agreed that "something ought to be done" about land reform but did not share Wallace's affinity for books on political economy, which twisted his brain more than it was already twisted. Fearing the distraction of yet another nonscientific obsession, Darwin lamented, "I hope you will [not] turn renegade to natural history; but I suppose that politics are very tempting."[64]

After Wallace completed his handbook on land reform in November 1881, calling it *Land Nationalisation; Its Necessity and Its Aims; Being a Comparison of the System of Landlord and Tenant with that of Occupying in their*

Influence on the Well-being of the People, he approached Alexander Macmillan with the idea of publication. Since this was his first attempt to publish a nonscientific work, he mustered all his marketing skills. One of his earlier essays on the subject, he told Macmillan, had won a number of important converts to the cause, who had underwritten the preliminary expenses of his nascent organization. To make the book easy to buy, he asked for a cheap paperback edition, all but promising to arrange for the acquisition of a thousand copies for distribution to working-class people. He considered the book to be the most important he had ever written; never before had anyone seen the facts he was about to bring forth. The publication of the book was urgent, and he expected a brisk sale.[65]

Wallace dedicated his book to the workingmen of England in the hope that revealing to them the chief cause of their misery would promote the radical reform necessary to give them a share of the wealth they had created for the privileged few. As he stated in the preface, his object was to teach the landless classes the nature of their rights and the way to gain those rights; it was thus a handbook of sorts, not an academic treatise. Unlike Marxists, he did not seek to overthrow the present system of government, nor did he expect the working class to seize power; he asked for justice for the mass of humanity. Without dwelling on the historical basis of land monopoly, he pointed to the fundamental error in the British social system, "the leading idea which has governed all social and industrial legislation for the last fifty years: that whatever favours and assists the production of wealth and the accumulation of capital by individuals, necessarily advances the well-being of the whole community." Wallace sympathized with the landlords and clearly understood that the loss of their hereditary entitlements would lead to the loss of their primary source of wealth. But he kept the common good uppermost in his vision of a just society. He also addressed the concerns of men like Spencer by proposing safeguards against the abuses of the state during the transition period.[66]

Although Macmillan agreed to publish *Land Nationalisation*, threats to the publisher compelled Wallace to switch to a new publishing house (Trübner). Wallace had vividly recounted the central role of a man involved in the infamous Sutherland evictions in Scotland, which began in 1807 and continued for some fourteen years. The man's sons were angered by Wallace's inflammatory report. He removed the name of the perpetrator but refused to omit the episode. According to his account, the marquis of Stafford, an English landowner, had purchased an estate of seven hundred thousand acres populated by fifteen thousand herdsmen and small farmers, who, in the course of a few years, were ordered to vacate their homes. Some relocated to the coast, where they were given small plots of land; others moved to Canada. In their place, Stafford and his agents brought in thirty-nine sheep farmers and their shepherds. Although the public believed that the process of relocation had been gentle and humane, one of the former tenants published an exposé in 1856 that cast an entirely different light. During their removal, he wrote, families suffered terribly. Most people were forced to sleep outside as they literally dismantled their homes, and some died from exposure or exhaustion. Crops were ruined through neglect. At first, people were allowed to move out at their own pace. But then the pace of relocation accelerated, and the evictions became more brutal. Houses were demolished with furnishings and personal items still inside. Crops were deliberately destroyed. Unknown numbers of people perished. Moreover, the new home sites were often worthless patches of moor or bog unsuitable for cultivation. Only the treatment of the Irish, Wallace declared, surpassed the brutality of the Sutherland evictions.[67]

In late 1881, Henry George came to Great Britain, stopping first in Dublin. From Dublin, he sent a letter to Wallace describing his initial impressions of English rule in Ireland. He had not intended to speak publicly on the issue of land reform before arriving in England itself, he wrote, but he could not keep silent about the "degrading tyranny now rampant." After making inflammatory statements in a public lecture about the British government, he was arrested and then released after a brief detention. In London, George spoke at several of the meetings of the Land Nationalisation Society, where he and Wallace became formally acquainted. Unlike Wallace, George was an accomplished public speaker, delivering his lectures in a slow, deliberate manner and underscoring his message with dramatic pauses. His lecture tour augmented the sale of his book, which by 1882 approached a hundred thousand copies in the United States and Britain. *Land Nationalisation* eventually went through five editions and, according to Wallace, had a large circulation, but its sales paled in comparison with those of *Poverty and Progress*. It is not clear why Wallace's book failed to create an equivalent stir, since it differed little from George's book. Historian Gertrude Himmelfarb has suggested that what differentiated George from Wallace was the specific mechanism—the single-tax scheme—that George proposed for carrying out nationalization, "much as the distinction and fame of Darwin and . . . Wallace came not from the theory of evolution as such but from the particular mechanism that explained the process of evolution."[68] But there may have been other reasons. Perhaps Wallace's style was too didactic. He did not mince words; he was clear and direct. George, abstract and idealistic, knew how to tug at the reader's heartstrings. George was also younger and more charismatic, and he loved the podium as an actor loves the stage. Wallace, shy and uncomfortable in large social forums, was undoubtedly less effective in communicating his message. But Americans also possessed a certain mystique. Epidemic delusions, whether spiritual or socialist, swept from west to east across the Atlantic. If Wallace's book had appeared first, it might have been more successful. The fact remains that George's book, not Wallace's, influenced the younger generation of English intellectuals and workingpeople.[69] Second-tier status never bothered Wallace, however. Dedicated to his cause, he took to the road with his fellow reformers in the Land Nationalisation Society, invading the towns and villages of England in yellow vans and preaching to anyone interested in their humanitarian message. By the mid-1890s, the society had sponsored more than three hundred "van meetings" and nearly five hundred lectures since its inception, modestly affecting public opinion while failing to achieve radical social change.[70]

Wallace had not altogether abandoned his science. During this period, he continued to publish reviews, articles, and letters in *Nature.* Some of them clarified and defended his theories on geographic distribution. He carried on a lively argument with his good friend Raphael Meldola on the coloration of arctic animals. In July 1881, he proposed a highly original theory of the origin of language in a review of *Anthropology: An Introduction to the Study of Man and Civilisation*, a new book by Edward Burnett Tylor. Wallace was especially critical of Tylor's remarks on the origin of human languages. Tylor doubted the sufficiency of the theory that emotional, imitative, and suggestive sounds were the basis on which all languages were founded. Language, he noted, is always growing, and new words are continually made "by choosing fit and proper sounds." Words once emotional or imitative had been modified to a great degree, concealing their original and less abstract characteristics. Yet Tylor still concluded that it would be "unscientific" to accept this theory of the origin of language because "other causes"—which he did not enumerate—may have had a greater impact. However, it seemed to Wallace that the imitative and emotional origin of language was "demonstrated by a body of facts almost as extensive and complete as that which demonstrates the origin of species by natural selection, and that the 'other causes' are in both cases exceptional and subordinate." He then proceeded to call attention to the wide and far-reaching character of imitative words, demonstrating the degree to which the force, expressiveness, and beauty of human languages depend on them. He gave examples of a number of onomatopoeic English words to prove his point. We see a *splash*; cold makes us *shiver*; we hear a *knock*; we feel a *stumble*: "How clearly do such words as slide, glide, and wave imply slow and continuous motion, the movement of the lips while pronouncing the latter word being a perfect double undulation. How curiously do the tongue and palate seem to be pulled apart from each other while pronouncing the words glue and sticky." Even in the motion of the breath, one could detect meaning. The words "in," "out," "up," "down," "elevate," and "depress" were all "pronounced with an inspiration and expiration respectively, the former being necessarily accompanied with a raising, the latter with a depression, of the head." We pronounce "come" with a closure and contraction of the lips during inspiration, and "go" with open and protruding lips during expiration: "When we name the mouth or lips we use labials; for tooth and tongue, dentals; for the nose and things relating to it, nasal sounds; and this peculiarity is remarkably constant in most languages, civilised and savage." Wallace noted that many savages pointed with the lips as we did with the finger.

As Wallace later boasted, he "enormously" extended the principle of onomatopoeia in the origin of vocal language.[71] In this, as in many other instances, his ingenuity astonished his contemporaries. Hyde Clarke, a noted English anthropologist, recognized a new avenue of study: "Indeed what Mr. Wallace gives us is very little, but when it comes to be applied it acquires the highest importance." He carried Wallace's idea into the realm of psychology, symbology, and mythology—to the intimate connection between sign languages and speech languages, noting that the latter evolved from but did not completely eradicate the former. "I have derived particular advantage from Mr. Wallace in being enabled to understand my own work," he confessed. In 1895 Wallace expanded his review into a longer article. After reading this article, William Gladstone, in his other guise as classical scholar, informed Wallace that there were many thousands of illustrations of Wallace's ideas in the *Iliad* and the *Odyssey*.[72]

Many such gems are buried in the reviews that Wallace churned out to earn a few extra shillings. With the fecundity of a Mozart, he created little masterpieces in minor art forms. He rarely produced hackwork and could not write unless inspired—and there was nothing more inspiring than the nature of the human species, a problem that kept him awake at night and stimulated his imagination. During his travels, he had amassed extensive vocabularies, notes, and artifacts with the apparent intention of preparing a full anthropological treatment of *Homo sapiens* in the tradition of ethnologists James Cowles Prichard and William Lawrence, but his materials from Brazil had been destroyed as a result of the fire aboard the *Helen*. He had greater success in the Far East. In an appendix of *The Malay Archipelago*, he provided a list of 117 words in 33 distinct languages. But he never got around to writing a treatise on the subject of the human species, either because other projects distracted him from this monumental task or because others—Darwin, Huxley, Sir Charles Lyell, and John Lubbock—covered the same ground. Instead, his novel ideas crop up in letters to colleagues and in a number of shorter essays. His anthropological theories are not appreciated today in part because they appeared in a variety of publications over a period of decades, just as his early writings on evolution had appeared piecemeal in obscure journals read only by a specialized audience.

Darwin sent his last letter to Wallace in July 1881. He expressed a weariness of life that Wallace probably ignored, since matters of health were common in letters from Darwin. But this time Darwin was speaking the truth. Over the next several months, his health declined. In March 1882, he suffered chest pains and developed an irregular pulse following even the slightest physical effort, which prevented him from taking his favorite stroll along the Sandwalk. Even at rest, he felt exhausted and faint. When he realized that he would never work again, he fell into a deep depression. His personal physician, whose office was in London, could make house calls only on rare occasions. Huxley wanted Darwin to move to London, but Darwin preferred to stay at Down. In the middle of the night on April 18, he developed severe nausea and "a distressing sense of weakness." Several times he lost consciousness. He remained in a state of anxiety until fifteen minutes before his death, which occurred at 4:00 P.M. on April 19, the result of complications related to heart disease.[73]

Darwin had expressed a wish to be buried at Down, but William Spottiswoode, president of the Royal Society, and a number of government dignitaries, including twenty members of Parliament, persuaded Darwin's family to have him buried in Westminster Abbey alongside other illustrious Englishmen; his crypt was placed only a few feet away from that of Sir Isaac

Newton. In granting permission for the placement of the crypt, the dean of Westminster, the Reverend George Granville Bradley, had elected to ignore Darwin's agnosticism. The funeral service began precisely at noon on April 26. It was an imposing gathering, Francis Galton observed in a letter to his sister Emma later that day. The whole "family" of scientific men filled the pews. Meanwhile, the Darwin family had selected ten men as pallbearers, among them the duke of Argyll, who as a cabinet minister and man of science represented the government. The duke and Wallace brought up the rear—Wallace having been asked by George Darwin (at the prompting of Huxley) only at the last minute, which narrowly averted a serious faux pas. Spottiswoode, Lubbock, Huxley, and Hooker, representing English science, also helped carry the casket. Several cabinet ministers paid their final respects. Gladstone and the queen did not. "Thus all shades of opinion and station were merged," Galton wryly noted. Except for the tallest pallbearers, the solemn procession up the nave to the chancel was invisible to most of the congregation and sank out of sight in "a trough of crowded humanity." A hymn specially composed for the occasion included some lines from Proverbs: "Happy is the man that findeth wisdom, and the man that getteth understanding . . . " (3:13). The rest of the service seemed perfunctory to Galton, like a graduation ceremony.[74]

Huxley wrote the most famous eulogy, in which he declared in his characteristically histrionic style that Darwin had "found a great truth, trodden underfoot, reviled by bigots, and ridiculed by all the world; he lived long enough to see it, chiefly by his own efforts, irrefragably established in science, inseparably incorporated with the common thoughts of men, and only hated and feared by those who would revile him but dare not."[75] Wallace's, written a year later for the *Century Magazine* and reprinted in the 1895 edition of *Tropical Nature*, was equally moving. "However much our knowledge of nature may advance in the future," Wallace predicted, "it will certainly be by following in the pathways he has made clear for us; and for long years to come the name of Darwin will stand for the typical example of what the student of nature ought to be. And if we glance back over the whole domain of science, we shall find none to stand beside him as equals."

Two months after Darwin's death, on June 29, Trinity College, University of Dublin, awarded Wallace the honorary degree of Doctor of Laws based on the recommendation of the Reverend Samuel Haughton, the senior lecturer in geology. Haughton had long ago forgiven Wallace for his spirited rebuttal of his article on beehives. Even a heated argument about geologic climates carried on between Wallace and Haughton in several issues of *Nature* in late 1880 and early 1881 had done nothing to lessen Haughton's estimation of him.[76] Wallace, who had an aversion to ceremonies, nevertheless accepted the honor. He later recalled that a short but "flattering Latin speech" by the public orator introduced him as "the friendly rival of Darwin"; it continued: "Equally familiar to both are the different species and varieties of animals. Darwin, indeed, was the first to pluck the golden laurel-branch. Yet through this did Wallace suffer no eclipse; for as Virgil sang—'One branch removed, another was to hand: Another, bright and golden as the first.'" Before leaving Dublin, Wallace had breakfast with Haughton and other members of a committee from the Zoological Gardens, whose "instructive and witty" conversation he enjoyed. "The brilliant midsummer morning, the cosy room and friendly party assembled rendered this one of the many pleasant recollections of my life," he wrote in his autobiography.[77] The award was an important step in Wallace's scientific rehabilitation, and he seemed grateful for the recognition.

Throughout these difficult decades, Wallace had rarely been ill. Now and then he suffered from bouts of malaria, but otherwise he enjoyed unusually good health. The assault on his reputation, his indefatigable efforts to defend himself and promote his causes, and his unquenchable thirst for knowledge preserved rather than sapped his youthful vigor. In late 1883, however, he developed a strange and apparently painful affliction of his eyes, which he attributed to long hours of reading and writing at night with a poor light source. An eye specialist diagnosed an "inflammatory process of the retina" due to eyestrain and advised confinement to a darkened room for several weeks and no reading or writing for several months. Wallace was told that with care his eyes would be as strong as ever in two or three years, and he was forbidden from doing any literary work. He appears to have followed the doctor's orders almost to the letter, though two short essays, "The Morality of Interest—The Tyranny of Capital" for the *Christian Socialist* and "How to Experiment in Land Nationalisation" for the Land Nationalisation Society, managed to escape someone's watchful notice. Otherwise, 1884 was a "lost year," a period of time that he called a "tremendous trial" and of which he recorded nothing in his autobiography except a few terse sentences.[78]

By early 1885, Wallace had recovered sufficiently to resume his intellectual pursuits. One of the first things he did was to enter a competition for the best paper proposing remedies for the depression of trade, which continued to afflict the industrialized world. His essay did not win the £100 prize—it was far too radical—but the judges were sufficiently impressed to ask him whether he would allow them to publish the first part of the essay and leave out the part with which they disagreed. Indignantly, Wallace declined the offer and talked Macmillan into publishing an expanded version. He called his book *Bad Times*. Its unwieldy subtitle—*An Essay on the Present Depression of Trade, Tracing It to Its Sources in Enormous Foreign Loans, Excessive War Expenditure, the Increase of Speculation and of Millionaires, and the Depopulation of the Rural Districts; With Suggested Remedies*—was a compendium of what he considered the root causes of the depression: loans to undeserving despots, rampant militarism, the rapid rise of millionaires at the expense of the working poor, and mass migration to the cities as a result of the inequitable distribution of land. England and Europe had become enormous casinos managed by ruthless speculators protected by unjust laws, who bilked the credulous and ignorant multitude of their life savings.

Bad Times was darker and more pessimistic than *Land Nationalisation*. Wallace questioned the morality of European society and condemned capitalism—even European civilization itself—as an utter failure. He proposed four simple solutions: (1) England ought not to lend money to foreign despots, since its commercial prosperity rested on the well-being of its trading partners, not the wealth of tyrants; (2) the working class should be enfranchised and educated to vote for representatives who authorized military expenditures solely for defense; (3) a graduated income tax should be instituted to limit the fortunes of millionaires and their heirs and to redistribute wealth for the public good; and, most important, (4) land reform should be imposed to prevent rural flight. Wallace concluded: "It is . . . by applying the teachings of a higher morality to our commerce and manufactures, to our laws and customs, and to our dealings with all other nationalities, that we shall find the only effective and permanent remedy for Depression of Trade."[79] Except for one or two positive reviews, *Bad Times* was both a critical and a commercial failure. Macmillan returned a hundred unsold copies to him, which he distributed to his friends.[80]

Undaunted by the lack of enthusiasm in his own country, Wallace—now aged sixty-three and at a time of life when his colleagues were winding down their careers—prepared for an ambitious lecture tour of the United States.

CHAPTER 16

To the Land of Epidemic Delusions

IN HIS AUTOBIOGRAPHY, Wallace states that he was invited to Boston to give a series of lectures in the autumn of 1886.[1] But this was a half-truth. The trip to the United States was his own doing. He had received an invitation to lecture in Sydney, and friends advised him to go by way of America and perhaps give some lectures there as well. In January 1886, he sought advice about the prospects of making such a tour "a financial success" from Othniel Charles (O. C.) Marsh of Yale University, the most famous paleontologist in the United States, whom he had met in London some years back and who had vaguely suggested coming to America to lecture.

Wallace had several reasons for going to the United States, but the most pressing was financial; in fact, he underlined the words "financial success" in his letter to Marsh. The civil pension that Darwin had procured for him provided some security, but it was not enough to live on comfortably. Wallace told Marsh that he would do anything in his power to make more money to support his family, even if it meant risking a sea voyage despite his age and "somewhat precarious" health. (At home were two teenagers, Violet, aged sixteen, and Will, aged fourteen, and he was determined that they would enjoy every advantage he had been denied in his adolescence.) He would go only if Marsh thought that the prospect for financial success was good. Less important reasons, which he did not mention to Marsh, included visiting leading American spiritualists and progressive political and social thinkers. Moreover, a journey to the United States offered a rare opportunity to examine the North American flora and fauna and indulge his other passion—geographic distribution. He also hoped to travel to California to visit his brother John, whom he had not seen in forty years.[2]

Wallace planned to arrive on the East Coast in the early autumn. After lecturing in Boston, Baltimore, and Washington, D.C., he would proceed to California in early winter. He would remain there until the spring and then cross the Pacific to New Zealand and Australia, where he had relatives. From Australia, he would sail to South Africa and return to London in December

1887 or early January 1888. It was an ambitious itinerary, one to which he looked forward with dread. He had not sailed on the open seas for more than twenty-five years, but the memories of the miseries he had suffered were still fresh.

Marsh consented to write to a number of associates at various American universities on Wallace's behalf. One of them was Daniel Coit Gilman, first president of the newly established Johns Hopkins University, to whom he sent a copy of Wallace's letter. Marsh told Gilman that anything Wallace was likely to say on natural history would be "good." As for the land question and depression of trade—two other subjects that Wallace told Marsh he would be willing to discuss—he offered no opinion.[3] Wallace had also contacted Carl Wilhelm (C. W.) Ernst, editor of the *Beacon*, a magazine similar in content to the *Athenaeum* but more radical, being a beacon for socialists. Ernst praised Wallace's political writings to Gilman: "I think almost as much [of him] as the most eminent men of science think of his work in natural history and animal geography." Moreover, Ernst had received an enthusiastic response from the Lowell Institute in Boston, where Wallace was already booked to give a series of lectures. But an opportunity to hear Wallace speak was not something people jumped at. When Thomas Huxley had visited the United States in the 1870s, he attracted crowds of admirers. His fame as an orator and as Darwin's bulldog had reached beyond English shores, and he was so well received that for a while he contemplated accepting a position at an American university. But there were questions about Wallace. Ernst knew nothing of his "social graces," for example, and planned to contact a cousin in London to learn more about Wallace's character; as he told Gilman, he was "quite unwilling to recommend even so eminent a gentleman without previous inquiries."[4]

It is not clear what Ernst was concerned about, but he seemed to have some reservations. Gilman also had reservations, but for different reasons. (Ernst was a spiritualist who edited *Banner of Light*, a spiritualist periodical.) Gilman asked another Johns Hopkins faculty member to solicit an opinion from Huxley, who let the cat out of the bag. "The substance of what he has to say is sure to be worth listening to," Huxley replied in March 1886, "even if it should be about spirit rapping and writing (though I presume that he will keep clear of that topic)." He had "grave doubts" about Wallace's style of speaking. Although he had never heard Wallace address a large audience, in smaller meetings Wallace was not a dynamic lecturer. Huxley qualified this judgment, however, with his usual dry humor. "I hate listening to lectures," he confessed to Gilman, "and have often said I would not hear my own if I could help it. . . . As soon as a man begins to speak I have an irresistible temptation to think of something antipodal to his subject."[5] But Wallace would surprise many of his American listeners, for he was later praised as a "more effective speaker than most of the eminent Englishmen who have lectured in the United States."[6]

Gilman put the matter aside for a couple of months. In early May, the agent Wallace had hired to arrange his lecture tour reminded Gilman of his client's planned visit. Wallace had sent an outline of his lecture series, which seemed agreeable to Gilman, who passed his recommendation on to the appropriate university committee. Gilman informed the agent that a better venue than Johns Hopkins was the Peabody Institute, established in 1857 by the American philanthropist George Peabody as an educational institution for the working classes to improve themselves—similar in spirit to the Mechanics' Institutes of Great Britain—which could accommodate larger audiences and pay a larger stipend.[7]

With the first part of his lecture tour secured, Wallace directed his attention to natural history. The topics that people wanted to hear him speak about were zoogeography and his theories on the origin of coloration. He added a lecture on the "Darwinian" theory since natural selection had been coming under increasing attack, especially in the United States. He composed eight lectures, to be given twice weekly during November and December, a little later than he would have liked. The preparation involved more time, trouble, and expense than he had anticipated. He created a number of slides to illustrate protective and warning coloration and mimicry, which required the costly services of an artist. He also encumbered himself with several maps and diagrams that, in their waterproof case, formed an awkward package six feet long.[8]

On October 9, 1886, Wallace bade farewell to Annie, Will, and Violet, and began his ocean journey on a slow steamer that left from London. The days of the many-masted sailing vessels had passed, but voyage on the open seas was no less stomach-churning. By the time he disembarked in New York City two weeks later, on October 23, he had given up any thought of going to Australia and South Africa and decided to limit his travels to the continental United States.

He spent his first night in a hotel. The next morning, a Sunday, Albert G. Browne, an editor for one of the New York newspapers who had called on him at Godalming earlier that summer, picked him up at his hotel. Browne had come with Henry George, and the three men drove through Central Park to Browne's house on East Nineteenth Street. At the time, George was campaigning for mayor of New York City as the Labor Party candidate. (He would lose to the Democrat, Abram S. Hewitt, though he pulled in more votes than Theodore Roosevelt, the Republican candidate who was favored to win the election.) Wallace attended one of George's political rallies at the Brooklyn Bridge and was asked to speak on his behalf, but being no populist barn burner he failed to impress his audience, who wanted to be whipped into a frenzy in support of their man. A few days later, he boarded the train to Boston, where he would live in a simple business-class hotel, the Quincy House, for the next two months.[9] Wallace's agent had done a good job advertising his client's lecture series. He billed Wallace as "England's Great Naturalist" and "the most distinguished living naturalist in the world" and included clippings of favorable reviews of Wallace's most recent books.[10]

The eight lectures in Boston were well received. One reporter called his lecture on Darwinism "a masterpiece of condensed statement . . . a most beautiful specimen of scientific work" and complimented Wallace on his modest and straightforward, if somewhat lackluster, manner and style.[11] Although Wallace steered clear of spiritualism per se, he could not resist inserting some remarks about his belief in a guiding Mind and a reference to human beings' true nature, which he said was spiritual, not material. He had abandoned his earlier suggestion that certain aspects of the human form were derived by some process other than natural selection, introducing in its place a teleological explanation of human origins:

> I can not only believe that [the human] body has been derived, by gradual development, from a lower animal type, but that the very reason and purpose of the existence of the otherwise meaningless animal and vegetable kingdoms, and even of our whole material world, has been, the ultimate production of the noble and perfect human form by and through which the spirit of man—the man himself—might be developed, and be prepared, through *struggle* and *effort*, and by *increasing warfare* against physical and moral evil, for a higher and more permanent existence.[12]

Wallace enjoyed the intellectual ferment of an American city with the pretensions and class consciousness of London. He made the formal acquaintance of Darwin's American confidant, the Harvard botanist Asa Gray;

met with O. C. Marsh; and participated in the meetings of the National Academy of Sciences. At times, the amount of attention he received disconcerted him. When asked by Marsh to make some remarks about geographic distribution at the academy's annual meeting, held at the Massachusetts Institute of Technology, he said in his autobiography that, having been caught off guard, he had mumbled something about the phenomenon of seed dispersal by the wind. In fact, he made a more formal presentation. Standing before America's leading scientific savants, he read "The Wind as a Seed-Carrier," which focused on "one of the most significant problems in geographical distribution." A large number of arctic plants, he said, were now so widely distributed throughout the Southern and Northern Hemispheres that plants in New Zealand, Australia, and the extreme southern tip of South America were nearly identical to those on the high mountains of more temperate zones. Darwin had theorized that the effects of a glacial epoch might explain this phenomenon, but no such epoch was known to have occurred in the tropics. Wallace's studies of the fauna of oceanic islands offered another possible explanation. Under favorable circumstances, new species of plants could appear in out-of-the-way places thanks to the transfer of seeds by wayward birds, humans, or the wind. An occasional storm, perhaps of a magnitude that occurred only once in a century, could transport light seeds for great distances. Wallace later noted that this was a novel theory to the participants at the meeting, and no doubt it stimulated considerable discussion.[13]

At one of Gray's evening gatherings, Wallace was cajoled into talking about the circumstances leading up to his famous discovery of natural selection and the penning in 1858 of "On the Tendency of Varieties to Depart Indefinitely from the Original Type," a story by then almost legendary.[14] He also spent several evenings with the essayist and novelist Oliver Wendell Holmes, who expressed a keen interest in spiritualism when he had a few moments alone with Wallace. During the day, Wallace visited the various libraries and museums and found the quality of the museums high in comparison with similar institutions in England. The superior organization of the zoological and archaeological collections at Harvard stimulated him to address one of his pet issues—"museums for the people"—in a series of articles that he published the following year in the *Fortnightly Review*.[15] He also made several excursions to other parts of the Northeast to lecture at local colleges and universities. At Yale, he met James Dwight Dana, whom he described as "the first of American geologists." From New Haven, he proceeded to Vassar College, in Poughkeepsie, New York, where he was impressed by the novelty of higher education for women, with half the faculty being "lady professors"—a concept alien to the British—and noted that the curriculum was as rigorous as the one for men.[16] He spent ten days in Baltimore speaking on Darwinism and zoogeographic distribution at the Peabody Institute and lectured at Johns Hopkins University on November 30 and December 2, 7, and 9. A reporter for the American journal *Science* attended these lectures and noted that Wallace delivered them "in a clear and easy manner, and [with] that indefinable attractiveness which comes from many years of original research." Concerning the evolution of the human mind, the reporter politely observed that "the lecturer expressed [a] view peculiar to himself."[17] Wallace returned to Boston and divided his time during the last three weeks of December between scientific and spiritualist gatherings, feeling equally at ease in both worlds.

Spiritualism had attracted a number of prominent Bostonians, among them William James, professor of philosophy and religion at Harvard and brother of the novelist Henry James. For years, William James had been fascinated with trance mediums, thought transference, clairvoyance, and extrasensory perception. He had helped found the American Society for Psychical Research in 1884, shortly after meeting Wallace's friend the physicist William Barrett, who had come to the United States in August 1884 with a contingent of English researchers to encourage American scientists to undertake a serious investigation of psychic phenomena.[18] Barrett had been instrumental in organizing the British Society for Psychical Research in 1882, after the failure of his efforts to win recognition by members of the British Association for the Advancement of Science. Wallace was elected one of its honorary members, but he declined, citing his reputation as a "crank." His name remained on the list of honorary members anyway, and in the early twentieth century he was invited to become its president, an offer he also refused. Wallace kept his distance from the society because it had ignored his and Sir William Crookes's early work and approached the subject too cautiously and skeptically for his tastes.[19]

Wallace spent the afternoon of December 12 with James and other Harvard faculty, with whom he discussed both spiritualism and psychic research; two days later, he dined with James and several of these psychic researchers. In his American journal, Wallace reported that James invited him to a "remarkable" séance at which a beautiful woman materialized. This woman took Wallace by the hand, identified herself as having joked with him at a London séance, and allowed him to feel her ear, as he had done back then. Another figure also appeared—an elderly gentleman with a vague resemblance both to his father and to Darwin but who answered to the name of a deceased Australian cousin.[20] But Wallace and James did not always see eye to eye on spiritualist issues. In the 1880s James approached the subject as a skeptic, whereas at this point in his life Wallace accepted almost everything without reservation. Twenty years later, James would sound more Wallacean in his attitude toward the paranormal. In a 1909 essay he wrote:

> When . . . a theory gets propounded over and over again, coming up afresh after each time orthodox criticism has buried it, and each time seeming solider and harder to abolish, you may be sure that there is truth in it. [Lorenz] Oken and Lamarck and [Robert] Chambers had been triumphantly dispatched and buried, but here was Darwin making the very same heresy seem only more plausible. How often has "Science" killed off all spook philosophy, and laid ghosts and raps and "telepathy" away underground as so much popular delusion. Yet never before were these things offered us so voluminously, and never in such authentic-seeming shape or with such good credentials. The tide seems steadily to be rising, in spite of all the expedients of scientific orthodoxy. It is hard not to suspect that here may be something different from a mere chapter in human gullibility. It may be a genuine realm of natural phenomena.[21]

On December 29, his final evening in Boston, Wallace was the honored guest at a dinner at the posh Parker House that included James, Holmes, Gray, and about a dozen other distinguished American scientists and savants. The dinner, "luxurious in the extreme," was clearly meant to impress Wallace; the tables were decorated with a profusion of ferns, daffodils, violets, and roses. The conversation was at first a bit pedantic and obscure. The arrogance of a few of his hosts astonished him. The poet James Russell Lowell, for example, directed a remark to Wallace in Latin, evidently intending to test his scholastic abilities; by now, Wallace had forgotten the little Latin he once knew, and he waffled, failing the test. As the wine flowed, the bonhomie increased and the conversation became relaxed and wide-ranging, covering politics, travel, Sir James Brooke, and spiritualism—but the snobbery had spoiled the evening for Wallace. "I was not so much impressed by the Boston celebrities as I ought to have been," he recollected in his autobiography.[22]

The following day, during a snowstorm, he boarded a train for Washington, D.C., without any set itinerary. He lodged for four days with Charles Valentine Riley, chief of the United States Entomological Commission, and then moved to the Hamilton Hotel. Except for a ten-day stint in Canada, where he stayed with the parents of his colleague and sometime critic Grant Allen, he was stranded in the American capital for four months because his agent had failed to book any lectures there in advance. Not until March did he have his first official speaking engagement in the city, a lecture to the Women's Anthropological Society on the origin of the human races and the nature of language.

The failure of his agent notwithstanding, Wallace enjoyed Washington more than Boston, especially its intellectual life. It must have been Riley who arranged to get him free access to the Cosmos Club and its library. Founded in 1878, the club was composed of men and women who had done meritorious work in science, literature, and the arts or had distinguished themselves for their public service. Riley took him to one of its evening receptions, where he met most of Washington's leading scientific and political figures. One of the first men he met there was Major John Wesley Powell, the one-armed Civil War hero who headed the United States Geological Survey. Powell and Wallace, both self-taught scientists and self-reliant individuals, took an instant liking to each other. In 1869, at the age of thirty-five, Powell had led an expedition with nine other men to explore the Colorado River. It was a daring and dangerous journey, never previously attempted, but he fulfilled his mission, which was to prove that the Colorado River had carved out (in line with Sir Charles Lyell's uniformitarian doctrine) the Grand Canyon. Wallace considered Powell to be not only a first-rate geologist but also a great anthropologist and psychologist, but on what basis he made such a claim is unclear. It appears that Powell was an "advanced thinker," as Wallace liked to call people who shared—or at least were open to—his radical political ideas. He introduced Wallace to Charles Nordhoff, a journalist who in 1875 had published *The Communistic Societies of the United States*, a study of the Shakers, some Native American tribes, and other communal groups. Wallace spent a good deal of time in the Geological Survey's library, reading about the history of glaciation and the antiquity of man in America. He frequently lunched with Powell and two other members; while dining "in an informal way" on bread, cheese, fruit, cakes, and tea, they had many "interesting conversations [on] all kinds of subjects."[23]

What Powell thought of a lecture that he had arranged for Wallace is not known. Presumably he had been forewarned about its content, for it was given at his request before the all-male Anthropological Society of Washington on February 15. The lecture, "Social Economy versus Political Economy," offended many of Wallace's listeners by denigrating the "blind" acquisition of wealth so characteristic of the American capitalist culture and championing the rights of the oppressed and underprivileged working classes. As Wallace would later recall:

> It was an attempt to show how and why the old "political economy" was effete and useless, in view of modern civilisation and modern accumulations of individual wealth. Its one end, aim, and the measure of its success, was the accumulation of wealth, without considering who got the wealth, or how many of the producers of the wealth starved. What we required now was a science of "social economy," whose success should be measured by the good of all. . . . [Until] this is produced there must be no labour expended on luxury, no private accumulations of wealth in order that unborn generations may live lives of idleness and pleasure.[24]

"It is astounding," marveled a reporter from the *Washington Post*, "that a man who really possesses the power of induction and ratiocination, and who in physical synthesis has been a leader in his generation, should express notions of political economy [that] belong only or mainly to savage tribes."[25]

Wallace made two enduring friendships in Washington: one with the noted ornithologist Elliot Coues and the other with the equally noted paleobotanist Lester Frank Ward. Wallace usually spent his Sunday evenings at the home of Coues (pronounced "cows"), a nonpracticing physician who, as both a founder of the American Society for Psychical Research and a leader in the American theosophical movement, was as unorthodox as Wallace. The theosophical movement, which originated in 1875 when Helena Petrovna Blavatsky (more commonly known as Madame Blavatsky, or HPB), who had been born into a noble Russian family, and Colonel Henry Steel Olcott founded the Theosophical Society, was an outgrowth of spiritualism, but it was inspired by Eastern religions (Olcott was a Buddhist). Like the most progressive spiritualists, theosophists denied the existence of a personal deity, denigrated the role of priests, preached that each person controlled his or her destiny, and renounced affiliation with any particular religious organization. Theosophy was thus a universal religion. Unlike spiritualists, theosophists believed in reincarnation. Moreover, HPB and her disciples repudiated one of the fundamental principles of spiritualism: the belief that the living could communicate with the dead through the intercession of a medium. The souls of good people, they said, were unreachable; only those souls bent on mischief appeared at séances in order to deceive credulous sitters. Wallace understandably wrote very little about Blavatsky, but he must have found Coues's brand of theosophy more congenial than hers, for the two men became good friends. One evening, at a meeting of the American Society for Psychical Research, a paper on "atoms, molecules, force, matter, organism, spirit, etc." generated a lively discussion on the "atom-force theory," spiritualism, and theosophy. Coues, Wallace wrote, "glorified" Madame Blavatsky and theosophy, while Wallace claimed a higher position "for our spiritualism." Nevertheless, Wallace noted, it was a good discussion.[26]

Coues, an assistant director of the Smithsonian Institution, was highly respected in his field and had been elected to the National Academy of Sciences. He also had helped found the American Ornithological Union. In 1879 he had earned "one of the highest compliments paid of recent years to American science" when he was presented with a document—or memorial, as the English liked to call such an honor—signed by the leading English naturalists of the day, including Thomas Huxley, Charles Darwin, John Gould, Alfred Newton, and Wallace, commending his *Birds of the Colorado Valley: A Repository of Scientific and Popular Information Concerning North American Ornithology.* Coues's book was the most comprehensive treatise on the bird life of the Colorado River basin to date; the English naturalists singled out its remarkably thorough bibliography for special praise. At thirteen hundred pages (excluding indexes), it catalogued every observation ever published by other ornithologists, professional and amateur, of some four hundred species of birds. Moreover, for each bird Coues described the geographic distribution, anatomy, plumage, migration route, behavior, affinities, courtship, food habits, nest, egg, and songs. Not even Charles-Lucien Bonaparte, the author of the *Conspectus Generum Avium*, had been so encyclopedic.[27]

Coues introduced Wallace to the leading spiritualists of the nation's capital and took him to the séances of a public medium with the odd name of P. L. O. A. Keeler. Keeler conducted his sittings in front of a large curtain through which a number of items—pencils, paper, tambourines, pocket watches—were thrust, as though the fabric were as permeable as air. Wallace found Coues to be an ideal companion—as brilliant as Huxley, in his opinion, but much less intimidating (and less incredulous).[28]

The other man Wallace befriended, Lester Ward, shared Wallace's predilection for natural history and political philosophy. Although he had earned degrees in medicine and law, he was largely self-educated in science—perhaps another reason for Wallace's attraction to him. At the time of Wallace's visit, Ward was a paleobotanist at the Smithsonian. His careful study of the venation of leaves of fossil plants provided Wallace with material that he referred to in his chapter on the geologic evidence of evolution in his book *Darwinism*, published in 1889. Later Ward would leave a career in science to become a prominent sociologist, one of the pioneers in that field in the United States. The two men took long walks on Sunday afternoons in the country around Washington, D.C., where Wallace collected a number of interesting botanical specimens, which he shipped back to England to plant in his garden. While breaking for lunch, they launched on their favorite subjects. Ward was a liberal thinker—in fact, he was a socialist. "In that respect," Wallace wrote in 1905, "he was in advance of me." In 1906 Ward published a theory of planned progress called telesis, which hypothesized that humans, through education and development of intellect, could direct social evolution. It was an idea that Wallace undoubtedly shared. But when it came to metaphysical issues, Ward remained an "absolute agnostic" or monist, a person who believes that mind and matter are one rather than distinct entities, as spiritualists preached. In this respect the two men differed. Although Wallace later became a committed socialist, Ward never became a spiritualist. Nevertheless, Wallace and Ward maintained a cordial relationship and exchanged letters for years. When Ward came to England in 1894 to attend the British Association meeting in Oxford, Wallace invited him to spend a few days with him and Annie.[29]

During his first walk with Ward on February 13, Wallace noticed at once the differences and similarities between English and American flora; this was to become a theme that would pervade his travels. Although it was winter, the woody country northwest of Washington, D.C., contained a good number of flowers. On dry banks, he found the "beautiful little May-flower (*Epigoea repens*)" and the "pretty spotted wintergreen (*Chimaphila maculata*)," both members of the heath family and both genera peculiar to the United States, except for a few allied species in Japan. The forest slopes were covered with the attractive mountain laurel (*Kalmia latifolia*), another wholly American genus. In fact, of fourteen flowering plants that he discovered on this ramble, only three belonged to British genera and four or five to European genera; the majority were either endemic to the United States or found elsewhere only in Japan, eastern Asia, and the Pacific Islands.[30]

Wallace filled his time with visits to museums, art galleries, libraries, and other cultural institutions. He got a taste of American-style politics when he spent a day at the House of Representatives. One evening, he went (much to his chagrin) to a "large and brilliant" reception in his honor at the home of Stilson Hutchins, the founder and editor of the *Washington Post*, and hosted by Isabella Beecher Hooker, a leading spiritualist, suffragette, and trance medium, who was better known as the sister of the novelist Harriet Beecher Stowe. Beecher Hooker made a point of introducing Wallace to her guests—mainly congressmen and their families—and installed him at the head of a receiving line, where he was obliged to shake hands with fifty or sixty people who knew nothing about him. This reception attracted the attention of the press; a reporter described Wallace as a "tall, large, well-preserved man of sixty-five, with white hair and a full white beard, having a slight resemblance to the late poet [William Cullen] Bryant."[31]

On April 24, Wallace finally boarded a westbound train and rid himself of his winter accoutrements. Despite the significant dent in his budget and the considerable loss of time, he had few regrets about his stay in Washington. He had developed several important friendships and felt comfortable in both the scientific and spiritualist communities—more comfortable, in fact, than he had been in London, where the scientific and the spirit world failed to mix. He crossed the American continent slowly, taking notes on the geology, geography, and vegetation en route. What struck him most, he later wrote, was the newness and rawness of the country, the almost universal absence of that harmony between wild nature and human cultivation that charmed visitors to England. As he would later write in an article for the *Fortnightly Review,* "In these North-Eastern States, the native forests have been so ruthlessly destroyed, that fine trees are comparatively rare, and such noble elms, beeches, oaks, and sycamores as are to be found arching over the lanes and shading the farmhouses and cottages in a thousand English villages, are only to be seen near a few towns in the older settled states, or as isolated specimens which are regarded as something remarkable."[32]

He made an early stop at Coalburg, West Virginia, where the naturalist William Henry Edwards, author of *A Voyage up the River Amazon*, awaited him. It was Edwards who had persuaded him to embark with Henry Bates on the Brazilian adventure that became a defining event in their lives. Wallace had corresponded with Edwards intermittently during the ensuing forty years, mostly on entomological subjects. Edwards had retired to this picturesque, impoverished region twenty-five years earlier. He lived in a "nice house with a broad verandah" in the midst of an orchard that extended to the foot of a steep, forested mountain. During Wallace's four-day visit, he and Edwards took walks in the Appalachians. West Virginia—with its high and sloping hills, tramways, coal trucks on railways, and coal-mine engines—reminded him of some of the valleys of southern Wales. The backwoods life of rural America—with its "ugly snake fences," "queer little wooden huts," and "ragged, dirty children"—evidently did not impress him. Wallace was impressed, however, with his host's extensive collection of North American butterflies and his elaborate drawings of the larvae. Edwards had accurately depicted the larvae "at every moult, from their first emergence from the egg up to the pupa stage, which often served to determine otherwise too-closely allied species."[33]

After Coalburg, Wallace continued by train through the heartland, stopping in Cincinnati, Ohio. Here, in late April, the woods were "carpeted" with the flower called spring-beauty, while elsewhere there were "sheets" of Dutchman's breeches (*Dicentra cucullaria*), small patches of "exquisite little" blue-eyed Mary (*Collinsia verna*), the "handsome" celandine poppy, the "elegant" purple phlox, and the blood-root (*Sanguinaria canadensis*), with its white star-shaped flowers—all endemic species. But there were more "homely-looking" species, like creeping yellow buttercups and blue, white, and yellow violets, which were "utterly insignificant" compared with the many new and strange forms he encountered.[34]

He continued onward to Indianapolis, Indiana; St. Louis, Missouri; Sioux City, Iowa (where he met Bandusia Wakefield, "a fine botanical artist and amateur botanist"); and several towns in Kansas—including Lawrence, where he examined the University of Kansas's fine collection of Cretaceous plants, and Manhattan, where Edwin Alonzo Popenoe of Kansas State Agricultural College showed him the "peculiar" prairie flora.[35] As he traveled west, it grew hotter and dustier, and his nights aboard the train, usually in a sleeping compartment he shared with a stranger, often were uncomfortable. His diagrams, which could not be stored with ordinary luggage, had been left behind in Cincinnati, and he was forced to lecture without them, substituting impromptu drawings. The novelty of transcontinental travel quickly palled, and he considered cutting short his journey and returning home, perhaps concerned about his health in such a miserable climate. But he changed his mind and on May 23, 1887, finally arrived in Oakland, California, where he boarded a ferry to San Francisco.

John, who had come down from his home in Stockton, met him at the train station. Wallace says virtually nothing more about the reunion with the brother he had not seen in four decades. They checked into the Baldwin Hotel that evening, but had little time to talk. For the next two days, they were besieged by callers, including an astronomer from the Lick Observatory; a geologist from the Geological Survey; Joseph Le Comte, a geologist and zoologist who had studied with Louis Agassiz and was now president of the University of California at Berkeley; and a few reporters. John had arranged for his brother to give a lecture on "Darwinism" and on "Colour," but to whom and where these lectures were given Wallace does not say. He did, however, manage to squeeze in a séance on Friday, May 27, even coaxing John to accompany him. Undoubtedly at Wallace's instigation, John brought along a blank folding slate. It was a bright, sunny morning, and he, John, and the medium sat at a small table close to the window, while two other men—J. J. (James Jerome) Owen, former editor of the *San Jose Daily Mercury* and now the editor of the spiritualist newspaper the *Golden Gate*, and a physician and friend of Owen's—observed the proceedings from across the room. At one point, John's folding slate was set on the floor a foot or two away from the table. Although the slate was always in sight as they conversed, a few minutes later it was found to have writing on both inner faces. "I wish I could describe to you my spirit home," the message read. "But I cannot find words suitable in your earthly language to give it the expression it deserves. But you will know all when you join me in the spirit world. . . . Your loving sister, Elizabeth Wallace. Herbert is here." (Elizabeth, the firstborn child, died in infancy in 1808; Wallace might have meant Eliza, a sister who died of tuberculosis in 1832 at the age of twenty-two.) It occurred to Wallace to ask the medium whether writing could be produced on paper placed between slates. After a moment's pause, "as if asking the question of his guides," the medium told Wallace to take a paper pad, tear off six sheets, and place them inside a folding slate chosen randomly from a pile of slates nearby. This Wallace did, and he and John placed their hands over the folded slate. In a few minutes, the slate was opened and they found six portraits "in a peculiar kind of crayon drawing." Among the six portraits were simple but recognizable drawings of the early spiritualist and physician Benjamin Rush, one of the co-signers of the Declaration of Independence and confidant of President John Adams; the medium Daniel Dunglas Home; and a girl, who identified herself as "The spirit of Mary Wallace," probably (according to Wallace) his sister who had died at the age of eight, in 1822, a year before he was born.[36]

On the day of one of his lectures in San Francisco, Wallace accompanied Henry Gibbons, a professor of field botany at the University of California at Berkeley, and the soon-to-be-famous conservationist John Muir to the remnants of the grand redwood forests in the foothills north of San Francisco. When their carriage reached one particular clump of young redwoods, Gibbons directed Wallace to stand inside a circle of saplings, each twenty to thirty feet high, and showed him that they all arose from the outer edge of a huge charred trunk of an old tree destroyed by fire forty years earlier. The stump was thirty-four feet in diameter, as large as the biggest of the more celebrated sequoias. It was within a hollowed-out area of this mammoth stump that the three men ate lunch, but what they discussed Wallace did not find worthy of recording. He mentioned in his autobiography only that he admired Muir's book *The Mountains of California*, which he called a "beautiful volume."[37] Their mutual friend Asa Gray, Muir's mentor in botany, probably helped arrange the meeting. Ten years younger than Wallace, Muir had tried to emulate his great traveling-naturalist predecessors by walking from Indianapolis to South America, an expedition that ended prematurely on the Gulf of Mexico, where he contracted malaria. He then headed west to California and spent another five years exploring the Sierras from his base in Yosemite Valley. Wallace and Muir shared a passion for preserving the natural environment, and for the next twenty years they occasionally corresponded (their letters appear not to have survived), though they never developed a close relationship. Muir gravitated toward Joseph Hooker, most likely because of Hooker's greater expertise in botany—or perhaps because Wallace was too unorthodox even for the unorthodox Muir. It has been suggested that Muir's consultations with Wallace, which may have included a visit to his home in England in 1893, influenced his concept of conservation and his efforts to establish a system of national parks in the United States.[38]

Wallace joined John and his American family in Stockton for the Decoration Day weekend and then returned briefly to San Francisco on Sunday, June 5, to give a lecture on spiritualism before an audience of more than a thousand people at the Metropolitan Theater. He entitled his lecture—or homily, for it was written more in the spirit of a sermon than an academic discourse—"If a Man Dies, Shall He Live Again?" A reporter noted that those present "were evidently anxious to hear an answer to Job's significant question," and Wallace gave it to them.[39] It was the question of questions, Wallace said, one that the ancients never solved and modern scientists ignored. Yet he believed that the ultimate decision arrived at, whether negative or affirmative, not only was of vital interest to each individual but also determined the future welfare or misery of mankind. If Job's question was answered in the negative, it left the human condition "utterly hopeless," destroying the dream of reward for justice, truth, and unselfishness and placing no restraint on our evil tendencies. The greatest good for the greatest number, a noble philosophical belief of the utilitarians, could not, he felt, sufficiently motivate those who were concerned solely with their welfare. But the universal lesson of science—that all humanity and the world itself would eventually end—also would have no positive influence on the actions of human beings. In such a world, might alone would constitute right and the most selfish and brutal would dominate the world. Happily, he said, such a hell on earth would never exist "because there are causes now at work which forbid a disbelief in man's spiritual nature and his continued existence after death."[40]

After briefly outlining the evidence of spiritualist phenomena throughout human history, "proving" that they were natural phenomena, Wallace criticized scientific materialism. Science, he said, had penetrated so far into the mysteries of nature that it could not believe that spirit existed, while physiologists were unable to believe in the possibility of any mind without a corresponding material brain. "It is in the most materialistic epoch of the earth's history, in the midst of a society which prides itself on discarding all superstition and basing its belief on the solid foundation of physical science, that this new and unwelcome visitor [spiritualism] has intruded itself and maintained a vigorous existence for more than thirty years." Spiritualism's entire history, he noted, proclaimed it to be not imposture or delusion or the survival of savage beliefs, but a great and all-important truth:

> The essential teaching of Spiritualism is that we are all of us, in every act and thought, helping to build up a mental and spiritual nature which will be far more complete after the death of the body than it is now. . . . Spiritualism also teaches that every one will suffer the natural and inevitable consequences of a well- or ill-spent life; and the believer receives certain knowledge of these facts regarding a future state. . . . The struggle against material difficulties develops the qualities of patience and perseverance and courage, and undoubtedly the fruits of the

> ages, mercy, unselfishness and charity, could not possibly be exercised and trained except in a world where wrong and oppression, misery and pain and crime called them into action. . . . An imperfect world of sin and suffering may be the best and perhaps the only school for developing the highest phase of a personified spiritual existence.

His $146 payment for this unexpected invitation to speak surpassed his earnings for any single scientific lecture before or after his American tour.[41]

On Wallace's return, John and his daughter, May, accompanied him on a trip to Yosemite Valley, traveling for two hours by train and then for a leisurely two days, with frequent stops, by stagecoach in unbearable heat and dust. Despite the intense discomfort, it was a wonderfully instructive journey from the perspective of botany. The least interesting part was the lower foothills, up to about three thousand feet, which had been defaced by gold miners. At first there was scant vegetation, and the only conifer was the spiritless scrub pine (*Pinus sabiniana*). At about a thousand feet, a "coarse, unornamental" tree, the ponderosa pine (*P. ponderosa*), took root. Higher up were stands of sugar pine (*P. lambertiana*), a handsome species with large clusters of pinecones. Only above twenty-five hundred feet did the forests become picturesque, with the appearance of the elegant Douglas fir (*Pseudotsuga menziesii*) and red cedar (*Thuja plicata*). As they steadily climbed to four thousand feet, they began to encounter the most beautiful of all California firs, the white (*Abies concolor*) and noble (*A. nobilis*), whose symmetrical growth and dense horizontal branches "adorned with the most delicate of blue-green tints" delighted the eye. In the forest understory were the occasional small oaks and maples, but the most conspicuous plants were the stately white azaleas and the California dogwoods, with their six-inch white blossoms; the forest floor itself was bare except for some creeping herbaceous plants. The true "big trees," *Sequoia gigantea,* inhabited the belt from four thousand to the summit at seven thousand feet. Wallace admired the pines, firs, and cedars, but he found them "far exceeded by two others inhabiting the same country, the two Sequoias—*S. gigantea* and *S. sempervirens* [the redwood]." No other natural feature on earth was more impressive in its "display of the organic forces of nature" than these two giant species. Their increasing scarcity pained him, for at one time they had covered almost all the coastal and central mountains of California:

> Unfortunately these alone are within the power of man totally to destroy. Let us hope that the progress of true education will so develop

> the love and admiration of nature, that the possession of these altogether unequalled trees will be looked upon as a trust for all future generations, and that care will be taken, before it is too late, to preserve not only one or two small patches, but some more extensive tracts of forest, in which they may continue to flourish, in their fullest perfection and beauty, for thousands of years to come, as they have flourished in the past, in all probability for millions of years and over a far wider area.[42]

After reaching the summit, they zigzagged downward past Bridalveil Fall, the grand precipice of El Capitan always in view, entering Yosemite Valley with its rushing river. They spent two nights at a costly hotel in the valley, during which the two brothers took daily hikes, including a thousand-foot ascent to a point overlooking Upper Yosemite Fall. Wallace, who had lost little of his stamina, took a solitary six-mile walk to a grove of sequoias. The wild beauty of Yosemite captivated him, and he spent many hours examining the wildflowers and trees, especially admiring pink pentstemon (*Pentstemon newberryi*), a dwarf shrub with deep red flowers, and the brilliant scarlet Indian pink (*Silene californica*). He also pondered the valley's geologic origin, concluding that it had been formed by weather and the incursions of vegetation, which together had sculpted it from the crumbling granite over a period of time hard to conceive in human terms but now familiar to him from the work of Lyell. In an essay originally published in *Nineteenth Century* in 1893, he would write that Yosemite did not owe its exceptional physical features—the Cathedral Spires, the "subquadrangular mass" of El Capitan, the rounded summit of Half Dome—to any catastrophic origin, as some geologists still asserted. They were all, he wrote, "fully explained by that simple theory of earth sculpture by atmospheric agency which has been found applicable to the solution of similar problems in all other parts of the world," thus proving "the efficiency of causes now in action in producing the varied contours of the earth's surface."[43]

After spending a little more than a week exploring Yosemite, at 4:00 A.M. Wallace, John, and May left for Stockton, which they reached shortly after noon the same day. Waiting for Wallace was an invitation from Senator Leland Stanford, railroad magnate, politician, and spiritualist, whom he had met while in Washington, D.C. Stanford asked Wallace to be his guest at his estate in Menlo Park, located thirty miles south of San Francisco. The two men spent the day touring the estate's beautifully manicured grounds, which were tended by thirty Chinese gardeners, and a good part of the rest of Stanford's eight thousand–acre ranch in neighboring Palo Alto, studded with redwoods and eucalyptus groves. In a few years, the ranch would become the site of the university that the Stanfords were building in memory of their only son, Leland Junior, who had died at age fifteen of typhoid fever. Wallace may not have approved of Stanford's extravagant lifestyle or conservative politics, but he welcomed the proposed university's stated mission to "[inculcate] love and reverence for the great principles of government as derived from the inalienable rights of man to life, liberty and the pursuit of happiness." He hoped that this clause would be taught "in its spirit as well as in its letter. Never has a grander memorial been raised by parents to a beloved son."[44]

Wallace lingered in Stockton through the Fourth of July holiday, having been laid up in John's home for a week with an infected upper lip that required lancing and bandaging before it finally healed. His enforced few days of rest, he later wrote, gave him a rare opportunity to enjoy "that singular America festival," Independence Day. The firecrackers, Roman candles, and parade consisting of firemen, soldiers, politicians, animals, and clowns, plus the noisy crowds, brought back memories of a Brazilian carnival, though the Stockton display was a good bit smaller and less flamboyant.[45] Letters confirmed his next lecture schedule, and he made ready to depart, but not before giving his niece, May, his obvious favorite, a writing case and an amber brooch. On July 7 he said good-bye to John, May, and his two nephews, Herbert and Percy—none of whom he would ever see again—and boarded the 9:00 A.M. train to Sacramento.[46]

Wallace's next scheduled destination was Michigan, but no one expected him there until the end of the month, so he spent three days in the town of Summit in the Sierra Nevada, exploring hidden valleys and collecting flowers he had never seen. In small rocky valleys laced with drifts of snow and rivulets of meltwater, he admired the "grand examples" of ice action, huge granite blocks that were cracking, splitting, and moving slowly down the valley slopes. Frost, snow, vegetation, and sun all aided this gradual process of denudation and degradation. He made a brief excursion by stagecoach to Lake Tahoe, whose wooded, mountainous shores he found "less imposing" than the forested mountains surrounding Loch Lomond in Scotland and Lake Windermere in the English Lake District, and the lake itself far less grand than the lakes of Switzerland and northern Italy. "The rounded forms of the granite rocks are here (Sierra Nevada July 1887 Lake Tahoe) plainly due to glaciation," he recorded in his journal, "& have quite a different character to the globular or dome form at the Yosemite & elsewhere due to structural exfoliation. Here they have all the character of rugged weathered peaks & pinnacles, worn down smooth in rounded hummocks."

He continued on to Reno and across hot and arid Nevada into northwestern Utah, noting the appearance of stratified, slaty rocks in an ancient lake basin that was once part of the Great Salt Lake, 250 miles west of the town of Ogden. On July 18, he reached Denver, where Alice Eastwood, a twenty-eight-year-old botanist and expert in the flora of Colorado, acted as his guide. Wallace had met Eastwood in May, having briefly stopped in Denver on his way to California. At the time, Eastwood taught ancient history at a high school in Denver. The principal introduced her to Wallace, who was looking for a local botanist who could show him where to collect alpine plants, and Eastwood had agreed to take Wallace into the nearby mountains on his return journey. In the early 1890s she would move to California, where she would distinguish herself as that state's leading female botanist (until her death in 1953 at the age of ninety-four). She also developed an important friendship with John Muir.

In the company of Eastwood, Wallace made some interesting discoveries relating to the geographic distribution of the alpine flora of Europe and the United States. They set out on Tuesday, July 19, by rail to the town of Graymount, the railway terminus, situated at 9,500 feet at the junction of two valleys. After checking into a hotel, they took a short stroll and immediately discovered two rare British plants, the wintergreen (*Pyrola rotundifolia*) and the twin-flower (*Linnea borealis*). The next day, they hiked to an outpost known as Kelso's Cabin, where some miners had houses above the timberline at 11,000 feet. Along the way, they found a number of species that had European affinities, like an alpine form of the columbine (*Aquilegia coerulea*) and *Omphalodes nana*, var. *aretiodes*, whose tufts of blue flowers were embedded in the clumps of the common moss campion (*Silene acaulis*). But it was not until they passed the miners' houses on the way to the summit of Gray's Peak that they found "some of the chief gems of the alpine flora of the Rocky Mountains." Bordering a stream were "fine clumps" of the handsomest American primrose (*Primula parryi*), with its whorled crimson-purple flowers and yellow eye. In boggy places was an arctic species, the Greenland lousewort, and in rocky crevices grew a moss that was abundant in Scotland and Wales. The following morning, they explored Grizzly Gulch, a valley not far from their hotel, where they met two miners, who offered them lunch and showed them a good place to search for flowers. There Eastwood and Wallace "luxuriated" in the finest alpine flower garden they had yet seen, though Eastwood had visited the region numerous times. What especially attracted Wallace's attention was the number of species identical to those in Great Britain and Europe. That night, they slept in a miner's cabin at about a 12,500-foot elevation. The next morning, they ascended to the top of Gray's Peak, 14,500 feet above sea level, evidently without difficulty. Among the alpine plants allied to arctic species also found in Europe they noted pretty yellow flowers in the short turf just below the summit that, on closer examination, proved to be two distinct species shaped alike and flowering at the same time—an uncommon finding. It occurred to him, as a natural selectionist, that at such a high and exposed site the flowering season was short, pollinating insects were scarce, and the combined simultaneous display was advantageous to both species. Their mutual conspicuousness would attract whatever insects might visit at this great altitude, ensuring the survival and proliferation of both species in the struggle for life.[47]

In an 1891 two-part article published in the *Fortnightly Review*, "English and American Flowers," Wallace summarized his conclusions about the relationship of North American and European alpine plants. During his excursion with Eastwood, he found ninety-five species either identical to or allied to species in the high alpine or arctic regions of Europe, whereas only thirty were strictly American species. To understand how this came about, he said, it was necessary to go back to the glacial epoch, when a milder climate prevailed in what is now the arctic region. The present arctic flora, or its immediate ancestors, was then confined to the highest latitudes around the North Pole and its ring of mountains in northern Europe, Greenland, North America, and northeastern Asia. At that time, the Rockies, the Alps, and some of the Scandinavian mountains supported only alpine forms of plants in the surrounding lowlands. As the climate became colder and the ice sheet crept farther south across the two continents, the true arctic plants were driven southward, displacing the indigenous flora and occupying all the great mountain ranges south of the glaciers as well as the peaks that rose above them. Later, as the ice sheet retreated, these hardy plants kept close to the gradually retreating ice, moving up into the higher peaks of many mountains from which the ice had disappeared. Thus, he concluded, there were now many species common to the Rockies and the mountains of Wales, Scotland, and northern Europe. What made the flora appear different to the casual visitor were the proportions of plants of the endemic species to European-allied species. The showiest flowers were not those seen in Europe—like the white, purple, and yellow anemones; the rosy and purple primulas; or the white and yellow buttercups—but the blue and white columbines, the scarlet or crimson-bracted castilleias, and a host of purple or yellow composites.[48]

On July 26, he left Eastwood and his alpine paradise for Chicago, happy to have his own private sleeping car, and arrived two days later. The city, enveloped in a cold lake mist and formidable pollution, was still recovering from its devastating fire seventeen years earlier. The "irregularity" of its architecture—massive granite buildings towering above rotting wooden hovels—and the appalling spectacle of its lakefront, which was hideously disfigured by eight parallel lines of rails on which trains belched black smoke, distressed him during the few hours he wandered about. Soot killed the grass and blackened the moribund trees. Following an abysmal meal in the shabby railroad restaurant, he recalled the words of an American writer who had wryly observed of his homeland that "a whole huge continent has been so touched by human hands that over a large part of its surface it has been reduced to a state of unkempt, sordid ugliness."[49]

Wallace gave two lectures on animal coloration and Darwinism in Trowbridge, Michigan, following which he hurried on to Kingston, Ontario, to spend several days with Grant Allen's parents. From Kingston, he boarded the steamer *Vancouver* and sailed up the Saint Lawrence River, stopping at Montreal and Quebec before proceeding across the Atlantic to Liverpool. While at sea, he developed an upper-respiratory infection that conspired with his general state of dizziness and nausea to make for a terrible voyage. A thoughtful passenger sent him some grapes, which rats devoured during the night. On August 20, 1887, after an odyssey of ten months and ten days, he was finally reunited with Annie and the children in Godalming.[50] He had kept a meticulous account of his finances on the trip and estimated that his living expenses had been $3 a day. Thirty-eight lectures and two articles had earned him $3,167. His total outlay was $2,140, leaving him a net profit, after a few miscellaneous expenses, of $780, or £160.[51] From an economic standpoint, his trip had been a modest success, though it did little to augment his position in the annals of science. However, he derived more from the tour than he had expected. He had spent nearly a year indulging his two great passions, natural history and spiritualism, and he had had ample time to reflect on humanity's place in the universe, the conundrum that had occupied and would continue to occupy a large part of his intellectual life.

CHAPTER 17

The New Nemesis

WALLACE'S CELEBRATED OPPONENT William Benjamin Carpenter died in March 1885 after suffering severe burns from an explosion in his laboratory, but his place as major adversary was soon taken by George John Romanes. The battle between Wallace and Romanes arose on a different front. Romanes aspired to inherit Darwin's crown, a position that Wallace had all but abandoned but clung to with surprising tenacity when he felt it was threatened. Since Darwin's death in 1882, the theory of natural selection had once again come under assault, this time by evolutionists themselves, many of whom remained unconvinced that it was the primary force driving organic evolution. Although Wallace had shifted his focus from natural history to political theory after the publication of *Island Life*, he was drawn back into the conflict by Romanes and other scientists, who presented challenges he could not resist addressing.

Romanes, a lesser figure in the history of evolutionary theory, was born in Kingston, Ontario, in 1848, the year that Wallace and Bates set out for Brazil. After inheriting a considerable fortune, the family moved to London, where the four children lived in the fashion typical of the British upper classes. Romanes attended Cambridge, intending to become an Anglican minister, but after discovering the sciences he decided to devote his life to scientific research. In 1873, following the death of his father, he returned to London to live with his mother and transferred to University College, London, where he continued his experimental studies in the physiology laboratory of William Sharpey and John Burdon Sanderson. A letter submitted to the editor of *Nature* in June of that year, containing a somewhat abstruse discussion of the origin of coloration in an unusual specimen of a flounder at the London Aquarium, attracted the attention of Darwin, who no doubt perceived in Romanes a potential protégé whose work in experimental biology was worth encouraging.[1] Soon afterward, Romanes developed a close, almost sycophantic relationship with his mentor, which lasted until Darwin's death. Not long after his first contact with Darwin, he discovered the rudiments of a central nervous system in jellyfish (medusae), demonstrating that, even in this primitive form, nervous tissue is fundamentally the same throughout the animal kingdom, an important confirmation of the theory of common descent. He also wrote extensively on animal psychology and the mental evolution of the human species. Darwin helped him get elected to the Royal Society in 1879, though he had twice been denied admission. Egotistical and aggressive, Romanes rarely hesitated to publish his personal views in the form of verbose and pompous letters to the editor of *Nature,* which tried the patience of friends and foes alike. Although his wife later wrote that he was "absolutely incapable of anything but fairness," he could be pugnacious and indecorous.[2] It was this aspect of his personality that provoked clashes with Wallace.

Wallace reports in his autobiography that he first met Romanes in 1880, but he seems to have forgotten that they had exchanged a volley of letters on animal instinct in 1873. Romanes had accused Wallace of neglecting the significance of a "sense of direction" (homing instinct) in some animals. Wallace peremptorily replied that he would make no further remarks on the subject because he believed that "nothing satisfactory can be arrived at till experiments of the nature indicated in my last letter have been systematically carried out."[3] If Wallace thought nothing more of that episode, the twenty-five-year-old unknown and aspiring scientist would have remembered it clearly. They first met in person seven years later, in a "curious way," as Wallace described it. A letter by M (the biologist St. George Mivart) in *Nature* in February 1880—which theorized that thought, or "brain vibrations," could be conveyed across space to other brains, thus explaining the phenomena of clairvoyance and mesmerism—was answered by FRS, an interested skeptic. FRS asked for assistance from anyone who might help him to investigate thoroughly these so-called facts of psychic phenomena. He added that he had taken a good deal of trouble to investigate spiritualism and, while encountering a vast amount of "humbug," had also met with one or two things that he could not explain. He would "prosecute" his researches without bias or prejudice and would declare M's beliefs as facts if these facts were reproducible. He depicted himself as a "man of science" ready to take on "the powers of darkness"— as though his battle would end all battles. Wallace replied, reminding FRS that this ground had been covered many times before and warning that no amount of evidence would ever satisfy inveterate skeptics.[4]

A few days later, FRS wrote privately to Wallace, revealing himself as Romanes. In the letter, which was deferential and polite, he admitted to having been "staggered" by one or two facts that Wallace might find commonplace. Since the two men were not formally acquainted, he had been reluctant to make an unsolicited approach to discuss the subject, but now an opportunity had arisen. He was not as cynical as Wallace had become about the importance of his and Sir William Crookes's inquiries. "[Pray] do not suppose that I am blind to the importance of the testimony already accumulated," he wrote. "I should rather infer it is you who are blind to that importance; I think you underrate the impression which your own publications and that of your few scientific co-operators have produced. I know that this impression is in many minds profound." Wallace consented to a meeting, which prompted Romanes to ask for an appointment. His chief end was to discover what the "prophets" Crookes, Wallace, and others had already discovered. Romanes and Wallace met twice, first at Croydon in late March or early April 1880, and later in London. Romanes told Wallace that a close member of his family was a medium, and that the usual rapping messages had been spelled out—some true, some not, but all believed to have been produced by some other "intelligence" and not by any of the participants. Wallace gave Romanes his best advice but heard nothing more from him, assuming that he had abandoned his investigations after they failed to produce the evidence he sought.[5]

But Romanes was not interested in further investigations. Instead, he reported to Darwin on his visit with Wallace, like a spy who had made a successful incursion into enemy territory:

> [A]lthough I passed a very pleasant afternoon with him, I did not learn anything new about Spiritualism. He seemed to me to have the faculty of deglutition too well developed. Thus, for instance, he seemed rather queer on the subject of astrology! and when I asked whether he thought it worthy of common sense to imagine that, spirits or no spirits, the conjunctions of planets could exercise any causative influence on the destinies of children born under them, he answered that having already "swallowed so much," he did not know where to stop!![6]

Six years later, in July 1886, Romanes published an article in the *Journal of the Linnean Society* entitled "Physiological Selection: An Additional Suggestion on the Origin of Species," with which Wallace vehemently disagreed. The few responses to Romanes's argument were too weak, Wallace told his friend Raphael Meldola, so he would write one of his own "to expose the great *presumption* and ignorance . . . in declaring that Nat[ural] Select[ion] is *not* a theory of the origin of species,—as it is calculated to do much harm. See for instance the way the Duke of Argyll jumped at it like a trout at a fly!" An editorial in the *Times* on August 16 called Romanes "the biological investigator upon whom in England the mantle of Mr. Darwin has most conspicuously descended," which appalled Wallace. Moreover, the anonymous commentator had remarked that if physiological selection was accepted as valid by evolutionists, Romanes's paper would constitute "the most important addition made to the theory of evolution since the publication of 'The Origin of Species.'" As the greatest living exponent of the theory of natural selection in its pure form, Wallace felt that he was the only man who could come to the theory's defense. Romanes was posing as Darwin's successor. "This should be stopped before the press and the public finally adopt him as such," he told Meldola.[7]

Wallace's unusually heated remarks were no doubt a reaction to what he perceived as a veiled personal attack. He responded in the *Fortnightly Review*, in "Romanes *versus* Darwin: An Episode in the History of the Evolution Theory," published shortly after his departure for the United States. His review was intended to destroy Romanes's objections at their roots before they could invade the garden he had so lovingly cultivated.

Romanes's theory of physiological selection sought to solve a problem that Wallace had pointed out to Darwin in 1868: how varieties developed into species without a physical barrier separating the incipient species from its parental form. At the time, Wallace had suggested that physiological isolating mechanisms that prevented hybridization between a variety and its parental species evolved through natural selection; that is, the hybrids would have proved less successful in the struggle for survival than the pure offspring of either the variety or the parent species, and thus a variety sharing the same range as its parental form might evolve into a separate species. Darwin could find no evidence that natural selection had anything to do with what he called this disinclination to cross, but he had no other explanation for the phenomenon except to state that it was a by-product of evolutionary change. The failure to explain the disinclination to interbreed was a potential weakness in their theory, Wallace had warned at the time, one that their enemies would exploit. Now, some twenty years later, that enemy had arrived in the guise of Romanes.

According to Wallace's understanding of Romanes's theory—and there would be considerable confusion among naturalists over the years about what Romanes was trying to say[8]—"physiological variations" occurred first and were the actual starting point for the formation of new species. Natural selection therefore would not be a theory of the origin of species because it would not account for what Romanes maintained was the primary characteristic feature of species: the almost invariable infertility when two allied species were crossed. Moreover, individual species were not distinguished from one another solely by distinctive adaptations to the environment but by trivial, superficial, and altogether useless characteristics. To account for these characteristics, Romanes said, Darwin and his followers had suggested additional causes, such as sexual selection, Lamarckian use and disuse, and, most important, the prevention of intercrossing with parent forms. Prevention of intercrossing would occur when varieties were geographically isolated from the parental species. During that isolation, independent variability would be sufficient for the evolution of new species, for there would be an inherent tendency to develop variations, which would become fixed and be passed on to succeeding generations of animals. But, Romanes argued, most species must have originated in the midst of the parental form (sympatric speciation). Only in extreme cases, as in the Galápagos Islands, did speciation occur as a result of geographic isolation. Unless mutual infertility appeared at the outset, an incipient species would become extinct through intercrossing ("swamping effect").

After summarizing Romanes's theory, Wallace proceeded to deconstruct the three "great" objections to natural selection: the evolution of characteristics apparently unrelated to survival, the evolution of varieties into species even when they occupied the same range as the parent species, and the "disinclination to interbreed" among species in the wild. Wallace pointed out that an enormous number of specific characteristics that seemed useless to the observer often were not useless at all: "[T]his argument from our ignorance is a very bad one when we consider how recently whole groups of specific differences, formerly looked upon as useless, have been brought under the law of utility." Protective coloration was one example, and he provided ample evidence to prove that trivial characters distinguishing one species from another both were useful and had been "fixed" by natural selection: "I believe, therefore, that the alleged 'inutility of specific characters' claimed by Mr. Romanes as one of the foundations of his new theory, has no other foundation than our extreme ignorance . . . of the habits and life-histories of the several allied species, the use of whose minute but numerous differential characters we are therefore unable to comprehend."

As for the second objection, Wallace quoted Darwin, who believed in microgeographic isolation. According to Darwin, each newly formed variety is at first local, separated from the rest of the species. Similarly modified individuals exist in small local populations and interbreed. If this new variety is successful in the battle of life, it will slowly spread from a central habitat, competing with and conquering the unchanged parent species on the margins of the expanding habitat. Romanes challenged this "very large" assumption: that the same variation occurs simultaneously in a number of individuals inhabiting the same area. Wallace insisted "that which Mr. Romanes regards as 'a very large assumption' is . . . a very general fact, and, at the present time, one of the best-established facts in natural history." There was considerable variability within a population around a mean, as measurements of body parts and examination of color and markings in a number of animal species had shown. Species could roughly be divided into two groups, with significant divergence from the mean condition:

> We must also remember that at least 90 or 95 per cent of the offspring produced each year are weeded out by natural selection . . . so that, during any change of conditions necessitating readjustment to the environment, an ample supply of "simultaneous favourable variations" would occur calculated to bring about that readjustment. And since we have every reason to believe . . . that the slight specific differences of which these variations are the initial steps are in most cases utilitarian in character, we may feel sure that all useful variations, occurring so frequently, would be preserved and rapidly increased without any danger from the "swamping-effects of intercrossing."

Finally, Wallace dispatched the third objection, Romanes's assumption that it was almost a universal rule for natural species to be incapable of interbreeding. Wallace cited the experiments of the botanist Dean Herbert, who had produced hybrids from a number of distinct genera, and these hybrids could be propagated indefinitely. "The popular ideas as to the sterility of hybrids are derived from crosses between certain domestic animals by no means closely allied," Wallace pointed out, "such as the horse and ass, the canary and goldfinch, or the domestic fowl and the pheasant." It was difficult to cross close species of animals under circumstances approximating natural conditions—it was even difficult to cross two members of the same species in zoos—though he knew of animals classed in distinct families, like the pheasant and the grouse, producing hybrids in a state of nature. "[This] mere fact . . . is in itself an argument against there being any constant infertility between the most closely allied species, since if that were the case we should expect the infertility to increase steadily with remoteness of descent till when we came to family distinctions absolute sterility should be invariable."

While admitting that variations in fertility were highly probable, Wallace noted, "Mr. Romanes speaks of this physiological variation as if it were a simple instead of a highly complex form of variation, and as if it might occur sporadically within the limits of a species like some change of colour or modification of form." Physiological selection necessitated large leaps, some sort of major mutation, which violated the maxim that *Natura non facit saltum* (Nature does not make leaps), the uniformitarian principle underlying the Darwin–Wallace conception of the evolutionary process. If one individual, or a dozen individuals, within a large species was somehow incapable of breeding with the bulk of the members of its species, the chances of that individual mating with a "physiological complement" of the opposite sex without some other distinguishing characteristic to identify the potential mate was highly improbable. Even if such a mating took place, the variation would be a rare occurrence and would not be preserved by natural selection:

> Yet [Romanes] has arrived at a diametrically opposite conclusion, for he claims as the special feature of these variations that "they cannot escape the preserving agency of physiological selection." . . . This most extraordinary statement . . . seems to me to have been reached by ignoring altogether the cardinal fact of the tremendous struggle for existence, and the survival in each generation of only a small percentage of "the fittest." Mr. Romanes's argument almost everywhere tacitly assumes that his "physiological variations" *are* the fittest, and that *they* always survive! With such an assumption it would not be difficult to prove *any* theory of the origin of species.[9]

In a letter to *Nature* on September 2, Francis Darwin corroborated Wallace's remarks. He had evidence besides allusions in *The Origin of Species* that his father had been "familiar with the principle of physiological selection and, moreover, did not regard it with any great favour." In the same journal, Romanes reproached Francis Galton and Raphael Meldola for having made similar objections. Stung by Wallace's review, comments by Galton, Meldola, and Francis Darwin, plus other critiques of his theory that were "flowing through channels other than the pages of *Nature*," Romanes defended his paper, claiming that it was merely a preliminary statement of principles that he hoped would stimulate naturalists to cooperate with him in investigating and verifying his theory (though the "preliminary" statement had taken up seventy-five pages of the Linnean Society's journal). But no one seemed interested in verifying this theory, only in refuting it. After allowing these critiques to "exhaust themselves," Romanes published a lengthy rebuttal in the January 1887 issue of *Nineteenth Century*. Summarizing the main points of this rebuttal in a letter to *Nature* on January 13, he singled out Wallace, referring to his criticisms as "feeble." Moreover, Wallace and other critics had misunderstood him. He was dumbfounded, he said, that sentences that were straightforward and good examples of plain English could have confused such competent readers as Wallace. Here Romanes was simply wrong; one might easily lose one's way amid his long-winded and convoluted rhetorical style.

Wallace, who was growing tired of Romanes's tirades, replied to his letters in *Nature* from the United States. Physiological selection was a weak and unsatisfactory theory, he said, and he now left the question to the consideration of those naturalists who had the time and inclination to study the problem.[10] Reacting to Wallace's and other evolutionists' unrelenting criticisms, Romanes backpedaled. Physiological selection, he said, stood in the same relation to natural selection as did sexual selection: it was a supplementary theory, referring to only nonadaptive characters. Altering his tactics, he accused Wallace of ostracizing him from the community of evolutionists as an "arrogant heretic" who rebelled against the "highest authority." He claimed that there were authorities elsewhere in Europe who supported his theory. He agreed with Wallace that the only way to prove or refute his theory was through further experimentation. It would take him at least another three years to amass such evidence. "[It] will be a very long time before I shall have occasion to trouble [my critics] with the theory of physiological selection," he concluded.[11]

After returning from the United States in August 1887, Wallace spent the rest of the year tending his garden and wading through the numerous letters and piles of journals and books that had accumulated during his absence of ten months. Travel by rail across a vast continent had exhausted him, and he firmly resolved to lead a quiet country life. His friend Edward Bagnall (E. B.) Poulton, a professor of zoology at Oxford University, invited him to give a series of lectures in Oxford's university extension series, but Wallace preferred to lecture occasionally and then only to local scientific and literary societies.[12]

But he had not tired of writing. The many letters from Americans praising his "Darwinism" lectures, as well as the increasing number of attacks on the theory of natural selection, inspired him to write a popular book on the subject. He decided to "refashion" Darwin's *Origin of Species*, which many found difficult to understand, in order to shed light on those areas most commonly misunderstood by both scientists and the public. A defense of natural selection, he believed, was long overdue.[13]

Wallace spent much of 1888 putting together his book, which he intended to call "Popular Sketch of Darwinism" but whose title he later shortened to *Darwinism*.[14] The theory of natural selection alone explained the origin of species, he wrote. The publication in 1885 of August Weismann's theory on the "continuity of the germ-plasm"—or "panmixia," as Weismann called it—was a boon to Wallace, for it removed one of the major obstacles against the theory of natural selection by proving that acquired characteristics could not be inherited. In his youth, Weismann had been an enthusiastic naturalist who studied medicine in Frankfurt before switching to laboratory work. A serious eye disease, however, forced him to give up the microscope and devote himself to theoretical work, which eventually laid the foundation for modern genetics. Like Wallace, he was almost a "pure" natural selectionist and rejected the Lamarckian idea of the inheritance of acquired characters.[15]

Before Weismann enunciated his theory, no one had developed a satisfactory hypothesis to explain the phenomenon of heredity. Most naturalists believed in "blending inheritance," but the mechanism remained a mystery. Darwin thought that he had hit on a solution and called his theory "pangenesis," which he published in 1868 as *The Variation of Animals and Plants Under Domestication*. At the end of the first volume of this book, he struck a distinct Lamarckian note: "It is often sufficient for the inheritance of some peculiar character, that one parent alone should possess it. . . . But the power of transmission is extremely variable. In a number of individuals descended from the same parents, and treated in the same manner, some display this power in a perfect manner, and in some it is quite deficient. . . . The effects of injuries or mutilations are occasionally inherited; and . . . the long-continued use and disuse of parts produces an inherited effect." Darwin postulated the existence of "gemmules," or microscopic germs, "cast off" into the circulation by the almost infinite number of cells that compose an organism and then concentrated in the generative cells, primarily the ova and sperm. If properly nourished, he argued, any gemmule could reproduce a portion of the organism or even the whole organism itself. These gemmules were inconceivably small—perhaps the size of atoms—since all could be contained in an ovum or a spermatazoon. Sometimes the gemmules remained dormant or undeveloped in an individual, only to be activated somehow in future generations. Gemmules also could be modified by external conditions. If multiplying sufficiently, they were capable of replacing the older gemmules and developing new structures. For example, a mutilation in an adult theoretically could be passed to its offspring. Pangenesis also explained the persistence of rudimentary and useless organs.[16] Wallace initially embraced pangenesis as the best available theory to explain heredity. But the confusion the theory wrought among scientists caused Darwin to pronounce it "stillborn" and to predict, like a dejected Cassandra, that it eventually would reappear under some other name.[17] He was wrong. One of the first experimentalists to disprove the theory of pangenesis was his cousin Francis Galton, who pioneered the study of heredity and performed blood transfusions on rabbits, replacing the entire blood volume of one animal with the blood of another and then breeding the transfused rabbits. The offspring were not in the least bit altered and resembled their biological parents, not the donor rabbits. Thus the blood did not carry the hereditary material from the body to the reproductive organs, as Darwin had speculated.[18]

When Weismann published his views on heredity nearly twenty years later, Wallace abandoned Darwin's "ponderous" theory for one that he deemed far superior. He distilled Weismann's fundamental question as follows: "How is it that in the case of all higher animals and plants a single cell is able to separate itself from amongst the millions of the most various kinds of which an organism is composed, and by division and complicated differentiation to reconstruct a new individual with marvelous likeness, unchanged in many cases even throughout whole geological periods?"[19] That the union of spermatazoon and ovum reproduced not only the general character of the species but many of the individual characteristics of the parents and their remote ancestors could be explained only if the germ cells arose directly from the parent germ cells themselves. Weismann had stated that at birth a portion of the "germ-plasm," a special molecular substance contained in the germ cells derived from the parents, was reserved unchanged to produce the germ cells of the next generation. But this was not all.

Through sex, the germ-plasms of the two parents were united, intermingling the characters of many generations and creating unique individuals whose variability was precisely what natural selection worked on. This concept—the continuity of the germ-plasm—had also been postulated by Galton some years earlier.

Weismann arrived at his conclusions through induction, based on his extensive knowledge of cytology, embryology, and natural history. If his hypothesis were true, then the transmission of acquired characters could never occur because the material to be transmitted would have been segregated into germ cells at such an early stage in fetal development that climate, habits, or any other external factors could not affect it.[20]

Weismann's rejection of "soft" inheritance was received with great hostility by neo-Lamarckians—the majority of evolutionists at the time—and was not universally accepted for fifty years. Although Weismann did not arrive at a correct idea of the precise means of inheritance—that would have to wait until the rediscovery, at the turn of the twentieth century, of Gregor Mendel's work; the subsequent identification of the genetic material; and James Watson and Francis Crick's discovery in 1953 of the replicating mechanism—his general views on the segregation and transmission of genetic material are still valid. Wallace was among the first to recognize Weismann's genius and actively promote his ideas. After reading the essays (translated from the German by his friend Meldola) in *Nature*, Wallace, like the duke of Argyll, jumped at germ-plasm like a trout at a fly, for it proclaimed natural selection as the only mechanism of evolution. Weismann's work was not easy reading, but Wallace immediately understood his triumphant rebuttal of Lamarckian inheritance.[21]

Romanes and his theory of physiological selection remained a formidable obstacle. When Poulton suggested that Romanes review *Darwinism*, Wallace was outraged. "I think it would be almost *indecent* for [him] to review [it]!" he exclaimed.[22] Wallace had made Romanes a major target in the course of answering several of the most prominent objections to natural selection, and this renewed antagonism had resurrected the epistolary battle that ended a few months earlier. His book also singled out the Reverend John Thomas Gulick, an American missionary, naturalist, and friend of Romanes, who in "Divergent Evolution Through Cumulative Segregation" had maintained that geographic isolation alone produced speciation, without inter- and intraspecific competition. According to Gulick, there was an inherent tendency to variation in certain divergent groups within a species (now referred to as genetic drift). When one part of a species was isolated, this tendency to variation created a persistent divergence from the rest of the population despite identical conditions, and the variant line diverged further and further from the original type; if the two groups were prevented from interbreeding, a new species would then form. Gulick used as his example terrestrial mollusks in the Hawaiian Islands, where the various species, derived from a common ancestor, inhabited "identical" though geographically separate habitats. Natural selection could not have been operating, he said, because the food, climate, and enemies were the same.[23] Strongly disagreeing, Wallace stated, "It is an error to assume that what seem to us like identical conditions are really identical to such small and delicate organisms as these land mollusks, of whose needs and difficulties at each successive stage of their existence, from the freshly-laid egg up to the adult animal, we are so profoundly ignorant."[24]

According to Wallace, the two greatest arguments against natural selection as the sole driving force behind speciation were hybridization and sexual selection. He therefore devoted several chapters to these difficulties. Although he had treated sexual selection in other publications and offered nothing new on the subject, his discussion of hybridization was truly novel.

The reality of hybridization called into question the very definition of species. Before 1858, when Darwin and Wallace announced their joint theory of natural selection, the defining difference between a variety and a species was the degree of fertility. In general, the varieties of any one species, no matter how different they might be in external appearance, could freely interbreed and produce offspring that could continue to freely interbreed. But distinct species, no matter how closely they might resemble each other, usually were unable to interbreed, and if they were, their offspring always were sterile. Wallace pointed out that these distinctions were once considered a "fixed law of nature" and an absolute test and criterion of a true species versus a variety. As long as species were regarded as special creations, and the origin of species and that of varieties were viewed as different, the law could have no exceptions. If two species produced a fertile hybrid and the offspring of that hybrid also were fertile, the parent species were relabeled as varieties. Likewise, if two varieties produced sterile hybrids, the varieties were considered to be separate species. "Thus the old theory led to inevitable reasoning in a circle," Wallace said, "and what might be only a rather common fact was elevated into a law which had no exceptions." In the *Origin*, Darwin had proved the fallacy of such a law by demonstrating numerous exceptions.[25]

Unlike most of their predecessors and their fellow naturalists, Darwin and Wallace ceased to view a species as a distinct entity created by God, considering it instead as an assemblage of individuals modified to adapt to new conditions of life—an assemblage that also happened to interbreed and produce similar offspring. In a state of nature, animals preferred their own. Even slight differences of form or color were sufficient to deter them from interbreeding, in effect resulting in isolation of groups even when habitats overlapped. Wallace called this phenomenon "isolation of varieties by selective association" and felt that "the great stumbling-block of many naturalists will be completely obviated."[26]

Wallace posed the same hypothetical question he had put to Darwin long before: Could natural selection actually promote the failure of varieties of a species to interbreed successfully—that is, promote the development of reproductive isolating mechanisms? It was a question that turned Romanes's theory of physiological selection on its head. Relegating natural selection to a minor role in evolution, Romanes had suggested that physiological selection was not only a fact but also the true starting point of speciation. Given the fuller knowledge of the facts of variation uncovered since 1858, Wallace felt that he could show that natural selection sometimes was able to accumulate variations in an incipient species' ability to interbreed with another incipient species or with its parental species. His demonstration was based on a number of assumptions inferred from known facts presented by other naturalists, especially Darwin.

He first asked his readers to imagine two varieties of a species occupying an extensive area, each in the process of adapting to somewhat different conditions of existence. If the two varieties freely crossed and their offspring also were fertile, then further speciation of the two forms would be retarded or entirely prevented. He next asked his readers to suppose that the hybrids of these same two varieties were somehow less and less able to produce fertile offspring when crossed, a phenomenon correlating with the increasingly differing modes of life and the slight external or internal peculiarities that gradually arose between them. Wallace already had presented evidence from Darwin that such peculiarities were indeed real causes of the failure of two varieties of a domesticated species to hybridize. On one of the Faroe Islands, the half-wild native black sheep did not readily interbreed with imported white sheep; flocks of white and Chinese geese generally did not frequently hybridize; the differently colored herds of cattle in the Falkland Islands also tended to remain apart. Moreover, external appearances often signaled fitness or the lack thereof: white color and blue eyes in male cats correlated with deafness; pigeons of certain colors produced naked offspring; in Virginia, only black varieties of a certain pig could eat a plant called the paint-root (*Lachnanthes tinctoria*) without having their bones turn pink and their hooves fall off; and so forth. Although these were all examples in domesticated or partly domesticated animals, Wallace saw no reason that the tendency to mate with like-appearing members and the correlation of color with sometimes injurious constitutional peculiarities should not also occur in nature. Given the partial sterility of hybrids, it would follow that even if the two varieties freely crossed, the population of hybrids would increase less rapidly than those of the two pure forms. The offspring of the pure forms would doubtless be better adapted to their respective conditions of life. And when the struggle for existence became exceptionally severe, the smaller population of hybrids would die out before the offspring of the parent forms did, leaving the varieties pure. The greater the inability of the hybrid forms to produce fertile offspring, the likelier the pure forms of the varieties would outbreed the hybrid forms and eventually outcompete the hybrids for resources, causing their extinction:

> It must particularly be noted that this effect would result, not by the preservation of the infertile variations on account of their infertility, but by the inferiority of the hybrid offspring, both as being fewer in numbers, less able to continue their race, and less adapted to the conditions of existence than either of the pure forms. It is this inferiority of the hybrid offspring that is the essential point; and as the number of these hybrids will be permanently less where the infertility is greatest, therefore those portions of the two forms in which infertility is greatest will have the advantage, and will ultimately survive in the struggle for existence.

Wallace concluded that "specialisation to separate conditions of life, differentiation of external characters, disinclination to cross-unions, and the infertility of the hybrid [products] of these unions, would all proceed *pari passu*, and would ultimately lead to the production of two distinct forms having all the characteristics, physiological as well as structural, of true species."[27] Wallace's theory—which holds that as two populations within a species diverge to the extent that the hybrids between them are less well adapted than either parent form, natural selection will tend to eliminate the hybrids—is recognized as valid by some evolutionary biologists and is known as the Wallace Effect.[28]

In the fifteenth and final chapter of *Darwinism*, which represented nearly thirty years of reflection on the subject, Wallace examined the application of natural selection to human beings. Natural selection, he said, was sufficient to explain most of the facts of organic life, such as the origin of species, genera, families, orders, classes, and even kingdoms. But it was insufficient to explain certain major events in the history of evolution of life on earth. He therefore introduced a bit of catastrophism into his otherwise uniformitarian vision of the earth's history. Echoing Sir Charles Lycll in *The Geological Evidences of the Antiquity of Man*, he pointed to three stages in the development of the organic world "when some new cause or power must necessarily have come into action." The first stage was the change from inorganic to organic matter, when the earliest cell appeared. "[It] has been well said," Wallace noted, "that the first vegetable cell was a new thing in the world, possessing altogether new powers—that of indefinite reproduction, and, still more marvelous, the power of variation and of reproducing those variations till endless complications of structure and varieties of form have been the result." The second stage was the introduction of sensation or consciousness, "constituting the fundamental distinction between the animal and vegetable kingdoms." No explanation thus far had been intellectually satisfying, he said, and the phenomenon remained a mystery. The third and final stage was the existence in man "of a number of his most characteristic and noblest faculties, those which raise him furthest above the brutes and open up possibilities of almost indefinite advancement." These faculties, he believed, could not have been developed according to the same laws that determined the general and progressive evolution of the organic world. He was forced to conclude that these three distinct stages pointed to an unseen universe, "a world of spirit, to which the world of matter is altogether subordinate." The spiritual nature of humanity, therefore, was not inconsistent with his and Darwin's theory of evolution by natural selection; it depended on the same fundamental laws and causes that provided the materials for the evolutionary process to unfold. But this view of human nature should relieve us of the "crushing mental burthen" of the materialistic and fatalistic belief in our helplessness in the face of the implacable, blind forces of nature. Humankind's struggles were not pointless. In the near future, the earth, which for untold millions of years had been slowly developing the forms of life that would culminate in human beings, would be as it had never been. Darwin must have rolled over in his crypt in Westminster Abbey at *Darwinism*'s final dreamy paragraphs. Had he been summoned to a séance, he would have rapped out, "No! No!"

By the time Wallace had finished the manuscript of *Darwinism*, he was convinced that he had demolished every one of his adversaries, including the Darwin of sexual selection. Macmillan published the book in 1889, agreeing to a few conditions that Wallace had set. He wanted his book issued with "cut edges," not the uncut pages that resulted in ragged edges that accumulated dust and dirt over time. His concerns were economic: there seemed to be a prejudice among readers against uncut books. Darwin's own books, he said, were neatly cut by machine at his own insistence. In 1867 Darwin had written a letter to the *Athenaeum* advocating the abolition of the practice of leaving pages uncut, thereby requiring separation with a knife or finger. (This would "earn the gratitude of children who have to cut through dry and unillustrated books for the benefit of their elders," Darwin pointed out.)[29] Why could Wallace not demand the same conditions? "I know I shall . . . obtain the *gratitude* of all *readers* of my book even if you despise my bad taste," Wallace told Alexander Macmillan.[30]

The book was moderately successful and would go through three editions by the turn of the century. Herbert Spencer complimented Wallace on the book but chided him for a misleading modesty. "I regret that you have used the title 'Darwinism,'" Spencer said, "for notwithstanding your qualification of its meaning you will, by using it, tend greatly to confirm the erroneous conception almost universally current."[31] Wallace thanked his friend Grant Allen for reviewing the book favorably in the *Academy*. In a letter dated July 22, 1889, he wrote that he disagreed with one point to which Allen had alluded:

> You are right in your *hint* that my spiritualism *led* me to my views as to man, but I deny altogether that this is an *a priori* view, since the *facts* of spiritualism are to me just as real and certain as the facts of organic nature, and I am bound to bring the two into harmony. But you are wrong again to *this* view having had any influence in my rejection of sexual selection. That arose solely from the absence of evidence for it, and to me [the] enormously improbable assumption that the mating of butterflies depends on the *choice* of the female and that that choice is determined by small differences of colour![32]

Much to Wallace's chagrin, the editor of the *Contemporary Review* hired Romanes to write a critique, which appeared in the August 1889 issue. Romanes's principal objection to *Darwinism* was Wallace's exclusive adherence to natural selection. "He will not have any other 'factor,' and therefore says natural selection must eat up sexual selection like the lean kine have the fat kine," he told Francis Darwin before the review appeared. Pure Darwinians like Wallace and Weismann were harming the theory by discarding the "Darwinian recognition of use and disuse." "Wallace's jealousy . . . is foolish and inimical to natural selection theory itself, by forcing it into explanations which are plainly false."[33] Romanes contended that despite its title, the book was not an exposition of Darwinism but of "pure Wallaceism." Except for the origin of certain faculties of man, he said, the major difference between Wallace and Darwin was the scope they gave to the doctrine of natural selection. Wallace saw no limits to the mechanism, where Darwin saw several. Romanes also contended that Wallace had validated the hypothesis of physiological selection rather than disproving it.

Edwin Ray Lankester, once Wallace's impudent opponent in the trial of the medium Henry Slade but now more or less an ally, came to Wallace's defense in the October 10 issue of *Nature*. Calling the book "admirable," he stated that it contained "an exposition of highly important and interesting views . . . on subsidiary matters, which have either not been published previously or have appeared in a scattered and more or less inaccessible form." Indeed, the abundance of new facts and arguments, convincing or not, were of "extreme value and full of interest." But he also felt that Wallace's tendency to speculate and his failure to muster sufficient evidence to support some of his conclusions weakened the book. "With Darwin, one becomes accustomed to see no speculation put forward, no step of an argument advanced, unless there is an overwhelming weight of testimony in its favour," he said. "'Darwinism' can never take the place of the 'Origin of Species,' but may well serve as an introduction to the study of that and the other works of Darwin—the value of which, not only as storehouses of fact and suggestion, but as classical models of scientific discussion, cannot be over-estimated and will probably never be surpassed." Wallace's criticism of Romanes appeared to Lankester to entirely destroy all that was novel in "that laborious attack" on Darwin's theory of the origin of species. But Lankester lost patience with the concluding homage to the spiritualist doctrine. He could only remark—with greater diplomacy than he had exhibited in the past—that it remained an interesting problem for the "future student of human faculty" to reconcile Wallace's "wonderful ingenuity" and reasoning skills in the field of zoology with his views on the "so-called 'manifestations' of spiritualists."

Angered by Wallace's and Lankester's criticisms, Romanes recruited Gulick to support his cause. Gulick had hoped to reply to Wallace in *Nature*, but Romanes advised against it, having been chided by editor Joseph Norman Lockyer for using the magazine as his soapbox. Moreover, a rebuttal published in a periodical would have a transient effect, whereas a book would advance the truth of evolution. "If only 100 copies were sold and 100 more distributed [by] yourself," Romanes told Gulick, "they would find their way into every mind much worth influencing in our own generation."[34] In the meantime, in an article written for the *Fortnightly Review*, Romanes charged Wallace with plagiarism: "He presents an alternative theory to explain the same class of facts. Yet this theory is purely and simply without any modification whatsoever, a restatement of the first principles of physiological selection, as these were originally stated by myself." He added in reference to Wallace's concluding chapter that here "we encounter the Wallace of Spiritualism and astrology, the Wallace of vaccination and the land question, the Wallace of incapacity and absurdity."[35]

Wallace made no public reply to these inflammatory remarks for two months, convinced that people whose opinion he valued would condemn them. But he had an ace up his sleeve, and now he played it. Several years earlier, when he was in Kingston, Ontario, a woman who had attended one of his lectures told him that Romanes was a spiritualist and had tried to convert Darwin! Stunned, Wallace found it hard to believe that Romanes would have confided such interests to Darwin, knowing his attitude toward spiritualism. But the woman had proof. She was a good friend of Romanes's brother, an avowed spiritualist then living in Canada. He and Romanes had written to Darwin jointly, and she still had copies of those letters in her possession. The next morning, she brought them to Wallace and gave him carte blanche to make of them what he wished. He copied them out and kept them for future reference.[36]

"Whether or no it was good taste for you to appeal to the political and medical prejudices of your readers in a matter purely scientific . . . I leave others to judge," Wallace wrote to Romanes. "But as to your appeal to popular scientific prejudice by referring to my belief in Spiritualism and astrology (which latter I have never professed my belief in), I have something to say." He revealed that he had copies of two letters detailing Romanes's experiences of spiritual phenomena and his conviction of the truth of these facts and the existence of spiritual intelligences:

> Formerly you had thought there were two mental natures in Crookes and Wallace—one sane, the other lunatic! Now (you said) you belonged to the same class as they did. Tell it not in Gath! There are, then, two Romanes as well as two Wallaces. There is a Romanes "of incapacity and absurdity!!" But he keeps it secret. He thinks no one knows it. He is ashamed to confess it to his fellow-naturalists; but he is not ashamed to make use of the ignorant prejudice against belief in such phenomena, in a scientific discussion with one who has the courage of his opinions, which he himself has not.[37]

The revelation did not faze Romanes. He replied that it had never occurred to him that he had hit Wallace below the belt. If Wallace had called his critique of *Darwinism* an example of "incapacity and absurdity," he would not have objected. He was only making "fair comment" on Wallace's "different lines" of thought. But he had erred and would not refer to such notorious matters again. Regarding the early letters to Darwin, he had forgotten the details after fourteen years. Those letters, however, were meant to be strictly confidential and to be regarded as "provisional" only. Since that time, he had changed his mind about spiritualism. He wondered how Wallace had got hold of the letters. If by some occult process, he hoped that Wallace would publish them as evidence of a spirit world. If from a member of the Darwin family, he did not know whether Wallace had a right to read them.[38]

Wallace then challenged Romanes to deny the phenomena he had personally observed. What other explanation did Romanes have? Once again, he challenged him to tell it to the world. He himself had nothing to hide—his writings were public property, and people were free to reflect on his facts and arguments and criticize them if they wished. "After the way you have referred in print to my belief in such phenomena," Wallace said, "most persons would think I was quite justified in making known the fact of the existence of these letters and their general tenor." The honorable thing to do was for Romanes himself to publish the letters, with full details of the discovery of the imposture that had induced him to change those convictions so "earnestly and solemnly" expressed to Darwin. Wallace hoped that he would not be forced to reveal the letters' existence himself.[39] Now Romanes was genuinely alarmed. How Wallace had obtained copies of these letters mystified him. Wallace briefed the Darwin family about the nature of the feud with Romanes and assured them that no one in the family had given him access to private material. A year later, Romanes held up a white flag of sorts, requesting a photograph of Wallace for his forthcoming book on the Darwinian theory. Wallace politely declined.[40]

Wallace's somewhat underhanded tactic had succeeded: Romanes never published another criticism of him. But Wallace neither forgave nor forgot their rancorous interchange. In 1893 Romanes developed an illness that would prove fatal. William Thistleton-Dyer, the head of the Royal Botanic Gardens at Kew and a good friend of Wallace's, conveyed the news to him. Through Thistleton-Dyer, Romanes had made the strange request to speak with Wallace privately. Wallace was surprised. They hardly knew each other, he told Thistleton-Dyer; they had met face to face only once or twice, and only a half dozen or so letters had passed between them. The "very gross misstatements and personal attacks" on him for allegedly plagiarizing Romanes's theory of physiological selection were still offensive to him: "This accusation he supported by such a flood of words & quotations and explanations, as to obscure all the chief issues and render it almost impossible for the ordinary reader to disentangle the facts. I told him then that unless he withdrew this accusation as publicly as he had made it I should decline all future correspondence with him, & should avoid referring to him in any of my writings." Moreover, he concluded,

> [When] a man has made an accusation of *literary and scientific dishonesty,* and has done all he can to spread this accusation over the whole civilised world my only answer can be—after showing as I have done . . . that his accusations are wholly untrue—to ignore his existence. I cannot believe that he *can* want any sympathy from a man he says has wilfully & grossly plagiarized him, *unless* he feels that his accusations were unfounded. If he does so, & will write to me to that effect (for publication if I wish after his death), I will accept it as full reparation & write him such a letter as you suggest.[41]

But no written retraction was forthcoming, and Romanes died in 1894 of kidney failure, his theory of physiological selection dying with him. Despite its idiosyncratic ending, Wallace's book had dealt a fatal blow to Romanes's attempt to subvert the theory of natural selection as the primary explanation for the origin of species. From time to time over the next two decades, Romanes's hypothesis was revived in different guises. But other investigators after Wallace have proved that "sympatric speciation," at least in the non-adaptive, saltationist sense that Romanes proposed, does not occur in nature. Isolating mechanisms of some sort (spatial, temporal, or behavioral), reinforced by natural selection, are essential to and responsible for the formation of separate species from divergent varieties, a conviction that Wallace maintained steadfastly.[42]

CHAPTER 18

Thoroughly Unpopular Causes

SOMETIME IN 1884, a fellow spiritualist named William Tebb asked Wallace to write a pamphlet condemning smallpox vaccination. An ardent anti-vaccinationist and a participant at the annual meetings of the International Congress of Anti-Vaccinators, which had been founded in 1880, Tebb needed someone more eminent to legitimize the anti-vaccination campaign. He knew that he had come to the right man, for Wallace was instinctively attracted to a cause or an idea by its novelty and the degree of opposition to it. "The whole history of science shows us that, whenever the educated and scientific men of any age have denied the facts of other investigators on a priori grounds of absurdity or impossibility, the deniers have always been wrong," Wallace wrote in the preface to the third edition of *Miracles and Modern Spiritualism*, which was published in 1896. After a lifetime of experience fighting uphill battles, he was immediately inclined to favor the anti-vaccinationists' claims. Tebb laid before him some startling statistics confirming this inclination, and Wallace decided to take on the British medical establishment, which had all but unanimously endorsed smallpox vaccination.

Long regarded as one of the great scourges of humankind, smallpox provoked the same terror as polio would in the first part of the twentieth century and AIDS in the century's closing decades. The fatality rate varied widely, from 20 to 60 percent, and most survivors suffered permanent disfigurement. In 1717 Lady Mary Wortley Montagu, the wife of the English ambassador to Turkey, described the procedure of "ingrafting," used at the Ottoman court, which successfully prevented disease. Renowned for her beauty, two years earlier she had developed smallpox, which permanently disfigured her face; her twenty-year-old brother had died of the disease a few months before it ravaged her. In 1718 she ordered the embassy surgeon, Charles Maitland, to inoculate her five-year-old son; three years later, after returning to England, she insisted that Maitland inoculate her four-year-old daughter. This early form of variolation ("variola" means "pustule" or "pox"), as it was also called, involved making several scratches on the arm of a previously uninfected patient and introducing material taken from the pustules of individuals suffering from a mild case of the disease. Sir Hans Sloane, president of the Royal Society and the king's personal physician, witnessed the inoculation of Montagu's daughter and convinced the king to grant Maitland a license to perform experiments on prisoners. Following variolation, one of these prisoners was exposed to a child with smallpox and proved to be immune. After the two daughters of the Princess of Wales were inoculated, the procedure gained more widespread acceptance.

Approximately 3 percent of those variolated died from complications related to the "ingrafting" and became a source for new epidemics, but mortality rates declined tenfold among those inoculated. In eighteenth-century England, the prevailing belief among milkmaids was that exposure to cowpox, a far more benign disease, conferred protection against smallpox. This belief, considered mere folklore, interested a Gloucestershire physician named Edward Jenner, who proved their observations correct. His first experiment took place in 1796, when a milkmaid developed cowpox after exposure to an infected cow. Jenner extracted fluid from a pustule on the young woman and inoculated a healthy eight-year-old boy through two half-inch incisions on his arm—the same method as variolation. Six weeks later, he variolated the child with material from a person with a full-blown case of smallpox. Inoculation with cowpox fluid protected against the development of smallpox. Several months later, he repeated the variolation on the same child, with equal success. At the end of the year, in a paper presented to the Royal Society, he described thirteen cases of individuals who had contracted cowpox and did not develop smallpox after variolation. Jenner also demonstrated that person-to-person transmission of cowpox prevented smallpox. Thus "vaccination," as the procedure came to be called (*vacca* being the Latin word for "cow")—that is, the inoculation of an individual who never had had cowpox or smallpox with material from the pustule of someone with active cowpox—conferred protection against one of humankind's most dreaded diseases.

Although members of the Royal Society initially resisted his findings, Jenner persisted with his experiments, and those of other investigators confirmed the validity of his technique of vaccination. Within two years of the publication of his *Inquiry into the Causes and Effects of Variolae Vaccinae* in 1798, his methods were enthusiastically embraced and more than a hundred thousand people submitted to vaccination worldwide.[1] Two hundred years later, the World Health Organization declared the eradication of smallpox by vaccination one of the great triumphs of twentieth-century science. Wallace thus opposed a measure that ultimately saved untold millions of lives. From the perspective of history, his objections seem absurd and irresponsible, but—as is so often the case with Wallace—they were not unreasonable when considered in the context of his era.

Wallace had not always been an anti-vaccinationist. In the Malay Archipelago, he had praised the efforts of the Dutch to safeguard the health of the local inhabitants by administering the smallpox vaccine.[2] At some point in his life—perhaps in childhood, though he does not say—he had submitted to smallpox vaccination. He also ensured that all three of his children were vaccinated. In "An Answer to the Arguments of Hume, Lecky, and Others Against Miracles," a paper he had read to the members of the Dialectical Society in 1870, he cited Jenner along with Galileo, Benjamin Franklin, and the physician William Harvey as examples of men whose groundbreaking discoveries were "violently opposed by their scientific contemporaries." Twenty years later, in a revised version of the paper, Wallace expunged Jenner's name from the list. In the 1870s his fellow spiritualists had criticized him for his pro-vaccination views. Most agreed with the anonymous author of the article "Scientific Fallibility," who wrote that "vaccination is one of the fashionable quackeries of the present day, and there is just as much orthodoxy and fashion in 'science' as there is in religious opinions and ladies' head gear."[3]

But it was English politics more than pressure from spiritualists that changed Wallace's mind. The devastating impact of the smallpox epidemics of the eighteenth and early nineteenth centuries, combined with the success of Jenner's procedure, gradually led lawmakers to enact regulations aimed at universal vaccination. In 1840 the British Parliament passed the first Vaccination Act, which gave power to local boards to vaccinate everyone in their districts. Its aim was universal vaccination, but the law permitted individuals to decline. Persistent outbreaks of the disease and the zealous urging of medical practitioners, however, forced Parliament to amend the act in 1853 to make vaccination compulsory. The act commanded that all newborns be vaccinated within three months of birth and required physicians to provide a certificate justifying any delay. The certificate remained in force for two months, at which point the parents were legally bound to have their child vaccinated. Once a child was vaccinated, the parents were instructed to meet with their physician eight days later to verify that vaccination was successful. If the vaccination failed to take, the child was revaccinated. Sometimes overzealous vaccination disfigured a child's arm. Public-health officers enforced the vaccination law with routine inspections, and any parent refusing to comply was fined and forced to pay for the costs of prosecuting the case. Some deemed the law too lenient, since the parent could be fined only once.

Despite these early earnest efforts by the British government, smallpox refused to go away. The number of deaths in London alone from smallpox between 1850 and 1861 was 7,150, and an unknown number of people had been infected. In 1867 the Vaccination Act was repealed and replaced with more draconian measures. Every six months, a list of unvaccinated children had to be submitted to the local boards. A justice of the peace was then authorized to order the vaccination of all unvaccinated children. A parent could be prosecuted for noncompliance every six months until the child reached the age of fourteen. But hard-liners did not consider even this law severe enough. Another amendment mandated the prosecution and imprisonment of any board members who did not rigorously enforce the vaccination laws.[4] The vaccination laws afflicted the noncompliant poor the most. Any wealthy objector could skirt legal action by paying the necessary fines or bribing the prosecutors or local public vaccinators, whereas the poor could not afford the fines and were treated like criminals. The acts were regarded by some not only as a gross infringement on personal liberty but also as morally unjust.[5] In "Sanitation, Not Vaccination the True Protection Against Small-Pox," a paper presented at the Second International Congress of Anti-Vaccinators, held in Cologne in October 1881, Tebb stated that "it is no exaggeration to say that a more wretched and obnoxious edict has not been foisted on the English-speaking race since the passage of the atrocious Fugitive Slave Law in America thirty-five years ago."[6]

Since Jenner's day, there had always been opponents of vaccination, but opposition did not become organized into a movement in Great Britain until the 1870s. Anti-vaccinationists fell into two categories: those who did not believe in the efficacy of the procedure and those who believed that compulsory vaccination violated individual rights. But the categories were not mutually exclusive; some anti-vaccinationists objected on both grounds. At first, Wallace dismissed the claims of the vaccine's inefficacy as unconvincing; he was originally attracted to the anti-vaccinationist cause as a human-rights issue. But after studying some statistics that Tebb laid before him, he decided that the facts against vaccination were compelling. He put aside his other work and began a study of the existing statistics both before and after the introduction of vaccination, poring over an immense number of bills of mortality and other government documents. He also examined the statistics for measles, scarlet fever, diphtheria, and pertussis (whooping cough) and discovered that the incidence of all these diseases paralleled one another over time. The data from continental Europe, where smallpox vaccination was also compulsory, showed a similar pattern. Wallace suspected that something other than vaccination was affecting the fluctuations in the incidence of all these infectious diseases; smallpox vaccination did not seem to make an appreciable difference. In fact, between 1871 and 1880 the number of deaths from smallpox in London increased over that of the previous decade by 80 percent, to 15,551. These data contradicted government figures, which indicated that the death rate from smallpox in all groups was 576 per million inhabitants in the years 1838 to 1842, before the enforcement of vaccination; 223 per million in the years 1854 to 1871, when vaccination was not rigidly enforced; and only 111 per million after 1872, when vaccination of children was made compulsory. Like his friend Tebb, Wallace concluded that vaccination had no impact on the disease; his evidence indicated a rise in the incidence of smallpox since the vaccination campaign began, not a fall, as the British government suggested. (However, pro-vaccinationists pointed out that the apparent rise in cases of smallpox between 1871 and 1880 was due to improved reporting of cases.)[7]

Tebb also alerted Wallace to the work of J. Thomas Biggs, a sanitary engineer who sat on the Leicester town council. Leicester had long been known as a community of dissenters on many issues—for example, it was a hotbed of Unitarianism—and it took its strongest stance against vaccination. Biggs became the leader of the anti-vaccinationist movement after he was forced to liquidate some household goods to pay a fine for having violated the vaccination laws. He was instrumental in establishing the so-called Leicester method of dealing with smallpox, which involved strict quarantine of suspected cases and their contacts, along with immediate disinfection of the premises. Biggs gathered data into fifty-one tables, demonstrating his view of the value of sanitation over vaccination. His tables showed an increase in infant mortality during smallpox-vaccination campaigns—deaths that, according to Biggs, reflected both direct and indirect effects of vaccination and were due to a broad range of diseases, including erysipelas (a streptococcal skin infection), syphilis, diarrhea, convulsions, and bronchitis or pneumonia. However, Biggs failed to mention that although the number of unvaccinated persons in Leicester was large, that number amounted to no more than 10 or 15 percent of the entire population. C. T. Ritchie, president of the

Local Government Board overseeing vaccination, observed several years later, "While [medical authorities] are able now to seize upon and strangle the disease when it breaks out, it will be infinitely more difficult to do so when the number of unvaccinated persons has increased—as they will if Leicester continues to pursue the policy it is now pursuing."[8]

Some of the claims of anti-vaccinationists were legitimate. At the beginning of the anti-vaccinationist movement, the germ theory of disease had not yet been postulated and the origin of epidemic diseases was still being fiercely debated. Many outstanding scientists and physicians were anti-contagionist; they believed that the environment, like changes in weather, or some sort of internal derangement of the sick individual caused diseases like smallpox, cholera, and yellow fever. Contagionists, on the contrary, believed that various diseases could be transmitted from person to person. But even contagionists were uncertain about how a disease was transmitted. Until the turn of the twentieth century, nearly everyone thought that people could acquire yellow fever from the clothing and bedding of yellow-fever victims. Others, noting that the poor suffered from diseases at a higher rate than the wealthy, pointed to unhealthy living conditions as a source of disease and promoted reforms in sanitation. It was not a coincidence that the passage of a stricter Vaccination Act followed the enactment of the Public Health Act of 1848, which established the Board of Health in recognition of the state's duty to clean the streets, construct sewers, ensure proper ventilation of buildings, and prevent river pollution, among other measures.

Nevertheless, not until the early 1880s had Louis Pasteur and Robert Koch convinced some scientists that living agents caused epidemic diseases. In 1880, in a series of elegant experiments, Pasteur proved that a filterable agent smaller than a bacterium (he called it a parasite) could be cultured outside the body of an animal and then introduced into another animal, causing disease. Eventually the term "virus"—which means "poison" in Latin and was used by contagionists to describe the material passed from a person infected with smallpox or cowpox to an uninfected individual—was adopted to describe this filterable organism. Whether or not one accepted the germ theory, adequate treatments for common diseases did not exist. Disposable needles and sterile instruments were practically unknown. Practitioners used the same lances on every patient, transmitting all sorts of diseases. The vaccination fluid itself might contain contaminated material—the agents causing syphilis, erysipelas, or tuberculosis. When the clamors of anti-vaccinationists forced the medical community to find uncontaminated vaccination fluid, "pure" calf lymph was introduced, but the skeptics were no less skeptical. Why were cows any purer than humans, they asked. Who could be certain that diseases were not being transmitted from animals to people? "The new prescription is termed 'pure lymph from the calf'—a sweet name with a savour of Daphne [*sic*] and Chloe that would give credit to Barnum," one critic wrote in the debut issue of the *Vaccination Inquirer and Health Review* in 1879. "Nice words, however, often cover nasty realities; and thus it happens that 'pure lymph from the calf' is a euphism for pox taken from a cow at Beaugency [a slaughterhouse] and propagated by inoculation on calves."[9]

Wallace believed—correctly—that improvements in sanitation and personal hygiene would curb the spread of many diseases far better than other measures. A cleaner water supply, pollution-free air, better waste management, uncontaminated food, and the elimination of overcrowding diminished morbidity and mortality and increased longevity by reducing premature death—a fundamental dictum of modern public-health practice. "[The] remarkable contribution to the mass of evidence in the 'Reports' [the government's statistics on smallpox mortality] which brings out this fact most clearly receives no notice whatever," he complained.[10] This view echoed that of his friend Tebb, who at the Second International Congress of Anti-Vaccinators in Cologne in 1881 had complained that the smallpox epidemic raged "amongst the poor, ill-fed, uncleanly, intemperate, overworked populations . . . amongst those who live in the courts and alleys, in old and decayed habitations, and in the miasmatic atmosphere in which the neglected residuum of this immense city are reduced to dwell."[11]

Wallace and his anti-vaccinationist allies faced a war of propaganda by organized medicine. As technology advanced, doctors claimed greater credit for the reduction of deadly diseases, which set the stage for the exalted view of the medical profession held today. Although doctors did much good, attention to simple rules of public health improved their efforts. The various vaccination acts distracted people from the root causes of disease and its transmission; rather than investing in measures that would improve public health, the government instituted regressive laws that increased the misery of the average Englishman. Public opinion and popular sentiment against compulsory vaccination began to mount; horror stories about unfair treatment abounded. Conscientious objectors—often the sole source of subsistence for their families—were hauled off to jail and treated as common criminals.

Moved by the plight of the lower classes, whose fate rested in the hands of the more powerful upper classes, Wallace scoured the statistics for fallacies. During his late-night analyses of the data, Wallace also noted the absence of controlled experiments that might show how long the protection afforded by the vaccine lasted. Practitioners and government officials alike had jumped on the Jenner bandwagon in their quest to halt a deadly epidemic disease, but the duration of protection was unknown. Since no one knew why vaccination seemed to confer protection—the concept of antibodies lay decades in the future—some authorities guessed four, five, perhaps ten years of benefit. But no one had studied the problem from that angle. The number of people properly vaccinated but no longer protected could not be ascertained.

Wallace wrote up his findings in a tract entitled *Forty-Five Years of Registration Statistics, Proving Vaccination to be both Useless and Dangerous,* published in 1884 or early 1885, which he addressed to members of Parliament. (The second edition was published in 1889.) The objections of the anti-vaccinationists—and of Wallace in particular—outraged physicians. In a review published in the March 22, 1888, issue of *Nature* of two books that held diametrically opposite views on vaccination, the reviewer could not contain his anger. "Dr. Wallace's writing on this subject, whom we are ashamed as scientific men to find in the anti-vaccination ranks, are especially shown up and gibbeted," he said, commenting on the triumphant arguments of the pro-vaccinationist book. In a not very pleasant allusion to Wallace's infamous spiritualist beliefs, the commentator added:

> The book is throughout so carefully and faithfully written, and deals so well with the tactics of the anti-vaccinators, that it ought to have a wide circulation among those interested in the question. Many are interested in seeing conjuring tricks, and in witnessing optical delusions. So long as the tricks are not understood, there is an inclination in the minds of some to regard these tricks as much more difficult of performance than they really are, and some may even attribute them to supernatural agency. The same kind of tricks are played by such men as Dr. Wallace on our literary pursuits.

Wallace, of course, had become inured to this kind of criticism, replying coolly:

> [A] few preliminary misconceptions must be dealt with. One of these is that, as vaccination is a surgical operation to guard against a special disease, medical men can alone judge of its value. But the fact is the very reverse. [They] are interested parties, not merely in a pecuniary sense, but as affecting the prestige of the whole profession. In no other case should we allow interested persons to decide an important matter. [There] is much evidence to show that doctors are bad statisticians, and have a special faculty for misstating figures.[12]

His eye affliction, brought on in part by the intense scrutiny of a mass of data published in eye-straining type size, had interrupted his anti-vaccination crusade midstream. He wrote little on the subject again for two years, distracted by other pressing issues and his trip to the United States.

The vaccination issue reached its apogee in April 1889, when the anti-vaccination coalitions—spearheaded by J. A. Picton, who represented the city of Leicester in the House of Commons—pressured Queen Victoria's government to appoint a royal commission to evaluate the efficacy and methods of enforced vaccination.[13] Although Wallace was invited to serve as a member of the commission, he declined, ostensibly on the grounds that he could not devote the necessary time at this stage in his life. In fact, he objected to the composition of the commission—mostly physicians who, he was convinced, would dominate and prejudice the inquiry. He could do more good, he felt, by acting as witness rather than judge.[14] Other anti-vaccinationists were also wary. Although there were anti-vaccinationists on the fifteen-man commission, they were anti-compulsion, not necessarily anti-vaccination. "Anti-vaccinationists impugn vaccination itself," one commentator wrote in the *Vaccination Inquirer and Health Review.* "They have mastered its history and explored its fallacies, and they cannot therefore be represented by those whose opposition goes no deeper than dislike of the enforcement of the practice by fine and imprisonment."[15]

The Royal Commission on Vaccination was chaired by Lord Farrer Herschell, lord chancellor under Gladstone, and among its members were some of the most famous British medical men of the day, including Sir James Paget (reputed to be the best diagnostician in Britain), Sir William Scovell Savory (president of the Royal Society of Surgeons), and John Syer Bristowe (a pathologist at St. Thomas's Hospital in London). The anti-vaccination side was most prominently represented by Charles Bradlaugh, a famous atheist and member of Parliament from Northampton. Between February 19 and August 7, 1890, the commission held twenty-one meetings and examined ten witnesses, the majority of them anti-vaccinators.[16] Wallace was the first called to testify and was deposed four times. He had come well prepared, with many notes and professionally rendered charts. He began by stating that the efficacy of vaccination was not a medical but a statistical question. He knew nothing about medicine, he admitted, but he knew something about statistics. Erroneous methodology invalidated the comparisons of vaccinated and unvaccinated mortality rates. Before the implementation of vaccination in the eighteenth century, the disease's mortality rate was only 18 percent. But the contemporary average mortality from smallpox, according to medical authorities, was an "astonishing" 30 to 60 percent among the unvaccinated population, this despite improvements in sanitation and other quality-of-life measures—a striking difference that defied logic.

Wallace held up the first of his several diagrams, which graphed smallpox mortality rates in London between 1780 and 1820 and showed a steady and rapid decrease in mortality—too rapid, he insisted, to be explained by vaccination alone. The next few diagrams examined statistics from Scotland, Sweden, Prussia, and Vienna and demonstrated the failure of vaccination to prevent smallpox epidemics. One of the major difficulties of an analysis of this sort, Wallace pointed out, was the absence of a control population: in general, there were no comparable populations, one vaccinated, the other not. Statistics usually were drawn from countries with compulsory vaccination laws. He lectured the commission on the vital importance of arriving at true conclusions using strict controls, as John Tyndall and Darwin had done in their own meticulous experiments. He then laid before the commission what he considered to be his most important diagram, which related to smallpox in France. The French statistics were unique, he said, because they approximated those rigid conditions and showed that vaccination had done nothing to alter mortality rates. He also produced fifteen pages of statistics, arranged and written in his own hand, for the perusal of the commission following his presentation.

The commission's members were at first impressed with the voluminous evidence that Wallace had set before them. But in the period before his next deposition, they scrutinized the data carefully and found a number of egregious errors. The statistics from one French province, for example, were recorded twice and stood side by side in the same diagram! For another province, Wallace had recorded no smallpox deaths over many years, though officials recorded a total of 1,100 deaths in two years in a single town of the same province. A reporter from the *Lancet* remarked that Wallace's theory "was based upon a series of blunders that one would hardly have expected from a school-boy."[17]

The revelation of these mistakes and others caught Wallace by surprise. His haughty words returned to haunt him. Embarrassed, he offered to withdraw the tables, but his adversaries on the panel would not let him off the hook so easily. Sir Guyer Hunter, a member of Parliament, was particularly merciless, attacking him "in a style and tone which would have been the envy and admiration of the commonest and vulgarest" of barristers. In his defense, Wallace said that he regretted the inaccuracies, but he assured the panel that he had not manufactured the data but had only made use of what was available to him. Hunter ended his examination by asking Wallace in a derisive tone if the data were worthy of a Tyndall or a Darwin. Savory asked Wallace if he wanted the tables published as something worthy of his scientific reputation. Herschell asked Wallace if he expected the members of the commission to accept the statistics "blindfolded." Wallace bore the obloquy with his usual equanimity and was forced to admit that the statistics from France were valueless.[18]

Paget then asked Wallace to explain the great diminution of smallpox in London in the nineteenth century. Wallace attributed the decrease to improved sanitation and healthier lifestyles. Paget pressed him to state whether sanitation or vaccination had reduced the death rate from smallpox. Wallace did not know, but he insisted that his evidence showed that vaccination carried to its fullest extent did not prevent the spread of smallpox. How, then, Paget asked, could he explain the decrease of smallpox early in the century and its sudden increase decades later if sanitation had improved. Once again, Wallace said that he did not know, but he suggested that smallpox would have had to become more virulent after 1820 if the number of deaths of the unvaccinated was accurate. Paget directed one of the clerks to place a new diagram before Wallace, using Wallace's own data. "Why does this graph look so different from yours?" he asked. Wallace studied Paget's diagram and his own charts for a few moments before grasping what had been done. Paget had drawn his diagram on a greatly exaggerated vertical scale, with dramatic peaks and troughs, while his own, with a greatly reduced vertical scale, approximated an irregular curve. Wallace pointed out that the apparent differences were only illusory, and that the two graphs were identical. Paget gave up in exasperation.[19]

Angered by this attempt to humiliate him, Wallace stood his ground. However, he had lost his credibility and could do nothing to redeem himself. The *Lancet* professed mock sympathy for his plight, picturing him as a victim of persons of no standing who were either ignorant or unworthy of credence. They pointed out that although Wallace's evidence was mainly statistical, Wallace and his advisers avoided England's own ample evidence on smallpox mortality, instead selecting statistics from a past century and distant countries, the verification of which was "well-nigh" impossible. As for the statistics that Wallace had prepared himself, only he was responsible. Anti-vaccinationists had another opinion of the proceedings. It was reported in the *Vaccination Inquirer and Health Review* that "Dr. Wallace stood an examination . . . which proved, if it proved nothing else, that anyone who gives anti-vaccination evidence before this commission must be prepared to keep his temper through insults which most . . . would resent, and in which few . . . would indulge."[20]

Wallace concluded his testimony on May 21, the final day of the hearing. He took the opportunity to summarize his position. During the course of his examination, Wallace said, he had been asked questions which implied that he had undertaken his study without the full and accurate data a scientist required. This was true, to some extent, but he had never regarded the vaccination question as one of pure science: "If it had remained a question of medical science and practice, I should not have troubled myself about it, and certainly not have written on it." Once penal law enforced a medical dogma, however, the issue became one of politics and personal liberty; he thus felt impelled, in his small way, to aid in repealing a cruel and unjust law: "I could not wait years to study the question in all its intricacies and obscurities while men were being daily punished, as I believed, unjustly. Liberty is in my mind a far greater and more important thing than science." A few errors had been pointed out, but they were unimportant and did not affect his main argument. When set against the errors of politicians, his were "molehills compared with mountains." Moreover, he had good reason to distrust medical authorities. During the Franco-Prussian War, for example, smallpox mortality was given as 23,469 for unvaccinated French soldiers and only 263 for the vaccinated Germans. For years, these statistics were upheld as a justification for vaccinating against smallpox. But it had been shown that no statistics from that period existed—someone had manufactured them. As another example, he pointed out that one expert had taken the average mortality rate from smallpox in the eighteenth century and multiplied it by four to reflect the increase in population in order to compare it with statistics from the nineteenth century. These and other gross errors were important in "showing how untrustworthy are some medical men, how careless in a matter where accuracy is of vital importance, this carelessness, and even recklessness of statement, [extending] to official statements." Referring to the mass of statistics he had compiled, he reiterated his belief that vaccination was "absolutely powerless" either in preventing attacks of smallpox or in diminishing smallpox mortality.[21]

In late December 1890, the commission issued a preliminary report, but its inquiry dragged on for several years and the final report was not completed until 1896. In the meantime, Wallace continued to write on the subject, and his opponents continued to rebut his arguments. A pamphlet entitled *Vaccination a Delusion; Its Penal Enforcement a Crime: Proved by the Official Evidence in the Reports of the Royal Commission*, published in January 1898 following the commission's final report, was his answer to the commission's conclusions. He wrote:

> A Royal Commission, which one would have supposed would have striven to be rigidly impartial, has presented a Report which is not only weak, misleading, and inadequate, but is also palpably one-sided. . . . Whether we examine the long-continued records of London mortality, or those of modern registration for England, Scotland, and Ireland; whether we consider the "control experiment" or crucial test afforded by unvaccinated Leicester . . . the conclusion is in every case the same: that vaccination is a gigantic delusion; that it has never saved a single life; but that it has been the cause of so much disease, so many deaths, such a vast amount of utterly needless and altogether undeserved suffering, that it will be classed by the coming generation among the greatest errors of an ignorant and prejudiced age, and its penal enforcement the foulest blot on the generally beneficent course of legislation during our century.[22]

In a war of attrition, Wallace would always remain unvanquished. The *Lancet* reacted predictably to Wallace's anti-vaccination credo: "Mr. Wallace, despite the manner in which he has suffered pulverisation . . . in his cross-examination before the Royal Commission, has written yet another pamphlet, in which . . . he reiterates all his old arguments, all his old fallacies, and, be it added, all his old sins of omission."[23]

An amendment to the Vaccination Act was read in the House of Commons several days after the publication of the commission's final report in March 1896. The most radical of the commission's recommendations was the abandonment of person-to-person vaccination, which resulted in the transmission of a variety of other diseases, in favor of vaccination with sterilized calf lymphatic fluid. This recommendation was adopted in the proposed amendment. The other significant recommendation—that conscientious objectors be exempted from punishment by making a "statutory declaration" justifying their objection—was not adopted. Instead, conscientious objectors would still be fined, but only once; they would not be subjected to repeated penalties for failure to vaccinate the same child. The *Lancet* criticized the proposed amendment as "compulsory vaccination without compulsion." In its noble quest for universal vaccination without serious infringement on personal liberty, the government had, in the journal's opinion, inadvertently undermined its own cause. It was therefore essential that medical authorities educate the public and dispel erroneous doctrines about smallpox and smallpox vaccination.[24]

In 1898 a clear majority finally passed a new vaccination bill after protracted debate. Anti-vaccinationists like Wallace continued to proclaim the worthlessness of vaccination, but the new act considerably weakened the force of their protests. They could not charge that they had been ignored; many of their complaints had been seriously considered and appropriate remedies implemented; penalties for noncompliance were significantly reduced. The medical profession praised the legislation on the whole but did not feel that the act went far enough. Not included were provisions for revaccination, which it felt was crucial for ensuring lifelong immunity to the disease.

The legislation, however, proved to be a victory for democratic principles. The nation had come a long way since the beginning of the century, when a patronizing ruling elite often disregarded individual rights. Private liberty and the common good were balanced, and the outcome was as equitable as possible for an issue that had incited so much rancor and symbolized the tension between government dictates and individual rights. By throwing the weight of his scientific reputation behind a controversial subject, Wallace had played a significant role in the advancement of the cause of justice and freedom in his nation. He also forced the medical community to reexamine its scientific methodology and improve a procedure that was flawed even if its results were far more successful than Wallace and his anti-vaccinationist allies were willing to concede.

Wallace officially declared himself a socialist in the summer of 1889 after reading *Looking Backward*, a utopian novel by the American writer Edward Bellamy. Although Bellamy's tale was fictional, his realistic style and detailed conception of a socialist state—classless, egalitarian, and devoid of the excesses of capitalism—convinced Wallace that socialism was the highest form of political and social organization that human beings could achieve. Bellamy's socialist society bore a close resemblance to Robert Owen's New Lanarck. Wallace told his friend Grant Allen (also a socialist sympathizer and a member of the Land Nationalisation Society) that the book was a "work of true genius."[25] He was not alone. Many other intellectuals of the era considered *Looking Backward* second only to Marx's *Das Kapital* as a seminal achievement in the field of political science.[26]

Bellamy inspired Wallace to write "Human Selection," an article that he published in the September 1, 1890, issue of the *Fortnightly Review*. Darwin, who was among the first to raise the issue of human selection in *The Descent of Man*, expressed pessimism about humanity's future. He felt that those who succeeded in the race for wealth were not always the best or the most intelligent. The lower classes outbred the upper and threatened to overwhelm society with their inferior progeny. Although he offered no solutions, others did. Most solutions that scientists and social philosophers had proposed Wallace found morally and ethically reprehensible. Lumped under the umbrella term "social Darwinism," they advocated either state interference in reproduction to eliminate the "unfit" or state subsidies to encourage the more fit to marry young and outbreed everyone else. Eugenics, the invention of Darwin's cousin Francis Galton, was the science of social Darwinism, and ultimately led to the most egregious institutions of the twentieth century: fascism and the Final Solution. Herbert Spencer, another eugenicist, believed that under the beneficial influences of education, hygiene, and social refinement, civilized man could be improved, an assumption based on the Lamarckian notion of the inheritance of acquired characteristics.

Wallace had once praised Galton's work. In 1870 he had reviewed *Hereditary Genius*, in which Galton held that inheritance determined humans' natural abilities by means of the same law that governed the rest of the organic world: natural selection. *Hereditary Genius* was one of the first statistical analyses of human intelligence. Galton deduced intelligence from a study of biographical dictionaries and obituaries of eminent men and the grades of examination papers at Cambridge University. Like the great majority of contemporary critics, Wallace ignored the bias inherent in Galton's analysis, which was based on an upper-class, Anglo-Saxon, male-dominated ethos, focusing instead on Galton's conclusion—harmonious with his own—that the intellectual ability of the Athenians in the age of Pericles was at least two grades higher than the ability of nineteenth-century Europeans. Galton had used the "law of the deviation from an average" (today's standard deviation) to determine the intellectual status of his nation. "With all our boasted civilisation and the vast social and scientific problems [with] which we have to grapple; with all our world-wide interests, our noble literature, and accumulated wealth," Wallace declared, "the intellectual status of the most civilised modern nation is actually lower than it was more than two thousand years ago!"[27]

As usual, Wallace had his own solution, formulated after an additional twenty-five years of contemplation. Social Darwinism, a term he never used, was repugnant to him. Spencer's ideas held no weight either, since, as August Weismann had demonstrated, qualities acquired after birth could not be inherited. In "Human Selection," Wallace would disagree with all current proposals to elevate the human condition, because the present state of society, which was "vicious and rotten to the core," made it impossible to determine which individuals were the best or most "fit." These proposals smacked of classism, for the majority of men and women lived and worked in the most degrading of circumstances. Only after sweeping out the "Augean stables" of existing social organizations and re-creating a society based on justice and equality of opportunity would social, scientific, and political leaders be qualified to judge the fitness or unfitness of the members of their society.[28]

Strands of Malthus, Owen, and Bellamy can be detected in his essay, all woven into a new fabric on the loom of natural selection. Wallace placed the education of women at the top of his scheme. Educated women tended to marry at a later age or not at all. He predicted that with the increased education of women, the overall fertility rate of society would diminish, resulting in an equilibrium between the birth and death rates and the elimination of "the bugbear" of overpopulation. In his idealized society, women would become the agents of natural selection. They would marry for love, not necessity, and would abstain from marriage until they could find a worthy and sympathetic husband. Idle and selfish men would be almost universally rejected; the diseased or weak of intellect would remain unmarried; those who suffered from mental illness or hereditary diseases would have difficulty finding partners. State interference was unnecessary because a natural checks-and-balances system would operate. It may seem strange to hear Wallace advocating a form of sexual selection, but he had never argued that it did not exist, only that it was a part of natural selection in another guise:

> When we allow ourselves to be guided by reason, justice and public spirit in our dealings with our fellow-men, and determine to abolish poverty by recognizing the equal rights of all citizens of our common land to an equal share of the wealth which all combine to produce—when we have thus solved the lesser problem of a rational social organization adapted to secure the equal well-being of all, then we may safely leave the far greater and deeper problem of the improvement of the race to the cultivated minds and pure instincts of the men, and especially of the Women of the Future.

Wallace moved from Godalming in 1889, after eight years' residence. Since returning from the United States, he had suffered from persistent upper-respiratory problems. The least amount of exertion brought on violent palpitations, and asthma or bronchitis (his diagnosis) followed every cold. Hoping that a larger garden and a milder, sunnier climate would improve his health, he acquired a cottage at Parkstone, in Dorset, a four-hour train ride from London's Waterloo Station. The property had particularly attracted him because of its abundance of large evergreen shrubs and stands of purple veronica, with their bountiful spikes of beautiful flowers. Large eucalyptus trees provided shade and privacy. There were also two gardens, one with a southern and the other with a northern exposure. Oak and pine trees covered a hill on the eastern side of the property. Inspection of the soil convinced him that he could grow the rhododendrons and heath he had long wished to plant. The cottage itself, though small and cramped, was big enough for himself and Annie. Violet and Will, who were twenty and eighteen, respectively, no longer lived at home; Violet was a kindergarten teacher in Liverpool, and Will—or Willie, as his father called him—was studying electrical engineering at a local college.[29]

In May 1889 Wallace declined an honorary doctorate from Oxford University. He told Edward Poulton that he had a profound distaste for public ceremonies, that he was too much of an amateur in natural history and too "ignorant" to receive honors from a great university. Moreover, in addition to being preoccupied with the move into his new home, the ceremony coincided with his annual obligation to grade examination papers—a valued source of income. The greatest kindness his friends could do for him was to leave him in "peaceful obscurity."[30] Poulton changed the date to November to meet Wallace's objections. Trapped by Poulton's maneuver, he reluctantly accepted.

The ceremony was the usual cap-and-gown affair, with congratulatory speeches given in Latin. Dressed in the scarlet robe and black bonnet that he had had to purchase for the occasion, Wallace listened to praise for his early, seminal work in natural history, with only a passing reference to his other publications. The degree of Doctor of Civil Law (D.C.L.) was conferred on him, and from that day forward he was known as Dr. Wallace, a title he accepted with dignity and displayed with some pride as a validation of his lifework, achieved despite a formal education that ended at age thirteen.[31] Poulton gave Wallace the standard tour of the ancient campus, but Wallace, never content with a straightforward approach to anything, wanted to see Oxford through the eyes of the young Grant Allen. As an undergraduate, Allen had written a poem called "In Magdalen Tower," a magical evocation of Oxford at midnight, as viewed from a window somewhere in the venerable college. Since those early days, Allen had become a popular writer of fiction, with his literary success overshadowing his work in evolutionary biology. Wallace counted himself among Allen's admirers. He therefore was determined to find the room that had inspired Allen's poetical musings. Poulton enlisted the aid of a few of Allen's old classmates, and the small group embarked on a wild-goose chase through Magdalen College, visiting one room after another until they concluded that the lookout had been a figment of the author's overheated imagination.[32]

Although the move to Dorset afforded Wallace many hours of desired solitude, he maintained contact with the outside world through his ongoing correspondence and the piles of journals and magazines he received on a regular basis. From 1889 onward—his opposition to vaccination notwithstanding—he enjoyed a steady rise in reputation. On December 1, 1890, the Royal Society of London awarded him the first Darwin Medal "for his independent origination of the theory of the origin of species by natural selection." The annual meeting, held at the luxurious Hotel Metropole in Westminster, was "very numerously attended." The speeches were so long, a reporter for *Nature* stated, that he could not reproduce them in their entirety, but he wished to put on record a comment by one of the attendees, who praised Wallace "as an example of that chivalrous feeling which one would wish men of science should always exhibit."[33] In May 1891, Macmillan reissued *Contributions to the Theory of Natural Selection* (1870) and *"Tropical Nature" and Other Essays* (1878) in a single volume entitled *Natural Selection and Tropical Nature.* Although it never approached the popularity of some of his other books, it sold well enough to go through a second edition in 1895. In the updated introduction, Wallace related for the first time the unusual circumstances surrounding the composition of his most famous essay as he recovered from malaria on the island of Ternate, prompting Adolf Bernhard Meyer of the Zoological Museum in Dresden—who had been inspired by Wallace's example to make his own journey to New Guinea in search of birds of paradise—to note that "ordinary mortals dream nonsense in their fits of fever, a philosopher of Dr. Wallace's standing conceives original ideas!"[34]

Awed by his remarkable work in evolutionary biology, the younger generation of academically trained biologists considered Wallace one of the Grand Old Men of Victorian science. These men and women eagerly sought his opinions and thoughtfully considered his responses. When he spoke on evolutionary theory, everyone listened. When he spoke on other issues equally dear to him, his views were politely dismissed—at least by the science community—as eccentric or quaint. Meanwhile, letters poured in; books arrived at his doorstep for his perusal and critique; requests were made for contributions to magazines on both sides of the Atlantic. In the spring of 1892, he was awarded the Founder's Medal of the Royal Geographical Society, though he begged to be excused from the celebratory dinner.[35] From his redoubt in Parkstone, he continued to contemplate the wonders of nature and humanity's place in it. "The older I get," he remarked to his friend J. W. Marshall, "the more marvelous seem to me the mere variety of form and habit in plants and animals, and the unerring certitude with which from a minute germ the whole complex organism is built up, true to the type of its kind in all the infinitude of details! It is this which gives such a charm to the watching of plants growing, and of kittens rapidly developing their senses and habitude!"[36]

It was Wallace's position as England's leading evolutionary biologist that spurred a young Finnish social anthropologist named Edward Alexander Westermarck to ask his advice on a book that would soon become famous. *The History of Human Marriage* was an elaboration of Westermarck's 1890 doctoral dissertation. He had visited England in 1887 to study at the British Museum and had learned English in order to read the great English evolutionists and anthropologists in the original language. (He would later teach sociology at the University of London.) Two ideas in Westermarck's manuscript attracted Wallace's attention. First, Westermarck postulated that incest taboos arose as a result of natural selection to prevent the deleterious effect of inbreeding. Second, on the subject of sexual selection Westermarck disagreed with Darwin, adopting some of Wallace's views. "[The] sexual colours, scents, and sounds in the animal kingdom are complementary to each other *in the way that is best suited to make the animals discoverable*," Westermarck hypothesized after presenting a formidable array of evidence from the animal world:

> It is the way in which the sexual colours, sounds, and odours are distributed among the different animal species that, in my opinion, is the strongest evidence against Darwin's theory of sexual selection, because this distribution is invariably such as to make the animal distinguishable at a distance. And now I am prepared to go a step further. It seems to me probable that these secondary sexual characters are useful to a species not only because they facilitate reproduction by making it easier for the sexes to find each other, but also because they tend to prevent inbreeding by attracting individuals from a distance.[37]

"I have not studied the question myself," Wallace wrote to Westermarck on January 29, 1890, regarding the universal repugnance to incest, "but I have long felt and believed that the theory of primitive promiscuity of [John Ferguson] McLennon, [John] Lubbock, [Edward Burnett] Tylor and others was untrue, and I am very much pleased at the admirable and thorough way in which you have exposed its fallacy. Your facts and your arguments seem to me quite conclusive and must, I am sure, carry conviction to most persons who have not committed themselves to the opposite view."[38] In his introduction to the original edition of Westermarck's work, Wallace commented on those conclusions "diametrically opposed" to the views of some of the most eminent thinkers of the day, including Darwin, Spencer, Conwy Lloyd Morgan, and Lubbock. "With such an array of authority on the one side and a hitherto unknown student on the other," Wallace wrote with perhaps his own youthful iconoclasm in mind, "it will certainly be thought that all the probabilities are against the latter. Yet I venture to anticipate that the verdict of independent thinkers will, on most of these disputed points, be in favour of the newcomer who has so boldly challenged the conclusions of some of our most esteemed writers."[39]

Wallace met with Westermarck in late December—he put off the visit for a few weeks because his house was in the midst of being renovated, and the builders had "hardly left" him and Annie a room to inhabit—and agreed to proofread the entire book. In March 1891, he critiqued Westermarck's ninth chapter, a "full and ingenious discussion of the origin of sexual modesty." Wallace believed that the origin of the concealment of the sexual organs was not the same as that of the rest of the body by clothing. Nine-tenths of all "savages," he noted, use some sort of clothing, and total nudity is rare. Sexual intercourse occurs mainly at night and is rarely practiced in the open without some form of concealment. "This may have arisen partly from the helplessness against attack of both parties, and also because the females, *even of animals*, require a considerable amount of solicitation or courting to obtain consent." The upright position in humans—especially males—exposes the sexual organs more to accidental blows or wounds, or to be seized by an enemy, "and this would naturally lead to the bandaging of the penis, especially among people who did not constantly carry shields to protect the body. Even the completely naked women of the Uaupés showed a great sense of modesty in their *attitudes*, always turning sideways upon meeting a man and, when sitting, so disposing the legs as to well conceal the pudenda."[40] Wallace also gave Westermarck some insightful editorial advice. He disliked the final chapter, on the "comparatively unimportant subject [of] *The duration of Marriage*," feeling that few but the most "professed Anthropological students" would read it thoroughly. The final chapter—which would be the only popular part of the book, and one that everyone would read—should provide a summary of facts and conclusions presented in the book, especially how those conclusions differed from those of previous writers: "Such a chapter will ensure *good* reviews, and in all probability double the sale of the book."[41] *The History of Human Marriage* was published in 1892 by Macmillan (thanks to Wallace's influence) and became an immediate scientific success. It anticipated Freud's work *Totem and Taboo*, though Freud's conclusion about the origin of the incest taboo, based on his psychoanalytic theory of human development, differed from Westermarck's biological explanation.

In 1893 the Royal Society of London nominated Wallace for membership, an honor some thought long overdue. After the award of the Royal Medal in 1868, Wallace had made no effort to join, while lesser men were admitted year after year. A few men, like Herbert Spencer, begged to join the ranks but were denied because of a lack of scientific credentials. Henry Bates, whose claims were high but not as high as Wallace's, finally obtained admission in 1881, though he passed through the review process twice before his election. George Romanes was elected after his third attempt on the basis of a single paper. Spiritualism did not prevent Sir William Crookes from becoming a member or winning the Royal Medal in 1876, nor would it have presented an insurmountable obstacle to Wallace's membership.

But Wallace had no use for the society. Membership provided no tangible benefits beyond prestige—and prestige had never motivated him. The soireés and social snobbery repelled him. In 1877, when someone erred by appending "FRS" to his name in *Fraser's Magazine*, he notified Joseph Norman Lockyer at *Nature* of the mistake, apologizing for his "involuntary appearance under false colours."[42] Attendance at an occasional lecture was the only support he could muster for an organization he deemed unsympathetic to the needs of the common people. Wallace was not alone in his opinion. By the late 1880s, the society was criticized for having become a gentlemen's club, with disproportionately few true discoverers in its ranks. Thomas Huxley complained to Joseph Hooker that in their youth their mission had been to "exorcise the aristocratic flunkeyism" from the society, but in their old age they had to rid it of the wealthy engineers, businessmen, and self-styled savants who had "sold their souls for a good price" to parade as men of science.[43]

In October 1892, William Thistleton-Dyer wrote to inform Wallace that he was prepared to put his name forward for membership with his consent. Hooker planned to spearhead the petition, which would then be endorsed by Huxley, Poulton, Raphael Meldola, Galton, Francis Darwin, Edwin Ray Lankester, Thistleton-Dyer, and others. Even the obligatory £3 fee had been waived; all he had to do was sign on the dotted line. Wallace's reflex response was to refuse: "The fact is, that I have—for various reasons—left it so long that at this late period of my life it seems hardly worthwhile. Therefore, while thanking you very much for your kind interest in the matter, I really think it will be better to take no further steps toward it."[44]

Three months passed. Thistleton-Dyer consulted Huxley, who approved of the nomination and expressed his "strongest opinion" that Wallace should yield unless his objections were insuperable. So Thistleton-Dyer tried again: "All of us who belong to the RS have but one wish, which is that it should stand before the public as containing all that is best and worthiest in British Science. As long as men like you stand aloof, that cannot be said. . . . Huxley went the length of saying that to him it seemed a plain duty. . . . To dissociate yourself from the RS really amounts nowadays to doing it injury."[45] The impassioned appeal overwhelmed Wallace. He was mystified that anyone should care about his becoming a fellow. "I have for many years felt almost ashamed of the amount of reputation & honour that has been awarded me," he wrote. "[All] the work I have done is more or less amateurish & founded almost wholly on other men's observations." However, not wishing to appear ungrateful for the honor now being forced on him, he placed himself in Thistleton-Dyer's hands.[46] He went ahead, and Wallace was formally inducted into the Royal Society of London in June 1893.

Between 1889 and 1896, Wallace led the defense of natural selection. In 1893 he wrote "Are Individually Acquired Characters Inherited?" for the *Fortnightly Review*, having been inspired by Herbert Spencer's article "The Inadequacy of Natural Selection." Although Spencer had coined the phrase "survival of the fittest," he firmly believed in the inheritance of acquired characteristics on a priori grounds.[47] Attacking flaws in his logic, Wallace showed that Spencer did not understand the theory of natural selection. Wallace's pointed critique "dreadfully disturbed" Spencer, for his entire philosophy was based on the transmissibility of improvements in moral and mental characteristics. Without the possibility of the inheritance of acquired characteristics, Huxley said, Spencer's psychology went to pieces.[48] Despite his upset, Spencer was not converted by Wallace's article. How, then, does one explain his extraordinary reaction to the address of Lord Salisbury (Robert Arthur Talbot Gascyone-Cecil) to the 1894 meeting of the British Association for the Advancement of Science?

Salisbury, a former prime minister, was chancellor of Oxford University. His selection as president of the British Association provoked controversy because he was a politician rather than a scientist. A debate raged in the pages of *Nature* about the propriety of a politician leading a major scientific organization. Politics won out. The theme of Salisbury's presidential address was science and religion. By calling attention to the last meeting of the British Association at Oxford in 1860, Salisbury placed the legendary debate between Thomas Huxley and "Soapy Sam" Wilberforce in the proper historical perspective. In the popular mind, Salisbury said, the 1860 meeting had manifested the deep division between science and religion, but religion was only a cover for the resentment of the old guard toward the claims of younger rivals—that is, the yielding by adherents of the old paradigm to promulgators of a new one. Yet despite the triumph of the new thinking, certain inscrutable mysteries remained. In Salisbury's opinion, humankind had come no closer to Truth. Darwin had effected a revolution in the methods of scientific research and "disposed" of the concept of the immutability of species; nevertheless, "the deepest obscurity" still hung over the origin of the infinite variety of life. Support for the theory of natural selection was by no means unanimous, he felt. Not wishing to place himself in the line of fire by intervening in the controversy between neo-Darwinians and their opponents, he sympathized with those who had returned a verdict of "not proven" with respect to the ingenious hypothesis.

After describing the production of domestic breeds by artificial selection, Salisbury asked, "[I]n natural selection, who is to supply the breeder's place? . . . What is to secure that the two individuals of opposite sexes in the primeval forest, who have been both accidentally blessed with the same advantageous variation, shall meet, and transmit by inheritance that variation to their successors?" Nothing ensured that step except chance, which was something he could not accept. Appealing to teleology, he raised the odious specter of Lord Kelvin, who had said, "Overpoweringly strong proofs of intelligent and benevolent design lie around us, and if ever perplexities, whether metaphysical or scientific, turn us away from them for a time, they come back upon us with irresistible force, showing to us through nature the influence of a free will, and teaching us that all living things depend on one everlasting Creator and Ruler."[49]

Salisbury's address infuriated Spencer. "His nonsensical representation of the theory ought to be exposed," he complained to Wallace, "for it will mislead very many people. . . . It behooves you of all men to take up the gauntlet he has thrown down."[50] Wallace would later refer to Salisbury's conception of natural selection as "strange and almost incredible," arising from the belief that advantageous variations occurred singly, at long intervals, and in one or two individuals rather than constantly and everywhere within populations of millions, which could roughly be divided into two parts—the less and the more adapted.[51] But he brushed off Spencer's concerns, which astounded him. The whole world knew that both Wallace and Darwin were linked to the revolutionary doctrine of natural selection, Spencer said. "Everybody will look to you for a reply, and if you make no reply it will be understood that Lord Salisbury's objection is valid."[52] Wallace did nothing. A year later, still ruminating about the address, Spencer grumbled that it was left to him to take on Salisbury.[53]

Wallace's reticence was calculated: Spencer had failed him in the past. Not only had he refused to support the Land Nationalisation Society, but he had recanted his own socialist convictions, so eloquently espoused forty years earlier in *Social Statics*. Wallace idolized the author of *Social Statics*, not the man who now pleaded the case for Darwin. He abandoned the battle against Salisbury to Spencer, who would have to fend for himself. He had written more than enough on the subject, which anyone could verify if he or she so desired.

Wallace's greater concern was the attack from fellow scientists—the real danger to the public mind, in his opinion. He considered the threat from Salisbury, a politician, to be insignificant. The front against the Darwin–Wallace doctrine of natural selection had continued to widen. "Almost every educated man who can write good English but who cannot understand Darwin's theory of Natural Selection, seems to feel compelled to explain his difficulties and to offer his own preferable theory in the form of a volume on Evolution," Wallace wryly remarked in a book review for *Nature* in 1894.[54] Less than a year later, as if on cue, the British biologist William Bateson—who would later win fame as the founder of the science of genetics (a term he coined, along with the words "allele," "heterozygote," and "homozygote")—proposed his theory of "discontinuous variation," which held that the formation of species resulted from large, sudden mutations, not the gradual changes proposed by natural selectionists. After "wading through" Bateson's book *Material for the Study of Variation, Treated with Especial Regard to Discontinuity in the Origin of Species*, published in 1894, Wallace told Poulton that Bateson did not appear "to have any adequate conception of what Natural Selection is, or how impossible it is to escape from it. . . . [He] seems to think that, given a stable variation, Natural Selection must hide its diminished head! . . . He is so cocksure he has made a great discovery—which is the most palpable of mares' nests."[55] In a formal review of Bateson's work for the *Fortnightly Review*, Wallace characterized Bateson's views as "wholly erroneous" and a "backward step in the study of evolution."[56]

As usual, Wallace opened his review with a concise summary of the theory of natural selection, noting that "it is for want of giving full weight to the essentially intermittent nature of the struggle for existence that so many writers fail to grasp its full significance and continually set forth objections and difficulties which have no real importance." Darwin, he said, distinguished two classes of variations, which he termed "individual differences" and "sports." The former were small but numerous; the latter, large but rare. And it was these "sports" that Bateson referred to as "discontinuous variations." Neither Darwin nor Wallace discounted discontinuous variations, but they considered such variations to be unimportant and rarely, if ever, serving as the origin of species. While species formed a discontinuous series, Bateson said, the diverse environments on which they primarily depended shaded into one another imperceptibly and formed a continuous series. Bateson could not see how such modest environmental differences could thus be a directing cause of distinct specific differences; he therefore concluded that those differences were in some unknown way intrinsic to the nature of every organism and not directly dependent on natural selection.

Wallace believed that Bateson's argument rested on a fallacy: that in each locality, the environment of every species found there was the same, "and that all change of environment, whether in space or time, is continuous."

> [N]othing can be more abrupt than the change often due to diversity of soil, a sharp line dividing a pine or heather clad moor from calcareous hills; or to differences of level, as from a marshy plain to dry uplands; or, for aquatic animals, from the open sea to an estuary, or from a nontidal stream to an isolated pond. And when, in the course of geological time, an island is separated from a continent, or volcanic outbursts build up oceanic islands, the immigrants which reach such islands undergo a change of environment which is in a high degree discontinuous.
>
> Even more important, perhaps, is the fact that everywhere the environment as a whole is made up of an unlimited number of subenvironments, each of which alone, or nearly alone, affects a single species, as familiarly included in the term "their conditions of existence."[57]

Wallace dismissed Bateson's book as being "of small value" because it devoted a large amount of space to the various monstrosities of human hands and feet and of some mammalia that Bateson considered to have a direct bearing on the problem of the origin of species. Such abnormalities as six-, seven-, and eight-fingered, double-handed, and double-footed children; six- and seven-toed cats; double-footed birds; and so forth were also seen in nature and, Bateson believed, occurred as frequently among wild as among domesticated animals. Yet in no single instance, Wallace pointed out, had any similar monstrosity ever established itself in a successful race or local variety, even though such mutations must happen frequently. Such mutations were therefore disadvantageous to a high degree, and those animals possessing them could not compete with normal members of their species. Moreover, the most stable of all characteristics in the higher types of life were the number and relative position of serial parts, like the number of limbs or the number of parts of the limbs. At a very early point in geologic history, the main numerical relations of the essential parts of the higher organisms had become more or less fixed and stable, remaining relatively unchanged. The four limbs of vertebrates were already established in fishes in the Devonian period, as were the four wings and six legs of true insects: "We are thus irresistibly led to the conclusion that, among all the possible forms of variations now occurring, those affecting the number of important serial parts among higher organisms are those which have the least possible relation to whatever modification of species may now be going on around us. . . . Yet it is to variations of this nature . . . that the bulky and learned volume we are discussing has been devoted."[58]

Although Bateson's theory of discontinuous variation led to groundbreaking work in the field of genetics, contemporary evolutionists side with Wallace, calling Bateson's thesis "erroneous."[59] Despite Wallace's effective reply to Bateson and other critics, the theory of natural selection continued to lose ground to the non-Darwinian evolutionists. Wallace therefore next used the Linnean Society as a platform for his lifelong crusade. On June 18, 1896, he gave a lecture entitled "The Problem of Utility: Are Specific Characters Always or Generally Useful?" England's leading naturalists and biologists crowded the lecture hall to hear Wallace's most refined views to date on the species question: "Our final conclusion is that, whether we can discover their use or no, there is an overwhelming probability in favour of the statement that every truly specific character is or has been useful, or, if not itself useful, is strictly correlated with such a character."

In a letter to *Nature*, Thistleton-Dyer remarked that the problem of utility was the heart of the "Darwinian" theory. He was responding to Lankester's version of remarks that he himself had made at the Linnean Society meeting after Wallace's talk. Thistleton-Dyer, a botanist, had commented that he was increasingly convinced that specific characters in flowering plants were utilitarian. About animals, of which he knew little, he could not believe that all variations distinguishing one species from another were also not useful or correlated with useful traits, as both Darwin and Wallace had said. If time had allowed, he might have strengthened his position by referring to the large amount of important and convincing work supporting his belief that had been done in other countries. But in England, such research was being neglected. The result was that "the Darwinian theory of organic evolution seems hardly to have a convinced supporter left except for Mr. Wallace. In its place we have the 'Physiological Selection' of Dr. Romanes, the 'Discontinuous Variation' of Mr. Bateson, and, last of all, the extended 'Correlation Principle' of Prof. Lankester. A common feature of each is their more or less definite rejection of the principle of utility as accounting for specific characters."[60]

The lecture at the Linnean Society would be Wallace's last public appearance at an ordinary meeting of a scientific society. Although Lankester remarked that Wallace appeared to be in "full health and vigor," Wallace himself felt otherwise.[61] He believed that his health had broken down. "Fumigations"—exposure either to hot water vapor or to the nicotine in cigarette smoke, with its stimulating effect on respiration—and cigarettes themselves, the prescribed treatment of the times, were failing to improve his respiratory condition. His literary work had come to an end, and he did not expect to live much longer. He contented himself with gardening and the comforts of home.

Wallace may have been oppressed by the deaths of three important people in his life. In February 1892, Henry Bates had died; in September 1893, he had lost Fanny to cancer; and only three months later, he had mourned the death of Richard Spruce. Although he had been estranged from Bates for almost twenty years, he could not have remained unmoved by the death of the man who had profoundly changed the course of his early life. But the deaths of the latter two mattered more. Reflecting on the death of Fanny, he wrote to a friend:

> [It] seems unnatural and incredible that the living self with its special idiosyncrasies you have known so long can have left the body, still more unnatural that it should (as so many now believe) have utterly ceased to exist and become nothingness! With all my belief in, and knowledge of, Spiritualism, I have, however, occasional qualms of doubt, the remnants of my deeply ingrained scepticism; but my reason goes to support the psychical and spiritual phenomena in telling me that there must be a hereafter.

Three years later, Fanny communicated with him through a medium. "She loved you more than any one in the world," he was told. He knew that this was true, and he had not fully realized it until after her death.[62] His memories of Spruce were less morbid. He recalled an "intelligent, amusing and lovable companion." His only regret was that Spruce had never written up his travels; he hoped one day to edit Spruce's journals and make known to the world the name of one of the greatest but least recognized naturalists of the century.[63] Of Bates, he could find little to say that was truly heartfelt in the obituary he composed for *Nature*. Moreover, he refused the Royal Geographical Society's request to memorialize Bates in its journal. "I am afraid I cannot give you anything special about Bates," he wrote, "as I really know less than *scores* of others as to his life and work, though we went to the Amazon together. . . . Owing to my having lived in the country for the last 25 years I very rarely met him, and therefore know very little of his life and work in London," adding, "Even as to personal reminiscences I could give hardly anything, as my memory of the details of that long ago time are but dim."[64]

Wallace's health at the time is difficult to judge. He was not too ill to tend to his orchids in his greenhouse or to strike up a friendship with a doctor who shared his passion—even though the physician was a public vaccinator![65] His health did not deter him from traveling to Switzerland and collecting plants at high altitudes in July 1895.[66] But he was sufficiently worried to consult a medium, who visited him for a private séance at Parkstone and introduced him to two spirits, a Scottish physician and an Indian girl named Sunshine, both of whom made positive predictions about his future. Sunshine predicted a third chapter in his life, which would be characterized by satisfaction, retrospection, and work. He thought that both spirits were mistaken.[67]

Not long after the auspicious visit of the medium, an English travel entrepreneur by the name of Henry Simpson Lunn invited Wallace to lecture in Switzerland, offering to pay all his expenses. Instead of Darwinism, Lunn asked him to speak about the greatest scientific achievements of the nineteenth century. Wallace accepted Lunn's invitation, and in September 1896 he traveled to Davos, taking Annie and Violet with him. After much contemplation, he had narrowed down those achievements to thirteen major inventions and twelve theoretical discoveries. The thirteen inventions were the railway, steam navigation, the telegraph, the telephone, matches, gaslight, the electric light, photography, the phonograph, X rays, spectrum analysis, anesthetics, and antiseptics. The twelve theories included the conservation of energy; the molecular theory of gases; the direct measurement of the velocity of light and experimental proof of the earth's rotation, which "are put together, because hardly sufficient alone"; John Tyndall's discovery of the function of dust in nature, which explained the blue color of the sky and the ocean and the "gorgeous tints" at sunrise and sunset; the periodic table; the meteoritic theory of the universe, which postulated that "the nebulae of various forms and intensity represent . . . the early stages in the development of stars, suns, and planetary systems out of diffused meteoritic matter; while the stars themselves are of various temperatures, the heat increasing when the meteoritic matter is more rapidly aggregating"; the proof of a glacial epoch and its effects on the earth's surface; the proof of the great antiquity of humans; the theory of organic evolution; the cell theory and recapitulation theory in the embryo; the germ theory; and the nature and function of the white blood cell. Few in his audience would have disagreed. If one surveyed ancient Greece and Egypt, he said, one could find only eight theories that had revolutionized the world, compared with twelve theories in the nineteenth century alone. But in typical Wallace fashion, he went out on a limb, pointedly rejecting vaccination and predicting that phrenology would eventually attain general acceptance and prove to be a true science of the mind.[68] Attending his lecture were many physicians, who must have been as scandalized by his rejection of vaccination as they were puzzled by his advocacy of phrenology. Wallace spent a week in Davos, which offered much to amuse him. He took several walks through the nearby woods, meadows, and stony passes during the day, collecting some rare specimens for cultivation in his garden. In the evening, he attended a concert or lecture. He then met up with Annie, Violet, and five of their friends, who had gone to a "newly-discovered" place called Adelboden, near the village of Kandersteg.

When he returned to England, nocturnal bouts of his asthma plagued him. Unable to sleep, he was too exhausted during the day to work productively. In the summer of 1897, he graded his last examination papers, which he had been doing for twenty-seven years. While sitting for Albert Bruce-Joy, an Irish sculptor from the Royal Academy in London, who had been commissioned by an organization that Wallace did not bother to name to create a medallion for an unmentioned honor, he described the state of his health. Bruce-Joy, who apparently had suffered from a similar respiratory affliction, told Wallace that he had cured himself with drastic dietary changes. He had discovered a diet recommended by an American physician who believed that too much starch led to disease. To improve one's health, one ought to abstain from potatoes, bread, and "watery" vegetables, which "overloaded" the stomach and prevented proper digestion. In their place, one was advised to substitute well-cooked meat, fruits, nuts, eggs, and milk—all in moderation. In desperation, Wallace changed his eating habits; for years he had been a vegetarian, avoiding meat for ethical and humane reasons. Within a week, he felt better. In a month, he was "quite well," and he never again suffered another episode of asthma.[69] Wallace's miraculous cure raises a question about the underlying cause of his respiratory difficulties. In the nineteenth century, it is doubtful that a European vegetarian could hope to eat a well-balanced, nutritious diet. In Europe and the United States, vegetables were boiled to death, and only the natives of tropical countries enjoyed a year-round supply of fresh fruits. Moreover, foods were not fortified with vitamins, iron, and other minerals, as they are today. Although it is mere speculation, it is possible that Wallace suffered from some sort of vitamin deficiency—possibly a lack of thiamine—that led to a condition known as beri-beri. Or perhaps he suffered from the physiological effects of iron deficiency. (There is the so-called wet form of beri-beri, which is characterized by an enlargement of the heart and eventual heart failure, followed by the subsequent accumulation of fluid in the lungs and extremities. The lack of dietary iron, essential for generating hemoglobin, can also result in anemia, another potential cause of heart failure. Either condition can cause shortness of breath following the slightest activity or produce an irritating cough.) Whether it was charlatanry, true inductive reasoning, or plain common sense, the fact remains that Bruce-Joy's diet worked.

The end of his illness marked a turning point, the third chapter of his life, which the spirit Sunshine had predicted. With renewed vigor, he embarked on his first major literary enterprise in years, an expansion of the themes of his Davos lecture. His book was called *The Wonderful Century: Its Successes and Its Failures*, less than half of which was devoted to the outstanding achievements of the nineteenth century, the remainder addressing the century's mistakes. Its publication was timed to coincide with the vote in Parliament on the vaccination issue in the summer of 1898. According to Wallace, the major social failures, in addition to vaccination and the neglect of phrenology, were the penal system and the existence of private insane asylums. With regard to the latter, simply upon certification by two physicians, a person could be "kidnapped" and held without the knowledge or consent of his or her nearest friends or relatives. He urged that the fact of insanity be decided by a patient's acts, not by his or her opinions, and proved before a jury on the basis of medical evidence. He had in mind the case of a spiritualist friend who had been incarcerated in a private institution for several months for holding views similar to the ones he espoused. Eventually she had been rescued by the concerted efforts of her family. He felt that asylums run by public authorities, rather than private individuals motivated by economic gain, would minimize such abuses. A half century would pass before his recommendations were heeded.

Wallace also singled out militarism as a "curse of civilization." Most of the European wars of the nineteenth century were either dynastic squabbles or perpetrated for national aggrandizement, and were never waged to free slaves or protect the oppressed based on purely altruistic motives. Kings and kaisers, ministers and generals, nobles and capitalist millionaires were the "true vampires" of civilization, always seeking greater opportunity to suck the lifeblood of the people. The whole world had been turned into a "gambling table of the six Great Powers of Europe": England, France, Prussia, Russia, Italy, and Austria. This struggle for wealth was accompanied by the reckless destruction of nature and was among the greatest crimes of the century. The "demon greed," which resulted in the inequitable distribution of wealth, also was responsible for the enormous extent of urban poverty at the time. Wallace estimated that one-third of the population of London—more than 1 million people—lived below the poverty level. The charitable organizations that had been established in midcentury made noble efforts but were not sufficiently powerful to remedy this tragedy.[70]

Macmillan did not publish *The Wonderful Century*. Wallace preferred Swan Sonnenschein, the publisher of the second edition of *Land Nationalisation*. But he demanded no less than what he had demanded of Macmillan. He asked that a portion of the copies be cut to the size of his other books and produced with a pattern on the front and back covers "wholly in gold." He was thinking of those readers who already owned any of his other books and wanted matching sets. Uncut copies could be sent to libraries, which would "ultimately" have them rebound before they disintegrated. But he was even more emphatic about the appearance of the "cut" copies. The covers ought to be "exactly" the size of those of his earlier works. And he also "very much" wanted the top edges of these copies to be gilt, "which would agree with the gold cover and make the book suitable for gifts—and I am sure increase the sale." If that was too expensive, then he thought that the top edges could be marbled or colored to "take off the rawness"—but gilt "would be very much better."

The Wonderful Century received no reviews in the pages of *Nature* for reasons that Lockyer did not make public. About this lapse, which Wallace felt was deliberate—every one of his other scientific works had been reviewed in that journal—Wallace was philosophical. "They [*Nature*] never gave a word of notice to my book," he reported to William Barrett. "Probably . . . out of kindness to myself as one of their oldest contributors, since they would have had to scarify me, especially as regards the huge Vaccination chapter, which is nevertheless about the most demonstrative bit of work I have done."[71]

In June 1898, Wallace was asked to chair the International Congress of Spiritualists, to be held in London. He selected as the subject of his opening address justice, not charity, as the fundamental principle of social reform. During the past forty years, he told the congress, the view of the spiritual nature of human beings and the relation of the present life to the afterlife had undergone a change. The old doctrine of an afterlife of either reward or punishment, dependent on one's acceptance of dogmatic beliefs and ceremonial observances, had given way to the teachings of modern spiritualism, which preached moral and mental continuity from life through life after death. According to this new theory, which had been adopted even by non-spiritualists, the development of the body would save the soul. A healthy body was essential for a sound mind, which, in turn, was necessary for a sound soul. "Inasmuch as we have fully utilized and developed our faculties—bodily, mental, and spiritual—and have done all in our power to aid others in a similar development," he told his audience, "so have we prepared the future well-being of ourselves and for them." It was the moral duty of the more privileged to work strenuously to improve social conditions so that all could live a full and happy life—on earth and afterward. "If the accounts we get of the spirit world have any truth to them," he said, "the reclamation and education of the millions of undeveloped and degraded spirits which annually quit this earth, is a sore though cheerfully accepted burden, a source of trouble and sorrow to those more advanced spirits who have charge of them."

Wallace firmly believed that the very poorest classes of people were as good morally and as elevated intellectually as the middle and upper classes, "who look down upon them as in every way their inferiors." Social convention, not defects in character, had forced them to live in degraded conditions. He asked his audience to consider a real and fundamental remedy, not a mere palliative one. "Charity has increased enormously," he observed, "and has completely failed. Now it is time for us to try Justice." True justice meant equality of opportunity, and only through socialism, or socialist principles, could such justice be effected: "For our present state of society is not true individualism, because the inequalities of opportunity in early life are so great that often the worst are forced to the top, while many of the best struggle throughout life without a chance of using their highest faculties, or developing the best part of their nature." Equality of opportunity would put all this right, he said. This is "but bare justice." Life on earth was the school for the development of the spirit. For true philanthropists, the prevailing misery and degradation "*must* be and *can* be got rid of." Moreover, it was a philanthropist's duty and business to help to eliminate it. Although various religious sects were working in their respective fields to alleviate individual cases of want and suffering, this method had utterly failed to diminish the mass of human misery because it dealt with symptoms and not causes. Now the time had come to demand true social justice. Let justice, not charity, be our watchword, he proclaimed.[72]

Wallace thus entered the new century with the same indomitable spirit that had enlivened him in the old. He once told a friend that nothing stimulated him more than novelty and opposition. "An uphill fight in an unpopular cause, for preference a thoroughly unpopular one, or any argument in favour of a generally despised thesis," he proudly professed, "has charms for [me] that [I cannot] resist."[73] The opening decade of the twentieth century would provide him with a wealth of such charms.

CHAPTER 19

Satisfaction, Retrospection, and Work

By the turn of the century, Wallace felt the itch to move again. Parkstone had lost its rustic appeal, with villas being erected on all available land in his vicinity; now he had to walk almost two miles to reach open country. In 1901 he had hoped to put together a group of investors to buy an estate of from one hundred to three hundred acres and build a small number of houses for people who shared his environmental concerns. The idea was to secure a healthy and picturesque tract only one or two hours from London, with part of its woods and wilder areas to be reserved as a natural park for the enjoyment of the residents. He mailed a pamphlet outlining the plan to the various scientific societies and proposed to act as agent and surveyor, but no one took him up on the offer.[1]

After abandoning this fantasy of a naturalist's version of New Lanarck, he looked for something more modest for himself and Annie. He looked at a number of estates thirty or forty miles from London and was about to give up in despair when he found a spot only four miles from Parkstone, in Broadstone, a small town on the southern coast and an ideal retreat from the world, remote enough to ensure privacy but not too far from civilization to isolate him from English intellectual life. After some tough negotiating, he acquired three acres, having surveyed the property himself with his old sextant. The most charming part of the property, he told his son, Will, was an old grassy orchard with a score of trees, including apple, pear, plum, and cherry. From the top of the orchard was a view over moor and heather to Poole Harbour, and beyond the harbor to some hills and the sea. He found the land perfectly suited for his favorite shrubs. "I expect bulbs of all kinds grow well," he wrote, "and I mean to plant a thousand or so of snowdrops, crocuses, squills, daffodils, etc., in the orchard, where they will look lovely."[2] Will, now an electrical engineer and more earthbound than his father, asked several pertinent questions, but his father had already thought of everything: connections to water mains, arrangements for electricity, and revamping the old road, among other matters.[3] Wallace next hired an architect, who designed a many-gabled "very picturesque" but unpretentious two-story brick house.[4]

Wallace acted as his own contractor, estimating the cost of his new house at £1,000. By February 1902, he was visiting his property three days a week to supervise the gardeners, who were already at work preparing the loamy soil for his exotic plants. The following month, ground was broken and a basement excavated. In April he coordinated the repotting and planting of fifty young trees and shrubs that a friend had sent from Italy as a gift. To his horror, horses and cows from neighboring farms raided his property and devoured a number of them.[5] All sorts of problems delayed completion of the house until the end of the year. Meanwhile, expenses piled up, including wages of £10 to £12 a week and various bills averaging £30 to £50 a month. In December, eight men were still hard at work, another man was painting, and still another was constructing a road and carriage driveway to the front door. By Christmas, the house he called Old Orchard was ready for occupancy.

The house had set him back financially. He was £150 in debt to the bank and verging on bankruptcy, he wrote nonchalantly to Will, despite an advance from Macmillan for the revision of three of his books. He mischievously admonished Will not to tell his mother. He was not worried, however, for he had just been commissioned to write an article on any subject he desired for the *New York Independent*. It did not take him long to dream up a topic. While adding four new chapters on astronomy for another edition of *The Wonderful Century*, presumably with the acquiescence of (and advance payment from) Swan Sonnenschein, he suddenly felt inspired. He would show that our solar system, and therefore the earth, occupied a central position in the universe. If true, this central position could be no coincidence: it was further evidence, he believed, of intentional design by some supreme Intelligence or controlling Mind. Curtis Brown, his American agent, loved the idea, and in a few days Wallace completed his article. The *New York Independent* had offered him £20 for 2,500 to 3,000 words, but his article came to 8,000 words and he hoped to get £60 for it, the highest amount the newspaper had ever paid to a contributor. "But what is even better," he told Will, who was probably surprised by the subject, "Mr. Brown suggests (after reading it) that I make a *book* of it, of about 70,000 words, and he will arrange for its publication, and get me a good Royalty and a good *sum paid in advance*."[6] The article also appeared in the March 1, 1903, issue of the *Fortnightly Review*. As always, for Wallace the fragile state of his finances served as the primary source of inspiration. As he noted in his autobiography, "I feel that without the spur of necessity, I should not have done much of the work I have done."[7] Although no one believed him when he confessed to being constitutionally lazy, if he had had a choice and the financial resources, he undoubtedly would have enjoyed the simple country life of a gentleman farmer and the occasional excursion to the Continent to examine the volcanoes of southern Italy or the archaeological treasures of Pompeii.

For six months, Wallace immersed himself in astronomy, an entirely new discipline for this octogenarian. Every day he worked at his desk in his library, with its lovely view of nearby Poole Harbour. He enlisted Will to proofread the manuscript and make any corrections in the physics and math computations as he went along. Will objected to certain passages—especially to tangential topics like pollution. "The bit about the pure air came to me while writing and I let myself go," Wallace replied in self-defense. "Why should I not try and do a little good and make people think a little on such matters, when I have the chance of perhaps more readers than all my other books!" Will also objected to his father's treatment of the subject of *Homo sapiens*. "[Of] course that is the whole subject of the book!" Wallace exclaimed. "And I look at it differently from you, because I know facts about him you neither know nor believe yet." The new book, which he entitled *Man's Place in the Universe* as a conscious allusion to Thomas Huxley's *Man's Place in Nature*, was unusual by any standard. In it, Wallace combined astronomy with geology, physics, organic evolution, and a dash of spiritualism to argue powerfully against the probability that human beings existed anywhere else in the universe. "A great deal is speculative," he told Will, "but any reply to it is equally speculative. The question is, which speculation is most in accordance with the known facts, and not with prepossessions only."[8]

Wallace challenged an idea then raging throughout Europe and the United States that intelligent life existed on other planets in the solar system, most especially Mars. In the late 1870s, the Italian astronomer Giovanni Schiaparelli had observed lines on Mars that he called *canali* (channels), which the American astronomer Percival Lowell interpreted as a sign of intelligent life. In 1898 H. G. Wells had captured the public's imagination with his story *The War of the Worlds*, which depicted the invasion of earth by the technologically superior Martians. Wallace was unimpressed. Based on his readings and his correspondence with some of the foremost English astronomers and physicists of the day—including Joseph Norman Lockyer, George Darwin, Lord Kelvin, and Sir John Herschel—he had reduced current astronomical lore to six propositions. First, the stellar universe, though of enormous extent, was finite, with determinable limits, bounded in its extremity by the circular Milky Way. Second, the solar system was situated in the plane of the Milky Way and not far removed from its center; therefore, our planet was nearly at the center of the universe. Third, the universe consisted of the same matter and chemical elements throughout and was subject to the same physical and chemical laws; conditions similar, if not identical, to those on earth had to obtain on a planet if organic life was to develop there. Fourth, no other planet in the solar system, Mars included, had the properties or conditions that would permit the development of anything but the most primitive form of life. Fifth, the probability of there being any other sun with inhabited or habitable planets was inconceivably small. Sixth, the nearly central position of our sun was probably a permanent one, and had been specially favorable—perhaps absolutely essential—to the development of life on earth.[9]

One of the greatest inventions of the nineteenth century was spectrum analysis, which for the first time enabled scientists to analyze the composition of the sun, planets, and stars. Wallace stated that, broadly speaking, the entire stellar universe was constructed of the same series of elementary substances as those found on our planet. The whole material universe was essentially one. If the elements were the same, then the laws that governed these elements must also be the same. If any organized beings existed elsewhere in the universe, then they must also be the same. Although he admitted that life could exist under altogether diverse conditions in universes differently constructed, life in our universe would have had to arise under the same conditions as those existing on our planet. The conditions on Mars, Mercury, or any other planet in the solar system could not support the development of higher forms of life. If life existed in those places at all, it would be of the lowest type. The adaptability of a planet for the full development of life depended primarily on its size and, more directly, its mass. For example, Mars was not massive enough to retain water vapor; moreover, given the narrow temperature ranges in which life could survive on earth, Mars was too cold to sustain life.

Wallace portrayed the universe as a vast, integrated system, from the tiniest molecules to the largest stars. No one of its constituents could exist without the other. The existence of the earth was as important to the existence of the universe as the existence of the universe was to the existence of the earth. In short, the universe was a single organism composed of infinite and complex substructures. The most controversial aspect of the book was, of course, its conclusion. Contrary to the reigning dogma of the time, Wallace abandoned the materialist explanation of the origin of life in favor of the spiritualist one. He believed that his interconnected argument, founded wholly on facts and principles enunciated by modern science, led by deduction to one "great and definite" conclusion: that human beings, the culmination of conscious organic life, had arisen on our planet alone in the whole vast material universe. If human beings were the unique and supreme product of this vast universe, then some controlling Mind or Intelligence had conceived us for this very purpose. The immensity of the stellar universe, the long and slow and complex progress of nature, and the vast aeons of time that had passed before our development served as the raw materials and the spacious workshop for a Mind that "produced" the planet that eventually resulted in humankind. As Wallace wrote:

> All nature tells us the same strange, mysterious story, of the exuberance of life, of endless variety, of unimaginable quantity. All this life upon our earth has led up to and culminated in that of man. It has been, I believe, a common and not unpopular idea that during the whole process of the rise and growth and extinction of past forms, the earth has been preparing for the ultimate—Man. Much of the wealth and luxuriance of living things, the infinite variety of form and structure, the exquisite grace and beauty in bird and insect, in foliage and flower, may have been mere by-products of the grand mechanism we call nature—the one and only method of developing humanity.

The conception of the universe during Wallace's time neatly fit into his theory of man's place in the cosmos. In fact, one of his critics, Herbert Hall Turner, Savilian Professor of Astronomy at Oxford, acknowledged that Wallace drew his facts from the best sources of information available to him. But Turner was quick to point out that much of astronomical knowledge in the first years of the twentieth century was already "ancient history"; data from astronomers were pouring in at a logarithmic rate. The discovery in 1901 of "dark nebulae" that both reflected light and partly obscured it suggested to Turner that there probably were more stars than could be seen with the telescopes of the era. Moreover, he called Wallace's belief in the sun's central position in the universe irrelevant. In his critique of Wallace in the *Fortnightly Review*, Turner asked:

> Does it matter very much whether the Sun is at or near the centre of the visible universe if no better reason can be given for assigning any great significance to this position? Without the tremendous inference, the fact itself, if fact it be, can only invite our polite attention as a curious coincidence. Even as a coincidence it does not take high rank; for it can in any case only be temporary. If there *is* a centre of the visible Universe, and if we occupy it to-day, we certainly did not do so yesterday, and shall not to do so tomorrow.[10]

To his fellow scientists, Wallace's theory may have looked like anthropocentrism dressed up in evolutionary clothing, but to those dissatisfied with cold scientific reductionism it offered solace. Likewise for Wallace, who refused to believe that humanity, with its faculties, aspirations, and powers for good and evil, was a simple by-product of random forces—that human beings were merely animals of no importance to the universe and requiring no great preparation for their advent.[11]

In 1904 Francis Darwin suggested to his friend the artist William Rothenstein that he make a portrait of Wallace. Rothenstein, Oxford educated and a young portraitist of note, enjoyed painting unpopular eminent men or eminent men espousing unpopular ideas. In the late 1890s, he had published a four-volume series entitled *English Portraits* that included portraits of Thomas Hardy, Henry James, and George Bernard Shaw—though not everyone was a celebrity. As the caricaturist Max Beerbohm noted in his introduction to a volume of the painter's portraits and drawings, distinction was what Rothenstein liked, even if coupled with obscurity. Celebrities generally left Rothenstein cold unless they were something more than celebrated. Wallace was therefore the perfect choice.[12] Upon Darwin's recommendation, Wallace readily consented and invited the artist to spend a weekend with him. Rothenstein traveled to the village of Broadstone in late January, arriving by train in the evening. A cab took him along the main road past several houses, and then down a hill to Wallace's home, marked by a lantern hung specially for him on the gate. As instructed, he brought the lantern with him to the front door, where Wallace politely waited.

Wallace apparently was a disappointment to his portraitist. Having read his political and social writings, Rothenstein expected to find an expansive and warmhearted humanitarian but instead encountered someone rather distant and cold. ("Perhaps age had dulled him, and the years had damped down a once brightly burning fire," Rothenstein would later note in his memoirs.) Wallace's appearance reminded him of a Nonconformist preacher, not a scientist—whatever such a person was supposed to look like. He may have had in mind someone like Sir Joseph Hooker, who had charmed him more. "How handsome he must have been as a young man I saw from . . . photographs in the Darwin's [*sic*] house," he recalled of Hooker. "In the Hooker's [*sic*] house hung an entertaining picture of a very Victorian young Hooker, with side whiskers and sun helmet, receiving, in a tropical landscape, the fruits of the earth from kneeling savages." Old Orchard disappointed Rothenstein as much as Wallace's appearance did: no menagerie of glorious specimens from the tropics; no striking artifacts from Brazil or the Dutch East Indies to fill out the image of a great explorer. He was amused to note that an old faded photograph of Darwin in a cheap frame served as Wallace's only memento of their long and historic association. The young artist's eye discovered nothing of beauty even in the wild Dorset landscape. Old Orchard was the home of a "schoolmaster," he remarked, someone with plebeian tastes. His portrait, a simple pencil drawing like his other English portraits, reflects this impression. "The face . . . seems to me rather too *delicate* in feature, too small in *mouth*, too *young-looking!* and with that very *bored* expression," Wallace wrote in alarm to Rothenstein when he received the proof of the portrait a few weeks later. He attributed his dull countenance to having felt unwell that day and invited Rothenstein back in the spring, when the fresh air would enliven his complexion. Rothenstein declined the offer.[13] If Wallace had failed to impress Rothenstein, Rothenstein had failed to impress Wallace. The encounter was not mentioned in his later writings. Shy and reserved around strangers, Wallace was best seen when "brought out" by a sympathetic companion. The young artist evidently had found nothing with which to sympathize in the old philosophical naturalist.

Man's Place in the Universe, published at the end of 1903 in London (Chapman & Hall) and New York (McClure, Phillips), was a notable success (it quickly went through seven editions), prompting Curtis Brown to commission an autobiography with Chapman & Hall and Dodd, Mead in New York, the American publisher of *The Wonderful Century*. The publication of two new books by other publishing houses sent Macmillan into a panic. Wallace defended his decision by stating that he had been approached to write both books; had he originated the ideas himself, he would have offered Macmillan the option to publish them; at the same time, he did not lose the opportunity to point out Brown's help in easing every step along the way. Not only had Brown conceived ideas for the two books and all but promised publication in both England and the United States on "very liberal terms," but he had arranged for translations into several languages, which saved Wallace much trouble and correspondence. As a type of consolation, Wallace noted in a postscript that if his autobiography attracted attention, the sale of his books with Macmillan undoubtedly would increase.[14]

My Life; A Record of Events and Opinions, which Chapman & Hall brought out in October 1905, is Wallace's surprisingly candid account of his own successes and failures. Like a number of his later works, it intermixes facts with sermons from the bully pulpit—especially on those two bêtes noires: spiritualism and land nationalization. In it, Wallace looks back at his later life and recent literary output from a spiritualist perspective: "Now it seems to me a very suggestive fact that my literary work during the last ten years should have been so completely determined by two circumstances which must be considered, in the ordinary sense of the term, and in relation to my volition, matters of chance." If Henry Simpson Lunn had not invited him to Davos and suggested the subject of science in the nineteenth century, he would not have written *The Wonderful Century*. If he had not written *The Wonderful Century*, he would never have been directed to great astronomical problems, never have chosen as the subject of an article the sun's central position in the universe, never have undertaken the laborious task of researching and writing *Man's Place in the Universe*, and thus never have been asked to write his autobiography. If he had not met the sculptor Albert Bruce-Joy, who had recommended a diet that miraculously cured his asthma, he would not have had the stamina to embark on any of these demanding projects.[15] The spirit Sunshine's prophecy of a third chapter in his life, characterized by "satisfaction, retrospection, and work," had been fulfilled. The wholesome activity of mind and body, the beauty of Old Orchard, with its pure air and splendid views—this was satisfaction. The writing of his autobiography, which covered a period spanning three-quarters of a century—this was retrospection. The entire past ten years—this was work.

Wallace's autobiography sprawled over two volumes, each more than four hundred pages long. It contained numerous illustrations and two fold-out maps (one a reproduction of his original map of the Rio Negro and Upper Uaupés, and the other depicting the Malay Archipelago). It went through a second printing in 1906—including a version for Bell's Indian and Colonial Library, which printed books for circulation in India and the other British colonies—and was condensed and revised in 1908 with help from his son, Will. In the preface, Wallace stated that it was difficult to write a record of a life that spanned some eighty years without subjecting himself to the charge of "diffuseness or egotism." And from this charge, he admittedly could not altogether escape. But his experiences had been varied, "if not exciting," and he hoped that the frequent change of scene and occupation, as well as the diversity of his interests and of the persons with whom he had associated, would render his story "less tedious than might have been anticipated." The book is marked by its author's renowned eccentricities. For example, he analyzed himself phrenologically, shunning even the slightest reference to the new theory of the unconscious. He had dismissed the possibility of an unconscious mental life as early as 1896: in the preface to the third edition of *Miracles and Modern Spiritualism*, he had stated that "the 'second' or 'subconscious' self, with its wide stores of knowledge, how gained no one knows, its distinct character, its low morality, its constant lies is as purely a theoretical cause as is the spirit of a deceased person or any other spirit." Moreover, he believed that the course of his life proved the thesis of his idol, Robert Owen, that the development of one's character was due less to heredity than to the influences of one's environment.

He admitted to Arabella Buckley Fisher that the reviews of his autobiography were generally fair in their assessment of his so-called fads. Because of his frequent heterodoxy, one reporter called him an "anti-body." Another gave him a backhanded compliment, stating that he was the only man who believed in spiritualism, phrenology, anti-vaccination, and the centrality of the earth in the universe whose life was worth writing. "Then it points out a few things I am capable of believing but which everybody else knows to be fallacies, and compares me to Sir I Newton writing on the prophets!" he complained of this review. "Yet of course he praises my biology up to the stars—there I am wise—everywhere else I am a kind of weak, babyish idiot!"[16] The reviewer for *Nature*, J.A.T. (probably J. Arthur Thomson, professor of natural history at the University of Aberdeen), focused on Wallace's wisdom in natural history, having been urged by Joseph Norman Lockyer to confine his "attention . . . to what he tells us of his work as naturalist and biologist, though it is difficult, if not altogether legitimate perhaps, to abstract off one aspect of a life in this fashion." The review glowed with admiration and affection for its subject, referring to Wallace as the "Nestor of the evolutionist camp," a man whose long life had been "full of work, rich in achievement, and starred with high ideals."[17] The anonymous reviewer in the *Lancet* was equally reverential, though he could not help ending the review with the remark that "possibly some readers will find the lengthy disquisition upon spiritualism the most interesting part of Dr. Wallace's autobiography, not because the rest of these numerous pages pale in interest by contrast but because of the, to many of us, astounding fact that a man of Dr. Wallace's undoubted achievements in science and clearness in argument should have been convinced by what is unquestionably . . . said to be . . . largely imposture."[18]

For the next year, Wallace took a break from large literary projects and made only a handful of contributions to various publications, mainly on political, social, or spiritual issues: he continued to criticize government policy on smallpox vaccination; he lent his support to the Society for the Abolition of Capital Punishment; he gave practical advice on how to nationalize the railroads and socialize the government; and he urged South Africa to embrace racial equality.[19] Every morning, he spent two hours or so reading or writing in his study, surrounded by shelves filled with books, magazines, and journals. He responded to the many letters he received, including requests to write on any subject—even those he knew nothing about. Next he might wander into the adjoining conservatory. Having bought or received as gifts plants and seeds from Australia, New Zealand, the Himalayas, the mountains of Ceylon, the South African veld—including the far reaches of the vast British Empire—more than two hundred subtropical shrubs flourished on his premises. Every day he went outdoors into his garden, donning the massive pair of boots with inch-thick wooden soles in which he used to tramp through Wales or the mountains of Switzerland.

While his various occupations distracted her husband, Annie preoccupied herself with some domestic chore or worked in her part of the garden. She was fond of roses and primroses, which she nurtured as assiduously as her husband did his orchids, and she succeeded at the difficult art of hybridization. Her memory for botanical names was excellent; often when Wallace had difficulty recalling the name of a plant for a visitor, she would rescue him. At lunch they joined each other again, perhaps with Violet. In 1903, Violet, now thirty-four years old and unmarried, had returned from her schoolteacher's job in Liverpool to live with her parents for reasons that are unclear. She supported herself by tutoring three or four children in the neighborhood, which delighted her father, who treated her pupils as if they were his grandchildren.

In the afternoon, Wallace took a nap or interrupted his studies and writing to contemplate the beautiful view from his study window. In the evening, after supper, he read or wrote for a few more hours before going to bed. His reading was as eclectic as ever: Mark Twain, Leo Tolstoy, Pyotr Kropotkin. He loved novels, both contemporary works and the older classics, having a special affection for Charles Dickens and Walter Scott. But his favorite form of literature was poetry; over the decades, he had collected more than fifty volumes. One of his more recent discoveries was the great Hindu epic *Mahabharata*, which he placed in the same exalted league as Homer and the Psalms of the Old Testament. Quotations from some of his favorite authors are sprinkled throughout his later works—in part because he hated to see a blank page in any book.[20] He spent months investigating and debating the origin of a four-stanza poem called "Leonanie," which he attributed to Edgar Allan Poe or a spirit in his name:

> Leonanie—Angels named her
>   And they took the light
> Of the laughing star and framed her
> In a smile of white
>   And they made her hair gloomy
>   Midnight, and her eyes of bloomy
>   Moonshine, and they brought her to me
> In the solemn night . . .

The American poet James Whitcomb Riley, famous as the originator of "Little Orphan Annie," claimed to have concocted this poem as a youthful prank in 1877. At that time, Riley had been a struggling poet whom no one would publish; he wished to show the world that "the critics of verse would praise from a notable source what they did not hesitate to condemn from an emanation opposite," and, indeed, the poem had been praised as genuine by the highest authorities. Wallace was aware of the stunt but did not think that Riley had completely cleared up the mystery of the poem's origin; in his opinion, Riley was incapable of producing anything "with the same exquisite musical rhythm" as "Leonanie." Two versions of the poem were extant, Wallace argued, and his brother John had sent him the "better" version before his death in 1895: "The one issued by the alleged *composer* is not only inferior to the other, but contains such *incongruities* and verbal *errors*, as to seem to show that he did not realise the meaning of the poem, or appreciate its musical rhythm, as a whole, unless Riley *wrote* the poem under the *spiritual influence* of Poe—in an access of *inspiration* which has never recurred—(which I think quite possible) and then in his normal state altered and *spoilt* it."[21]

Wallace had not altogether abandoned his efforts on behalf of natural selection. Attacks on the Darwin–Wallace theory never failed to raise his hackles. "What a miserable abortion of a theory is 'Mutation,' which the Americans now seem to be taking up in place of Lamarckism," he wrote to his friend Edward Poulton.[22] The rediscovery in 1900 of Gregor Mendel's work by three European botanists and its promulgation in Great Britain by William Bateson seemed to divide evolutionists into two ideological camps: the naturalists and the experimentalists. Bateson, an experimental biologist with little experience in field biology, had seized on Mendel's rules of inheritance as proof of his belief in discontinuous variation; that is, every change in evolution was due to a major mutation. Although Mendel's laws explained only heredity, they were appropriated by Bateson to explain speciation. To a naturalist like Wallace, who had experienced firsthand the vast diversity of life, with its infinite gradations and variations, "mutationism" was special creation under another name.[23] Wallace appreciated the novelty of Mendel's work, but he did not grasp its full implications for his own theory. (Few at the time did.) The debate between the two opposing camps clouded the picture. Another three decades would pass before Mendelism and the theory of natural selection were synthesized into a single grand theory and no longer perceived as mutually exclusive. But Wallace placed Mendel's discoveries in the proper perspective. As he wrote to Poulton in July 1907, "'Mutation' as a theory is absolutely nothing new—only the assertion that new species originate *always* in sports, for which the evidence adduced is the most meagre and inconclusive of any ever set forth with such pretentious claims! 'Mendelism' is something new, and within its very limited range, important, as leading to conceptions as to the causes and laws of heredity, but only misleading when adduced as the true origin of species in nature, as to which it seems to me to have no part."[24]

Wallace could never remain idle for long. Without some great project, his life had no purpose. Perhaps perceiving that his time was running out, in

March 1906 he wrote a letter to Matthew B. Slater, the executor of Richard Spruce's estate, resurrecting the idea of editing Spruce's notes and journals for publication. Hooker, now eighty-nine, and Sir Clements R. Markham, a distinguished geographer and former president of the Royal Geographical Society, enthusiastically supported the project. Slater sent Wallace one of Spruce's journals. Although it brought back vivid memories of events long past, Wallace feared that it was too small a part of his journey to interest a publisher. In the nine years that had elapsed since Wallace had casually suggested publishing Spruce's papers, Slater had examined hardly anything in his possession. There must be much more, Wallace thought. He was certain that Spruce had kept a continuous journal of his fifteen years of travel in South America and copious notes on the various botanical questions that interested him from time to time. If he could see all of Spruce's notes and the entire journal, he would then be able to put together a book that would be of considerable interest to botanists, naturalists, and the educated reader alike. It was the voyages to the Orinoco and the Andes that were the most novel and most interesting, in his opinion. He asked Slater for a list of Spruce's papers first, fearing that Slater would send him boxes of unorganized material.[25]

Slater sent Wallace a list of papers and manuscripts, but none that made any reference to Spruce's journeys, extending over a two- or three-year period, up the Rio Negro and the Uaupés, Casiquiare, and Orinoco Rivers—regions that had not been explored botanically since Alexander von Humboldt and Aimé Bonpland had visited them. Surely there must be some journals and notebooks describing this extensive exploration. Slater suggested that Wallace come to York to examine the papers himself, but Wallace did not feel that he could leave Broadstone without endangering his health. "Though very well on *the whole*," he wrote, "I can only keep so by strict attention to *diet*, regularity of *habits*, and by means of those *conveniences* and *remedies* I can only have at home. For the last three years I have not been a night away and feel it is best for me *not* to leave home, as I feel many of the disabilities of old age." Wallace extended an invitation to Slater to visit his home to discuss the matter further. Spruce had been a meticulous worker, and Wallace wanted the botanical names of plants and all Indian or Portuguese terms written out very distinctly, exactly reproducing Spruce's spelling, to prevent publishing errors. Wallace was also fussy about the kind of manuscript paper onto which transcriptions were to be made, and he sent along a sample to Slater, who apparently accepted Wallace's imperious demands without resentment and agreed to do what he could to ensure the successful publication of such an important work.[26]

Slater visited Wallace in late June or early July; in August, he informed Wallace that he had found the missing journal. Relieved, Wallace declared himself ready to make a proposal to Macmillan, with which he had had a rapprochement. He was appalled, however, when the "journal" arrived a few days later, and he demanded to know how Slater could have thought "that the *copies* of *letters* to Mr. John Teasdale [one of Spruce's "botanizing" friends] are the missing 'Journal.'" The offending letters, which were long and gossipy, merely alluded to the journal, which Wallace insisted must therefore exist somewhere. Moreover, Spruce had drawn up an index of all nonbotanical subjects referred to in the "Journal" and in a "Little Square Book," the former index comprising 194 pages and the latter 187, indicating that both were substantial volumes. He urged Slater to carry on the search. "Until these *two important books* are found," he informed Slater, "nothing can be done in regard to making arrangements for publication." After once more looking through the material Slater sent him, Wallace could find nothing describing Spruce's eighteen-month residence in the town of Tarapoto, in Peru, or any references to the period from July 1857 until his departure for England in May 1864. "It is evident therefore that you must still have a large quantity of 'Journals'—either in separate small books, or bound together," he wrote to Slater on August 26. "As it is these 'Journals' *alone* that are of any use for giving an account of his whole Travels, please make *another* search for them."[27]

Two weeks later, Slater finally sent what appeared to be a missing journal, but he was once again mistaken. Wallace told him to keep looking. "It is most unfortunate you did not, at the time of Spruce's death, put all his MSS into *boxes* by themselves, so that they could be had when wanted," he complained. He had only half of what he needed to complete the project, and he fairly exploded into italics: "It is now *six months* since I offered to do what I could to get them published, and the *essential portions*, without which all the rest are useless, are still not forthcoming! Cannot you get a strong woman to help your daughter make a *thorough search* from *attic* to *cellar*, in *cupboards, boxes and bundles*, and *everywhere else*, till they are found?"[28]

Slater was ready to give up, though Wallace was not. He held out hope that Slater would eventually find the crucial journals. After another futile search, Slater told Wallace that the essential material was nowhere to be found. As a last resort, Wallace wrote to Markham to ask if he knew anything about the mislaid papers. Markham knew nothing and said that their loss was "most deplorable." Slater apologized to Wallace, insisting that these papers had never been in his possession.[29] Disappointed, Wallace nevertheless sent an outline of his proposed book to Macmillan, along with endorsements from Hooker and Markham. The outline promised an 80,000-word manuscript covering Spruce's explorations of the Lower Amazon, with a "very interesting" chapter on botany and geology and a discussion of the warlike women of the Amazon. There was also to be an account of Spruce's voyages on the Rio Negro and the Uaupés, Casiquiare, and Orinoco Rivers, some 120,000 words in length. For the years 1855 to 1857, he had a large number of Spruce's notes and letters to Sir William Hooker and George Bentham at Kew, as well as an 8,000-word narrative of Spruce's travels from Tarapoto to the town of Baños in Ecuador. Unfortunately, for the final five years of Spruce's South American travels, Wallace had only thirty four-page letters to Hooker and Bentham. Wallace advised Macmillan that Spruce was a naturalist of the highest rank and that as one of Spruce's most intimate friends he was willing to act as editor of the work, though it would involve much labor and some expense in the copying and typesetting of letters and journals. Slater would also require some share in the work, he said. He asked Macmillan for the same royalty arrangement as for the other books he had published with the firm.[30]

Although expecting little profit, Macmillan agreed at once to Wallace's conditions and generously offered to advance him £50, nearly half of which Wallace proposed to split with Slater, reserving for himself whatever might be earned from royalties—or, if Slater preferred, Wallace would agree to give him one-fifth of the royalties and £5 from the advance payment. "Even at the best I do not expect [the advance] to pay me for the time and work I must give to [the book]," he told Slater "but I greatly wish that Spruce's work should be made public."[31] In the same letter, he wrote excitedly about a new clue to the whereabouts of the missing journals. A close friend had consulted a clairvoyant who, after examining a sample of Spruce's handwriting, described a house with a Gothic porch and circular drive standing in wooded grounds. On the lower floor, entered by way of a stone corridor, he saw a large cupboard filled with bundles of old manuscripts and newspapers; but he could determine if the papers bore any connection to Spruce only if he had a lock of Spruce's hair. Wallace wished to know whether Slater recognized the house. Did it perhaps belong to one of Spruce's close friends, like Teasdale? "As a matter of curiosity I shall be glad to know if *that house* in any way corresponds to the description I sent you. I suppose you do not happen to have kept a lock of Spruce's hair? If so please send me a fragment."[32] Slater's response has not been preserved, but the answer was apparently "no" on all counts.

Wallace continued the laborious task of compiling the book, which involved copying out by hand Spruce's journal and the various notes and letters; some of this he did himself, and others were transcribed by Slater and his daughter. Wallace could have saved himself trouble by visiting Slater and personally selecting the material he needed, as Slater had first suggested, but he opted for the more cumbersome procedure of having Slater send items one at a time, a request with which Slater dutifully complied. By February 1907, Wallace had gathered and arranged enough material to fill two volumes, bridging gaps in the narrative with his own editorial comments.[33] He also had obtained from Teasdale a series of crude paintings by a native artist, which Spruce had collected and saved. After consulting Annie and his children, Wallace decided that these paintings were highly characteristic works of primitive art and worth printing in the text, but Macmillan refused on the grounds of expense. Wallace wrote to Teasdale to ask if any of Spruce's friends in York might be willing to bear the cost of the reproduction of the paintings "as a kind of tribute to the memory of one they highly esteemed."[34] The request annoyed Teasdale, who was prompted to complain to Slater, "You will see he throws out a suggestion as to money being raised. . . . I may say at once that the matter is not of sufficient interest to me to induce me to contribute more than a trifle to such an object."[35]

Wallace took a break from this tedious work to testify in London at the trial of the famous magician John Neville Maskelyne, who had been sued for libel by the Reverend Thomas Colley. Colley, an Anglican minister and an outspoken proponent of spiritualism, had offered £1,000 to anyone who could prove that a certain phenomenon he had witnessed at a séance thirty years earlier, in 1876, was the result of trickery. The phenomenon, described in a pamphlet he distributed, consisted of the materialization of a figure from the body of a medium and Nonconformist minister named Francis Ward Monck during a trance state. Wallace had mentioned Monck and this phenomenon in *My Life.* In the late 1870s, in order to study Monck's production carefully and preserve him from harm related to repeated séances, four men, including Darwin's cousin Hensleigh Wedgwood, had secured the medium's exclusive services for an entire year. Wedgwood had invited Wallace to see the phenomena associated with Monck, and one bright Sunday afternoon, in full daylight, Wallace witnessed what was certainly one of the most spectacular sights he had ever seen. After Monck, who was dressed in clerical black, went into a trance, he suddenly stood up and pointed to his left breast. From the left side of Monck's coat emanated a faint white patch, which brightened, seemed to flicker, and extended upward and downward in a "vaporous filament" until it formed a cloudy pillar from Monck's shoulder to his feet while remaining attached to his body. Monck then moved to one side and passed his hand through the connecting band, severing the apparition from his body. A thickly draped female figure gradually emerged. Monck turned toward the figure, said "Look," and clapped his hands. The figure clapped her hands in response, producing a faintly audible clapping sound. She then moved slowly back to him, grew fainter and shorter, and was apparently absorbed into his body at the same point from which she had emerged.[36] Colley, for his part, allegedly had seen an Egyptian "mahedi" who walked around the room, wrote messages, levitated Colley, and then was reabsorbed into Monck's body just as Colley attempted to embrace it and found Monck in his arms instead. The psychic researcher Edmund Podmore, coeditor of *Phantasms of the Living*, a ten-volume series documenting all forms of psychic phenomena, called Colley's impressions a hallucination, a charge that Colley emphatically denied.[37]

Monck was later imprisoned for three months as a "rogue and vagabond" after authorities found a "conjuring apparatus"—consisting of nothing more than a pair of "stuffed gloves"—in his room. This was not the first time that Monck's honesty had been questioned. William Barrett caught Monck engaging in a gross bit of fraud during a séance when he discovered that the medium had draped a piece of white muslin on a wire frame with a black thread attached to it, which he used to simulate a materialized spirit.[38] Having attended numerous séances with Monck under conditions that convinced him that imposture was unlikely, Colley proclaimed Monck's innocence, most recently at the Anglican Church's 1905 congress.[39] When Maskelyne read Colley's pamphlet, he jumped at the opportunity to prove fraud. For forty years, he had remained an unflinching critic of spiritualists, believing them all to be fools preyed on by rascals. He continued to pack his performance hall with spectators eager to see his unrelenting exposures of mediumistic tricks. He accepted Colley's challenge, and on October 1, 1906, he undertook to reproduce the "psychic parturition," as he called it, having advertised his intention in a pamphlet of his own. Colley, who was offered a front seat at the performance, chose to remain outdoors in the rain, distributing more copies of his own pamphlet. Onstage, Maskelyne reenacted the materialization to a remarkably successful degree, though he did not attempt reabsorption; as he would later explain to the court, "[T]he public would not [have been able to] stand the whole thing."[40]

Colley's charge of libel was based on three or four sentences in Maskelyne's pamphlet that accused him of having misrepresented himself as an archdeacon when he went to South Africa in the late 1870s to work with John William Colenso, the bishop of Natal. According to Maskelyne, Colley had had no right to call himself an archdeacon, not having received a degree from a legitimate institution, and had represented himself falsely without his archbishop's approval. Colley vehemently denied the accusations and sought to set the record straight, whereas Maskelyne's lawyers sensationalized the case as one more rebuttal to the spiritualists.[41]

At the trial, which lasted less than a week, Wallace volunteered as a witness and, in a curious case of déjà vu, came face to face with his unrepentant nemesis Maskelyne. Introduced as "Dr. Alfred Russel Wallace, F.R.S., D.C.L., L.L.D.," he had come to London not to defend Colley's reputation but those of Monck and spiritualism. Although he said that he had never met Colley until the day of the trial, he had met Monck in 1877 or 1878. At séances in a small house in Bloomsbury, he had observed exactly the same phenomena described by Colley and was as convinced as Colley of their genuineness. He had also attended Maskelyne's recent performance, and he pronounced the legerdemain "perfectly ludicrous." The conditions were entirely different, he said, the one on a lighted stage, the other in a small room in broad daylight. Maskelyne had failed to reproduce the phenomenon in its entirety. No white patch had emerged from Maskelyne's breast and grown before the audience's eyes, and at the end the "materialized" young lady walked off the stage instead of being gradually reabsorbed into Maskelyne's body. "It was an absurd travesty," Wallace testified, "while the other was a most marvelous sight to see." During cross-examination, Wallace denied that all mediums were eventually exposed as charlatans. He had rarely heard of true exposures and, contrary to Maskelyne's testimony, had never personally met a medium who was a scoundrel. Indeed, the magician's panoply of tricks failed to impress the jury. Nor were they amused by a frivolous attack on a member of the clergy—even if he was a spiritualist. They ruled in Colley's favor, fining Maskelyne £75 for libel and denying him the £1,000 reward. Thus ended Wallace's last major public contribution to the spiritualist cause.[42]

Wallace also interrupted his work on Spruce's papers to write a short book called *Is Mars Habitable?*—a response to Percival Lowell's 1906 book *Mars and Its Canals.* As a result of observations through his telescope at his observatory in Flagstaff, Arizona, Lowell, an amateur astronomer, had become convinced that the intricate lines on the surface of Mars were the artifacts of intelligent beings. The receding and advancing of the Martian polar caps led Lowell to affirm the presence of water, which was channeled in summer to irrigate the equatorial regions and was sufficient to supply the needs of a complex civilization. His book caused a minor sensation, attracting the attention of almost every magazine and newspaper in the United States and Great Britain. Since no reputable astronomer or physicist refuted Lowell's thesis, Wallace took it upon himself to give a careful, thorough, and "popular" exposition of the facts disproving Mars's capability of supporting advanced life, expanding on his arguments in *Man's Place in the Universe.* Macmillan agreed to publish the work—a somewhat unusual decision, since the firm had published Lowell's book in Great Britain, but perhaps it was eager to keep Wallace from straying to other publishing houses. It bowed to his request ("the *one* thing I ask") to publish this brief manuscript in a volume of the same size as his other books, and left the terms of his royalties unchanged.[43]

Although paying homage to Lowell's "admirable work," which had earned him the just praise of the scientific community, Wallace disagreed with his interpretation of his observations. Contrary to Lowell's assertions, he said, spectroscopic analysis did not confirm the presence of water vapor in the Martian atmosphere. It also had been proved that aqueous vapor could not exist on a planet whose mass was less than one-quarter that of earth. Stymied by Lowell's complex mathematical formulas demonstrating that Mars was as warm and habitable as earth even though it received less than half as much of the sun's heat as earth did, Wallace still believed that Lowell was wrong. For three months, he puzzled out a response. He consulted a number of physicists and mathematicians, including George Darwin, but no one agreed to search for errors and fallacies in Lowell's mathematics. "However I think I have done it myself by the rules of common sense," Wallace wrote confidently to his friend Fred Birch, a young naturalist preparing for an entomological exploration of South America whom Wallace was mentoring at the time.[44] In the latter half of the book, Wallace proposed his own theory of the formation of Mars from the aggregation of meteorites into a "cold mass," whose surface had been heated and liquefied by the further impact of more meteorites and then had cracked during the cooling process between impacts. Further cracking and crumbling of these fissures over long periods of time had led to what appeared to be canals and oases.[45] Although his theory could not be proved at the time, he attempted to replace speculation with an explanation based on known scientific principles. Underlying Wallace's philosophical premise was an idiosyncratic view of the origin of the universe—one might call it teleological evolution, the belief that a controlling Mind or Intelligence manipulated natural laws for distinct ends. It was this belief, more than anything else, that had made so many of his scientific contemporaries uncomfortable. Essential to his theory was his anthropocentrism—a belief in humanity's unique position in the universe. Nothing, barring the discovery of intelligent life elsewhere in the universe, could change his mind on that issue.

With Mars out of his system, Wallace returned to the Spruce book. By the end of 1907, he had nearly finished his manuscript. He was fastidious about every aspect of the book: he knew what he wanted, showed no hesitation in imposing his wishes on others, and was quick to express displeasure when his orders were not followed. In his obsession for perfection, he was tyrannical. His endless lists must have tried the patience of Slater and his editors alike. Age had made him more rigid and entitled but had not weakened the drive and determination of his youth. Wallace acted as his own ruthless editor, cutting his manuscript in half after having pored over it four or five times—and still the book came to two volumes. But when it was finished, he felt confident that it would appeal to "any intelligent person" and be of lasting interest to botanists and lovers of nature. He sent Macmillan a list of picturesque and highly descriptive chapter headings. He also included every one of Spruce's drawings that he felt was worth reproducing and old photographs of Ecuadorian and Amazonian scenery that, "if reduced to page size, will add greatly to the value of the book—which will be one of *permanent interest*." He had an additional recommendation: "May I suggest that in deciding on the size and form of the book, you keep in view the possibility of a cheap *one vol* edition—which I am inclined to think would have a permanent sale as a prize and gift book among 'classical' travels [*sic*]."[46] Wallace left no doubt that he managed the orchestra: he was composer, performer, and conductor. Macmillan was left to underwrite the concert.

The mapmaker, who displayed special incompetence by ignoring all of Wallace's detailed notes, drove him to the point of madness. Important towns or landmarks were omitted; roads were erased; imaginary rivers traversed imaginary valleys; cities and mountains were put in the wrong places; fictitious rivers drained into the Pacific. Arbitrarily, the cartographer put in the longitude of Quito as "o." "What possible use, but to confuse, is this on an English map?" Wallace complained. Consulting atlases and references that should have been at the fingertips of any professional cartographer, he missed nothing. Time and time again, he returned his most important map for revisions. "I have been more worried by this map than by all the rest of the book," he confessed. "Any fourth form schoolboy would be ashamed to send in such a map—and it is sent to me by a *Map-maker* with a *well-known* name for my *approval*!"[47]

There were other, more minor problems. Dissatisfied with the weight of the paper that Macmillan was using, Wallace wanted something lighter that would feel more pleasant to the touch. He also asked that the botanical passages, of interest only to the specialist, be offset from the rest of the text by smaller print.[48] Will, who worried about the stress of all this exasperating work, advised his father to draw up a statement detailing the financial arrangements of the book following publication in order to save the respective families time and trouble in the event of his death. It was a request that Wallace did not find strange or self-serving and to which he readily agreed.[49]

It is easy to forget that the person who edited *Notes of a Botanist on the Amazon & Andes*, which Macmillan published in 1908, was an octogenarian. Despite physical infirmities, he still possessed the energy and intellect to condense, organize, and critically assess a great mass of disparate material. What he produced is of enduring importance to botanists and to the history of science, and it remains a remarkable homage to the memory of a man whose extraordinary work might otherwise have been forgotten. As Mark R. D. Seaward, a noted British botanist, observed some ninety years later (partly in gratitude to Wallace's herculean efforts), "[N]o botanist studying tropical plants and their use can afford not to consult Spruce's specimens and detailed notes. Their value for those working in the fields of phytochemistry, ethnotoxicity, hallucinogens, narcotics and economic botany is immense."[50]

CHAPTER 20

A National Treasure Celebrated

WHEN ASKED WHY he was chosen by the president of the United States, William Jefferson (Bill) Clinton, as a recipient of the 1997 National Humanities Medal, Studs Terkel, historian of the American common man and woman, replied, "If you hang around long enough, anything is possible."[1] Wallace would no doubt have agreed with Terkel's sentiments. By 1908, when he was eighty-five years old, he had hung around long enough to become a national treasure. He had outlived most of his peers and critics. His heresies were now viewed as eccentricities, and his countrymen hailed him as a great man, awarding him the highest honors he could have received short of a knighthood. On July 1, 1908, he received the first Darwin–Wallace Medal from the Linnean Society in commemoration of the joint publication of their revolutionary papers fifty years earlier. In the first week of November, he was awarded both the Copley Medal and the Order of Merit.

The Darwin–Wallace Medal had been specially struck for the occasion. On each side of the gold medal were embossed the august bearded busts of Darwin and Wallace, the former in profile to show off the craggy brow overshadowing deeply recessed, brooding eyes; the latter in three-quarter view, with a less forbidding aspect. At the celebration, reported in the press throughout Europe, the United States, and the British Empire, silver copies were presented to Sir Joseph Hooker, Ernst Haeckel, Edouard Strasburger, August Weismann, Francis Galton, and Sir Edwin Ray Lankester—the six greatest living exponents of Darwinism in England and on the Continent. Wallace made a rare public appearance and gave an eloquent speech, typically downplaying the significance of his role in spearheading one of the greatest intellectual revolutions of all time. As he recalled:

> The idea came to me, as it had come to Darwin, in a sudden flash of insight: it was thought out in a few hours—with such a sketch of its various applications and developments as occurred to me at the moment—then copied on thin letterhead and sent off to Darwin—all

> within one week. Such being the actual facts . . . I should have had no cause for complaint if the respective shares of Darwin and myself in regard to the elucidation of nature's method of organic development had been thenceforth estimated as being, roughly, proportional to the time we had each bestowed upon it when it was thus first given to the world—that is to say, as 20 years' work is to one week. For, had he already made it his theory, after 10 years' work—15 years'—or even 18 years' elaboration of it—*I* should have had no part in it whatever, and *he* would have been at once recognised, and should ever be recognised, as the sole and undisputed discoverer and patient investigator of the great law of "Natural Selection."[2]

The award of the Copley Medal—"the ancient olive crown of the Royal Society," as Sir Humphrey Davy, president of the society in the 1820s, called it—came as no surprise to anyone except Wallace, who now joined the ranks of Michael Faraday, Charles Darwin, and the handful of others who had received both the Royal and Copley Medals for their distinguished contributions to science.[3] But the Order of Merit was completely unexpected. King Edward VII had established the Order of Merit in 1902, one year after acceding to the throne, for exceptionally meritorious service in either the military or the arts and sciences, limiting its award to twenty-four people annually, each appointed by the sovereign himself.[4] One of the first recipients of the Order of Merit was Florence Nightingale. Herbert Asquith, who became prime minister in 1908, had suggested Wallace's name. Asquith headed a Liberal government whose most significant achievement was social reform on a scale unprecedented in English history. Although Asquith characterized himself as a moderate, others in his cabinet were more radical. When the Liberals came to power in 1905, Wallace was delighted. "The most Radical Government yet known in England . . . gives me a new interest in life and the hope to live a few years longer to see what comes of it," he wrote to Raphael Meldola.[5] Old-age pensions, workers' compensation, unemployment and health insurance, eight-hour day for miners, greater taxation of the wealthy—in short, the enactment of several items on the socialist agenda—were among the many new laws passed during the party's ten-year administration.[6]

Asquith thought the selection of Wallace would be a "graceful and generally appreciated mark of [His] Majesty's interest in the progress of Natural Science." If the king had any doubts, Asquith wrote, "[His] Majesty might desire to be fortified by the highest scientific opinion, in which case [I] would . . . communicate confidentially with Lord Rayleigh, the President of the Royal Society."[7] The king knew little of Wallace or his radical political views; otherwise, he might have had misgivings. Rayleigh's and Asquith's unqualified endorsements seemed to be sufficient. Although deeply moved by the plight of the impoverished, the king was content with the status quo and instinctively distrusted the socialists, despite their assurances of loyalty to the Crown. His reign was marked by the striking contrast of regal splendor and high living with the squalor and discontent of the urban and rural poor, whose clamor for social change would reach a climax in the aftermath of World War I.[8] Conservatives in the government were shocked at the selection of an avowed socialist for the Order of Merit, but no one was more astonished than Wallace himself. For once the stars were aligned in Wallace's favor, with the golden jubilee of the discovery of natural selection, the advent of a radical government, and a spiritualist sympathizer as the head of Britain's highest scientific body. Thus Wallace came to be honored with two of his country's most prestigious awards for private citizens.

Lord Knolly, the king's private secretary, requested Wallace's presence for investiture at Buckingham Palace on December 14, but Wallace dreaded the pomp and circumstance of such a formal ceremony. At the time, he was conveniently suffering from some sort of illness, which made him hesitant to leave home for any reason—even to meet the king. He composed a polite letter of thanks, but begged to be excused from attending the ceremony on the grounds of age and delicacy of health. Privately he confided to Fred Birch that he was deterred by the expense of the court dress, a "kind of very costly livery," which was obligatory for the occasion.[9] To his great relief, after a few anxious weeks the king granted his request and sent one of his equerries, a Colonel Legge, to Broadstone with the Order of Merit in hand. Wallace was pleased with the medallion, which he described to Birch glowingly as "a very handsome cross in red and blue enamel and gold—rich colours—with a crown above, and a rich ribbed-silk blue and crimson riband to hang around the neck." The personable colonel, who regaled the elderly couple with royal gossip, was treated to tea and showed Wallace how to wear the medal on public occasions. He also presented Wallace with two official documents signed by the king, the first appointing him a member of the Order of Merit, and the second certifying by royal decree that he had a right to the order though not personally invested by the king. Not long afterward, a letter from Windsor Castle informed him that a chalk portrait was to be taken for the library's collections. William Strang, the artist, who was also a renowned etcher, engraver, and illustrator, spent the night at Wallace's house and within four hours completed "a very good" life-size head in colored chalk.[10] Despite his usual dislike of attention, Wallace could not completely suppress his pride; he allowed Macmillan to append "O.M." before the other titles after his name on the frontispiece of his latest book.[11]

Wallace received congratulatory letters from around the world. In distant Singapore, the awards did not escape the notice of Mrs. A. J. Bidwell, a daughter of Charles Allen, his assistant in the Malay Archipelago, who had died in 1899 or 1900. "Although not personally known to you," Bidwell's husband wrote, "my wife feels that she knows you, as her father was never tired of telling his children of the happy times he spent with you. There is a feeling in the Straits that you do to a certain extent belong to us from the long time you spent here and your writings."[12]

In late October, Wallace received a letter from Sir William Crookes inviting him to give a lecture at the Royal Institution in January 1909 to celebrate the golden jubilee of the publication of *The Origin of Species*. Located at 21 Albemarle Street, in a town house that had been converted in 1799 into a scientific institute with ample lab space, a lecture hall, libraries, and offices, the Royal Institution had quickly become one of the world's premier research organizations and scientific think tanks. It was there that Sir Humphry Davy isolated chlorine and iodine and invented his famous miner's lamp (Davy lamp), Michael Faraday carried out his groundbreaking experiments in chemistry and physics, and Lord Rayleigh discovered argon. The Friday Evening Discourses, consisting of formal lectures given to members and their guests on the latest practical or theoretical scientific discoveries, attracted serious scientists and fashionable socialites alike, owing to the high quality of the lecturers and the novelty of the subject matter. It was a great honor to be invited to speak, yet Wallace's first inclination was to decline, having nothing new to add to this well-worn subject. But an idea suddenly came to him, containing new arguments to quash the Mendelians. "I think I can put Darwinism in a new light, so as to leave 'Mutationism' etc etc *nowhere*!" he boasted to Meldola. "Not a line written yet but subject sketched out."[13] Wallace replied to Crookes with his provisional idea, to which Crookes immediately consented. He also offered to read all or part of the lecture if Wallace did not feel that his voice would hold up under the strain. By mid-January, however, Wallace had regained most of his strength and was able to read his own text himself. His discourse, which he entitled "The World of Life: As Visualised and Interpreted by Darwinism," was held on January 22.

The Royal Institution was "packed to the roof," the mathematician Karl Pearson, a protégé (and, later, first biographer) of Francis Galton, wrote to his mentor, who could not attend the lecture. Wallace was quite audible "but not very original," Pearson noted, lacking the vivacity necessary to hold the attention of his audience. Nevertheless, it was "really worth" being present to hear Wallace speak.[14]

The natural world, not the artificial one of the laboratory, was Wallace's great subject. Darwinism was misconstrued because non-naturalists failed to appreciate the vast numbers, the great variety, and the incessant intermingling of earth's "life-forms." From measurements of hundreds of specimens, it could be shown that the individuals of a species varied around a mean, which meant there was plenty of material for natural selection to act on. If it were not for the power of natural selection, the planet would be overrun by a handful of species. "It is the fact of the adaptation of almost all existing species to a continually fluctuating environment—fluctuating between periodical extremes of great severity—that has produced an amount of adaptation that in ordinary seasons is superficially complete," he said. Mutationism was a minor force in his conception of evolution. Summing up, he declared,

> It is only by continually keeping in our minds all the facts of nature which I have endeavoured, however imperfectly, to set before you that we can possibly realise and comprehend the great problems presented by the "World of Life"—its persistence in ever-changing but unchecked development throughout the geological ages, the exact adaptations of every species to its actual environment both inorganic and organic, and the exquisite forms of beauty and harmony in flower and fruit, in mammal and bird, in mollusk and in the infinitude of the insect-tribes; all of which have been brought into existence through the unknown but supremely marvelous powers of Life, in strict relation to that great law of Usefulness, which constitutes the fundamental principle of Darwinism.[15]

The success of Wallace's lecture prompted him to contemplate another book, which he would call *The World of Life: A Manifestation of Creative*

Power, Directive Mind, and Ultimate Purpose. While researching the lecture, he had come on Lord Salisbury's 1894 address to the British Association for the Advancement of Science. Although disagreeing with Salisbury's remarks on natural selection, he was intrigued by the overarching teleological theme. He had avoided teleology in his own lecture, but to Arabella Fisher he confided, "I am becoming more and more impressed with . . . a teleology of fundamental laws and forces rendering development of the infinity of life-forms possible (and certain) in place of the old teleology applied to the production of each species."[16]

As he approached his ninetieth year, Wallace was also becoming more mystical. This mysticism had crept into most of his later writings and finally was expressed as a grand philosophy in *The World of Life*, which was published by Chapman & Hall in December 1910 and went through five editions. Just as he now viewed earth as a single living organism, with its infinite variety of species adapting to an ever-changing climate and geography, so he viewed the universe as a single organism, with its infinity of stars surrounding a unique planet revolving around its unique sun, each of these elements inseparable from the others. What he once speculated about he now firmly believed: *Homo sapiens* was placed on earth for a reason. Wallace's thinking was undoubtedly influenced by the writings of Emanuel Swedenborg, whose life and works were intimately connected with the spiritualist movement. Wallace's spiritualist library, bequeathed originally to Oxford University but since transferred to the University of Edinburgh, contains several works by Swedenborg. Wallace never publicly acknowledged this influence, but the concluding chapter of *The World of Life* is almost incomprehensible without some knowledge of Swedenborg's mystical, intellectual, and utopian writings.

Born in Sweden in 1688, Swedenborg achieved fame as a mining engineer, astronomer, physicist, zoologist, and anatomist. In midlife, after a transforming visionary experience, he abandoned his scientific pursuits and cultivated his psychic powers, devoting himself to theology and philosophy. Spiritualists considered him the first modern medium, antedating by nearly a century Katie Fox, the American girl whose sensational trances in the late 1840s precipitated the nineteenth-century spiritualist movement.[17] Philosophers like Ralph Waldo Emerson cited him as a classic example of the mystic; Swedenborg's ideas inspired the transcendental movement in the United States.[18] Swedenborg's most extraordinary claim was traveling clairvoyance, the ability of the spirit to leave the body temporarily and witness events or phenomena before they occur, a power that first erupted in his youth but did not fully blossom until middle age. For twenty-seven years, from 1744 until his death in London in 1772, he was in constant touch with the otherworld. This world was inhabited by spirits of the dead with whom he claimed he could converse, and he made it his mission to reveal heaven and hell as he saw them during his extracorporeal journeys. However, his journeys were as much an exploration of the inner world as of the spiritual world. They formed the centerpiece of his philosophical writings, which have appealed to writers as diverse as the English poet William Blake, the Russian novelist Fyodor Dostoyevsky, and the master Japanese Zen scholar D. T. Suzuki.[19] In his travels, Swedenborg found limitless diversity in heaven, full of angels and spirits in infinite gradations, whose relative positions in the universe were determined by the degree of goodness, love, and wisdom they had exhibited during life on earth. Invisible to humans through the ordinary senses, angels were seen only through the eyes of the spirit within. When a person's "spirit sight" was opened, Swedenborg said, the angels and the countless communities they lived in materialized. Journeys in the "World of Spirit," as he termed it, were simply changes in state, not governed by time and space as normally conceived.[20] Living in the antiecclesiastical times of the Enlightenment, Swedenborg believed that Christ, the one true God, had entrusted him to bring fallen Christians back into the fold with a new, all-embracing (but nondogmatic) religion that claimed direct mystical communications between the material and the spiritual world.

It was perhaps Swedenborg's explorations of the spirit world rather than the specifics of his theology that appealed to Wallace. Intellectually Wallace had come full circle: from the mechanistic to the teleologic. But his teleology was like that of no other previous scientist or philosopher: all things on earth—animal, plant, and mineral—all things in the stellar universe, were coadapted. An unknowable Mind or controlling Intelligence guided that coadaptation. With sudden insight, Wallace grasped the nature of the First Cause. This vision, elaborately expressed in *The World of Life*, was a union of Swedenborgian mysticism and modern science: the vast whole of the universe reflected the Supreme Being, but the mechanics were executed by descending grades of intelligences and powers. He imagined an infinite number of angels working on every level: some guided the creation of the stars; some guided the formation of the earth and the other planets; some guided the development of life on earth; some guided the workings of the cell. Their tools were the forces that scientists were only just beginning to understand: gravitation; the velocity of light; the molecular behavior of gases; variation; heredity; the survival of the fittest; and the orderly process of cell division, differentiation, and growth. These "angels"—for want of a better term—provided the initial impetus, making certain that the universe developed along certain lines while adhering to strict laws. But that was not all. The production of the infinity of life-forms was beautifully coordinated for the ultimate development and education of an equally varied humanity. All gradations of human beings on the spectrum of good and evil had had a chance to survive and produce offspring. In the spirit world, death did not cut short the period of educational advancement; after death, in fact, individuals encountered the best conditions and opportunities to improve their spirit and thus their heavenly status.[21]

Wallace's philosophy was positivistic. Is nature cruel, he asked. No, but many of the leading philosophers of present and past generations had proclaimed the cruelty of life on earth. In the penultimate chapter of his book, Wallace noted:

> Huxley . . . spoke of the myriads of generations of herbivorous animals which "have been tormented and devoured by carnivores"; of the carnivores and herbivores alike as being "subject to all the miseries incidental to old age, disease, and over-multiplication"; and of the "more or less enduring suffering" which is the meed of both vanquished and victor; and he concludes that since thousands of times a minute, were our ears sharp enough, we should hear sighs and groans of pain like those heard by Dante at the gate of hell, the world cannot be governed by benevolence.

Thomas Huxley's conception was one of the strangest phenomena of the human mind, he said. He could not understand why so many intelligent people were more attracted to the belief in eternal suffering as promulgated by revealed religions than to the belief in a limited period of suffering on earth, to be replaced by the development of a race of spiritual beings who lived in eternity, free of the miseries of a mortal existence.[22]

Wallace believed that the vision of a world of tortured millions was nothing more than the projection of human sensations of pain onto the animal world: "The probability is, that there is as great a gap between man and the lower animals in sensitiveness to pain as there is in their intellectual and moral faculties." Most animals, from the lowest paramecia to the schools of fish and the herds of wildebeest, existed to be devoured. The vicious teeth and dagger-like claws of the cat family had evolved for the purpose of preventing the escape of captured food, not for the shedding of blood or the infliction of pain. The evidence that an animal devoured by a lion or puma suffered very little was conclusive: "The suddenness and violence of the seizure, the blow of the paw, the simultaneous deep wounds by teeth and claws, either cause death at once, or so paralyse the nervous system that no pain is felt till death very rapidly follows."[23] In a state of nature, the Carnivora hunt and kill to satisfy hunger, not for amusement; and "all conclusions derived from the house-fed cat and mouse are fallacious." If every evolutionist considered the utility of pain, he would see that animals had only as much sensitivity as was necessary for preserving the young from common dangers to life and limb before reaching maturity. Beyond that, animals were spared unnecessary suffering. Their relative lack of sensitivity, however, did not justify vivisection:

> The bad effect on the operator and on the students and spectators remains; the undoubted fact that the practice tends to produce a callousness and a passion for experiment, which leads to unauthorised experiments in hospitals on unprotected patients, remains; the horrible callousness of binding sufferers in the operating trough, after the experiment, by careless attendants, brutalised by custom, remains; the argument of the uselessness of a large proportion of the experiments, repeated again and again on scores and hundreds of animals, to confirm or refute the work of other vivisectors, remains; and, finally, the iniquity of its use to demonstrate already-established facts to physiological students in hundreds of colleges and schools all over the world, remains.[24]

Human beings were much more sensitive to pain, because of their bare skin; they had no protective armor or thick coat of hair with which to ward off blows. "And here I think I see the solution of a problem which has long puzzled me," Wallace said. "[W]hy man lost his hairy covering, especially from his back, where it would be so useful in carrying off rain. He may have lost it, gradually, from the time when he first became Man—the spiritual being, the 'living soul' in corporeal body, in order to render him more sensitive." After the invention of fire, he noted, our species exposed itself to thousands of self-made dangers. Those less sensitive to these dangers perished; those more sensitive survived. Hence sensitivity to pain developed through the agency of natural selection, but under the direction of a higher intelligence.[25]

To no one's surprise—including Wallace's—the concluding portion of his book elicited the most grumbles from the scientific community. "[It] is unfortunate that the author is never able to avoid the pitfalls of teleological speculation," the anonymous critic for *Nature* wrote.

> This tendency is still more strikingly manifested when the author proceeds to discuss such questions as the existence of pain in the lower animals, of the non-justifiability of vivisection. . . . On all these and similar questions Dr. Wallace writes confidently, sometimes intruding his speculative opinions in the midst of the treatment of purely scientific questions. . . . We are all familiar with the author's peculiar views on extra-scientific, social and political questions. Some of these tendencies to unbridled speculation seem to have reached an extreme limit in the twilight of a noble life. . . . But however much we may regret the intrusions by the author of these wild speculations . . . we recognise that they are inspired by the author's love of humanity and all living things, by a desire to ameliorate the sorrows and sufferings he sees around him, and by a hope—ill-founded though it may be—that such teachings may be of service to his fellow-men.[26]

Wallace complained to his friend William Thistleton-Dyer about being misunderstood. He recognized the action of natural selection as universal and capable of explaining "all the facts" of evolution from "amoeba to man." But natural selection had nothing to do with the basic mysteries of life. From the few simple elements available in air, earth, and water, innumerable structures, like bone, hair, and blood—"mere lumps of dead matter"—all organized to serve a definite purpose, were built up somehow to create life within every organism. No biologist or physiologist dared to grapple with the mysteries of the origins of life. "One and all they shirk it," Wallace said, "or simply state it to be insoluble. It is *here* that I state guidance and organising power are essential."[27] Thistleton-Dyer tried in vain to point out Wallace's philosophical inconsistencies. "[If] we admit that [Darwinism] is scientific," he replied, "then we are precluded from admitting a 'directive power.'" Science could explain nature only as it revealed itself to our consciousness. Thistleton-Dyer sympathized with anyone who sought an answer for these conundrums from nonscientific sources, but in his opinion scientific explanations and spiritual "craving" were wholly distinct and should be left that way. Alluding to the remarks of the great nineteenth-century German naturalist and embryologist Karl Ernst von Baer, he suggested that "the naturalist is not precluded from asking 'whether the totality of details leads him to a general and final basis of intentional design.' I have no objection to this, and offer it as an olive-branch which you can throw to your howling and sneering critics."[28]

But there were others, mainly nonscientists, who needed no olive branch, for they were neither howling nor sneering. Theodore Roosevelt, the new contributing editor to the *Outlook*, praised *The World of Life* and its author in an article published in the December 1911 issue, in which the former president examined the writings of several philosophers, including William James and Henri Bergson. "No one has criticized with greater incisiveness what [Wallace] properly calls 'the vague, incomprehensible, and offensive assertions of the biologists of the school of Haeckel.' He shows his scientific superiority to those men by his entire realization of the limitations of the human intelligence, by his realization of the folly of thinking that we have explained what we are simply unable to understand when we use such words as 'infinity of time' and 'infinity of space' to cover our ignorance." Roosevelt claimed (perhaps incorrectly) that Wallace, despite his advanced age, came nearer to the younger generation of scientific investigators, who were more ready to acknowledge that purely mechanistic explanations of evolution had broken down. Indeed, Roosevelt concluded, science furnished "an overwhelming argument for 'creative power, directional mind, and ultimate purpose' in the process of evolution."[29]

Wallace was a creature of habit until the very end of his life. At 6:00 A.M. he made a cup of tea on a gas stove in his bedroom, the exact quantity of tea and water having been measured out the night before and boiled for just the right amount of time the next morning. He then slept for another two hours, before calling in his gardener, who acted as valet, to help him dress. After preparing a cup of hot cocoa, he went into the study to read the newspaper or recently arrived letters. When his strength permitted, he strolled about his garden, supervising the planting of new beds of flowers; at other times, his gardener wheeled him about in a bath chair. He remained in touch with his favorite correspondents: Edward Poulton, Meldola, William Barrett, Thistleton-Dyer, and Arabella Fisher. He continued to read newspapers and fire off letters on issues that struck a nerve. When no protests were made against the development of "flying machines" and dirigibles for dropping bombs, he expressed his moral outrage to the *Daily News*. "If there ever was a time to call upon the Liberal government to dissociate itself from this proposed crime against humanity, it is now," he insisted. He also advocated open negotiations with other "civilized" nations to sign a treaty against the use of airplanes for the purposes of mass destruction, despite the assurances of military leaders that such a treaty was unnecessary.[30] When a railway strike threatened to paralyze the nation in 1911, he told Will that he had written to the two "strongmen" of the Liberal government, Asquith and David Lloyd George, recommending the immediate nationalization of the railways by royal proclamation, "on the ground of mismanagement for seventy years, and having brought the country to the verge of starvation and civil war."[31] The discovery in 1912 at Piltdown Common in Sussex of a skull said to be the long-sought missing link between man and ape caused a sensation but did not impress Wallace, though Lankester was completely taken in by the hoax. "The Piltdown skull does not prove much, if anything!" Wallace told Will.[32] He had never believed in a single missing link. In *The World of Life*, he had written that there was not, "as often assumed, one 'missing link' to be discovered but at least a score such links, adequately to fill the gap between man and apes; and their non-discovery is now one of the strongest proofs of the imperfection of the geological record."[33]

Even in his waning years, Wallace entertained visitors at Old Orchard. Scientists, writers, and politicians came to see him. Scientific travelers sought his advice, which he generously dispensed, or regaled him with tales of their exploits. He was accessible to anyone who might have something interesting to say, including "cranks" and "faddists" who wished to enlist his support for their causes—but he had little tolerance for "preposterous ideas" and politely pointed out the fallacies of those he disagreed with.[34]

Neither of his children had married, and both had returned to live with their parents. Violet would never marry. A chronic lung condition hindered Will from keeping a regular job, and he seems to have moved back sometime after 1911. Wallace worried about their future welfare, given that he possessed little of value beyond his house and property. The civil pension would be paid to Annie after his death; he counted on royalties to help sustain his son and daughter.[35] He therefore continued to write. By early 1913, he had finished two more books: *Social Environment and Moral Progress*, commissioned by his friend and future biographer the Reverend James Marchant, and *The Revolt of Democracy*, inspired by the recent strikes of the railway and other transport unions, miners, and London dock workers. "The condition of the workers as a whole is absolutely unbearable, is a disgrace to civilisation," he wrote in *The Revolt of Democracy*, "and fully justifies [their] most extreme demands. . . . Among the whole of the writers—whether statesmen or thinkers, capitalists or workmen—there is not one who has proposed any definite and workable plan by which the desired change of conditions will be brought about." And what demands were these workers making, he asked. All they wanted from their government were guarantees of a reasonably comfortable and civilized livelihood, a decent minimum of food and clothing, leisure and recreation, and houses fit for human beings.[36]

But that summer, a few months after completing *The Revolt of Democracy*, Wallace was convinced that he would never write another book. Severe eczema of his legs had plagued him all winter and now prevented him from walking or even standing up for any length of time. He also had developed a problem with his balance and fell down on several occasions. "Even moving about the room after books, etc., dressing and undressing, make me want to lie down and rest," he told an acquaintance.[37] Against the advice of his family, his gardener took him to a small cottage he recently had acquired, located a few miles southwest of Poole Harbour near the sea. The trip of two nights did him some good, temporarily counteracting his debilitating weakness. His mind remained lucid, and he did not lose interest in the social, spiritual, and scientific issues that had engaged him for much of his life. In October, he read W. L. Webb's *Brief Biography and Popular Account of the Unparalleled Discoveries of T. J. J. See*, which reprinted the astronomer's lectures on the origin of the universe.[38] He also was involved in a project with Marchant, who was planning a small book that would provide a popular account of Wallace's and Darwin's lives and works. Wallace believed that Marchant was the best person for the project, since he possessed "the *whole* of my books, and has read my *Darwinism seven* times!" And, indeed, the publisher John Murray already had commissioned Marchant to write the book. But Wallace sought advice from Macmillan about the issue of royalties. His contacts at Macmillan must have cringed at yet another lost opportunity; moreover, the house had not published his latest two books. Such a book as Marchant envisioned would involve a good deal of correspondence and proofreading, which at Wallace's age would place a great strain on him. "I wish to ask you . . . whether I could consistently with *fairness* and *custom* ask Mr. Murray to give me some share in the profits . . . for the sake of my family," he wrote to Macmillan in March 1913.[39] Macmillan seems not to have given an answer, for the project was eventually abandoned and never undertaken.

On Saturday, November 1, Wallace had a little burst of energy and walked around his garden unassisted. The following day, his appetite was hearty, but at 9:00 P.M., while he was at his desk reading or writing, he suddenly felt faint and shivered violently. Will sent for a Dr. Norman, who arrived an hour later and talked with Wallace for a long time in his study. Laughter could be heard at one point. "Wonderful man!" the doctor exclaimed, as he emerged from his elderly patient's office. His prognosis was grim, however: "I can do nothing for him." The next day, Wallace did not get up at the usual hour—lethargy compelled him to remain in bed. He sank into a near-comatose state, and on Friday, November 7, he passed away quietly at 9:25 A.M. without having regained consciousness.[40]

Friends suggested that Wallace be buried in Westminster Abbey beside Charles Darwin, but Annie, following her husband's wishes, preferred to bury him at the little cemetery in Broadstone, on a hill shaded by pines and cooled by breezes from the nearby sea. A monument consisting of a fossilized tree trunk from a local ancient geologic stratum and set on a base of Purbeck stone bore a modest epitaph that mirrored his modest life: "Alfred Russel Wallace, O. M. Born Jan. 8th, 1823. Died Nov 7th, 1913."

The bishop of Salisbury, Frederick Edward Ridgeway, conducted the funeral at the Dorset parish church on Monday, November 10. Compared with Darwin's funeral, Wallace's was a simple affair. Will, Violet, and some of Wallace's closest friends—including Raphael Meldola and Edward Poulton, representing the Royal Society; Henry Dukinfield Scott, representing the Linnean Society; and Joseph Hyder, representing the Land Nationalisation Society—attended the unpretentious ceremony. There were no more than a dozen formal wreaths, and only one from an organization (the Spiritualists' Society of Bournemouth sent a wreath of Madonna lilies and white chrysanthemums). The Bournemouth spiritualists also sent a spray of white roses given "in token of love for a very dear friend, whose passing is a loss to the cause of humanity, to which he gave his brilliant talents."[41] Annie, who was ill, remained at home; she would die a year later, in December 1914.

Numerous obituaries appeared in prominent publications. The *Lancet*, long a foe, stated that while some of his writings "were not marked by the discrimination that might have been expected from so famous a man, and his attitude towards one phase of scientific medicine was unfortunate . . . it would be wrong to be blind to the value of grand receptive intelligences like his, because in one or two points the conclusion which Wallace came to appears to us to be hasty. Medicine combines with the scientific world in deploring the loss of a leader, and in acknowledging the debt of the world to a pioneer in the interpretation of the universe."[42] The *New York Times* called him "the last of the giants belonging to that wonderful group of intellectuals that included, among others, Darwin, Huxley, Spencer, Lyell, and Owen, whose daring investigations revolutionized and evolutionized the thought of the century." Another commentator for the same newspaper added, "No apology need be made for the few literary or scientific follies of the author of that great book on 'The Malay Archipelago.'"[43] The *Times* of London praised his "restless, always creative, and original intelligence." Through everything he did, its reporter said, there "ran a thread of continuity; there was to the end the same earnest search for truth, the same wonderful industry garnering facts, the same wide outlook, and the same indifference to anything which might turn him aside from that quest to meaner things."[44]

The editors of *Light*, a spiritualist newspaper, passed over his scientific achievements because "these points are fully dealt with in the various newspapers," while spiritualism, which concerned Wallace for nearly sixty years, "receives but scanty notice at the hands of the Press." "Dr. Wallace," they wrote on the eve of World War I, "was not only a scientist, he was a humanitarian. His great heart was moved with compassion because of the struggles and sufferings of the people; his sympathy was ever with reformatory enterprises, and he never feared to use his pen and his influence on behalf of unpopular causes." Although Wallace was not a medium, "he was a prophetic seer in the best sense—for with forward-looking vision he saw the coming of the day of co-operation, of brotherhood, of altruism, and of spiritual emancipation." Wallace, they noted, had declared that spiritualism had made him a better human being and had given him the key to all that seemed dark and hopeless in human life. It made him more tolerant of the weaknesses of others and greatly improved his character. "We obtain the greatest happiness ourselves by doing all we can to make those around us happy," they remarked, quoting from one of Wallace's writings on the subject of spiritualism.[45]

Not long before Wallace's death, a movement had been under way to commission his portrait for the Royal Society. Meldola, Poulton, and the manager of Union of London and Smith Bank had begun to receive subscriptions. The effort would continue after the hiring of the artist J. Seymour Lucas to paint a posthumous portrait based on old photographs. Contributors included Sir William Crookes, the geologist Sir Archibald Geikie, the physician Sir William Osler, and William E. Darwin, one of Darwin's surviving sons.[46]

Soon after his death, a committee was formed, headed by Poulton and Meldola, to have a medallion placed in Westminster Abbey to honor Wallace's memory. Among those who agreed to underwrite the cost, which came to £200, were Geikie, Crookes, Lankester, Dukinfield Scott, and the astronomer Herbert Hall Turner. The dean of Westminister Abbey, Herbert E. Ryle, granted the request, and Albert Bruce-Joy, the sculptor whom Wallace had credited with changing the course of his life by recommending a radical alteration in diet, was commissioned to create the medallion. On November 1, 1915 (All Souls' Day), at the height of World War I, it was unveiled, along with medallions honoring the memories of two other great and recently deceased English scientists: Sir Joseph Hooker and Lord Joseph Lister. At the dedication ceremony, Ryle remarked:

> Alfred Russel Wallace was a most famous naturalist and zoologist. He arrived by a flash of genius at the same conclusions which Darwin had reached after sixteen years of most minute toil and careful observation. . . . It was a unique example of the almost exact concurrence of two great minds working upon the same subject, though in different parts of the world, without collusion and without rivalry. . . . Between Darwin and Wallace goodwill and friendship were never interrupted. Wallace's life was spent in the pursuit of various objects of intellectual and philosophical interest. . . . All will agree that it is fitting his medallion should be placed next to that of Darwin, with whose great name his own will ever be linked to the worlds of thought and science. . . . These are . . . men whose life work it was to utilise and promote scientific discovery for the preservation and betterment of mankind.[47]

The ceremony at Westminster Abbey was indeed a fitting tribute to Alfred Russel Wallace. In the worlds of thought and science, his ideas continue to live on; but in the popular imagination, his name is no longer linked as intimately with that of Charles Darwin. Although occasionally he assumed the role of heretic rather than that of an unswerving believer in the scientific revolution he helped to initiate, it was to restrain the excesses of science, not its progress. Even if his cautionary words do not always convince us, even if his vision strikes us as too utopian in these more cynical times, his accomplishments in evolutionary biology and his tireless crusade to improve the lot of his fellow human beings deserve our attention and admiration. As his friend Edward Bagnall Poulton eulogized his memory on the centenary of his birth in 1923, "there remains enough of mighty achievement to make sure for him a high place in the temple of fame."[48]

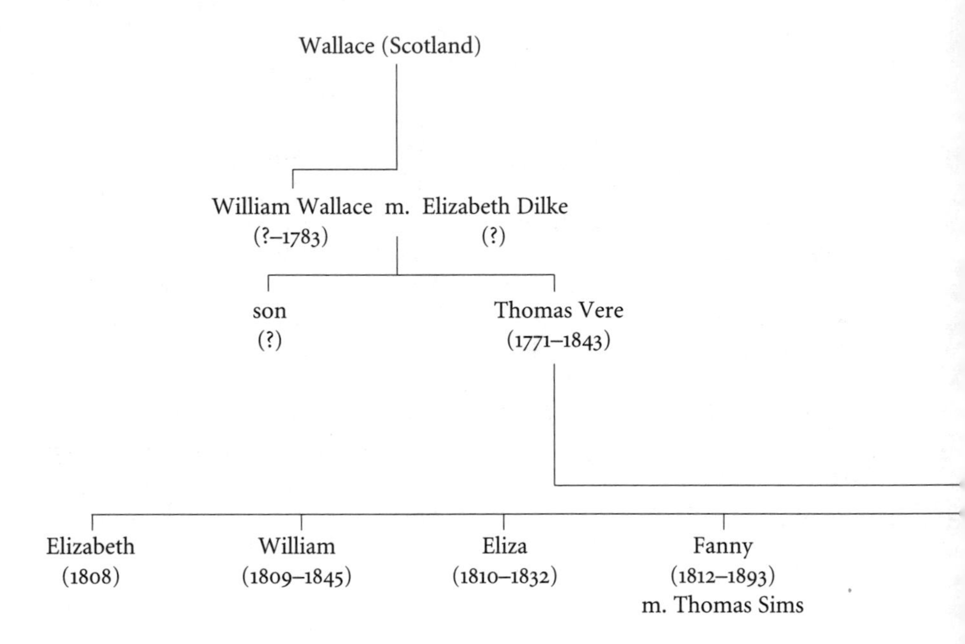

Wallace (Scotland)
William Wallace m. Elizabeth Dilke
(?–1783)
(?)
son
(?)
Thomas Vere
(1771–1843)
Elizabeth
(1808)
William
(1809–1845)
Eliza
(1810–1832)
Fanny
(1812–1893)
m. Thomas Sims

Genealogy of the Wallace Family

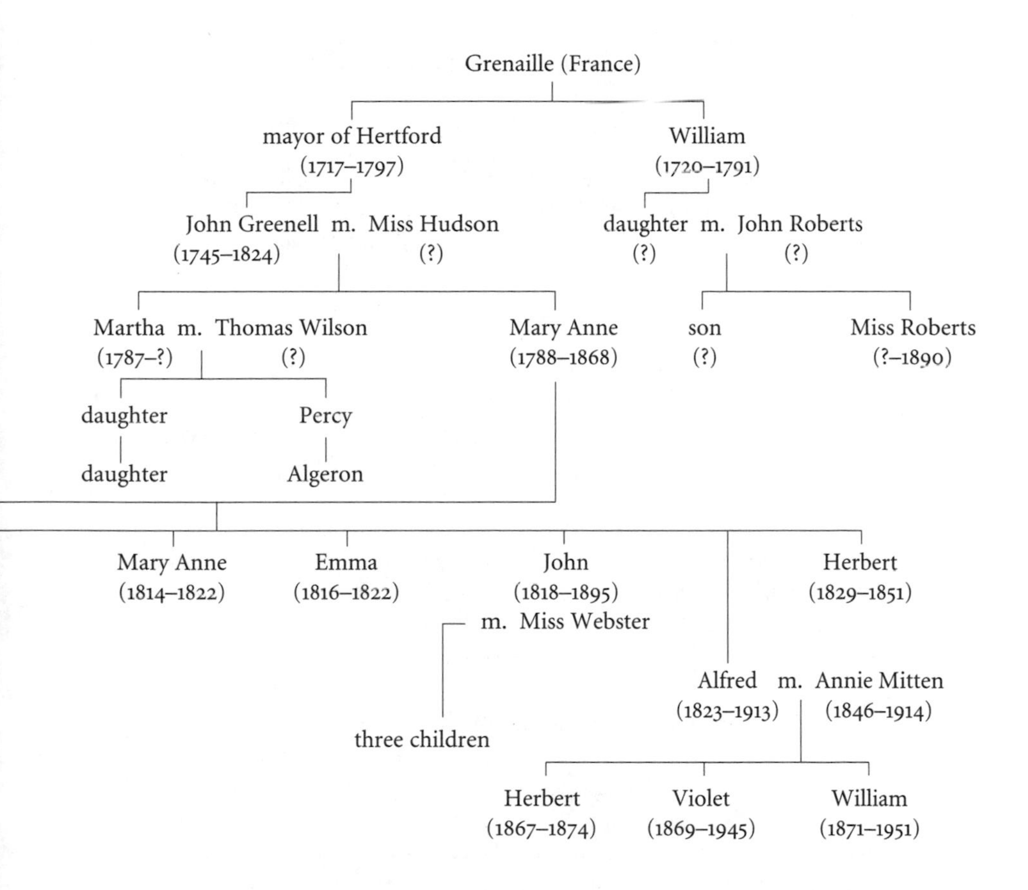

Notes

ABBREVIATIONS

Correspondents

AB	Arabella Buckley
AN	Alfred Newton
ARW	Alfred Russel Wallace
CD	Charles Darwin
CL	Charles Lyell
EPB	Edward Bagnett Poulton
FS	Frances Wallace Sims (Fanny)
GJR	George John Romanes
HWB	Henry Walter Bates
JH	Joseph Dalton Hooker
RM	Raphael Meldola
RS	Richard Spruce
SS	Samuel Stevens
THH	Thomas Henry Huxley
WBC	William Benjamin Carpenter
WTD	William Thistleton-Dyer

Printed Sources

AJ	Alfred Russel Wallace, American journal (Archives of the Linnean Society of London)
AMNH	*Annals and Magazine of Natural History*
CCD	Charles Darwin, *The Correspondence of Charles Darwin*, 9 vols., ed. Frederick Burkhardt and Sydney Smith (Cambridge: Cambridge University Press, 1985–1994)
DAR	Darwin archives (Cambridge University Library)
JLS	*Journal of the Proceedings of the Linnean Society of London*
JMA	Alfred Russel Wallace, journal from the Malay Archipelago (Archives of the Linnean Society of London)
JRGS	*Journal of the Royal Geographical Society*
LJH	Joseph Dalton Hooker, *The Life and Letters of Joseph Dalton Hooker*, 2 vols., ed. Leonard Huxley (New York: Appleton, 1918)

LLD	Charles Darwin, *The Life and Letters of Charles Darwin*, 2 vols., ed. Francis Darwin (1887; reprint, New York: Appleton, 1904)
LLH	Thomas Henry Huxley, *Life and Letters of Thomas Henry Huxley*, 2 vols., ed. Leonard Huxley (New York: Appleton, 1913)
LLL	Charles Lyell, *Life, Letters, and Journals of Sir Charles Lyell*, 2 vols., ed. K. Lyell (London: Murray, 1881)
LLW	Alfred Russel Wallace, *Alfred Russel Wallace: Letters and Reminiscences*, ed. James Marchant (New York: Harper, 1916)
MA	Alfred Russel Wallace, *The Malay Archipelago* (1869; reprint, Oxford: Oxford University Press, 1986)
ML	Alfred Russel Wallace, *My Life; A Record of Events and Opinions*, 2 vols. (New York: Dodd, Mead, 1905)
NRA	Henry Walter Bates, *The Naturalist on the River Amazons* (London: Murray, 1892)
PZL	*Proceedings of the Zoological Society of London*
SJ	Alfred Russel Wallace, notebook ("Species Journal," 1855–1859): MS 180 (Archives of the Linnean Society of London)
SSS	Alfred Russel Wallace, *Studies Scientific and Social*, 2 vols. (London: Macmillan, 1900)
TARN	Alfred Russel Wallace, *A Narrative of Travels on the Amazon and Rio Negro* (1853; reprint, New York: Haskell House, 1969)

INTRODUCTION

1. Alfred Russel Wallace, "On the Tendency of Varieties to Depart Indefinitely from the Original Type; Instability of Varieties Supposed to Prove the Permanent Distinctness of Species," *JLS (Zoology)* 3 (1858): 53–62, reprinted in *Contributions to the Theory of Natural Selection* (1870; reprint, New York: AMS Press, 1973), 26–44.

2. Thomas Bell, "Presidential Address to the Members of the Linnean Society," *JLS (Zoology)* 4 (1859): viii–ix.

3. ARW to HWB, December 24, 1860, *LLW*, 59.

4. Stephen Jay Gould, "Wallace's Fatal Flaw," *Natural History*, January 1980, 26–39.

5. Another contemporary adherent of spiritualism was Sir Arthur Conan Doyle, a physician and the creator of Sherlock Holmes.

6. Alfred Russel Wallace, "The Limits of Natural Selection as Applied to Man," in *Contributions to the Theory of Natural Selection*, 359–60.

7. Thomas S. Kuhn, *The Structure of Scientific Revolutions* (Chicago: University of Chicago Press, 1996), 90.

8. H. L. McKinney, *Wallace and Natural Selection* (New Haven, Conn.: Yale University Press, 1972), 138.

9. H. James Birx, introduction to *Island Life*, by Alfred Russel Wallace (1880; reprint, Amherst, N.Y.: Prometheus Books, 1998), xvii.

10. Ernst Mayr, *The Growth of Biological Thought: Diversity, Evolution, and Inheritance* (Cambridge, Mass.: Harvard University Press, Belknap Press, 1982), 498.

11. John Marsden, "On Human Credulousness," November 27, 1997 [available at: www.linnean.org].

12. Carl Jung, "The Psychological Foundations of Belief in Spirits," in *Psychology of the Occult*, ed. and trans. R. F. C. Hull, Bollingen series 20 (Princeton, N.J.: Princeton University Press, 1981), 109.

1. ORIGINS OF A HERETIC

1. The chronology of Wallace's early years is confusing. He had a poor memory for dates, especially early ones; he believed until the late 1880s that he had been born in 1822, but a family prayer book records the year as 1823 (ARW to EBP, February 23, 1903 [courtesy of the librarian, Oxford University Museum of Natural History]). The dates of other events are also uncertain until April 1848, when dated letters document his departure for Brazil.

2. English education at the time was no better for a child of privilege, like Charles Darwin, who looked back at his early school years as a blank (*LLD*, 29). The deplorable state of elementary English education prompted Thomas Huxley, an anatomist and a spokesman for evolution, to propose educational reform, citing his own early experiences. A child of his era, he said, could pass through school with honors and yet not know that the earth circled the sun ("A Liberal Education" [1868], in *Science and Education* [1901; reprint, New York: Citadel, 1964], 86). See also *ML*, 1:71.

3. Fanny Wallace was the only surviving daughter of the family. The oldest daughter died in infancy, two others died in early childhood, and the fourth died in 1832 at age twenty-two from tuberculosis.

4. Robert Owen, *"A New View of Society" and Other Writings*, ed. Gregory Claes (New York: Penguin, 1991), 16, 22, 27–32, 43.

5. Thomas Paine, *The Age of Reason* (1794; reprint, New York: Buccaneer Books, 1976), 18, 67–68, 181, 186.

6. Alfred Russel Wallace, "Personal Suffrage, a Rational System of Representation and Election," *Fortnightly Review*, January 1, 1907, 3.

7. *ML*, 1:108.

8. Ibid., 108–9; Jane Camerini, "Evolution, Biogeography, and Maps: An Early History of Wallace's Line," *Isis* 84 (1993): 700–727; James Moore, "Wallace's Malthusian Moment: The Common Context Revisited," in *Victorian Science in Context*, ed. Bernard Lightman (Chicago: University of Chicago Press, 1997), 290–311.

9. Swansea government archives, 2000 [available at: www.swansea.gov.uk/archives].

10. *ML*, 1:165.

11. Ibid., 110.

12. Ibid., 191. See also Joseph Ewen, "Tracking Richard Spruce's Legacy from George Bentham to Edward Whymper," in *Richard Spruce, Botanist and Explorer*, ed. M. R. D. Seaward and S. M. D. Fitzgerald (Kew: Royal Botanic Gardens, 1996), 45.

13. *ML*, 1:190–91, 192. See also John Burnett, *A History of the Cost of Living* (New York: Penguin, 1969), 263–64. In the 1830s the average regularly employed semi-skilled urban worker earned anywhere between 15 shillings and £1 a week.

14. *ML*, 1:192–93.

15. John Lindley, *The Elements of Botany, Structural and Physiological; A Fifth Edition of the Outline of the First Principles of Botany* (London: Bradbury & Evans, 1847), 125. Wallace states that he used Lindley's fourth edition, a copy of which is in the library of the Linnean Society of London (*ML*, 1:192).

16. *ML*, 1:133–38.

17. Ernst Mayr, *The Growth of Biological Thought: Diversity, Evolution, and Inheritance* (Cambridge, Mass.: Harvard University Press, Belknap Press, 1982), 251–60.

18. H. L. McKinney, *Wallace and Natural Selection* (New Haven, Conn.: Yale University Press, 1972), 6.

19. William Swainson, *A Treatise on the Geography and Classification of Animals* (London: Longman, Reese, 1835), 201–19, 350–52.

20. Wallace's annotated copy of Swainson's *Treatise* is in the library of the Linnean Society of London.

21. *ML*, 1:198–99.

2. THE STRUGGLE FOR EXISTENCE

1. George M. Trevelyan, *British History in the Nineteenth Century, 1782–1901* (London: Longman, Green, 1934), 144, 205, 271; John Snow, "On the Mode of Communication of Cholera," in *Snow on Cholera* (1854; reprint, New York: Hafner, 1965); K. Theodore Hoppen, *The Mid-Victorian Generation, 1846–1886* (Oxford: Oxford University Press, 1998), 10–11, 129–31; Max Beer, *A History of British Socialism* (London: Bell, 1929), 1:52–56.

2. According to Robert Owen, "Mr Malthus . . . is correct, when he says that the population of the world is ever adapting itself to the quantity of food raised for its support; but he has not told us how much more food an intelligent and industrious people will create from the same soil, than will be produced by one ignorant and ill-governed. It, however, is as one to infinity" (*"A New View of Society" and Other Writings*, ed. Gregory Claes [New York: Penguin, 1991], 86–87). See also Claes's introduction, xx.

3. Thomas Malthus, *An Essay on the Principle of Population, as it Affects the Future Improvement of Society with Remarks on the Speculations of Mr. Godwin, M. Condorcet, and Other Writers*, ed., with an introduction, Antony Flew (1798 and 1830; reprint, New York: Penguin, 1985), 71–72, 83–84, 169, 199, 250.

4. Ibid., 272–73.

5. Edward Clodd, "Memoir of the Author," in *NRA*, xix.

6. H. P. Moon, *Henry Walter Bates, FRS (1825–1892): Explorer, Scientist and Darwinian* (Leicester: Leicestershire Museums, Art Galleries and Record Service, 1976), 15–18. See also Clodd, "Memoir of the Author," xix–xx.

7. *ML*, 1:237. See also Alfred Russel Wallace, acceptance speech, in *The Darwin–Wallace Celebration held on Thursday, 1st July 1908 by the Linnean Society of London* (London: Longman, Green, 1909), 7–8.

8. In his autobiography, Wallace gives the year of William's death and his return to Neath as 1846, but subsequent events do not make sense chronologically if

William died in 1846. The copy of William's death certificate in the West Glamorgan, Wales, archives states that he died on March 15, 1845, and was buried in the Llantwit-juxta-Neath cemetery.

9. *ML*, 1:243.

10. Ibid., 251–53.

11. James A. Secord, introduction to *Vestiges of the Natural History of Creation*, by Robert Chambers (1844; reprint, Chicago: University of Chicago Press, 1994), xli–ii. According to Secord, Chambers never officially acknowledged its authorship.

12. Robert Chambers, *Vestiges of the Natural History of Creation*, ed. James A. Secord (1844; reprint, Chicago: University of Chicago Press, 1994), 20–26, 53, 72–75, 82–84, 147, 153–55.

13. Ibid., 195.

14. Ibid., 212–13.

15. Ibid., 199, 234–35.

16. Secord, introduction, xxx–xxxi.

17. Joel Schwartz, "Darwin, Wallace, and Huxley and *Vestiges of the Natural History of Creation*," *Journal of the History of Biology* 23 (1990): 131–51. See also CD to THH, September 2, 1854, *CCD* 5:212–13; and *LJH*, 1:168.

18. H. L. McKinney, *Wallace and Natural Selection* (New Haven, Conn.: Yale University Press, 1972), 11; *ML*, 1:254. Wallace gives the year of his letters as 1847, but in "Memoir of the Author," Clodd dates Wallace's letter of November 9 to 1845.

19. Alfred Russel Wallace, *The Wonderful Century: Its Successes and Its Failures* (New York: Dodd, Mead, 1899), 136–37, 255.

3. A DARING PLAN

1. Richard Burckhardt, introduction to *Zoological Philosophy*, by Jean-Baptiste Lamarck (1809; reprint, Chicago: University of Chicago Press, 1984), xv–xxvii.

2. Martin Rudwick, introduction to *Principles of Geology*, by Charles Lyell (1830; reprint, Chicago: University of Chicago Press, 1991), 1:viii–xvii.

3. Charles Lyell, *Principles of Geology* (1832; reprint, Chicago: University of Chicago Press, 1991), 2:20–21.

4. Ibid., 2, 35, 64–65, 141, 156–57, 169, 182.

5. Rudwick, introduction, xv–xxvii. See also Alfred Russel Wallace, *The Wonderful Century: Its Successes and Its Failures* (New York: Dodd, Mead, 1899), 110–16; and Archibald Geikie, "Sir Charles Lyell, Bart., F.R.S." [obituary], *Nature*, March 4, 1875, 341–42. In promulgating James Hutton's doctrine, Charles Lyell was no longer subject to charges of blasphemy. By now, most scientists—even those who were clergymen—discounted the biblical interpretation of the age of the earth as calculated by the seventeenth-century Irish archbishop James Ussher. The earth's age was not a few thousand years but many millions.

6. ARW to HWB, November 9, 1845, *ML*, 1:254–56; Edward Clodd, "Memoir of the Author," in *NRA*, vii.

7. William Swainson, *On the Natural History and Classification of Quadrupeds* (London: Longman, Rees, 1835), 263.

8. *ML*, 1:257.

9. Ibid., 255–57.

10. Alexander von Humboldt, *Personal Narrative of a Journey to the Equinoctial Regions of the New Continent*, trans., with an introduction, Jason Wilson (1814–1825; reprint [abridged], New York: Penguin, 1995), 5. See also Malcolm Nicolson, historical introduction to *Personal Narrative*, by Humboldt, ix–xxxiii; and Loren McIntyre, "Humboldt's Way," *National Geographic*, September 1985, 328.

11. Charles Darwin, *Voyage of the Beagle (Journal of Researches into the Geology and Natural History of the Various Countries Visited by the H.M.S. Beagle)*, with an introduction by Janet Browne (1839; reprint, New York: Penguin, 1989), 287–88.

12. *LLD*, 1:53.

13. *ML*, 1:255–56.

14. Clodd, "Memoir of the Author," vii, xxi; *ML*, 1:264.

15. William H. Edwards, *A Voyage up the River Amazon: Including a Residence at Pará* (New York: Appleton, 1847), 11–12, 17–18, 44–45, 62, 143, 177–81.

16. *NRA*, 111. See also Norman Moore, "Memoir," in *Wanderings in South America, the North-West of the United States and the Antilles in the Years 1812, 1816, 1820, & 1824*, by Charles Waterton (New York: Sturgis & Walton, 1909), 16–17.

17. Edwin Williamson, *The Penguin History of Latin America* (New York: Penguin, 1992), 167–251. See also Humboldt, *Personal Narrative*, 261.

18. Andrew Porter, *The Oxford History of the British Empire*, vol. 3, *The Nineteenth Century* (Oxford: Oxford University Press, 1999), 128–29.

19. Humboldt, *Personal Narrative*, 270–71, 210.

20. Alfred Russel Wallace, "White Men in the Tropics," in *SSS*, 2:99–101.

21. H. P. Moon, *Henry Walter Bates, FRS (1825–1892): Explorer, Scientist and Darwinian* (Leicester: Leicestershire Museums, Art Galleries and Record Service, 1976), 51.

22. *ML*, 1:262–63.

23. Janet Browne, "Biogeography and Empire," in *Cultures of Natural History*, ed. N. Jardine, J. A. Secord, and E. C. Spary (Cambridge: Cambridge University Press, 1997), 307–11.

24. Alfred Russel Wallace, "Museums for the People" (1869), in *SSS*, 2:2.

25. Browne, "Biogeography and Empire," 309.

26. *ML*, 1:264.

27. Ibid., 265.

28. Ibid., 266.

29. *LLD*, 1:35–37.

30. Anne Larsen, "Equipment for the Field," in *Cultures of Natural History*, 358.

31. Clodd, "Memoir of the Author," xxvi.

32. Samuel Stevens, "Directions for Collecting and Preserving Specimens of Natural History in Tropical Climates" (1855), Linnean Society of London Collection.

33. Ernst Mayr, *The Growth of Biological Thought: Diversity, Evolution, and Inheritance* (Cambridge, Mass.: Harvard University Press, Belknap Press, 1982), 170.

34. Henry Walter Bates, "Hints to Travelers," *JRGS* 34 (1864): 306–16.

35. ARW and HWB to William Hooker, April 13, 1848, Directors' Correspon-

dence, vol. 26, doc. 566–67 (March–April 1848, Bates and Wallace), Archives of the Royal Botanic Gardens, Kew.

36. ARW and HWB to William Hooker, March 30, 1848, Directors' Correspondence, vol. 26, doc. 566–67 (March–April 1848, Bates and Wallace).

37. *ML*, 1:193–94, 267.

38. *NRA*, 1; *ML*, 1:268. Bates notes the date of their departure as April 26, whereas Wallace gives April 20. Bates is probably correct since his journals survived his journey, whereas Wallace's journal did not and he was writing from memory.

4. TRAVELS ON THE AMAZON . . .

1. *NRA*, 1; *ML*, 1:268.

2. *TARN*, 1.

3. Ibid., 2.

4. *NRA*, 21.

5. *ML*, 1:269; *NRA*, 7–9; *TARN*, 3–4, 12.

6. *TARN*, 10, 306–7.

7. Alfred Russel Wallace, "Equatorial Vegetation," in *"Tropical Nature" and Other Essays* (1878; reprint, New York: AMS Press, 1975), 29–34.

8. *TARN*, 302.

9. *NRA*, 31.

10. ARW to SS, excerpt in *Zoologist* 8 (1850): 2839; *TARN*, 34; ARW to William Hooker, August 20, 1848, Directors' Correspondence, vol. 26, doc. 566–67 (March–April 1848, Bates and Wallace), Archives of the Royal Botanic Gardens, Kew.

11. *NRA*, 57; *TARN*, 35–36.

12. *TARN*, 53; *NRA*, 67–70, 326–30.

13. Samuel Stevens, "Advertisement to Collectors," *AMNH*, 2nd ser., 3, no. 13 (1849): 74–75.

14. Samuel Stevens, in *Transactions of the Entomological Society of London*, 1st ser., 5 (1847–1849): lviii.

15. Edward Bagnell Poulton, a close friend of Wallace in his later years, wrote a short biography of Wallace for the Royal Society, in which he denied that a quarrel had occurred ("Alfred Russel Wallace [1823–1913]," *Proceedings of the Royal Society of London* 95 [1924]: i–xxxv). But Richard Spruce, in a letter to William Hooker, wrote, "I forget to mention that we have several times seen Mr Wallace. . . . He and Bates quarreled and separated long ago" (RS to William Hooker, August 3, 1849, Letters to Spruce [1842–1890], no. 259, Archives of the Royal Botanic Gardens, Kew).

16. *TARN*, 58–59.

17. *ML*, 1:15, 282–83.

18. SS to Lord Palmerston, August 4, 1849, with an extract of Wallace's letter to him, Public Record Office, Foreign Office Archives, FO 13/271. Stevens asked for a response within ten days, but the Foreign Office did not reply until August 21, a week after the departure of a ship to Brazil.

19. *TARN*, 93–94.

20. ARW to SS, September 14, 1849, excerpt in *AMNH*, 2nd ser., 5, no. 26 (1870): 156–57.

21. RS to William Hooker, August 3, 1849, Letters to Spruce (1842–1890), no. 259.

22. Richard Spruce, *Notes of a Botanist on the Amazon & Andes . . . During the Years 1849–1864*, ed. Alfred Russel Wallace (London: Macmillan, 1908), 1:xxi–xxxiii; Alfred Russel Wallace, "Richard Spruce, Ph.D., F.R.G.S." [obituary], *Nature*, February 1, 1894, 318–19.

23. Spruce, *Notes of a Botanist*, 1:74–75.

24. *TARN*, 106. Wallace called this water lily *Victoria regia* [*sic*] (Spruce, *Notes of a Botanist*, 1:75–76).

25. *ML*, 1:279.

5. . . . AND THE RIO NEGRO

1. *TARN*, 111.

2. Ibid., 114.

3. Richard Spruce, *Notes of a Botanist on the Amazon & Andes . . . During the Years 1849–1864*, ed. Alfred Russel Wallace (London: Macmillan, 1908), 1:201–2.

4. *TARN*, 112–14.

5. Alfred Russel Wallace, "On the Umbrella Bird (*Cephalopterus ornatus*), 'Ueramimbe,' Lingoa Geral," *PZL* (1850): 206–7.

6. Ibid.

7. *TARN*, 120; *NRA*, 176.

8. *Strychnos toxifera* was named and classified by the botanist George Bentham. The English explorer Robert Schomburgk identified it as the source of curare in the 1830s and discovered—through bold experiments on himself and animals—that the poison killed only when it entered the bloodstream and not when ingested ("On the 'Urari,' the Arrow Poison of the Indians of Guiana; with a Description of the Plant from which It Is Extracted," *AMNH*, 1st ser., 7, no. 45 [1841]: 407–27). See also *NRA*, 295–96; *TARN*, 147–48. Wallace calls the blowpipe *gravatána.*

9. *NRA*, 296.

10. *ML*, 1:276–78; *TARN*, 133–35.

11. Alfred Russel Wallace, "Fishes of the Rio Negro and Its Tributaries," *PZL* (1905): 189. Wallace apparently described this fish fifty years before it was first brought to the attention of the Zoological Society of London in 1897.

12. Alfred Russel Wallace, "On Some Fishes Allied to *Gymnotus*," *AMNH*, 2nd ser., 14, no. 83 (1854): 398–99.

13. Charles Waterton, *Wanderings in South America, the North-West of the United States and the Antilles in the Years 1812, 1816, 1820 & 1824* (New York: Sturgis & Walton, 1909), 59.

14. *TARN*, 330.

15. Robert S. Ridgely and Guy Tudor, *The Birds of South America: The Suboscine Passerines* (Austin: University of Texas Press, 1994), 778–79.

16. *TARN*, 148–49.

17. Ibid., 155.

18. ARW to an unknown correspondent, December 1850, *ML*, 1:285.

19. Alfred Russel Wallace, "On the Rio Negro," *JRGS* 23 (1853): 212–17; *TARN*, 165–69.

20. *TARN*, 172.

21. Alfred Russel Wallace, Notebook: Fishes of the Amazon and Rio Negro, Descriptions (quoted by permission of the Trustees of the Natural History Museum, London).

22. *TARN*, 176–80.

23. Spruce, *Notes of a Botanist*, 1:211.

24. *ML*, 1:285.

25. Charles Darwin, *Voyage of the Beagle (Journal of Researches into the Geology and Natural History of the Various Countries Visited by the HMS Beagle)*, with an introduction by Janet Browne (1839; reprint, New York: Penguin, 1989), 171–73.

26. *TARN*, 190–91, 343.

27. Alfred Russel Wallace, "On the Insects Used for Food by the Indians of the Amazon," *Transactions of the Entomological Society of London*, n.s., 2, pt. 8 (1854): 241–44.

28. Edward Clodd, "Memoir of the Author," in *NRA*, xxv.

29. HWB to SS, December 23, 1851 [*sic*], excerpt in *Zoologist* 9 (1851): 3144. The date must be a misprint, since the next letter published in the journal is dated April 30, 1851, when Bates had arrived in Pará.

30. "Great Beauty of South-American Lepidoptera," *Zoologist* 9 (1851): 3059–60.

31. Clodd, "Memoir of the Author," xxvi.

32. *NRA*, 178.

33. HWB to Mary Anne Wallace, Pará, June 13, 1851 (estate of A. J. R. Wallace, Bournemouth).

34. Spruce, *Notes of a Botanist*, 1:267.

35. *TARN*, 236, 250.

36. Ibid., 243.

37. Ibid., 249–50.

38. *ML*, 1:316.

39. Elbert Hubbard, *Little Journeys to Homes of Great Scientists: Wallace* (New York: Roycrofters, 1905), 96. The number of specimens that Hubbard claims Wallace brought home was not corroborated by Wallace, who states only that he estimated the value of the lost specimens at approximately £500.

6. DISASTER AT SEA . . . AND A CIVILIZED INTERLUDE

1. *ML*, 1:303–12; *TARN*, 271–79; ARW to an unknown correspondent [about the fire], October 19, 1852, excerpts in *Zoologist* 10 (1852): 3541–43, 3643, and *Times* (London), October 2, 1852, 8.

2. *ML*, 1:307.

3. Ibid., 310.

4. Alexander von Humboldt, *Personal Narrative of a Journey to the Equinoctial Regions of the New Continent*, trans., with an introduction, Jason Wilson (1814–1825; reprint [abridged], New York: Penguin, 1995), 270.

5. John Burnett, *A History of the Cost of Living* (New York: Penguin, 1966), 192–93.

6. Quoted in George W. Stocking Jr., *Victorian Anthropology* (New York: Free Press, 1987), 216.

7. Henry Walter Bates, pocketbook, which includes an almanac from the period that names all the scientific societies active in London at the time (59–60), British Library, Department of Manuscripts, London.

8. See, for example, the By-Laws of the Entomological Society of London, revised in 1848, *Proceedings of the Entomological Society of London.*

9. *ML*, 1:324.

10. Alfred Russel Wallace, "The Dawn of a Great Discovery: My Relations with Darwin in Reference to the Theory of Natural Selection," *Black and White*, January 17, 1903, 78.

11. Edward Newman, "Preface," *Zoologist* 8 (1850): v–vi.

12. Ibid. See also obituary of Edward Newman, *Zoologist* 34 (1876): iii–xxii.

13. Edward Newman expresses his inability to accept the theory of evolution in "The Death of a Species," *Zoologist* 27 (1869): 1529–42.

14. "Advertisement, " in preface of *AMNH*, 1st ser., 2 (1838).

15. Alfred Russel Wallace, "On the Habits of the Butterflies of the Amazon Valley," *Transactions of the Entomological Society of London*, n.s., 2, pt. 7 (1854): 253–64.

16. Alfred Russel Wallace, "On the Monkeys of the Amazon," *AMNH*, 2nd ser., 14, no. 84 (1854): 451–54; *ML*, 1:325; *TARN*, 294–96. For clarification of New World primates, see Ronald M. Nowak, *Walker's Primates of the World* (Baltimore: Johns Hopkins University Press, 1999), 97–128.

17. *ML*, 1:377; Wallace, "On the Monkeys of the Amazon."

18. Reviews of *Palm Trees of the Amazon* and *Narrative of Travels*, by Alfred Russel Wallace, *AMNH*, 2nd ser., 13, no. 73 (1854): 56–57.

19. H. L. McKinney, introduction to *Palm Trees of the Amazon and Their Uses*, by Alfred Russell Wallace (1853; reprint, Lawrence, Kan.: Coronado, 1971). See also T. C. Archer, note to the editor, *Hooker's Journal of Botany* 7 (1855): 213–14.

20. RS to William Hooker, January 2, 1854, Spruce Correspondence, no. 108, Archives of the Royal Botanic Gardens, Kew.

21. William Hooker to RS, January 5, 1855, Spruce Correspondence, no. 372; McKinney, introduction. Four years later, Richard Spruce published an article that more strongly defended Wallace's identification of *Leopoldinia piassaba* as a new species of palm without undermining Carl von Martius's reputation, attributing the confusion to insufficient data ("On *Leopoldinia Piassaba*, Wallace," *JLS [Botany]* 4 [1859]: 58–60).

22. Edward Newman, "President's Address to the Entomological Society," *Zoologist* 12 (1854): 4223–24.

23. *ML*, 1:327.

24. ARW to AN, March 29, 1875, Newton Papers, no. 281, Cambridge University Library.

25. Alfred Russel Wallace, "Attempts at a Natural Arrangement of Birds," *AMNH*, 2nd ser., 18, no. 105 (1856): 206–7, 214.

26. Patricia T. Stroud, *The Emperor of Nature* (Philadelphia: University of Pennsylvania Press, 2000), 314–15.

27. Wallace's annotated copy of Charles-Lucien Bonaparte's *Conspectus Generum Avium* is in the library of the Linnean Society of London. See also *ML*, 1:328.

28. *ML*, 1:330.

29. Alfred Russel Wallace, "On the Rio Negro," *JRGS* 23 (1853): 212–17; *ML*, 1:318–20.

30. Richard Spruce, *Notes of a Botanist on the Amazon & Andes . . . During the Years 1849–1864*, ed. Alfred Russel Wallace (London: Macmillan, 1908), 1:329–30.

31. Clements R. Markham, *The Fifty Years' Work of the Royal Geographical Society* (London: Murray, 1881), 19–27.

32. Charles Lyell, *Principles of Geology* (1833; reprint, Chicago: University of Chicago Press, 1991), 3:384–88; "Roderick Impey Murchison," in *Dictionary of Scientific Biography*, ed. Charles Coulton Gillispie (New York: Scribner, 1981).

33. Clements R. Markham, unofficial history of the Royal Geographical Society, ca. 1901, 368, Archives of the Royal Geographical Society, London.

34. *ML*, 1:327.

35. ARW to Roderick Impey Murchison, June 30, 1853, Correspondence Block 1851–1860, Archives of the Royal Geographical Society.

36. S. C. Otway, envoy in Madrid, to Lord Addington, August 19, 1853, and Henry Elliott, envoy to The Hague, August 30, 1853, Foreign Office Correspondence, Archives of the Royal Geographical Society.

37. ARW to Norton Shaw, August 27, 1853, Archives of the Royal Geographical Society.

38. ARW to Norton Shaw, November 7, 1853, Archives of the Royal Geographical Society.

39. Foreign Office to Norton Shaw, January 11 and 13, 1854, Archives of the Royal Geographical Society; *ML*, 1:330.

40. ARW to Norton Shaw, February 4 and 6, 1854, Archives of the Royal Geographical Society.

41. ARW to Norton Shaw, February 8, 1854, Archives of the Royal Geographical Society.

42. Copy of certificate of election, Archives of the Royal Geographical Society (courtesy of A. Tatham).

43. K. Theodore Hoppen, *The Mid-Victorian Generation, 1846–1886* (Oxford: Oxford University Press, 1998), 167–73.

44. *ML*, 1:336.

7. THE MALAY ARCHIPELAGO

1. ARW to SS, May 9, 1854, *Zoologist* 13 (1854): 4395–97; *MA*, 35.

2. ARW to FS, June 25, 1855, *LLW*, 46–48.

3. *MA*, 41. See also ARW to William Hooker, October 10, 1854, *Hooker's Journal of Botany* 7 (1855): 200–203.

4. Alfred Russel Wallace, in *Hooker's Journal of Botany* 7 (1855): 206–9.

5. James Brooke published an outline of his mission to the Malay Archipelago in *Athenaeum*, October 13, 1838, 744–46.

6. Gertrude Jacob, *The Raja of Sarawak* (London: Macmillan, 1876), 1:240; Spenser St. John, *The Life of Sir James Brooke, Rajah of Sarawak* (1879; reprint, Kuala Lumpur: Oxford University Press, 1994), 76, 89–90, 315–16.

7. St. John, *Life of Sir James Brooke,* 219, 274–75.

8. H. L. McKinney, *Wallace and Natural Selection* (New Haven, Conn.: Yale University Press, 1972), 33. McKinney believes that this notebook contains reflections that were being compiled for a larger book, and Wallace himself corroborates this assertion at various points in the notebook. The notebook has been used to track the development of his thought, especially in regard to his famous essay "On the Tendency of Varieties to Depart Indefinitely from the Original Type; Instability of Varieties Supposed to Prove the Permanent Distinctness of Species," *JLS (Zoology)* 3 (1858): 53–62.

9. *MA*, 51–74; *SJ*, 9–30; *ML*, 1:341; Alfred Russel Wallace, "Some Account of an Infant Orang-Utan," *AMNH*, 2nd ser., 17, no. 101 (1856): 386–90; "On the Orang-Utan or Mias of Borneo," *AMNH*, 2nd ser., 17, no. 102 (1856): 471–76; "On the Habits of the Orang-Utan in Borneo," *AMNH*, 2nd ser., 17, no. 103 (1856): 26–32; and "A New Kind of Baby," *Chambers's Journal*, November 22, 1856, 325–27.

10. Edward Blyth to CD, August 4, 1855, *CCD* 5:400; Wallace, "On the Orang-Utan or Mias of Borneo." See also Alfred Russel Wallace, "Notes for a paper on the Orang-Utan or 'Mias,'" MS Notebook 1855–8 Aves and Mammalia (courtesy of the Linnean Society of London).

11. Wallace, "On the Habits of the Orang-Utan in Borneo."

12. For a comparison with Wallace's articles on the orangutan in the *AMNH*, see *Zoologist* 14 (1856): 5183–85, 5329–31.

13. *ML*, 1:354.

14. Ibid., 355.

15. ARW to HWB, January 4, 1858, *LLW*, 53–55.

16. Quoted in McKinney, *Wallace and Natural Selection*, 44; Edward Forbes, "On the Manifestation of Polarity in the Distribution of Organized Beings in Time," *Proceedings of the Royal Institution of London* 57 (1854): 332–37.

17. Alfred Russel Wallace, "On the Law Which Has Regulated the Introduction of New Species," *AMNH*, 2nd ser., 16, no. 93 (1855): 184–96, reprinted in *Contributions to the Theory of Natural Selection* (1870; reprint, New York: AMS Press, 1973), 1–25.

18. James Brooke to ARW, July 4, 1856, Additional MSS 46441, nos. 2–3, Department of Manuscripts, British Library, London.

19. *ML*, 1:355.

20. Barbara Beddall, "Wallace, Darwin and the Theory of Natural Selection: A Study in the Development of Ideas and Attitudes," *Journal of the History of Biology* 1 (1968): 271.

21. *Sir Charles Lyell's Scientific Journals on the Species Question*, ed. Leonard G. Wilson (New Haven, Conn.: Yale University Press, 1970), xxix–xxxiv.

22. Ibid., xxxix–xl, 3–5, 55–56.

23. Barbara Beddall, "Wallace, Darwin and Edward Blyth: Further Notes on the Development of Evolution Theory," *Journal of the History of Biology* 5 (1972): 153–58.

24. CD, note, *CCD* 5:522n.1.

25. ARW to SS, August 21, 1856, ADD 7339.234, Cambridge University Library.

26. CD to W. B. Tegetmeier, November 29, 1856, *CCD* 6:290.

27. *Sir Charles Lyell's Scientific Journals*, xlvii.

28. Ibid., xlvi.

29. CD to JH, May 11, 1856, *CCD* 6:108.

30. Thomas Henry Huxley, notes on Wallace's 1855 Sarawak Law, in "Darwiniana," Thomas Henry Huxley Papers, Huxley Archives, Imperial College of Science, Technology, and Medicine, London [catalogued in Warren R. Dawson, *The Huxley Papers: A Descriptive Catalogue of the Correspondence, Manuscripts and Miscellaneous Papers of the Rt. Hon. Thomas Henry Huxley . . . preserved in the Imperial College of Science and Technology, London* (London: Macmillan, 1946), 41:57].

31. Jacob, *Raja of Sarawak*, 1:vii.

32. *MA*, 95–97. Wallace's collections attracted the attention of his former mentor, Edward Newman. In his anniversary address to the Entomological Society in January 1856, Newman expressed his delight: "Considered as a whole the entomological publications of 1855 do credit to the country in which we live. [I]insects of all classes have been collected abroad by our corresponding members, Messrs. Wallace and Bates. There is, therefore, every ground for satisfaction at the progress we are making, every reason for believing that entomology is progressing and will progress amongst us, not perhaps to take a station by the side of astronomy, mathematics or general physics, neither to be lauded as eminently utilitarian; but to be cherished as affording healthful occupation both to mind and to body" ("Anniversary Address to the Entomological Society of London, Jan. 28, 1856," *Proceedings of the Entomological Society of London* [1855–1856]: 153).

33. Alfred Russel Wallace, "Observations on the Zoology of Borneo," *Zoologist* 14 (1856): 5113–17.

34. Ibid.

35. ARW to SS, May 10, 1856, ADD 7339.232, Cambridge University Library.

36. Alfred Russel Wallace, "Attempts at a Natural Arrangement of Birds," *AMNH*, 2nd ser., 18, no. 105 (1856): 193–216.

37. *ML*, 1:349–53.

38. *MA*, 167–68.

39. Ibid. See also ARW to SS, August 21, 1856, ADD 7339.234.

40. ARW to SS, August 21, 1856.

41. Alfred Russel Wallace, "On the Zoological Geography of the Malay Archipelago," *JLS (Zoology)* 4 (1859): 174.

42. ARW to SS, December 1, 1856, *Zoologist* 15 (1857): 5652–57.

43. ARW to FS, December 10, 1856, *LLW*, 52.

44. ARW to SS, March 10 and May 15, 1857, *Proceedings of the Entomological Society of London* (1856–1857): 91–92.

45. *SJ*, 46.

46. *JMA*, 2:70.

47. Ibid.

48. Ibid., 76–77.

49. Ibid., 85.

50. ARW to SS, March 10 and May 15, 1857, "Postscript," *Proceedings of the Entomological Society of London* (1856–1857): 92–93.

51. Alfred Russel Wallace, "On the Great Bird of Paradise," *AMNH*, 2nd ser., 20, no. 120 (1857): 411–16.

52. Alfred Russel Wallace, "On the Natural History of the Aru Islands," *AMNH*, 2nd ser., 20, supplement (1857): 473–85.

53. *JMA*, 2:87.

54. HWB to ARW, November 19, 1856, *LLW*, 52–53.

55. ARW to HWB, January 4, 1858, *LLW*, 53–54.

56. CD to ARW, May 1, 1857, *LLW*, 107–8.

57. Barbara Beddall, "Darwin and Divergence: The Wallace Connection," *Journal of the History of Biology* 21 (1988): 38.

58. Alfred Russel Wallace, "On the Entomology of the Aru Islands," *Zoologist* 16 (1858): 5889–94.

59. Minutes of the meeting of the Zoological Society of London, February 23, 1858, *Zoologist* 16 (1858): 6040–42.

60. Ibid., 6041. The ratlike marsupial was described as "*Myoictis wallacii.* Rusty-brown, with interspersed black longer hairs; head redder; throat, chest and belly pale reddish; side of the neck at the base of the ears bright reddish; ears, and the greater part of the tail bright red-brown; tip of the tail black . . . 'in-houses as destructive as rats to every thing eatable . . . '—Wallace" (John Edward Gray, "List of Species of Mammalia Sent from the Aru Islands by Mr. A. R. Wallace to the British Museum," *AMNH*, 3rd ser., 2, no. 9 [1858]: 222–23).

61. Exhibition of Wallace's lepidoptera and coleoptera by Samuel Stevens at the meeting of the Entomological Society of London, March 1, 1858, *Zoologist* 16 (1858): 6036.

62. Wallace, "On the Natural History of the Aru Islands," 481.

63. *SJ*, 46.

64. Alfred Russel Wallace, "Note on the Theory of Permanent and Geographical Varieties," *Zoologist* 16 (1858): 5887–89. See also the excellent discussion of this issue in McKinney, *Wallace and Natural Selection*, 67.

65. Alfred Russel Wallace, "On the Habits and Transformations of a Species of *Ornithoptera*, Allied to *O. Priamus*, Inhabiting the Aru Islands, near New Guinea," *Transactions of the Entomological Society of London*, n.s., 4 (1858): 272–73.

66. *SJ*, 37.

8. THE MECHANISM REVEALED

1. ARW to CD, December 22, 1857, *LLW*, 109–11. Only a fragment of this letter exists, but the contents can be inferred from Darwin's response.

2. ARW to HWB, November 19, 1856, *LLW*, 52–53.

3. In his "Species Journal," Wallace writes: "Lyell talks of the 'balance of species being preserved by plants, insects, & mammalia & birds all adapted to the purpose.' This phrase is utterly without meaning. Some species are very rare & others very abundant. Where is the balance? Some species exclude all others in particular tracts. Where is the balance? . . . If any state can be imagined proving a want of balance then a balance may perhaps be admitted, but what state is that? . . . [D]isprove all Lyell's arguments first at the commencement of my last chapter." For specific refutations of Charles Lyell's ideas, see *SJ*, 34, 37–39, 41, 43, 45, 49–51, 149–50 (courtesy of the Linnean Society of London).

4. *JMA*, 2:123; *MA*, 312.

5. ARW to HWB, November 19, 1856, *LLW*, 55; *ML*, 1:359–60.

6. The dates between Wallace's arrival in Ternate in early January and his departure for Dorey in late March are recorded imprecisely. This was the crucial period in which he formulated his principle of natural selection (he did not use that term, which was Darwin's invention). When one examines his "Species Journal," his second Malay journal, *The Malay Archipelago*, and his insect, bird, and mammal registries, one is frustrated by the lack of precise dates. There is a vagueness in these documents about dates that does not exist in others. Ternate or Gilolo? The question of where and when he formulated his theory will forever remain unclear. Wallace maintained that the idea had come to him in Ternate, but his itinerary suggests that he was in Gilolo. It would not matter except for the fact that H. L. McKinney, one of the foremost Wallace scholars, believes that Wallace deliberately deceived his audience by citing the locus of his discovery as Ternate, a more famous and exotic island; McKinney describes this as a foreshadowing of Wallace's later intellectual "aberrations." An accusation of lying is a serious charge. Wallace prided himself on his intellectual honesty. Although he had a sense of drama, he never deliberately altered the truth. There is probably a simpler explanation for this confusion. Wallace's address was Ternate; mail came and went from Ternate, not Gilolo. Gilolo was also a province of Ternate, and the distinction between the two islands, only ten miles apart, was not important for the purposes of correspondence.

7. The earliest of Wallace's recorded recollections of this insight was in 1869. The German naturalist Adolph Bernhard Meyer reported having received a letter from Wallace sometime in 1869 describing how he discovered the theory of natural selection (Letter to the Editor, *Nature*, August 29, 1895, 415). Wallace's letter to Meyer apparently has been lost; Meyer's correspondence was archived in Dresden and destroyed in the Allied bombing raids of World War II. The second account is in a letter from Wallace to Alfred Newton, of December 3, 1887, in the abbreviated edition of Darwin's biography: *The Autobiography of Charles Darwin and Selected Letters*, ed. Francis Darwin (1892; reprint, New York: Dover, 1958), 201. The rest are in Alfred Russel Wallace, introduction to *Natural Selection and Tropical Nature: Essays on Descriptive and Theoretical Biology* (London: Macmillan, 1891); *The Wonderful Century: Its Successes and Its Failures* (New York: Dodd, Mead, 1899), 139–40; *ML*, 1:361–62; and acceptance speech, in *The Darwin–Wallace Celebration held on Thursday, 1st July 1908 by the Linnean Society of London* (London: Longman, Green, 1909).

8. *ML*, 1:361.

9. Alfred Russel Wallace, "On the Tendency of Varieties to Depart Indefinitely from the Original Type; Instability of Varieties Supposed to Prove the Permanent Distinctness of Species," *JLS (Zoology)* 3 (1858): 53–62, reprinted in *Contributions to the Theory of Natural Selection* (1870; reprint, New York: AMS Press, 1973), 26–44.

10. *ML*, 1:363.

11. Barbara Beddall, "Wallace, Darwin, and the Theory of Natural Selection: A Study in the Development of Ideas and Attitudes," *Journal of the History of Biology* 1 (1968): 316. Beddall does not believe that Wallace had received Darwin's letter by early March 1858. Wallace states that a mail steamer arrived in Ternate on March 9 (*JMA*, 2:128). He mentions nothing about a letter from Darwin, only his extreme disappointment that he had not received a fresh supply of arsenic and "other necessaries" for his upcoming voyage to New Guinea, for which he had been waiting for two months. In his autobiography, however, he implies that he had received Darwin's letter of December 22, 1857, which describes Charles Lyell's and Edward Blyth's reception of "On the Law Which Has Regulated the Introduction of New Species," *AMNH*, 2nd ser., 16, no. 93 (1855): 184–96.

12. CD to ARW, December 22, 1857, *LLW*, 111.

13. *ML*, 1:359–60.

14. *MA*, 497.

15. ARW to Norton Shaw, September 28, 1858, Ternate, Archives of the Royal Geographical Society, London; Roderick Impey Murchison, "Presidential Address," *JRGS* 29 (1859): clxxvii–viii; Alfred Russel Wallace, "Notes of a Voyage to New Guinea," *JRGS* 30 (1860): 172–77.

16. ARW to Mary Anne Wallace, October 6, 1858, *LLW*, 57–58.

17. ARW to SS, excerpt in *Zoologist* 17 (1859): 6412.

18. Unfortunately, neither letter has been preserved, but their contents can be inferred from ARW to JH, October 6, 1858, *CCD* 7:166. (Wallace's reply to Darwin also has been lost.) Certainly Darwin was more distressed by Wallace's essay than he led Wallace to believe. During their long and ambivalent relationship, Darwin would never reveal to Wallace the depths of his despair. Not until after Darwin's death, in 1882, would Wallace get a true picture of Darwin's reaction to his essay, derived from parts of Darwin's correspondence and papers published in 1887 (*LLD*, 2:472–87).

19. *LLD*, 1:67.

20. Janet Browne, *The Secular Ark: Studies in the History of Biogeography* (New Haven, Conn.: Yale University Press, 1983), 203.

21. *LLD*, 1:43.

22. CD to William Darwin Fox, March 27, 1855, *CCD* 5:293–94.

23. Malcolm J. Kottler, "Charles Darwin and Alfred Russel Wallace: Two Decades of Debate over Natural Selection," in *The Darwinian Heritage*, ed. David Kohn (Princeton, N.J.: Princeton University Press, 1985), 382.

24. CD to Asa Gray, September 5, 1857, quoted in Browne, *Secular Ark*, 210–11.

25. Browne, *Secular Ark*, 212–16. Browne quotes from Darwin's journal documenting the step-by-step composition of "Natural Selection" and the development of his "principle of divergence." Without answering the question of priority, she remarks that the principle of divergence became a part of Darwin's account of evolu-

tion only two days before he received Wallace's letter, "a coincidence that still grips the imagination."

26. CD to CL, June 18, 1858, *LLD*, 1:473.

27. CD to CL, June 25, 1858, *LLD*, 1:474, postscript on 475.

28. Francis Darwin, quoted in Gavin de Beer, foreword to *Charles Darwin and Alfred Russel Wallace: Evolution by Natural Selection* (Cambridge: Cambridge University Press, 1958), 37.

29. For discussions of the early history of the Linnean Society and of the events leading up to the emergency meeting of July 1, 1858, see A. T. Gage, *A History of the Linnean Society* (London: Linnean Society, 1938), 1–13, 52–54.

30. CD to Erasmus Darwin, June 19, 1858, *CCD* 7:110; CD to CL, June 26, 27, and 29, 1858, *CCD* 7:119–20; CD to William Darwin Fox, July 12, 1858, *CCD* 7:124.

31. The minutes of the Linnean Society indicate that from 1854 until the meeting on July 1, 1858, Lyell did not attend any of the society's gatherings—though his name might have been omitted on occasions when he was present. The fact that he was given top billing on July 1 suggests that his presence was unusual. Since Lyell was one of the most famous scientific men in England, it would have been surprising if his name was ever omitted when he was present. Wallace attended the Linnean Society meetings several times in 1853 as a guest of Samuel Stevens, but never when Joseph Hooker, Lyell, or Darwin was present (courtesy of the Linnean Society of London).

32. Quoted in Beddall, "Wallace, Darwin, and the Theory of Natural Selection," 303–4.

33. De Beer provides a copy of the introductory letter from Hooker and Lyell, the extracts from Darwin's manuscript of 1844, Darwin's letter to Asa Gray, and Wallace's essay, all presented at the July 1, 1858, meeting of the Linnean Society (*Charles Darwin and Alfred Russel Wallace*, 91–254, 257–58, 264–67.).

34. Quoted in Beddall, "Wallace, Darwin, and the Theory of Natural Selection," 303. In a letter to Francis Darwin, George Bentham described his reaction to the presentation at the Linnean Society of Darwin's and Wallace's papers (*CCD* 9:101n). However, P. F. Stevens believes that Bentham "misremembered" his reaction to the meeting, confusing his reaction to *The Origin of Species*—recorded in his diary on December 24, 1859 ("I have now finished carefully reading Darwin's book on the Origin of Species which has considerably modified my former opinions on their immutability")—with his alleged reaction to the joint presentation ("George Bentham and the Darwin/Wallace Papers of 1858: More Myths Surrounding the Origin and Acceptance of Evolutionary Ideas," *Linnean* 11 [1995]: 14–16). In fact, in order to accommodate the request of Hooker and Lyell, he agreed to table his paper "On the New Genera of Brazilian Plants" because it would take more than an hour to read. Like the president of the society, Thomas Bell, and the majority of others present that day, Bentham seems not to have grasped the revolutionary nature of Wallace's and Darwin's papers.

35. Thomas Bell, "Presidential Address to the Members of the Linnean Society," *JLS (Zoology)* 4 (1859): viii–ix.

36. THH to JH, September 5, 1858, Thomas Henry Huxley Papers, Imperial College of Science, Technology, and Medicine, London [catalogued in Warren R. Dawson, *The Huxley Papers: A Descriptive Catalogue of the Correspondence, Manuscripts*

and Miscellaneous Papers of the Rt. Hon. Thomas Henry Huxley . . . preserved in the Imperial College of Science and Technology, London (London: Macmillan, 1946), 2:35].

37. CD to JH, July 13, 1858, *CCD* 7:129.

38. CD to CL, July 18, 1858, *CCD* 7:137.

39. H. L. McKinney, *Wallace and Natural Selection* (New Haven, Conn.: Yale University Press, 1972), 139.

40. John Langdon Brooks, *Just Before the Origin: Alfred Russel Wallace's Theory of Evolution* (New York: Columbia University Press, 1984), 241.

41. Arnold Brackman, *A Delicate Arrangement: The Strange Case of Charles Darwin and Alfred Russel Wallace* (New York: Times Books, 1980), xi.

42. Frederick Burkhardt and Sydney Smith, introduction to *CCD*, 5:xix.

43. Ernst Mayr, *The Growth of Biological Thought: Diversity, Evolution, and Inheritance* (Cambridge, Mass: Harvard University Press, Belknap Press, 1982), 296. See also Browne, *Secular Ark,* 181; and Kottler, "Charles Darwin and Alfred Russel Wallace," 371–72.

44. Kottler, "Charles Darwin and Alfred Russel Wallace," 376–77.

45. Ibid., 381. See also Charles Darwin, *The Origin of Species by Means of Natural Selection or the Preservation of Favored Races in the Struggle for Life* (1859; reprint of 6th ed., New York: Modern Library, n.d.), 87.

46. ARW to JH, October 6, 1858, *CCD* 7:166.

47. ARW to Mary Anne Wallace, October 6, 1858, *LLW*, 58.

48. ARW to SS, October 29, 1858, ADD 7339.23, Cambridge University Library.

49. *ML*, 1:365–66.

50. Ibid., 363.

51. ARW to JH, October 6, 1858, *CCD* 7:166.

52. *MA*, 335–36.

53. ARW to SS, Batchian, October 29, 1858, excerpt read at the March 22, 1859, meeting of the Zoological Society, *PZL* 27 (1859): 129.

54. Ibid.

55. Philip Lutley Sclater, "Note on Wallace's Standard-wing, *Semioptera wallacii*," *Ibis*, 1st ser., 2 (1860): 26–28. Sclater renamed Wallace's specimen *Semioptera wallacii*, the name by which it is still known. The birds of paradise comprise a family called the Paradisaeidae, which includes forty-five species in seventeen genera. The standard-wing is today considered a true bird of paradise. The genus *Semioptera* is closely related to the New Guinean genus *Paradisaea*, which includes the great, lesser, and king birds of paradise (B. Coates and K. D. Bishop, *A Guide to the Birds of Wallacea* [Alderley, Queensland, Australia: Dove, 1997], 416–17).

56. CD to ARW, January 25, 1859, *LLW*, 111–12.

57. *SJ*, 166–67; CD to ARW, January 25, 1859, *LLW*, 111–12.

58. Thomas Boyd, in *Zoologist* 17 (1859), 6357–59.

59. ARW to Thomas Sims, April 1859, *ML*, 1:367–68.

60. Philip Lutley Sclater, "The Geographical Distribution of Birds," *JLS (Zoology)* 2 (1858): 130.

61. Alfred Russel Wallace, "On the Zoological Geography of the Malay Archipelago," *JLS (Zoology)* 4 (1859): 172–84. The "baboon-monkey" that Wallace alludes to

probably is *Macaca nigrescens*, which is not a true baboon but a macaque. Baboons and macaques belong to the same family of monkeys, Cercopithecidae. *Cynopithecus* is no longer accepted as the name of a genus.

62. CD to ARW, August 9, 1859, *LLW*, 114–15.

63. *JMA*, 3:190. See also Alfred Russel Wallace, "The Ornithology of Northern Celebes," *Ibis*, 1st ser., 2 (1860): 140–47.

64. CD to ARW, April 6, 1859, *LLW*, 112–13.

65. Wallace, "Ornithology of Northern Celebes," 146.

66. George Robert Gray, "List of Birds by Mr. Wallace at the Molucca Islands, with Descriptions of New Species, etc.," *PZL* 28 (1860): 341.

67. William Wilson Saunders, in *JLS (Zoology)* 5 (1861): 93.

68. Alfred Russel Wallace, "On the Ornithology of Ceram and Waigiou," *Ibis*, 1st ser., 3 (1861): 283–84. See also ARW to SS, November 26, 1859, *Ibis*, 1st ser., 2 (1860): 305.

69. CD to ARW, November 13, 1859, *LLW*, 115–16.

70. ARW to Thomas Sims, March 15, 1861, *LLW*, 62.

71. ARW to HWB, December 24, 1860, *LLW*, 59.

72. ARW to George Silk, September 1, 1860, *ML*, 1:371–72.

73. CD to ARW, May 18, 1860, *LLW*, 117–18.

74. Quoted in Ernst Mayr, introduction to Charles Darwin, *The Origin of Species*, facsimile of 1st ed. (Cambridge, Mass.: Harvard University Press, 1975), xv.

75. Review of *The Origin of Species*, by Charles Darwin, *AMNH*, 3rd ser., 5, no. 26 (1860): 132–43.

76. Asa Gray, review of *The Origin of Species*, by Charles Darwin, *AMNH*, 3rd ser., 6, no. 35 (1860): 373–86.

77. "On the 'Origin of Species' . . . a Review," *Athenaeum*, November 19, 1859, 659–60.

78. *LLH*, 2:94.

79. "Objects and Rules of the British Association for the Advancement of Science," in preface to every volume of the *Report of the British Association for the Advancement of Science.*

80. Burlington House, *A History of the British Association* (London: Lund, Humphries, 1922), 83–84.

81. "Science: British Association," *Athenaeum*, July 7, 1860, 18–19.

82. "Section D—Zoology and Botany, Including Physiology," *Athenaeum*, July 14, 1860, 64–66. According to the *Oxford English Dictionary*, the Herald's College, founded as a Royal Corporation in 1483, was the government's official organization for proving pedigrees and conferring coats-of-arms on families.

83. JH to CD, July 2, 1860, *LJH*, 1:525.

84. For two other accounts of this meeting, see *LJH*, 1:520–27, and *LLD*, 2:113–18. Perhaps the most thorough account of this episode is in J. R. Lucas, "Wilberforce and Huxley: A Legendary Encounter," *Historical Journal* 22 (1979): 313–30. Hooker's role in the affair has not been widely publicized. In later years, Hooker was concerned about his own and his family's reputation and rarely spoke publicly in support of Darwin's doctrine. It was left to Huxley and, later, Wallace, who were less cautious, to do so.

85. *JMA*, 4:200. See also *MA*, 340.

86. ARW to George Silk, September 1, 1860, *ML*, 1:371–72.

87. Wallace, "On the Ornithology of Ceram and Waigiou," 287.

88. Ibid., 291. See also *AMNH*, 3rd ser., 10, no. 60 (1862): 468.

89. In 1853 Mary Anne Wallace was living with her daughter, Fanny; in 1862, when Wallace returned to London, she was still living with Fanny. But it appears that sometime between 1853 and 1862, she lived briefly at Columbia Cottage on Albion Road in Hammersmith, to which Wallace addressed some of his letters to her.

90. ARW to Mary Anne Wallace, July 20, 1861, *LLW*, 67–68.

91. *MA*, 114–15.

92. ARW to FS, October 10, 1861, *LLW*, 69.

93. Alfred Russel Wallace, "Narrative of Search After Birds of Paradise," read at the May 27, 1862, meeting of the Zoological Society, *PZL* 30 (1862): 153–59.

94. ARW to George Silk, n.d., *LLW*, 380–81.

95. *ML*, 1:557.

96. John Gould, introduction to *Birds of New Guinea and the Adjacent Papuan Islands: Including Many New Species Recently Discovered in Australia*, vol. 1 (London: Sotheran, 1875). See also *MA*, 557.

97. ARW to Philip Sclater, March 31, 1862, Archives of the Zoological Society of London.

9. BEAUTIFUL DREAMER

1. *MA*, viii.

2. Thomas Henry Huxley, *Evidences as to Man's Place in Nature* (1863; reprint, London: Watts, 1921), 22.

3. "Review of *Naturalist on the River Amazons*," *Athenaeum*, April 11, 1863, 488.

4. *ML*, 1:385–86.

5. ARW to CD, April 7, 1862, *LLW*, 119.

6. SS to Norton Shaw, June 17, 1862, Correspondence Block 1861–1870, Archives of the Royal Geographical Society, London.

7. Tim Severin, *Spice Island Voyage* (New York: Carrol & Graf, 1997), photo on 8.

8. HWB to JH, April 30, 1862, Letters to J. D. Hooker, vol. 2, no. 45, Archives of the Royal Botanic Gardens, Kew.

9. Edward Clodd covered up the embarrassment by moving up Henry Bates's wedding date to 1861 ("Memoir of the Author," in *NRA*). But later biographers discovered that the wedding certificate listed January 1863 as the actual date of Bates's marriage to Sarah Ann Mason (Robert Sprecher, "The Darwin–Bates Letters: Correspondence Between Two Nineteenth-Century Travellers and Naturalists," *Annals of Science* 25, no. 1 [1969]: 41; H. P. Moon, *Henry Walter Bates, FRS [1825–1892]: Explorer, Scientist and Darwinian* [Leicester: Leicestershire Museums, Art Galleries and Record Service, 1976]).

10. Sprecher, "Darwin–Bates Letters," pl. 1; photographic portrait of Bates in the office of the director of the Royal Geographical Society; oil painting attributed to Thomas Sims, which is in fact a photograph by Sims colorized by his wife, Fanny.

11. Henry Walter Bates, "Contributions to an Insect Fauna of the Amazon Valley," *Transactions of the Linnean Society* 23 (1862): 502.

12. CD to HWB, November 20, 1862, quoted in Sprecher, "Darwin–Bates Letters," 36. See also *LLD*, 2:185.

13. HWB to CD, May 2, 1863, quoted in Sprecher, "Darwin–Bates Letters," 44–45.

14. CD to HWB, December 3, 1861, quoted in Sprecher, "Darwin–Bates Letters," 18–20.

15. ARW to HWB, quoted in Edward Clodd, *Memories* (London: Watts, 1926), 64–65.

16. CD to HWB, December 3, 1861, quoted in Sprecher, "Darwin–Bates Letters," 20.

17. Wallace states that he presented the paper on May 11, but the *Proceedings of the Zoological Society of London* gives the date of the reading as May 27. For the entire paper, see *ML*, 1:387–94.

18. *LLD*, 1:123.

19. *ML*, 2:2–3.

20. *LLD*, 1:136. See also Thomas J. Barloon and Russell Noyes, Jr., "Charles Darwin and Panic Disorder," *Journal of the American Medical Association*, January 8, 1997, 138–41; and Jeremy Adler, "The Dueling Diagnoses of Darwin," *Journal of the American Medical Association*, April 23–30, 1997, 1275. For a discussion of Chagas' disease, see "Chagas' Disease," in *Manson's Tropical Diseases*, ed. Gordon C. Cook (Philadelphia: Saunders, 1996), 1198–1212.

21. *ML*, 2:11.

22. *The Autobiography of Charles Darwin*, ed. Nora Barlow (1958; reprint, New York: Norton, 1993), 115. After Darwin's death in 1882, the family was divided about the portions of Darwin's autobiography that should be suppressed. Emma Darwin was especially concerned about her husband's remarks on religious issues. She and several other family members also censored passages that might be deemed offensive to friends. Barlow, Darwin's granddaughter, restored the censored passages (introduction, 11–12). For some reason, the passage quoted here was felt to be offensive and was omitted from *The Life and Letters of Charles Darwin*, which contains the expurgated version of Darwin's autobiography.

23. ARW to CD, August 8, 1862, *LLW*, 120–21.

24. *ML*, 2:45–51. In his autobiography, Wallace states that he was invited to Devonshire in 1863 or 1864, but his letter of August 8 to Darwin indicates that he first visited in August or September 1862. Wallace visited James Brooke several times between 1862 and 1864, a period that coincided with a dynastic crisis in the Brooke household. Brooke's nephew, Captain Brooke Brooke, the heir apparent to the throne of Sarawak, had been disinherited, in part because of pressure from Lady Burdett-Coutts. "My unfortunate habit of speaking my thoughts too plainly broke off the acquaintance," Wallace wrote. "I thought the captain had been rather hardly treated, and one day, when the subject was mentioned . . . I ventured to say so." Although Brooke continued to correspond with Wallace until his death in 1868, Burdett-Coutts made sure that the two men never saw each other again.

25. *MA*, 210–18; Alfred Russel Wallace, "On the Physical Geography of the Malay Archipelago," *JRGS* 33 (1863): 217–34.

26. Roderick Impey Murchison, "Presidential Address," *JRGS* 39 (1869): clxii–clxv.

27. Ernst Mayr, "Wallace's Line in the Light of Recent Zoogeographic Studies," *Quarterly Review of Biology* 19 (1944): 1–14.

28. *ML*, 1:386.

29. Charles Darwin, *The Origin of Species Species by Means of Natural Selection or the Preservation of Favored Races in the Struggle for Life* (1859; reprint of 6th ed., New York: Modern Library, n.d.), 198.

30. Ibid., 202.

31. Samuel Haughton, "On the Form of the Cells made by various Wasps and by the Honey Bee; with an Appendix on the 'Origin of Species,'" *AMNH*, 3rd ser., 11, no. 77 (1863): 415–29.

32. Alfred Russel Wallace, "Remarks on the Rev. S. Haughton's Paper on the Bee's Cells, and on the Origin of Species," *AMNH*, 3rd ser., 12, no. 82 (1863): 303–9.

33. CD to ARW, January 1, 1864, *LLW*, 122–23.

34. ARW to CD, September 30, 1862, *LLW*, 122.

35. Herbert Spencer, *First Principles* (New York: Collier, 1902), 35, 50–52, 59–60.

36. *Autobiography of Charles Darwin*, 91–94.

37. *ML*, 2:23.

38. Charles Lyell, *The Geological Evidences of the Antiquity of Man, with an Outline of Glacial and Post-Tertiary Geology*, 4th ed. (1873; reprint, New York: AMS Press, 1973), 98–104.

39. Ibid., 104. For a thorough explication of the ideological underpinnings of this paradigm shift, see George W. Stocking Jr., *Victorian Anthropology* (New York: Free Press, 1987).

40. *ML*, 1:206–11.

41. Alfred Russel Wallace, *The Wonderful Century: Its Successes and Its Failures* (New York: Dodd, Mead, 1899), 167.

42. For a full account of the history of phrenology, see Roger Cooter, *The Cultural Meaning of Popular Science* (Cambridge: Cambridge University Press, 1984); Samuel Greenblatt, "Phrenology in the Science and Culture of the 19th Century," *Neurosurgery* 37 (1995): 792–94; George Combe, *The Constitution of Man Considered in Relation to External Objects* (1834; reprint, Delmar, N.Y.: Scholars' Facsimiles & Reprints, 1974); and "Mr. George Combe and the Philosophy of Phrenology," *Fraser's Magazine*, November 1840, 509.

43. Wallace, *Wonderful Century*, 175–77.

44. Charles Darwin, *Voyage of the Beagle (Journal of Researches into the Geology and Natural History of the Various Countries Visited by the HMS Beagle)*, with an introduction by Janet Browne (1839; reprint, New York: Penguin, 1989), 172.

45. *ML*, 1:287–89.

46. *SJ*, 63; Richard Owen, "On the Anthropoid Apes, and Their Relations to Man," *Proceedings of the Royal Institution of Great Britain* 2 (1854–1858): 35–36.

47. *JMA*, 1:51. See also *MA*, 421–22.

48. Obituary of James Hunt, *Anthropological Review and Journal of the Anthropological Society of London* 8 (1870): lxxix–lxxxiii. The *Anthropological Review* and the *Journal of the Anthropological Society of London* were bound together; articles in the *Anthropological Review* are referred to by volume and number, while those in the *Journal* are referred to by only the number of the bound volume.

49. James Hunt, "Opening Address to the Anthropological Society of London on 24 February 1863," *Anthropological Review* 1, no. 1 (1863): 1–20.

50. K. Theodore Hoppen, *The Mid-Victorian Generation, 1846–1886* (Oxford: Oxford University Press, 1998), 230–31.

51. Stocking, *Victorian Anthropology*, 43, 48–53.

52. *ML*, 1:254–55. See also William Lawrence, *Lectures on Comparative Anatomy, Physiology, Zoology and the Natural History of Man* (London: Bell & Daldy, 1866); and James Cowles Prichard, *Researches into the Physical History of Mankind*, vol. 1 (London: Houlston and Stoneman, 1851).

53. ARW to THH, February 26, 1864, Thomas Henry Huxley Papers, Imperial College of Science, Technology, and Medicine, London [catalogued in Warren R. Dawson, *The Huxley Papers: A Descriptive Catalogue of the Correspondence, Manuscripts and Miscellaneous Papers of the Rt. Hon. Thomas Henry Huxley . . . preserved in the Imperial College of Science and Technology, London* (London: Macmillan, 1946), 28:61].

54. Hunt, "Opening Address," 5–6, 10.

55. Alfred Russel Wallace, "The Origin of Human Races and the Antiquity of Man Deduced from the Theory of 'Natural Selection,'" *Anthropological Review and Journal of the Anthropological Society of London* 2 (1864): clviii–clxx.

56. "Discussion" [March 1, 1864, meeting], *Anthropological Review and Journal of the Anthropological Society of London* 2 (1864): clxx–clxxxvii.

57. Henry F. J. Guppy, "Notes on the Capabilities of the Negro for Civilisation," *Anthropological Review and Journal of the Anthropological Society of London* 2 (1864): ccxiii; "Discussion."

58. JH to CD, May 14, 1864, *DAR*, 101:219.

59. CD to JH, May 22, 1864, *More Letters of Charles Darwin*, ed. Francis Darwin (London: Murray, 1903), 2:31–32.

60. CD to ARW, May 28, 1864, *LLW*, 127–28.

61. Ibid.

62. ARW to CD, May 29, 1864, *LLW*, 128–30.

63. Ibid.

64. Lyell, *Geological Evidences of the Antiquity of Man*, 336. Lyell proposed the term "Pleistocene" in 1839 to denote the epoch that began at the end of the Tertiary period and preceded the Recent, or present, epoch. (The Recent epoch has been renamed the Holocene, and the Pleistocene and Holocene make up the Quaternary period.) In 1830 Lyell had subdivided the Tertiary into four epochs: the Eocene (the oldest), Miocene, Older Pliocene, and Newer Pliocene. He also had invented the concept of a quantitative geologic chronometer to estimate the relative ages of any two Tertiary deposits. He did this by comparing the number of extant marine mollusk species with that of extinct species in the various geologic strata. In the Newer Pliocene, for

example, 90 to 95 percent of the marine mollusks were extant; in the Eocene, only 3 percent were extant. In the mid-nineteenth century, there was no reliable method of dating geologic strata, and Lyell preferred vague terminology like "very long time" or "great duration." "We cannot ascertain at present the limits, whether of the beginning or the end, of the first stone period, when Man coexisted with the extinct mammalia," he wrote, "but that it was of great duration we cannot doubt" (422–23). He noted that the earliest date that scientists could pinpoint with certainty was 776 B.C., the year of the first Olympiad in Greece. Estimations of earlier dates were possible if one knew the rates of denudation or elevation of various formations. For example, the submergence of Wales to 1,400 feet would require 56,000 years if the rate of submergence were 2.5 feet a century: "I am aware that it may be objected that the average rate here proposed is a purely arbitrary and conjectural one" (334–35).

65. CL to ARW, May 22, 1864, Additional MSS 46435, nos. 13–15, Department of Manuscripts, British Library, London. A portion of the letter is in *ML*, 1:418.

66. ARW to CL, May 24, 1864, *ML*, 1:419.

67. *ML*, 1:433.

68. CD to JH, March 13, 1863, *LLD*, 2:196–97; CD to CL, March 6, 1863, *LLD*, 2:196.

69. Lyell, *Geological Evidences of the Antiquity of Man*, 551–52.

70. *ML*, 2:24–25.

71. Alfred Russel Wallace, "On the Progress of Civilisation in Northern Celebes," *Reader*, October 15, 1864, 485–86. See also Meeting of the British Association, September 19, 1864, Section E, Geography and Ethnology, *Report of the British Association for the Advancement of Science* (1864): 149–50.

72. "Anthropology at the British Association," *Anthropological Review* 2, no. 7 (1864): 332–34. The tense has been altered to reflect a live discussion.

73. *ML*, 1:382.

74. John William Colenso, "On the Efforts of Missionaries Among Savages," *Anthropological Review and Journal of the Anthropological Society of London* 3 (1865): ccxliviii–cclxxii. See also "Discussion" [May 16, 1865, meeting], *Anthropological Review and Journal of the Anthropological Society of London* 3 (1865).

75. John Stuart Mill, "Public Responsibility and the Ballot," *Reader*, April 29, 1865, 474–75; Alfred Russel Wallace, reply to John Stuart Mill, *Reader*, May 6, 1865, 517.

76. CD to ARW, 22 September 1865, *LLW*, 135.

77. ARW to CD, October 2, 1865, *LLW*, 136–37.

78. Review of *Prehistoric Times*, by John Lubbock, *Anthropological Review* 3, no. 10 (1865): 338–46.

79. Thomas Paine, *The Age of Reason* (1794; reprint, New York: Buccaneer Books, 1976), 6–10, 181–82.

80. JH to CD, October 6, 1865, *LJH*, 2:54.

10. A TURN TOWARD THE UNKNOWABLE

1. *ML*, 1:365.

2. *ML*, 2:399.

3. *ML*, 1:433, 2:399.

4. Alfred Russel Wallace, "On the Insects Used for Food by the Indians of the Amazon," *Transactions of the Entomological Society of London*, n.s., 2, pt. 8 (1854): 241–44.

5. *JMA*, 1:36. See also *MA*, 226–27.

6. *ML*, 1:262–63.

7. ARW to George Silk, January 20, 1862, *LLW*, 70.

8. *ML*, 1:409.

9. ARW to AN, February 19, 1865, Newton Papers, no. 186a, Cambridge University Library. See also *ML*, 1:409–11.

10. *ML*, 1:414.

11. *LLH*, 1:74–76.

12. RS to George Bentham, March 16, 1858, Richard Spruce, *Notes of a Botanist on the Amazon & Andes . . . During the Years 1849–1864*, ed. Alfred Russel Wallace (London: Macmillan, 1908), 2:203–4.

13. Clements R. Markham, unofficial history of the Royal Geographical Society, ca. 1901, 368–69; RGS Council Minute Book (1841–1865), April 11, 1864, 1:272, Archives of the Royal Geographical Society, London.

14. *ML*, 1:415.

15. *ML*, 1:410.

16. Ibid. Miss L. has been identified as Marion Leslie by Peter Raby, who also identified her father as Lewis Leslie, an auctioneer (*Alfred Russel Wallace: A Life* [Princeton, N.J.: Princeton University Press, 2001], 171, 181).The Leslies lived in fashionable Kensington; Wallace, his mother, and Thomas and Fanny Sims lived in the less fashionable Notting Hill.

17. ARW to CD, January 20, 1865, *DAR*, 2nd ser., 106–7:20–21.

18. CD to ARW, January 29, 1865, *LLW*, 132.

19. ARW to AN, February 19, 1865, Newton Papers, no. 186a.

20. The sources consulted for this brief history of spiritualism include Janet Oppenheim, *The Other World: Spiritualism and Psychical Research in England, 1850–1914* (Cambridge: Cambridge University Press, 1988), 217–18; Arthur Conan Doyle, *The History of Spiritualism* (1926; reprint, New York: Arno Press, 1975), 1:62–84; Alfred Russel Wallace, "The Scientific Aspect of the Supernatural" (1866), in *Miracles and Modern Spiritualism* (1896; reprint, New York: Arno Press, 1975), 152–59; and Carl Jung, "On Spiritualistic Phenomena," in *Psychology and the Occult*, ed. and trans. R. F. C. Hull, Bollingen series 20 (Princeton, N.J.: Princeton University Press, 1977), 92–93.

21. Alfred Russel Wallace, "A Defence of Modern Spiritualism" (1874), in *Miracles and Modern Spiritualism*, 157–58.

22. Wallace, "Scientific Aspect of the Supernatural," 132; *ML*, 2:294; *Athenaeum*, May 5, 1860, 612–13.

23. Alfred Russel Wallace, *The Wonderful Century: Its Successes and Its Failures* (New York: Dodd, Mead, 1899), 195.

24. Adrian Desmond and James Moore, *Darwin* (New York: Norton, 1994), 27.

25. A lengthy discussion of the controversy surrounding John Elliotson is in "Animal Magnetism: Third Report of Facts and Experiments," *Lancet* 1 (1837–1838): 441–46. In the same issue, see also "Letter to the Editor on Animal Magnetism,"

454–55; "The Life and Death of Animal Magnetism," 834–36; and "The Faculties of Elizabeth O'Key," 873–77. More on the subject is in "The Animal Magnetism Fraud and Humbug," *Lancet* 2 (1838–1839): 380–81. In the same issue, see also "The Immoral Tendency of Animal Magnetism," 450–51; "Resignation of Dr. Elliotson," 561–62; and "Respecting Dr. Elliotson's Resignation," 590–97.

26. *ML*, 1:234–35.

27. Ibid.

28. Ibid.

29. Wallace, "Scientific Aspect of the Supernatural," 132.

30. Ibid., 132–34.

31. Ibid., 134–35.

32. Ibid., 135–36.

33. Ibid., 136–37.

34. Ibid., 138.

35. Ibid., 127, 130.

36. James Hunt, "On Physio-Anthropology, Its Aim and Method," *Anthropological Review and Journal of the Anthropological Society of London* 5 (1867): cclvi–cclvii. See also "Discussion" [June 4, 1867, meeting], *Anthropological Review and Journal of the Anthropological Society of London* 5 (1867).

37. Hunt, "On Physio-Anthropology."

38. ARW to CD, October 2, 1865, *LLW*, 136–37.

39. RS to William Mitten, October 29, 1864, Letters to Spruce (1842–1890), no. 163, Archives of the Royal Botanic Gardens, Kew. See also Duncan Porter, "With Humboldt, Wallace and Spruce at San Carlos de Rio," in *Richard Spruce, Botanist and Explorer*, ed. M. R. D. Seaward and S. M. D. Fitzgerald (Kew: Royal Botanic Gardens, 1996), 61; CD to ARW, January 2, 1864, *LLW*, 123–24; Spruce, *Notes of a Botanist*, 2:320–21; David Robinson, *Historic Hurstpierpoint in Picture Postcards* (courtesy of M. A. Hayes, librarian, Sussex County, England); and M. R. D. Seaward and S. M. D. Fitzgerald, introduction to *Richard Spruce, Botanist and Explorer* (Kew: Royal Botanic Gardens, 1996), 8.

40. A copy of the marriage certificate states: "Marriage solemnized at the Parish church of Hurstpierpoint, County Sussex

CONDITION	*RANK OR PROFESSION*	*RESIDENCE*
Alfred Russel Wallace, 42	*bachelor gentleman*	*St Pancras, London*
Annie Mitten, 20	*spinster*	*Hurstpierpoint*

in the presence of Philip Smith and Rose Elizabeth Mitten—April 5, 1866" (Parish Register Transcripts of Hurstpierpoint [1608–1875], microfilm). According to the Baptismal Record of Hurstpierpoint, Annie was baptized on April 19, 1846 (microfilm no. 919093, p. 9, no. 72, International Genealogical Index, held by the Church of Jesus Christ of Latter-day Saints, Salt Lake City, Utah). Wallace, who was not accurate with dates, guessed that she was about eighteen when they married, but she was actually twenty.

41. Wallace, "Scientific Aspect of the Supernatural," 139–41.

42. Wallace, "Defence of Modern Spiritualism," 169–70.

43. Alfred Russel Wallace, account of séance, *Spiritualist Magazine*, February 1867, quoted in "Anthropological News: 'Science and Spiritualism,'" *Anthropological Review* 5, no. 17 (1867): 242–43.

44. Alfred Russel Wallace, preface to *On Miracles and Modern Spiritualism* (London: Burns, 1874).

45. *ML*, 2:296.

46. Edwin Ray Lankester, "Dr. Carpenter, C.B., F.R.S." [obituary], *Nature*, November 26, 1885, 83–85. See also B. G. Gardiner, "Picture Quiz: William Benjamin Carpenter," *Linnean* 16 (2000): 7–13.

47. *ML*, 2:42–43.

48. WBC to ARW, April 9, 1864, MS 46439, no. 3, Department of Manuscripts, British Library, London.

49. *ML*, 2:296–97.

50. *The Scientific Aspect of the Supernatural*, with its inscription, is in the Oxford University Museum of Natural History.

51. Wallace, *Miracles and Modern Spritiualism*, 42–43, 101. Unfortunately, James Challis's main claim to fame was his failure to discover the planet Neptune, the existence of which was predicted by two other astronomers. The night on which Challis was supposed to make his crucial observation. he was distracted by his wife's offer of tea. By the time he and his friend William Kingsley arrived at the telescope, the sky had clouded over. A German astronomer soon announced his discovery of the new planet.

52. Edward Burnett Tylor to ARW, November 26, 1866, MS 46439, no. 6.

53. *ML*, 2:298.

54. Ibid., 299.

55. "Anthropological News: 'Science and Spiritualism."

56. Robert Chambers to ARW, February 20, 1867, MS 46439, nos. 15–16.

57. *ML*, 2:302.

58. Ibid., 297.

59. ARW to the editor of *Pall Mall Gazette*, May 1868, MS 46439, nos. 35–43. For a synopsis of this letter, which was not published in the magazine, see *ML*, 2:300.

11. THE OLYMPIAN HEIGHTS AND THE BEGINNINGS OF THE FALL

1. ARW to CD, July 2, 1866, *LLW*, 140–43.

2. CD to ARW, July 5, 1866, *LLW*, 144–45.

3. British Association for the Advancement of Science, meeting of August 1866, *Quarterly Journal of Science*, October 1866, 541.

4. "Transactions of the Sections, Anthropology," *Report of the British Association for the Advancement of Science* (1866): 93–95.

5. James Hunt, "On the Application of the Principle of Natural Selection to Anthropology: In Reply to Views Advocated by Some of Mr. Darwin's Disciples," *Anthropological Review* 4, no. 15 (1866): 320–40.

6. "Discussion," *Anthropological Review* 4, no. 15 (1866): 391–408.

7. CD to ARW, February 23, 1867, *LLW*, 146–47.

8. Alfred Russel Wallace, "Mimicry, and Other Protective Resemblances Among Animals" (1867), in *Contributions to the Theory of Natural Selection* (1870; reprint, New York: AMS Press, 1973), 45–129.

9. *ML*, 2:3.

10. CD to ARW, February 26, 1867, *LLW*, 148–49.

11. ARW to CD, March 10, 1869, *LLW*, 193–94.

12. *ML*, 2:4–6.

13. RS to William Mitten, October 13, 1867, Letters to Spruce (1842–1890), Archives of the Royal Botanic Gardens, Kew; Wallace family prayer book (courtesy of Richard Wallace); CD to ARW, October 12 and 13, 1867, *LLW*, 157.

14. ARW to CD, February 7, 1868, *LLW*, 159.

15. Ernst Mayr, "Darwin, Wallace, and the Origin of Isolating Mechanisms," in *Evolution and the Diversity of Life: Selected Essays* (Cambridge, Mass.: Harvard University Press, Belknap Press, 1997), 129–30.

16. Charles Darwin, *The Variation of Animals and Plants Under Domestication*, 2nd ed. (London: Murray, 1888), 2:170–71.

17. CD and ARW correspondence, late February to April 6, 1868, *LLW*, 161–72.

18. ARW to CD, April 8, 1868, *LLW*, 172–73. See also Mayr, "Darwin, Wallace," 130–33. "In retrospect," Mayr says, "it is clear that in spite of the brilliance of analysis on both sides, the problem at the time was insoluble, because of lack of knowledge not only about sterility, but also about the other isolating mechanisms, genetics, and population structure. It was, though, while it lasted, a noble debate" (133). According to Mayr, recent work suggests that the greater part of the genetic basis of these isolating mechanisms is an incidental by-product of the genetic divergence of isolated gene pools and is acquired during this isolation, thus supporting Darwin's view. But Wallace would come back to the debate twenty-five years later, and new evidence supports his contention.

19. CD to ARW, February 22 and March 17, 1868, *LLW*, 160, 166.

20. ARW to CD, March 19, 1868, *LLW*, 167; CD to ARW, March 19 to 24 [*sic*] and April 30, 1868, *LLW*, 176–77.

21. Alfred Russel Wallace, "A Theory of Birds' Nests; Shewing the Relation of Certain Differences of Colour in Birds to Their Mode of Nidification" (1868), in *Contributions to the Theory of Natural Selection*, 231–63. Wallace announces his "law" on 240.

22. CD to ARW, April 15, 1868, *LLW*, 175.

23. Wallace later wrote that he had gone to Switzerland in June 1867 (*ML*, 1:412), but he actually went in June 1868, according to William Mitten to RS, June 7, 1868, Letters to Spruce (1842–1890).

24. Edward Lurie, *Louis Agassiz: A Life in Science* (Chicago: University of Chicago Press, 1960), 95–98.

25. ARW to William Mitten (now lost); William Mitten to RS, August 10, 1868, Letters to Spruce (1842–1890). See also *ML*, 1:412–14.

26. ARW to CD, August 16, 1868, *LLW*, 179–80.

27. British Association for the Advancement of Science, report, "Presidential Address," *Quarterly Journal of Science*, October 1868, 505–6.

28. Alfred Russel Wallace, "Creation by Law," *Quarterly Journal of Science*,

October 1867, 471–88, reprinted in *Contributions to the Theory of Natural Selection*, 264–301.

29. ARW to CD, August 30, 1868, *LLW*, 181–82.

30. ARW to CD, September 5, 1868, *LLW*, 182–83.

31. ARW to CD [incomplete], September 18, 1868, *LLW*, 183–85; CD to ARW, September 23, 1868, *LLW*, 185–86; ARW to CD [postscript], September 27, 1868, *LLW*, 188.

32. ARW to AN, September 27, 1874, Newton Papers, no. 307, Cambridge University Library.

33. Douglas McKie, "The Origins and Foundation of the Royal Society of London," *Notes and Records of the Royal Society of London* 15 (1960): 1.

34. Roy M. MacLeod, "Of Medals and Men: A Reward System in Victorian Science, 1826–1914," *Notes and Records of the Royal Society of London* 26 (1971): 81–108.

35. CL to ARW, January 24, 1868, MS 46435, nos. 70–72, Department of Manuscripts, British Library, London.

36. Minutes of the meetings of the Royal Society (courtesy of Mrs. Gill Jackson).

37. JH to CD, May 17, 1867, *DAR*, 102:163; CD to JH, May 21, 1867, *DAR*, 94:26–27.

38. CL to ARW, April 4 and October 22, 1867, MS 46435, nos. 37–39, 54–55.

39. Charles Lyell, "Theories as to the Nature of Species, and Darwin on Natural Selection," in *Principles of Geology, or the Modern Changes of the Earth and Its Inhabitants*, 11th ed. (New York: Appleton, 1872), 276–84.

40. Thomas Henry Huxley, "On the Classification and Distribution of the Alectoromorphae and Heteromorphae," *PZL* (1868): 313.

41. "Anniversary Meeting, President's Address, November 30, 1868," *Proceedings of the Royal Society of London* 17 (1869): 148.

42. *ML*, 2:35.

43. Death certificate of Mary Anne Wallace, General Records Office, United Kingdom, Registration District Horsham, PAS 704914/2000, number 602. I express my gratitude to both Francis Langley, West Sussex County Record Office, who reviewed records from Ifield, and Robert W. O'Hara, who patiently searched death records from throughout Great Britain, to find Mary Anne Wallace in the death registry.

44. *ML*, 2:377–80.

45. Alexander Macmillan to Harper Brothers, October 21, 1868, *Letters of Alexander Macmillan*, ed. George Macmillan (Glasgow: University Press, 1908), 248–49 (collections of the Newberry Library, Chicago).

46. Copy of contract between Wallace and Macmillan & Company, Archives of Macmillan & Company (courtesy of John Handford).

47. ARW to CD, January 20, 1868, *LLW*, 190–91.

48. CD to CL, March 20, 1869, APS 367, American Philosophical Society, Philadelphia.

49. CL to ARW, March 13, 1869, *LLW*, 287.

50. Canon Charles Kingsley to ARW, May 5, 1869, *LLW*, 287; ARW to Charles Kingsley, May 7, 1869, Special Collections and Archives, Knox College Library, Galesburg, Ill. Kingsley was a theologian and liberal thinker whose book *The Water-Babies: A Fairy Tale for a Land-Baby*, an indictment of the cruelties of the English class and economic system as affecting as Charles Dickens's *Oliver Twist*, was widely admired.

51. Richard Curle, *The Last Twelve Years of Joseph Conrad* (London: Sampson, Low, Marston, 1929), 120–21. See also Florence Clemens, "Conrad's Favorite Bedside Book," *South Atlantic Quarterly* 38 (1939): 305–15.

52. *ML*, 2:397.

53. *MA*, 596–99.

54. CD to ARW, March 27, 1869, *LLW*, 197.

55. CD to ARW, April 14, 1869, *LLW*, 198–99.

56. According to James Marchant, who examined Darwin's original copy of Wallace's essay (*LLW*, 197).

57. ARW to CD, April 18, 1869, *LLW*, 199–200.

58. CD to CL, May 4, 1869, *LLD*, 2:297–98; CL to CD, May 5, 1869, *LLL*, 2:442–43. Charles Lyell's letter to Wallace has not been preserved.

59. CL to ARW, January 24, 1868, MS 46435, nos. 70–72.

60. CL to ARW, July 21, 1868, MS 46435, nos. 68–69.

61. CL to ARW, March 13, 1869, MS 46435, nos. 101–2.

62. Roy M. MacLeod, "The X Club: A Social Network of English Science in Late-Victorian England," *Notes and Records of the Royal Society* 24 (1970): 309. The other members were George Busk, the secretary of the Linnean Society and a noted surgeon and anatomist; Thomas Archer Hirst, a mathematician at University College, London; the archaeologist John Lubbock; the chemist Edward Frankland; Herbert Spencer; John Tyndall; and William Spottiswoode, a mathematician, philosopher, and linguist.

63. *LLH*, 1:276–80; *LJH*, 1:540–41.

64. Walter Gratzer, *A Bedside Nature: Genius and Eccentricity in Science, 1869–1953* (New York: Freeman, 1997), xi.

65. Alfred Russel Wallace, "Government Aid to Science" [letter to the editor], *Nature*, January 13, 1870, 288–89.

66. "Government Aid to Science" [editorial], *Nature*, January 13, 1870, 279–80.

67. Alfred Russel Wallace, "Government Aid to Science" [letter to the editor], *Nature*, January 20, 1870, 315.

68. *ML*, 2:381.

69. *Field, the Country Gentleman's Newspaper*, February 12, 1870, 149; *ML*, 2:382.

70. *ML*, 2:382–86.

71. J. J. Walsh, "Experimental Proof of the Rotundity of our Earth," *Field, the Country Gentleman's Newspaper*, March 5, 1870, 199; *ML*, 2:382–83; Walsh, in *Field, the Country Gentleman's Newspaper*, March 26, 1870, 266.

72. John Hampden, letter to the editor, *Field, the Country Gentleman's Newspaper*, April 2, 1870, 305.

73. Parallax, letter to the editor, *Field, the Country Gentleman's Newspaper*, April 16, 1870, 347. It is not clear who Parallax was. He may have been Samuel Birley Rowbotham, who in 1832 founded the Universal Zetetic Society in the United States and Great Britain; the society promoted the belief that the earth was flat.

74. Alfred Russel Wallace, letter to the editor, *Field, the Country Gentleman's Newspaper*, April 16, 1870, 347.

75. *ML*, 2:387–88.

76. CD to ARW, July 12, 1871, *LLW*, 220–21.

77. ARW to Robert McLachlan, secretary of the Entomological Society, May 17, 1871, Oxford University Museum of Natural History; *ML*, 2:388.

78. ARW to Robert McLachlan, October 26, 1871, Oxford University Museum of Natural History.

79. ARW to Robert McLachlan, November 25 and 28, 1871, Oxford University Museum of Natural History.

80. *ML*, 2:388–90.

12. WALLACE AND *THE DESCENT OF MAN*

1. CD to ARW, March 31, 1870, *LLW*, 206.

2. Alfred Russel Wallace, preface to *Contributions to the Theory of Natural Selection* (1870; reprint, New York: AMS Press, 1973), iv–v.

3. Alfred Russel Wallace, "The Limits of Natural Selection as Applied to Man," in *Contributions to the Theory of Natural Selection*, 361. See also "British Association for the Advancement of Science, Norwich, Friday, August 21," *Times* (London), August 22, 1868, 4. The full text of John Tyndall's address was printed in the *Report of the British Association for the Advancement of Science (Norwich)* 38 (1868): 1–6.

4. Thomas Henry Huxley, *Darwinian Essays* (1871; reprint, New York: AMS Press, 1970), 161.

5. Wallace, "Limits of Natural Selection as Applied to Man," 364–71.

6. CD to ARW, January 26 and March 31, 1870, *LLW*, 205–6.

7. HWB to CD, May 20, 1870, quoted in Robert Sprecher, "The Darwin–Bates Letters: Correspondence Between Two Nineteenth-Century Travellers and Naturalists," *Annals of Science* 25, no. 2 (1969): 117.

8. R. E. Claparède, "Critique of A. R. Wallace's *Contributions to the Theory of Natural Selection* in 'Scientific Serials,'" *Nature*, August 11, 1870, 306; ARW to CD, July 6, 1870, *LLW*, 207.

9. CD to JH, July 8, 1870, *DAR* 94:177–78.

10. P. S., "Wallace on Natural Selection," *Nature*, October 13, 1870, 471–73.

11. Alfred Russel Wallace, "Man and Natural Selection," *Nature*, November 3, 1870, 8–9.

12. Anton Dohrn, "Contributions to the Theory of Natural Selection," *Academy*, February 15, 1871, 138–40; and "Contributions to the Theory of Natural Selection," *Academy*, March 1, 1871, 159–60.

13. Huxley, *Darwinian Essays*, 173–77, 165–69.

14. JH to CD, November 1871, *LJH*, 2:130.

15. Sigmund Freud, *Introductory Lectures on Psychoanalysis*, trans. James Strachey (New York: Norton, 1977), 285.

16. *The Autobiography of Charles Darwin*, ed. Nora Barlow (1958; Reprint, New York: Norton, 1993), 126.

17. *ML*, 2:7.

18. ARW to CD, November 24, 1870, *LLW*, 209.

19. Charles Darwin, *The Descent of Man, and Selection in Relation to Sex*, 2 vols.

(1871; reprint, Princeton, N.J.: Princeton University Press, 1981). Darwin referred to Wallace fifty-four times in his text. The next most frequently referenced individual is Edward Burnett Tylor, who is mentioned thirty-nine times.

20. Ibid., 1:26.

21. Ibid., 68.

22. Ibid., 104.

23. Ibid., 231–35.

24. Ibid., 279.

25. Ibid., 403–9.

26. Ibid., 2:93.

27. Ibid., 132–41.

28. Ibid., 155, 167, 171.

29. Ibid., 187–89.

30. Ibid., 200–201.

31. Ibid., 223–24.

32. Ibid., 302–15.

33. Ibid., 327–82, 376n.

34. ARW to CD, March 11, 1871, *LLW*, 213–14.

35. Alfred Russel Wallace, review of *The Descent of Man, and Selection in Relation to Sex*, by Charles Darwin, *The Academy*, March 15, 1871, 177–83.

36. This point is clearly laid out by Wallace in "The Origin of Species and Genera," *Nineteenth Century* 7 (1880): 95.

37. Darwin, *Descent of Man*, 1:70–71.

38. CD to Henrietta Darwin, March 28, 1871, and Erasmus Darwin to Henrietta Darwin, March [?], 1871, *Emma Darwin: A Century of Family Letters, 1792–1896*, ed. Henrietta [Darwin] Litchfield (New York: Appleton, 1915), 2:202–3.

39. Helena Cronin, *The Ant and the Peacock* (Cambridge: Cambridge University Press, 1994), 234, 246.

13. THE DESCENT OF WALLACE

1. ARW to CD, November 24, 1870, and March 11, 1871, *ML*, 2:212–13; abstract of transcript in the Thurrock Library Archives, Grays.

2. Advertisement for The Dell, 1876, Thurrock Library Archives, Grays (courtesy of Victor W. Tucker).

3. CD to ARW, July 27, 1872, *LLW*, 223.

4. Charles H. Smith provides a comprehensive list of Wallace's publications in *Alfred Russel Wallace: An Anthology of His Shorter Writings*, ed. Charles H. Smith (Oxford: Oxford University Press, 1991).

5. HWB to CD, November 15, 1873, quoted in Robert Sprecher, "The Darwin–Bates Letters: Correspondence Between Two Nineteenth-Century Travellers and Naturalists," *Annals of Science* 25, no. 2 (1969): 118.

6. ARW to CD, November 18, 1873, *LLW*, 231–32; CD to ARW, November 19, 1873, *LLW*, 232–33.

7. *ML*, 2:60–61. Augustus Mongredien also had radical political views, and his ideas about free trade influenced Wallace, who would not openly express them for another decade.

8. *ML*, 2:95.

9. ARW to AN, December 28, 1870, Newton Papers, no. 251a, Cambridge University Library.

10. Alfred Russel Wallace, preface to *Geographical Distribution of Animals, with a Study of the Relations of Living and Extinct Faunas as Elucidating the Past Changes of the Earth's Surface* (London: Macmillan, 1876), vii; *ML*, 2:94.

11. Joe D. Burchfield, *Lord Kelvin and the Age of the Earth* (Chicago: University of Chicago Press, 1990), 47, 52, 71, 80.

12. Adrian Desmond, *Archetypes and Ancestors* (Chicago: University of Chicago Press, 1982), 179; E. H. Jellinek, "Dr. H. C. Bastian, Scientific Jekyll and Hyde," *Lancet*, December 23, 2000, 2180–83; *ML*, 2:11; CD to ARW, July 12, 1871, *LLW*, 220–21.

13. Alfred Russel Wallace, "The Beginnings of Life," part 2, *Nature*, August 15, 1872, 299–303.

14. CD to ARW, August 28, 1872, *LLW*, 225–26; ARW to CD, August 31, 1872, *LLW*, 226–27.

15. Anton Dohrn to CD, August 20, 1872, *DAR*, 162:209.

16. Alfred Russel Wallace, letter to the editor, *Nature*, February 20, 1873, 303.

17. Responses to Wallace's letter to the editor arrived from Alfred Bennett (*Nature*, February 27, 1873); Charles Kingsley (March 6, 1873); the editor (March 7, 1873); Charles Darwin (March 13, 1873), who offered the example of turtles returning to Ascension, an island in the South Atlantic; W. H. Brewer (March 13, 1873); and J. T. (March 20, 1873). On March 20, the lead editorial argued against Wallace's "smell-hypothesis." Other letters "poured in" but were not published. Arthur Davis Hasler, a professor of biology at the University of Wisconsin, developed the hypothesis of "olfactory imprinting" in the 1940s to explain how salmon could journey thousands of miles to spawn in the precise stream of their birth (obituary of Arthur Davis Hasler, *New York Times*, March 31, 2001, A15). The salmon found their way back because of a finely honed, ingrained sense of smell. Hasler evidently was not aware of Wallace's hypothesis, developed eighty years earlier.

18. ARW to CD, May 16, 1871, *LLW*, 216–17.

19. ARW to AB, December 25, 1870, *LLW*, 419.

20. ARW to AB, February 28, 1874, *LLW*, 421.

21. AB to ARW, May 26, 1874, MS 46439, nos. 95–97, Department of Manuscripts, British Library, London.

22. E. E. Fournier D'Albe, *The Life of Sir William Crookes*, with a foreword by Sir Oliver Lodge (London: Fisher Unwin, 1923), 12; ARW to William Wallace, May 4, 1893 (courtesy of Richard Wallace).

23. Arthur Conan Doyle, *The History of Spiritualism* (1926; reprint, New York: Arno Press, 1975), 1:186–88.

24. Alfred Russel Wallace, "A Defence of Modern Spiritualism" (1874), in *Miracles and Modern Spiritualism* (1896; reprint, New York: Arno Press, 1975), 166–67.

25. William Crookes, "Spiritualism Viewed by the Light of Modern Science," *Quarterly Journal of Science*, July 1870, 316–20.

26. William Crookes, "Some Further Experiments on Psychic Force," *Quarterly Journal of Science*, October 1871, 472.

27. ARW to AB, June 1, 1871, *LLW*, 420.

28. Ibid.

29. Crookes, "Some Further Experiments on Psychic Force," 478–79, 490.

30. ARW to William Crookes, October 8, 1871, Darwin Collection of the Dittrick Medical History Center, Case Western Reserve University, Cleveland.

31. Crookes, "Some Further Experiments on Psychic Force," 482–83, 472.

32. "Spiritualism and Science," *Times* (London), December 26, 1872, 5.

33. Dialectical Society, *Report on Spiritualism of the Committee of the London Dialectical Society Together with the Evidence, Oral and Written, and a Selection from the Correspondence* (London: Burns, 1873), correspondence, iv, 229.

34. Ibid., 1–13.

35. "Spiritualism and Science."

36. William Benjamin Carpenter, "Epidemic Delusions," *Popular Science Monthly* 2 (1872–1873): 15–36.

37. Henrietta [Darwin] Litchfield, ed., *Emma Darwin: A Century of Family Letters, 1792–1896* (New York: Appleton, 1915), 2:216–17; CD to JH, January 18, 1874, *DAR*, 95:311.

38. THH to George Darwin, January 22 and 28, 1874, Thomas Henry Huxley Papers, Imperial College of Science, Technology, and Medicine, London [catalogued in Warren R. Dawson, *The Huxley Papers: A Descriptive Catalogue of the Correspondence, Manuscripts and Miscellaneous Papers of the Rt. Hon. Thomas Henry Huxley . . . preserved in the Imperial College of Science and Technology, London* (London: Macmillan, 1946), 13:87, 89].

39. *LLH*, 1:422–25; George Darwin to THH, January 30 and 28, 1874, Thomas Henry Huxley Papers [catalogued in Dawson, *Huxley Papers*, 13:91, 89].

40. CD to Francis Galton, April 21, 1874, Darwin Collection of the Dittrick Medical History Center.

41. CD to Francis Galton, March 29, 1874, Darwin Collection of the Dittrick Medical History Center.

42. Wallace, "Defence of Modern Spiritualism," 189n.

43. Ibid., 197–98n.

44. ARW to AB, April 24, 1874, *LLW*, 422.

45. International Genealogical Index, baptismal records for the United Kingdom, for Herbert Spencer Wallace, christened April 24, 1874, at Hurstpierpoint, collection of the Newberry Library, Chicago; death certificate of Herbert Spencer Wallace, General Records Office, Registration District Cuckfield, subdistrict of Hurstpierpoint, County of Sussex, no 482.

46. For Arabella Buckley's comment to James Marchant about Wallace's reaction to Bertie's death, see *LLW*, 422.

47. Wallace, "Defence of Modern Spiritualism," 146–52.

48. Ibid., 210–11.

49. Ibid., 228–29.

50. AB to ARW, May 26, 1874, MS 46439, nos. 95–97.

51. CD to ARW, December 14, 1874, Darwin Collection of the Dittrick Medical History Center.

52. CD to AB, February 23, 1875, *LLD*, 2:373–74; CD to JH, February 25, 1875, *LLD*, 2:374–75; *ML*, 1:435.

53. ARW to CD, July 23, 1876, *LLW*, 241–43; ARW to CD, December 6, 1874, *LLW*, 233.

54. Wallace, preface to *Geographical Distribution of Animals*, viii–ix.

55. *ML*, 2:95–98.

56. ARW to AN, February 14 and May 26, 1875, Newton Papers, nos. 276a and 290.

57. ARW to G. L. Craik, February 25, 1875, Macmillan–Wallace letters, nos. 1–4, Department of Manuscripts, British Library, London.

58. ARW to G. L. Craik, February 2, 1876, Macmillan–Wallace letters, nos. 1–4.

59. ARW to G. L. Craik, March 26, 1876, Macmillan–Wallace letters, nos. 1–4.

60. "Wallace's Geographical Distribution of Animals," part 1, *Nature*, June 22, 1876, 165–68; "Wallace's Geographical Distribution of Animals," part 2, *Nature*, June 29, 1876, 186–89.

61. Alfred Newton, in *Athenaeum*, September 16, 1876, 367.

62. Joseph Hooker, address to the Geography Section of the British Association for the Advancement of Science, excerpt in *Nature*, September 8, 1881, 448.

63. *ML*, 2:94; Wallace, preface to *Geographical Distribution of Animals*, xv.

64. CD to ARW, June 5, 1876, *LLW*, 234–35.

65. Philip J. Darlington, *Zoogeography: The Geographical Distribution of Animals* (New York: Wiley, 1957), 10–12, 14, 21, 29. See also Ernst Mayr, "What Is a Fauna Is a Collective Noun," in *Evolution and the Diversity of Life: Selected Essays* (Cambridge, Mass.: Harvard University Press, Belknap Press, 1997), 552–53.

66. ARW to Alexander Macmillan, March 30, 1876, Macmillan–Wallace letters, nos. 11–12.

14. THE WAR ON SPIRITUALISM

1. ARW to CD, June 7, 1876, *LLW*, 236–37.

2. *ML*, 2:413–15.

3. Michael Foster to ARW, March 6, 1875, MS 46435, no. 288, Department of Manuscripts, British Library, London.

4. Alfred Russel Wallace, "By-Paths in the Domain of Biology, Being an Address Delivered to the Biological Section of the British Association, (Glasgow, September 6th, 1876), as President of the Section" (1876), in *"Tropical Nature" and Other Essays* (1878; reprint, New York: AMS Press, 1975), 254–55.

5. ARW to AN, July 27, 1876, Newton Papers, no. 303, Cambridge University Library.

6. *Times* (London), quoted in *Spectator*, September 16, 1876, 1147.

7. For the complete text of the paper and subsequent discussion at the meeting

of the British Association for the Advancement of Science in Glasgow, see *Spiritualist and Journal of Psychological Science*, September 22, 1876, 84–88.

8. Edwin Ray Lankester, "A Spirit-Medium," *Times* (London), September 16, 1876, 7.

9. Alfred Russel Wallace, "A Spirit Medium," *Times* (London), September 19, 1876, 4.

10. J. Park Harrison, "A Spirit Medium," *Times* (London), September 20, 1876, 8.

11. Augustus Lane-Fox, "A Spirit Medium," *Times* (London), September 22, 1876, 10.

12. Lankester, "Spirit-Medium." See also "A Spirit Medium," *Times* (London), October 3, 1876, 9. Another scientific investigator, Searjant Cox, president of the Psychological Society of London, also had been touched by a hand in one of Henry Slade's séances that summer, but this hand, which appeared in broad daylight, was that of a woman—warm, soft, and moist like a living hand and half the size of Slade's. It emerged from the opposite side of the table, grabbed a pencil at Slade's request, and threw it on the floor ("The Slade Prosecution," *Times* [London], October 30, 1876, 11).

13. Lankester, "Spirit-Medium." See also Horatio Donkin, letter to the editor, *Times* (London), September 16, 1876, 7.

14. "Spirit Medium."

15. William Barrett, "A Spirit Medium," *Times* (London), September 22, 1876, 10.

16. Henry Slade, "A Spirit Medium," *Times* (London), September 23, 1876, 9.

17. "Slade Prosecution."

18. The account of the trial of Henry Slade is from the *Times* (London), October 3, 11, 13, and 21 and November 1, 1876; Arthur Conan Doyle, *The History of Spiritualism* (1926; reprint, New York: Arno Press, 1975), 1:280–83; "Spiritualism," *Spectator*, October 14, 1876, 1281–82; Janet Oppenheim, *The Other World: Spiritualism and Psychical Research in England, 1850–1914* (Cambridge: Cambridge University Press, 1988), 26–27; *Spectator*, November 4, 1876, 1360–61; and *Medium and Daybreak, a Weekly Journal Devoted to the History, Phenomena, Philosophy, and Teachings of Spiritualism*, February 2, 1877, 75–77.

19. Edwin Ray Lankester, "Charles Robert Darwin," in *Biographical Dictionary and Synopsis of Books, Ancient and Modern*, ed. Charles Warner (New York: International Society, 1896), 4385–93. I would like to thank Richard Milner for alerting me to this unusual source.

20. Emma Darwin to Leonard Darwin, July 22, 1876, *Emma Darwin: A Century of Family Letters, 1792–1896*, ed. Henrietta [Darwin] Litchfield (New York: Appleton, 1915), 2:223. The letter supposedly is dated July 22, 1876, but the allusion to the Slade trial seems clear.

21. *Medium and Daybreak*, February 2, 1877, 75–77.

22. *Times* (London), quoted in ibid.

23. *Spiritualist Newspaper*, August 30, 1878, 103.

24. *ML*, 2:299.

25. *Lancet*, September 23, 1876, 431–33. For comments on *The Geographic Distribution of Animals*, see *Lancet*, October 14, 1876, 537–38.

26. ARW to HWB, November 10, 1876, Correspondence Block 1871–1880, Archives of the Royal Geographical Society, London.

27. *ML*, 2:61.

28. ARW to CD, January 17, 1877, *LLW*, 244.

29. *ML*, 2:380–81.

30. THH to John Tyndall, October 27, 1876, Thomas Henry Huxley Papers, Imperial College of Science, Technology, and Medicine, London [catalogued in Warren R. Dawson, *The Huxley Papers: A Descriptive Catalogue of the Correspondence, Manuscripts and Miscellaneous Papers of the Rt. Hon. Thomas Henry Huxley . . . preserved in the Imperial College of Science and Technology, London* (London: Macmillan, 1946), 1:149].

31. Alfred Russel Wallace, "The Curiosity of Credulity," *Athenaeum*, January 12, 1878, 54.

32. "Spiritualism."

33. "Professor Barrett and Dr Carpenter," *Spectator*, October 28, 1876, 1343–44.

34. William Crookes, letter to the editor, *Nature*, November 1, 1877, 7–8.

35. William Benjamin Carpenter quotes the offending statements in Wallace's "Curiosity of Credulity" in "The Curiosities of Credulity," *Athenaeum*, December 22, 1877, 814.

36. WBC to ARW, October 5, 1877, MS 46439, nos. 144–46, Department of Manuscripts, British Library, London.

37. Alfred Russel Wallace, draft of response to William Benjamin Carpenter, October 7, 1877, MS 46439, nos. 148–49.

38. William Benjamin Carpenter, letter to the editor, *Nature*, November 1, 1877, 8–9.

39. William Benjamin Carpenter, letter to the editor, *Nature*, November 8, 1877, 26–27.

40. William Crookes, letter to the editor, *Nature*, November 15, 1877, 43–44.

41. Carpenter, "Curiosities of Credulity"; Wallace, "Curiosity of Credulity," 54.

42. ARW to William Barrett, December 9, 1877, *LLW*, 427.

43. Quoted in "The British National Association of Spiritualists: The Debate on Capt. Burton's Paper," *Spiritualist Newspaper*, December 13, 1878, 283–87.

15. PHOENIX FROM THE ASHES

1. Alfred Russel Wallace, "The Climate and Physical Aspects of the Equatorial Zone," in *"Tropical Nature" and Other Essays* (1878; reprint, New York: AMS Press, 1975), 20–21.

2. Alfred Russel Wallace, "Animal Life in the Tropical Forests," in *"Tropical Nature" and Other Essays*, 123.

3. Alfred Russel Wallace, "The Colours of Animals and Sexual Selection," in *"Tropical Nature" and Other Essays*, 161–62.

4. Ibid., 172–73.

5. Ibid., 187.

6. Ibid., 195.

7. Ibid., 197–98.

8. Ibid., 203.

9. Ibid., 210–11.

10. Edward Percival Wright, "Tropical Nature," *Nature*, June 6, 1878, 140–41.

11. CD to ARW, August 31, 1877, *LLW*, 245–46.

12. ARW to Grant Allen, February 17, 1879; CD to Grant Allen, "before" February 21, 1879, Darwin Collection of the Dittrick Medical History Center, Case Western Reserve University, Cleveland.

13. *ML*, 2:98.

14. ARW to Alexander Macmillan, July 14, 1878, Macmillan–Wallace letters, nos. 15–16, Department of Manuscripts, British Library, London.

15. ARW to JH, August 27, 1878, Directors' Correspondence, vol. 104 (1864–1899), Archives of the Royal Botanic Gardens, Kew.

16. ARW to CD, September 14, 1878, *LLW*, 248.

17. CD to ARW, September 16, 1878, *LLW*, 248.

18. ARW to CD, September 23, 1878, *LLW*, 249; ARW to JH, October 27, 1878, Directors' Correspondence, vol. 104 (1864–1899); Alfred Russel Wallace, "Epping Forest," *Fortnightly Review*, November 1, 1878, 628–45.

19. This anecdote is related by William Wallace (*LLW*, 352).

20. ARW to RM, November 27, 1879, Oxford University Museum of Natural History; AB to CD, December 16, 1879, *DAR*, 160:366.

21. ARW to CD, January 9, 1880, *LLW*, 250–51.

22. *ML*, 2:394; AB to CD, December 16, 1879, *DAR*, 160:366.

23. CD to AB, December 17, 1879, *DAR*, 143:180.

24. JH to CD, December 18, 1879, *DAR*, 104:136, 137.

25. CD to JH, December 19, 1879, *DAR*, 95:494; AB to CD, December 29, 1879, *DAR*, 160:368.

26. ARW to CD, January 9, 1880, *LLW*, 250–51.

27. *ML*, 2:11.

28. CD to ARW, January 5, 1880, *LLW*, 249.

29. ARW to Francis Darwin, July 3, 1901, *LLW*, 328. Darwin had written in *The Origin of Species* that he "cannot doubt that the theory of descent with modification embraces all the members of the same great class or kingdom. I believe that animals are descended from at most only four or five progenitors, and plants from an equal or lesser number. Analogy would lead me one step farther, namely, to the belief that all animals and plants are descended from some one prototype. But analogy may be a deceitful guide. . . . [On] the principle of natural selection with divergence of character, it does not seem incredible that, from such low and intermediate forms, both animals and plants may have been developed; and, if we admit this, we must likewise admit that all the organic beings which have ever lived on this earth may be descended from some one primordial form. But this inference is chiefly grounded on analogy and it is immaterial whether or not it be accepted. . . . [In] regard to the members of each great kingdom, such as the Vertebrata, Articulata, &c., we have distinct evidence in their embryological homologous and rudimentary structures, that within each kingdom all the members are descended from a single progenitor" (*The*

Origin of Species by Means of Natural Selection or the Preservation of Favored Races in the Struggle for Life [1859; reprint of 6th ed., New York: Modern Library, n.d], 371–72).

30. ARW to CD, January 9, 1880, *LLW*, 250–51.

31. Alfred Russel Wallace, *Island Life* (1880; reprint, Amherst, N.Y.: Prometheus Books, 1998), 229.

32. *ML*, 2:101, 335–36. See also Wallace, *Island Life*, 335.

33. ARW to RM, December 27, 1880, Oxford University Museum of Natural History; ARW to CD, October 11, 1880, *LLW*, 251.

34. CD to ARW, November 3, 1880, *LLW*, 252 and note, containing Wallace's summary of Darwin's objections. Wallace refers to Darwin's foolscap notes in *ML*, 2:12.

35. ARW to CD, January 1, 1881, *LLW*, 255–56. Wallace revisited this issue five years later. In a letter to A. Schuster, a geologist, he wrote: "I am desirous of obtaining evidence as to the *size* and *weight* of dust particles which can be carried great distances through the atmosphere. . . . Can you be so good as to furnish me with any particulars of this dust. . . . My object is to obtain evidence that *small seeds may* be carried over considerable widths of ocean by violent winds, and the only method available is to show that mineral particles of equal or greater specific gravity are thus carried. . . . I have some of the Krakatoa dust, but there was I believe no gale sufficiently strong at that time to carry more than the finest dust for considerable distances" (ARW to A. Schuster, July 18, 1886, Sc 382, Schuster correspondence, no. 3, Archives of the Royal Society of London).

36. CD to ARW, January 2, 1881, *LLW*, 256.

37. Archibald Geikie, review of *Island Life*, by Alfred Russel Wallace, part 2, *Nature*, February 24, 1881, 393.

38. JH to ARW, November 10, 1880, *LLW*, 289–90; JH to CD, November 24, 1881 [*sic*], *LJH*, 2:244. This book gives the year as 1881, which seems unlikely, since the letter refers to events in 1880 (after the publication of *Island Life* but before the award of the civil pension to Wallace).

39. THH to CD, November 14, 1880, *LLH*, 2:15.

40. CD to AB, October 31, 1880, *DAR*, 143:182.

41. Arabella Buckley quoted Wallace's remark in AB to CD, November 4, 1880, *DAR*, 160:370; CD to AB, November 9, 1880, *DAR*, 134:183.

42. JH to THH, November 26, 1880, Thomas Henry Huxley Papers, Imperial College of Science, Technology, and Medicine, London [catalogued in Warren R. Dawson, *The Huxley Papers: A Descriptive Catalogue of the Correspondence, Manuscripts and Miscellaneous Papers of the Rt. Hon. Thomas Henry Huxley . . . preserved in the Imperial College of Science and Technology, London* (London: Macmillan, 1946), 3:259].

43. CD to THH, November 26, 1880, Thomas Henry Huxley Papers [catalogued in Dawson, *Huxley Papers*, 5:349, plus attachment].

44. Henrietta [Darwin] Litchfield, ed., *Emma Darwin: A Century of Family Letters, 1792–1896* (New York: Appleton, 1915), 2:243.

45. CD to Francis Darwin, December 28, 1880, *DAR*, 211:69.

46. CD to HWB, December [?], 1880, and January 3, 1881, quoted in Robert Sprecher, "The Darwin–Bates Letters: Correspondence Between Two Nineteenth-Century Travellers and Naturalists," *Annals of Science* 25, no. 2 (1969): 123–24.

47. CD to THH, January 7, 1881, Thomas Henry Huxley Papers [catalogued in Dawson, *Huxley Papers*, 5:356]; Charles Darwin, quoted in *Emma Darwin*, 2:243.

48. CD to ARW, January 7, 1881, *LLW*, 257; ARW to CD, January 29, 1881, *LLW*, 259. See also *ML*, 2:394–95.

49. Arnold Brackman, *A Delicate Arrangement: The Strange Case of Charles Darwin and Alfred Russel Wallace* (New York: Times Books, 1980), 290.

50. *LLD*, 2:530. An authoritative account of Darwin's efforts to get Wallace a civil pension is in Ralph Cope, "'I Will Gladly Do My Best': How Charles Darwin Obtained a Civil List Pension for Alfred Russel Wallace," *Isis* 83 (1992): 3–26.

51. *ML*, 2:253–54.

52. Alfred Russel Wallace, "Reciprocity the Essence of Free Trade" (1879), in *SSS*, 2:167–83.

53. For comments about Robert Lowe's personality, see K. Theodore Hoppen, *The Mid-Victorian Generation, 1846–1886* (Oxford: Oxford University Press, 1998), 593. See also Robert Lowe, "Reciprocity and Free Trade," *Nineteenth Century* 5 (1879): 992–1002; and Alfred Russel Wallace, "A Few Words in Reply to Mr. Lowe," *Nineteenth Century* 6 (1879): 179–81.

54. *ML*, 2:257.

55. Alfred Russel Wallace, "How to Nationalise the Land," *Contemporary Review* 38 (1880): 716–36.

56. Hoppen, *Mid-Victorian Generation*, 595.

57. *ML*, 2:258.

58. *LLW*, 390.

59. Land Nationalisation Society, Land Nationalisation Tract, no. 1 (London: Land Nationalisation Society, 1881), shelfmark 8282ff.22, British Library, London.

60. *ML*, 2:258.

61. Land Nationalisation Society, Land Nationalisation Tract, no. 17 (London: Land Nationalisation Society, ca. 1887), shelfmark 8282ff.22.

62. *ML*, 2:261; ARW to CD, July 9, 1881, *LLW*, 260–61.

63. Henry George, *Progress and Poverty* (1882; reprint, New York: Robert Schalkenback Foundation, 1997), 140–42; Max Beer, *A History of British Socialism* (London: Bell, 1929), 2:243–44.

64. ARW to CD, July 9, 1881, *LLW*, 260–61; CD to ARW, July 12, 1881, *LLW*, 261–62.

65. ARW to Alexander Macmillan, November 29, 1881, Macmillan–Wallace letters, no. 19.

66. Alfred Russel Wallace, *Land Nationalisation; Its Necessity and Its Aims; Being a Comparison of the System of Landlord and Tenant with that of Occupying in their Influence on the Well-being of the People* (London: Trübner, 1882), 11–13.

67. *ML*, 2:276; Wallace, *Land Nationalisation*, 56–60.

68. Gertrude Himmelfarb, *Poverty and Compassion* (New York: Vintage, 1992), 315.

69. Beer, *History of British Socialism*, 2:242–45; *ML*, 2:274.

70. Land Nationalisation Society, *13th Report of the Land Nationalisation Society,*

1893–4 [report of April 9, 1894, meeting], Land Nationalisation Tract, no. 56 (London: Land Nationalisation Society, 1894).

71. Alfred Russel Wallace, "Anthropology," *Nature*, July 14, 1881, 242–45. See also *ML*, 2:406.

72. Hyde Clark, letter to the editor, *Nature*, August 25, 1881, 380–81; *ML*, 2:406.

73. *LLD*, 2:525–30, and appendix 1, 531–32; Francis Darwin to Francis Galton, April 20, 1882, quoted in *The Life, Letters and Labours of Francis Galton*, ed. Karl Pearson (Cambridge: Cambridge University Press, 1924), 2:197.

74. Francis Galton to Emma Galton, April 26, 1882, quoted in *Life, Letters and Labours of Francis Galton*, 2:198. See also the note quoting Galton's recollections of Darwin's funeral, as published in the *Times* (London), May 25, 1898.

75. Wallace quotes Huxley's eulogy of Darwin, as published in *Nature* (*ML*, 2:16).

76. ARW and Samuel Haughton correspondence, *Nature*, December 9, 1880, 124; January 13, 1881, 241–42; and January 20, 1881, 266–67.

77. *ML*, 2:88–89, 102.

78. *ML*, 2:230–31n.

79. Alfred Russel Wallace, *Bad Times: An Essay on the Present Depression of Trade, Tracing It to Its Sources in Enormous Foreign Loans, Excessive War Expenditure, the Increase of Speculation and of Millionaires, and the Depopulation of the Rural Districts; With Suggested Remedies* (London: Macmillan, 1885), 84–118.

80. *ML*, 2:104–5.

16. TO THE LAND OF EPIDEMIC DELUSIONS

1. *ML*, 2:105.

2. ARW to O. C. Marsh [copy of letter forwarded to Daniel C. Gilman], January 23, 1886, Daniel C. Gilman Papers, MS 1, Box 1.50, Special Collections, Milton S. Eisenhower Library, Johns Hopkins University, Baltimore; ARW to O. C. March [original], January 23, 1886, Othniel C. Marsh Papers, nos. 17705–8, Yale University, New Haven, Conn.

3. O. C. Marsh to Daniel C. Gilman, February 12, 1886, Daniel C. Gilman Papers, MS 1, Box 1.50.

4. C. W. Ernst to Daniel C. Gilman, February 2, 1886, Daniel C. Gilman Papers, MS 1, Box 1.50.

5. THH to H. N. Martin, March 4, 1886, Daniel C. Gilman Papers, MS 1, Box 1.50.

6. "Notes," *Nature*, January 20, 1887, 281.

7. B. W. Williams to Daniel C. Gilman, March 30 and May 10, 1886; Daniel C. Gilman to B. W. Williams, May 12, 1886, Daniel C. Gilman Papers, MS 1, Box 1.50.

8. *ML*, 2:105–6.

9. Ibid., 107–9.

10. A copy of one such advertisement from an unknown publication is in the Daniel C. Gilman Papers, MS 1, Box 1.50.

11. *ML*, 2:107–8.

12. Alfred Russel Wallace, excerpt of concluding passage of lecture on Darwinism, Daniel C. Gilman Papers, MS 1, Box 1.50.

13. A summary of Wallace's lecture is in "Fall Meeting of the National Academy of Sciences, Nov. 9–11," *Science*, November 19, 1886, 450. See also ARW to O. C. Marsh, November 23, 1886, Othniel C. Marsh Papers, no. 17713

14. *AJ*, November 17, 1886.

15. *ML*, 2:109–10; Alfred Russel Wallace, "American Museums: The Museum of Comparative Zoology, Harvard University," *Fortnightly Review*, September 1, 1887, 347–59; and "American Museums: Museums of American Pre-historic Archaeology," *Fortnightly Review*, November 1, 1887, 665–75.

16. *ML*, 2:113; *AJ*, November 26–28, 1886.

17. "Dr. Wallace on the Development Theory," *Science*, December 17, 1886, 560–63.

18. Linda Simon, *A Genuine Reality: A Life of William James* (New York: Harcourt, Brace, 1998), 189–91.

19. *ML*, 2:352–53.

20. *AJ*, December 28, 1886.

21. William James, "The Final Impressions of a 'Psychical Researcher'" (1909), in *Memories and Studies* (New York: Longman, Green, 1911), 178–79.

22. *AJ*, December 29, 1886; *ML*, 2:115–16.

23. *ML*, 2:188.

24. Ibid., 128–29.

25. Ibid., 129.

26. Janet Oppenheim, *The Other World: Spiritualism and Psychical Research in England, 1850–1914* (Cambridge: Cambridge University Press, 1988), 162–66; *AJ*, March 4, 1887.

27. *Nature*, June 26, 1879, 208. Elliot Coues, *Birds of the Colorado Valley: A Repository of Scientific and Popular Information Concerning North American Ornithology*, Miscellaneous Publication no. 8 (Hayden, Wash.: United States Geological and Geographical Survey of the Territories, 1878), as cited in *Dictionary of Scientific Biography*, ed. Charles Coulton Gillespie (New York: Scribner, 1981), 3:439. See also Paul Russell Cutright and Michael J. Brodhead, *Elliott Coues, Naturalist and Frontier Historian* (Chicago: University of Illinois Press, 1981), 179–80.

28. *ML*, 2:117, 358–59.

29. ARW to Lester F. Ward, June 7, 1894, quoted in Bernhard J. Stern, "Letters of Alfred Russel Wallace to Lester F. Ward," *Scientific Monthly* 40 (1935): 375–76; *ML*, 2:117.

30. Alfred Russel Wallace, "English and American Flowers," part 1, *Fortnightly Review*, October 1, 1891, 530.

31. *ML*, 2:122–24. See also "Reception to Prof. Wallace," *Washington Post*, February 5, 1887, 1. William Cullen Bryant, in a portrait photograph by Mathew Brady, does show a slight resemblance to Wallace.

32. Wallace, "English and American Flowers," part 1, 532.

33. *ML*, 2:139–40.

34. Wallace, "English and American Flowers," part 1, 531.

35. ARW to Lester F. Ward, September 25, 1887, quoted in Stern, "Letters of Alfred Russel Wallace to Lester F. Ward," 377.

36. *ML*, 2:363–65.

37. Ibid., 158.

38. C. Michael Hall, "John Muir: Travels in Australia, 1903–1904," in *John Muir: Life and Work*, ed. Sally M. Miller (Albuquerque: University of New Mexico Press, 1996), 303–4.

39. "The Life Hereafter, Future State Considered," *San Francisco Chronicle*, June 6, 1887, 8.

40. Alfred Russel Wallace, "If a Man Dies, Shall He Live Again?" *Banner of Light*, June 25, 1887, 1 (reprint courtesy of the American Society for Psychical Research).

41. *AJ*, June 6, 1887; *ML*, 2:160.

42. Alfred Russel Wallace, "English and American Flowers," part 2, *Fortnightly Review*, December 1, 1891, 810.

43. Alfred Russel Wallace, "Inaccessible Valleys" (1893), in *SSS*, 1:26–27.

44. *ML*, 2:165–69, 168n.

45. *AJ*, July 4, 1887.

46. *ML*, 2:170; *AJ*, July 6, 1887.

47. Wallace, "English and American Flowers," part 2, 798–801.

48. Ibid., 801–3.

49. Wallace, "English and American Flowers," part 1, 532–33.

50. *AJ*, August 17, 1887. 51. Wallace used a conversion rate of approximately $5 to £1. Unfortunately, the balance sheet in the American journal is difficult to follow. According to Peter Raby, who was given access to Wallace's private ledgers, he brought home £350, or $1,750 (*Alfred Russel Wallace: A Life* [Princeton, N.J.: Princeton University Press, 2001], 245).

17. THE NEW NEMESIS

1. George John Romanes, "Permanent Variation of Colour in Fish," *Nature*, June 5, 1873, 101.

2. For a history of Romanes's early life, see *The Life and Letters of George John Romanes*, ed. Ethyl Romanes (London: Longman, Green, 1908).

3. George John Romanes, "Perception and Instinct in the Lower Animals," *Nature*, August 7, 1873, 282–83; Alfred Russel Wallace, reply to George John Romanes, *Nature*, August 14, 1873, 302.

4. *ML*, 2:327–28; FRS, "A Speculation Regarding the Senses," *Nature*, February 12, 1880, 348.

5. *ML*, 2:329–33.

6. George John Romanes to CD, April 22, 1880, quoted in *Life and Letters of George John Romanes*, 101–2.

7. ARW to RM, August 28, 1886, Oxford University Museum of Natural History.

8. For example, William Thistleton-Dyer wrote: "I should be sorry to have misrepresented the views of Mr. Romanes. . . . But if I have done so, I must plead an extenuation that I know of no recent writer whose papers I find so difficult to thoroughly comprehend. With an appearance of lucidity there seems to me to be often an underlying obscurity of ideas by which I find myself as often completely befogged" ("Mr. Romanes's Paradox," *Nature*, November 1, 1888, 7–8).

9. Alfred Russel Wallace, "Romanes *versus* Darwin: An Episode in the History of the Evolution Theory," *Fortnightly Review*, September 1, 1886, 300–316.

10. Alfred Russel Wallace, letter to the editor, *Nature*, February 17, 1887, 366.

11. George John Romanes, letter to the editor, *Nature*, February 24, 1887, 390–91.

12. ARW to EBP, October 20, 1887, Oxford University Museum of Natural History.

13. *ML*, 2:201.

14. ARW to AB (Fisher), February 16, 1888, *LLW*, 295–96.

15. For a biographical sketch of August Weismann, see Ernst Mayr, *The Growth of Biological Thought: Diversity, Evolution, and Inheritance* (Cambridge, Mass.: Harvard University Press, Belknap Press, 1982), 698–703.

16. Charles Darwin, *The Variation of Animals and Plants Under Domestication*, 2nd ed. (London: Murray, 1888). For Darwin's in-depth explanation of his complicated theory of inheritance, see 2:349–99.

17. CD to JH, February 23, 1868, *LLD*, 2:259–61.

18. Peter J. Bowler, *The Eclipse of Darwinism* (Baltimore: Johns Hopkins University Press, 1983), 38–39.

19. Alfred Russel Wallace, *Darwinism: An Exposition of the Theory of Natural Selection with Some of Its Applications* (London: Macmillan, 1889), 438.

20. Ibid., 437–39.

21. H. N. Moseley, review of *The Continuity of the Germ-Plasma Considered on the Basis of a Theory of Heredity*, by August Weissman, *Nature*, December 17, 1885, 154; and "Dr. August Weismann on the Importance of Sexual Reproduction for the Theory of Natural Selection," *Nature*, October 28, 1886, 629–32.

22. ARW to EBP, January 5, 1888, Oxford University Museum of Natural History.

23. John T. Gulick, "Divergent Evolution Through Cumulative Segregation," *JLS (Zoology)* 20 (1888): 189–274, 312–80.

24. Wallace, *Darwinism*, 146–48.

25. Ibid., 152–53.

26. Ibid., 168–73.

27. Ibid., 175–76.

28. James Murray, *Genetic Diversity and Natural Selection* (New York: Hafner, 1972), 65.

29. *LLD*, 2:220.

30. ARW to Alexander Macmillan, March 10 and 12, 1889, Macmillan–Wallace letters, nos. 22–24, Department of Manuscripts, British Library, London.

31. Herbert Spencer to ARW, May 18, 1889, *LLW*, 30.

32. ARW to Grant Allen, July 22, 1889, Darwin Collection of the Dittrick Medical History Center, Case Western Reserve University, Cleveland. See also Grant Allen, review of *Darwinism*, by Alfred Russel Wallace, *Academy* 36 (1889): 41–42. Allen's review was more critical than Wallace admitted in his letter to him. The bulk of the review tore into Wallace's "losing battle" in favor of natural selection as the almost sole cause of the production of species. Allen predicted that "in another century . . . endless new factors in evolution which escape us now will be added to the three or four main ones—natural selection, sexual selection, use and disuse, direct action of environment—which we now possess."

33. George John Romanes to Francis Darwin, January 20, 1889, quoted in *Life and Letters of George John Romanes*, 214–17.

34. George John Romanes to John T. Gulick, October 20, 1889 (from letters presented to the Linnean Society by Dr. Addison Gulick, February 20, 1961, courtesy of the Linnean Society of London).

35. *ML*, 2:334–35.

36. Ibid., 125–26.

37. Ibid., 335–36.

38. Ibid., 337–38.

39. Ibid., 338–39.

40. Ibid., 342–43.

41. ARW to WTD, September 26, 1893, MS 46436, no. 300, Department of Manuscripts, British Library, London.

42. Ernst Mayr, "Karl Jordan on Speciation," in *Evolution and the Diversity of Life: Selected Essays* (Cambridge, Mass.: Harvard University Press, Belknap Press, 1997), 137–39. For a more detailed discussion of the modern debate about sympatric speciation, see Helena Cronin, *The Ant and the Peacock* (Cambridge: Cambridge University Press, 1994), 381–430.

18. THOROUGHLY UNPOPULAR CAUSES

1. Nicolau Barquet and Pere Domingo, "Smallpox: The Triumph over the Most Terrible of the Ministers of Death," *Annals of Internal Medicine*, October 15, 1997, 635–42.

2. *MA*, 359.

3. "Scientific Fallibility," *Medium and Daybreak*, March 29, 1872, 109. The article refers to remarks made by Henry Pitman: "In the admirable paper by Mr. Alfred R. Wallace on 'Miracles,' he refers to the 'great discoveries' of Gallileo [*sic*], Harvey, and Jenner. I think that nine Spiritualists out of ten will be of the opinion that Jenner made no discovery, that vaccination was practised before his day, and that no term is bad enough to denote the vile superstition" (*Medium and Daybreak*, March 15, 1872). Wallace's presentation to the Dialectical Society, "An Answer to the Arguments of Hume, Lecky, and Others Against Miracles," was reprinted as "No Antecedent Impossibility in Miracles: A Reply to Modern Objectors," *Spiritualist Magazine*, March 1872, which, in turn, he revised, and restored the original title, for *Miracles and Modern Spiritualism* (1896; reprint, New York: AMS Press, 1975).

4. H. N. Mozley, "Compulsory Vaccination in the British Isles," *Vaccination Inquirer and Health Review* 7 (1885): 79–80.

5. Alfred Russel Wallace, *The Wonderful Century: Its Successes and Its Failures* (New York: Dodd, Mead, 1899), 315. See also "To Our Readers," *Vaccination Inquirer and Health Review* 1 (1879): 1–2.

6. William Tebb, "Sanitation, Not Vaccination the True Protection Against Small-Pox," October 12, 1881, 2 [available at: www.whale.to/vaccine/tebb1.html].

7. *ML*, 2:368–69; Wallace, *Wonderful Century*, 259–70. The decline in rates of smallpox is noted in the testimony of C. T. Ritchie, president of the Local Government

Board, in "The Vaccination Acts," *Hansard's Parliamentary Debates*, 3rd ser., 334, no. 15 (April 5, 1889): 1741.

8. J. D. Swales, "The Leicester Anti-Vaccination Movement," *Lancet*, October 24, 1992, 1019–21; Wallace, *Wonderful Century*, 271–72; "Vaccination Acts," 1742.

9. *Vaccination Inquirer and Health Review* 1 (1879): 11.

10. Wallace, *Wonderful Century*, 299–300.

11. Tebb, "Sanitation, Not Vaccination," 3.

12. Wallace, *Wonderful Century*, 222–23.

13. Memorandum, Her Majesty's Government to ARW, April 1889, MS 46440, Department of Manuscripts, British Library, London.

14. *ML*, 2:369.

15. *Vaccination Inquirer and Health Review* 11 (1889): 74.

16. "Royal Commission on Vaccination," *Times* (London), December 25, 1890, 4; *Lancet*, January 3, 1891, 45; *Lancet*, January 10, 1891, 94–95; *ML*, 2:369–70; "Extracts from the Report of the Royal Commission on Vaccination," *Vaccination Inquirer and Health Review* 12 (1891): 164–68.

17. *Lancet*, January 10, 1891, 94.

18. Ibid., 95.

19. "Royal Commission on Vaccination."

20. "Extracts from the Report of the Royal Commission on Vaccination."

21. "Royal Commission on Vaccination"; "Extracts from the Report of the Royal Commission on Vaccination."

22. Wallace, *Wonderful Century*, 314.

23. "Mr. Alfred Russel Wallace and the Medical Profession," *Lancet*, March 12, 1898, 734–35.

24. *Lancet*, March 19, 1898, 799–800.

25. ARW to Grant Allen, July 22, 1889, Darwin Collection of the Dittrick Medical History Center, Case Western Reserve University, Cleveland.

26. Erich Fromm, foreword to *Looking Backward*, by Edward Bellamy (New York: Buccaneer Books, 1994), v.

27. Alfred Russel Wallace, review of *Hereditary Genius*, by Francis Galton, *Nature*, March 17, 1870, 501–3.

28. Alfred Russel Wallace, "Human Selection" (1890), in *SSS*, 1:509–26.

29. *ML*, 2:203, 228–29; ARW to EBP, May 28, 1889, Oxford University Museum of Natural History. In a letter to Edward Westermarck, Wallace states that one could catch a train at 8:50 A.M. from Waterloo Station in London and arrive in Parkstone a little before 1:00 P.M. (K. Rob Wikman, "Letters from Edward B. Tylor and Alfred Russel Wallace to Edward Westermarck," *Acta Acadamie Aboensis, Humanoria XIII* 7 [1940]: 15).

30. Wikman, "Letters," 15; ARW to EBP, June 2, 1889, Oxford University Museum of Natural History.

31. *ML*, 2:119–20. The conferral of the degree took place on Tuesday, November 26, 1889, at 2:00 P.M. (*Oxford Magazine*, November 1889, courtesy of Christine Mason, principal library assistant, Bodlein Library, Oxford University). Those awarded

the D.C.L. (a higher degree) wore scarlet or red gowns, with red silk facings six inches wide and red silk half-sleeves from the elbow to the wrist. The headgear was a black bonnet, not the "motor-board" conferred on lower-degree candidates (courtesy of Shepherd & Woodward, "Oxford-gown outfitters").

32. Edward Bagnett Poulton, "Alfred Russel Wallace (1823–1913)," *Proceedings of the Royal Society of London* 95 (1924): xxi–ii.

33. The announcement that Wallace was to be awarded the first Darwin Medal on December 1, 1890, was published in "Notes," *Nature*, November 20, 1890, 61; John Hopkinson, on Wallace's chivalrous behavior toward Darwin, quoted in *Nature*, December 4, 1890, 107.

34. Adolph Bernhard Meyer, letter to the editor, *Nature*, August 29, 1895, 415.

35. ARW to D. W. Freshfield, April 21, 1892, Correspondence Block 1881–1910, Archives of the Royal Geographical Society, London.

36. ARW to J. W. Marshall, September 23, 1892, *LLW*, 307.

37. Edward A. Westermarck, *The History of Human Marriage*, 5th ed. (New York: Allerton Books, 1922), 1:486, 488.

38. ARW to Edward A. Westermarck, January 29, 1890, Wikman, "Letters," 13–14.

39. Alfred Russel Wallace, introduction to *The History of Human Marriage*, by Edward A. Westermarck (London: Macmillan, 1892), 1:ix–x.

40. ARW to Edward A. Westermarck, March 12, 1891, Wikman, "Letters," 16.

41. ARW to Edward A. Westermarck, May 3, 1891, Wikman, "Letters," 18–19.

42. Alfred Russel Wallace, "Mr. Crookes and Eva Fay," *Nature*, December 6, 1877, 101.

43. THH to JH, May 26, 1889, *LLH*, 2:224.

44. WTD to ARW, October 23, 1892, *LLW*, 444; ARW to WTD, October 25, 1892, Directors' Correspondence (1864–1899), Archives of the Royal Botanic Gardens, Kew; List of Candidates (1888–1908), April 1, 1893, Archives of the Royal Society of London.

45. WTD to ARW, January 12, 1893, *LLW*, 444–45.

46. ARW to WTD, January 17, 1893, MS 46436, no. 281, Department of Manuscripts, British Library, London.

47. Thomas Huxley discusses Herbert Spencer's belief in Lamarckian inheritance in THH to W. P. Ball, October 27, 1890, *LLH*, 2:283–85.

48. THH to W. P. Ball, October 27, 1890, *LLH*, 2:283–85. Wallace addresses Spencer's belief in Lamarckian inheritance in "Are Individually Acquired Characters Inherited?" part 2, *Fortnightly Review*, May 1, 1893, 665–68. See also ARW to RM, June 10, 1893, *LLW*, 309.

49. Lord Salisbury [Robert Arthur Talbot Gascoyne-Cecil], "Presidential Address, the British Association, Oxford, August 8 [1894]," *Nature*, August 9, 1894, 338–343.

50. Herbert Spencer to ARW, August 10, 1894, *LLW*, 312.

51. Alfred Russel Wallace, "The World of Life: As Visualised and Interpreted by Darwinism" [excerpt from lecture presented at the Royal Institution], *Nature*, January 28, 1909, 386.

52. Herbert Spencer to ARW, August 19, 1894, *LLW*, 312–13.

53. Herbert Spencer to ARW, September 28, 1895, *LLW*, 317.

54. Alfred Russel Wallace, "Another Substitute for Darwinism," *Nature*, October 4, 1894, 541.

55. ARW to EBP, September 8, 1894, *LLW*, 313.

56. Alfred Russel Wallace, "The Method of Organic Evolution" (1895), in *SSS*, 1:348.

57. Ibid., 355.

58. Ibid., 362.

59. See, for example, Ernst Mayr, *The Growth of Biological Thought: Diversity, Evolution, and Inheritance* (Cambridge, Mass.: Harvard University Press, Belknap Press, 1982), 733.

60. William Thistleton-Dyer, letter to the editor, *Nature*, July 30, 1896, 293–94.

61. Edwin Ray Lankester, letter to the editor, *Nature*, July 16, 1896, 245–46.

62. ARW to J. W. Marshall, March 6, 1894, *LLW*, 436; *ML*, 2:415.

63. ARW to Philip Slater, December 31, 1893, Manchester Archives and Local Studies, Manchester Central Library.

64. ARW to John Scott Keltie, March 1, 1892, Correspondence Block 1881–1910, Archives of the Royal Geographical Society, London.

65. ARW to Violet Wallace, November 25, 1894, *LLW*, 358.

66. ARW to Annie Wallace, July [?], 1895, *LLW*, 358–59.

67. *ML*, 2:415.

68. Wallace, *Wonderful Century*, 150–55, 191–93.

69. *ML*, 2:229–30.

70. Wallace, *Wonderful Century*, 325–69.

71. ARW to William Barrett, October 30, 1899, *LLW*, 433–34.

72. Alfred Russel Wallace, "Justice, not Charity, as the Fundamental Principle of Social Reform: An Appeal to My Readers" (1898), in *SSS*, 2:521–28.

73. J. W. Sharpe, friend and neighbor in Godalming, reminiscence of Wallace, *LLW*, 354.

19. SATISFACTION, RETROSPECTION, AND WORK

1. Copy of pamphlet and letter from Wallace, May 22, 1901, archives of the Zoological Society of London.

2. ARW to William Wallace, October 26, 1901, *LLW*, 360–61.

3. ARW to William Wallace, November 28, 1901 (courtesy of Richard Wallace).

4. ARW to William Wallace, November 6, 1901, *LLW*, 361–62. Wallace tells Will: "Of course the house will be larger than we want, but I look to future value, and rather than build it smaller, to be enlarged afterwards, I would prefer to leave the drawing-room and bedroom adjoining with bare walls inside till they can be properly finished." For a description of Wallace's house, see Ernest H. Rann, "Dr. Alfred Russel Wallace at Home," *Pall Mall Magazine*, March 1909, 275–84.

5. ARW to William Wallace, April 6, 1902 (courtesy of Richard Wallace); ARW to William Wallace, March 2, 1902, *LLW*, 363–64.

6. ARW to William Wallace, December 19, 1902 (courtesy of Richard Wallace).

7. *ML*, 2:396.

8. ARW to William Wallace, July 8, 1902, *LLW*, 364–65.

9. Alfred Russel Wallace, *Man's Place in the Universe* (New York: McClure, Phillips, 1904), 306–14.

10. For a critique of Wallace's article "Man's Place in the Universe," *Fortnightly Review*, March 1, 1903, see Herbert Hall Turner, "Man's Place in the Universe: A Reply to Dr. Wallace," *Fortnightly Review*, April 1, 1903, 600, 602.

11. Wallace, *Man's Place in the Universe*, 317–18.

12. The portrait of Wallace (provenance unknown, 1904) is reproduced in Max Beerbohm, introduction to *The Portrait Drawings of William Rothenstein* (London: Chapman & Hall, 1926), pl. xlii.

13. William Rothenstein, *Men and Memories: Recollections of William Rothenstein*, vol. 2, *1900–22* (London: Faber and Faber, 1932), 30–34; ARW to William Rothenstein, January 13 and 26 and February 19, 1904, bMS Eng 1148 (1565), Houghton Library, Harvard University, Cambridge, Mass.

14. ARW to Macmillan, January 11, 1905, Macmillan–Wallace letters, nos. 52–53, Department of Manuscripts, British Library, London.

15. *ML*, 2:232–34.

16. ARW to AB (Fisher), November 7, 1905, *LLW*, 449–50.

17. J. Arthur Thomson, "A Great Naturalist," *Nature*, December 14, 1905, 145–46.

18. "Reviews and Notices of Books," *Lancet,* November 11, 1905, 1409–11.

19. *Alfred Russel Wallace: An Anthology of His Shorter Writings*, ed. Charles H. Smith (Oxford: Oxford University Press, 1991), 523.

20. Rann, "Wallace at Home"; Alfred Russel Wallace, "Home Life," in *LLW*, 366.

21. ARW to Paul Lemperly, March 28, 1904, Darwin Collection of the Dittrick Medical History Center, Case Western Reserve University, Cleveland. A full account of the hoax is given in Elizabeth J. Van Allen, *James Whitcomb Riley: A Life* (Bloomington: Indiana University Press, 1999), 100–110.

22. ARW to EBP, August 5, 1904, *LLW*, 328.

23. Ernst Mayr, *The Growth of Biological Thought: Diversity, Evolution, and Inheritance* (Cambridge, Mass.: Harvard University Press, Belknap Press, 1982), 548–49.

24. ARW to EBP, July 27, 1907, *LLW*, 334.

25. ARW to Matthew B. Slater, March 24 and April 8, 1906, Wallace–Slater correspondence, nos. 6–7, Manchester Archives and Local Studies, Manchester Central Library.

26. ARW to Matthew B. Slater, May 18 and 31, 1906, Wallace–Slater correspondence, nos. 8–9, plus addendum.

27. ARW to Matthew B. Slater, August 17 and 26, 1906, Wallace–Slater correspondence, nos. 14 and 16.

28. ARW to Matthew B. Slater, August 26 and September 6, 1906, Wallace–Slater correspondence, nos. 16–17; ARW to William Wallace, September 8, 1906 (courtesy of Richard Wallace).

29. ARW to Matthew B. Slater, September 13, 1906; Matthew B. Slater to ARW, October 9, 1906; ARW to Matthew B. Slater, October 11 and 23, 1906, Wallace–Slater correspondence, nos. 18–21.

30. ARW to Macmillan, October 26, 1906, Macmillan–Wallace letters, nos. 56–59.

31. ARW to Macmillan, November 7, 1906, Macmillan–Wallace letters, no. 60; ARW to Matthew B. Slater, November 7, 1906, Wallace–Slater correspondence, no. 22.

32. ARW to Matthew B. Slater, November 14 and 20, 1906, Wallace–Slater correspondence, nos. 23–24.

33. ARW to Matthew B. Slater, February 22, 1907, Wallace–Slater correspondence, no. 33; ARW to Macmillan, February 6, 1907, Macmillan–Wallace letters, nos. 62–63. See also ARW to John Teasdale, February 16, 1907, Wallace–Slater correspondence, no. 56.

34. ARW to Macmillan, February 6 and 9, 1907, Macmillan–Wallace letters, nos. 62–64; ARW to John Teasdale, February 22, 1907, Wallace–Slater correspondence, no. 34.

35. John Teasdale to Matthew B. Slater, February 19, 1907, Wallace–Slater correspondence, no. 57.

36. *ML*, 2:346–48.

37. Janet Oppenheim, *The Other World: Spiritualism and Psychical Research in England, 1850–1914* (Cambridge: Cambridge University Press, 1988), 70.

38. Arthur Conan Doyle, *The History of Spiritualism* (1926; reprint, New York: Arno Press, 1975), 1:304–5.

39. Oppenheim, *Other World*, 68–71. For a full account of the proceedings see "Colley vs. Maskelyne," *Times* (London), April 26, 27, and 29 and May 1, 1907.

40. "Colley vs. Maskelyne," *Times* (London), April 27, 1907, 14.

41. "Colley vs. Maskelyne," *Times* (London), April 26, 1907, 3; and April 27, 1907, 14.

42. "Colley vs. Maskelyne," *Times* (London), April 27, 1907, 14. This was the day on which Wallace testified.

43. ARW to Macmillan, September 25, 1907, Macmillan–Wallace letters, no. 73.

44. ARW to Fred Birch, September 12, 1907, *LLW*, 410.

45. Alfred Russel Wallace, *Is Mars Habitable? A Critical Examination of Professor Percival Lowell's Book "Mars and its Canals," with an Alternative Explanation* (London: Macmillan, 1907), 90–93.

46. ARW to Macmillan, January 17, 1908, Macmillan–Wallace letters, nos. 83–86.

47. ARW to Macmillan, July 11 and August 16 and 27, 1908, Macmillan–Wallace letters, nos. 102–7.

48. ARW to Macmillan, January 25 and March 26, 1908, Macmillan–Wallace letters, nos. 87 and 95.

49. ARW to Matthew B. Slater, January 21, 1908, Wallace–Slater correspondence, no. 44.

50. M. R. D. Seaward and S. M. D. Fitzgerald, introduction to *Richard Spruce, Botanist and Explorer* (Kew: Royal Botanic Gardens, 1996), 8–9.

20. A NATIONAL TREASURE CELEBRATED

1. Studs Terkel, quoted in *Chicago Sun Times*, September 30, 1997, 4.

2. Alfred Russel Wallace, acceptance speech, in *The Darwin–Wallace Celebration*

held on Thursday, 1st July 1908 by the Linnean Society of London (London: Longman, Green, 1909), 5–11.

3. The Copley Medal was beautifully designed. The face of each coin (the Copley consisted of two coins, one gold and the other silver) had the arms of the Royal Society, the name Copley, and the word "Dignissimo" and Wallace's name below. The reverse side displayed the royal arms (ARW to William Wallace, December 2, 1908, *LLW*, 370).

4. *Aspects of Britain: Honours and Titles* (London: Her Majesty's Stationery Office, 1996), 96. See also Keith Middlemas, *The Life and Times of Edward VII* (London: Weidenfeld & Nicolson, 1972), 202.

5. ARW to RM, December 22, 1908, Oxford University Museum of Natural History.

6. George M. Trevelyan, *A Shortened History of England* (New York: Viking, 1988), 529.

7. Herbert Asquith to Edward VII, October 29, 1908, RA VIC/R29/62, Windsor Castle Library Archives (courtesy of Lady de Bellaigue). Asquith refers to himself in the third person.

8. Middlemas, *Life and Times of Edward VII*, 206–7.

9. ARW to Fred Birch, December 30, 1908, *LLW*, 448. The full text of the letter is among the private papers of Richard Wallace.

10. ARW to William Wallace, December 2 and 17, 1908, *LLW*, 370. See also ARW to Fred Birch, December 30, 1908, *LLW*, 448.

11. ARW to Macmillan, November 10, 1908, Macmillan–Wallace letters, no. 115, Department of Manuscripts, British Library, London.

12. A. J. Bidwell to ARW, December 10, 1908, MS 46442, nos. 62–63, Department of Manuscripts, British Library, London.

13. ARW to RM, November 16, 1908, Oxford University Museum of Natural History.

14. Karl Pearson to Francis Galton, January 23, 1909, quoted in *The Life, Letters and Labours of Francis Galton*, ed. Karl Pearson (Cambridge: Cambridge University Press, 1924), 2:370.

15. Alfred Russel Wallace, "The World of Life: As Visualised and Interpreted by Darwinism" [excerpt from lecture presented at the Royal Institution], *Nature*, January 28, 1909, 384–86.

16. ARW to AB (Fisher), March 6, 1909, *LLW*, 337.

17. Arthur Conan Doyle, *The History of Spiritualism* (1926; reprint, New York: Arno Press, 1975), 1:15.

18. Jorge Luis Borges, "Testimony to the Invisible," in *Testimony to the Invisible: Essays on Swedenborg*, ed. James F. Lawrence (West Chester, Pa.: Chrysalis Books, 1995), 3.

19. A collection of essays that explore Emanuel Swedenborg's influence on modern philosophical thought is James F. Lawrence, ed., *Testimony to the Invisible: Essays on Swedenborg* (West Chester, Pa.: Chrysalis Books, 1995).

20. Emanuel Swedenborg, *Heaven and Hell*, trans. G. F. Dole (1758; reprint, West Chester, Pa.: Swedenborg Foundation, 1994), 50–52, 177, 333–34.

21. Alfred Russel Wallace, *The World of Life: A Manifestation of Creative Power, Directive Mind, and Ultimate Purpose* (London: Bell, 1911), 393–97.

22. Ibid., 371.

23. Ibid., 377.

24. Ibid., 381.

25. Ibid., 378.

26. Review of *The World of Life,* by Alfred Russel Wallace, *Nature*, June 8, 1911, 480–81.

27. ARW to WTD, February 8, 1911, MS 46438, nos. 193–94, Department of Manuscripts, British Library, London.

28. WTD to ARW, February 12, 1911, *LLW*, 342–44.

29. Theodore Roosevelt, "The Search for Truth," *Outlook*, December 2, 1911, 819–26.

30. Alfred Russel Wallace, "Flying Machines in War," *Daily News*, February 6, 1909, 4.

31. ARW to William Wallace, August 20, 1911, *LLW*, 398.

32. ARW to William Wallace, August 26, 1913, *LLW*, 346.

33. Wallace, *World of Life*, 246–47.

34. *LLW*, 373.

35. ARW to Macmillan, March 10, 1913, Macmillan–Wallace letters, nos. 121–22.

36. Alfred Russel Wallace, *The Revolt of Democracy* (London: Funk & Wagnall, 1914), 12, 14.

37. ARW to Dr. Littlefield, January 11, 1913, *LLW*, 346.

38. ARW to Sir Oliver Lodge, October 27, 1913, *LLW*, 411–12.

39. ARW to Macmillan, March 10, 1913, Macmillan–Wallace letters, nos. 121–22.

40. *LLW*, 378, 472.

41. Obituary of Alfred Russel Wallace, *Light: A Journal of Psychical, Occult, and Mystical Research*, November 15, 1913, 545.

42. "Alfred Russel Wallace" [obituary], *Lancet*, November 15, 1913, 1410.

43. "Alfred Russel Wallace" [obituary], *New York Times*, November 8, 1913.

44. "A Search After Truth," *Times* (London), November 8, 1913.

45. "The Promotion of Dr. A. R. Wallace," *Light: A Journal of Psychical, Occult, and Mystical Research*, November 15, 1913, 546–47.

46. "The Death of Dr. Russel Wallace: A Veteran of Science," *Times* (London), November 8, 1913.

47. Herbert E. Ryle, remarks at the dedication ceremony held at Westminster Abbey on November 1, 1915, *LLW*, 473–75.

48. Edward Bagnall Poulton, "Alfred Russel Wallace (1823–1913)," *Proceedings of the Royal Society of London* 95 (1924): i–xxxv.

Biographical Index

Agassiz, Jean Louis Rodolphe (1807–1873) Swiss-born American geologist and professor of natural history at Harvard University who was famous for his theory of glaciation.

Allen, Charles Grant Blairfindie (1848–1899) Canadian-born novelist and naturalist who was a friend of Wallace and Darwin.

Ansted, David Thomas (1814–1880) English professor of geology at King's College, London.

Asquith, Herbert (1852–1928) English politician who as Liberal prime minister (1908–1916) supported Wallace's nomination for the Order of Merit.

Barrett, William Fletcher (1844–1925) Irish professor of physics at the Royal College of Science, Dublin, who was a spiritualist investigator, friend and correspondent of Wallace, and founder of the British Society for Psychical Research in London.

Bastian, Henry Charlton (1837–1915) English neurologist who developed the theory of archebiosis (spontaneous generation), which briefly was supported by Wallace.

Bates, Henry Walter (1825–1892) English entomologist and assistant secretary of the Royal Geographical Society (1864–1892) who accompanied Wallace to Amazonia.

Bateson, William (1861–1926) English geneticist who was a proponent of the mutation theory of the origin of species, which Wallace opposed.

Becouer, Jean-Baptiste (1718–1777) French pharmacist and bird collector who invented arsenic soap, which was used to preserve animal specimens.

Bell, Thomas (1792–1880) English dental surgeon at Guy's Hospital, London, and professor of zoology at King's College, London, who was president of the Linnean Society when Darwin's and Wallace's papers were presented in 1858.

Bellamy, Edward (1850–1891) American novelist whose utopian socialist novel *Looking Backward* converted Wallace to socialism.

Bennett, John Joseph (1801–1876) English botanist who was secretary of the Linnean Society when Darwin's and Wallace's papers were presented in 1858.

Bentham, George (1800–1884) English botanist who served as agent for Richard Spruce during his travels in Amazonia, who was president of the Linnean Society (1861–1874), and whose classification of seed plants served as the foundation of modern systems of vascular-plant taxonomy.

Biggs, John Thomas (?) English anti-vaccinationist from Leicester.

Birch, Fred (?) English naturalist to whom Wallace provided guidance and advice for his collecting expedition to South America in the early twentieth century.

Blavatsky, Helena Petrovna (1831–1891) Russian-born American theosophist who, with Henry Steel Olcott, founded the Theosophical Society, an offshoot of spiritualism.

Blyth, Edward (1810–1873) English naturalist and curator of the Museum of the Royal Asiatic Society of Bengal, Calcutta, who was an important correspondent of Darwin.

Boisduval, Jean-Alphonse (1801–1879) French lepidopterist whose work on the Papilionidae influenced Wallace while he was in the Malay Archipelago.

Bonaparte, Charles-Lucien (nephew of Napoleon) (1803–1857) French ornithologist and taxonomist whose *Conspectus Generum Avium* was an indispensable resource for Wallace while he was in the Malay Archipelago.

Bradlaugh, Charles (1822–1891) English radical atheist and member of Parliament from Northampton who served on the Royal Commission on Vaccination as an opponent of vaccination.

Bristowe, John Syer (1827–1895) English physician and pathologist at St. Thomas's Hospital, London, who served on the Royal Commission on Vaccination as a proponent of vaccination.

Britten, Emma Hardinge (1823–1899) English early advocate of the modern spiritualist movement.

Brooke, James (1803–1868) English soldier and White Rajah of Sarawak who was admired by Wallace.

Bruce-Joy, Albert (1842–1924) Irish portrait sculptor whose dietary advice "cured" Wallace's lifelong asthma and who created a medallion with Wallace's bust that hangs in Westminster Abbey, near Darwin's tomb.

Buckley, Arabella (Burton) (later Fisher) (1840–1929) English secretary to Charles Lyell (1864–1875) who lectured and wrote on science (especially for children) and was a friend and confidante of Wallace.

Burton, Richard Francis (1821–1890) English explorer, Arabist, and spiritualist sympathizer.

Campbell, George Douglas (eighth duke of Argyll) (1823–1900) Scottish naturalist, Whig politician, and philosopher who was a believer in design in nature and opposed the Darwin–Wallace theory of natural selection, yet helped Wallace obtain a civil pension.

Candolle, Alphonse de (1806–1893) Swiss botanist and professor of botany in Geneva.

Carpenter, William Benjamin (1813–1885) English physician and physiologist who was professor of physiology at the Royal Institution (1844–1856) and professor of forensic medicine at University College, London, and who vehemently opposed Wallace's spiritualist beliefs.

Chambers, Robert (1802–1871) Scottish journalist, publisher, and philosopher whose *Vestiges of the Natural History of Creation*, which advocated evolution, created a sensation when published in 1844.

Claparède, Jean Louis René Antoine Édouard (1832–1871) Swiss naturalist and professor of comparative anatomy in Geneva, specializing in insects and spiders, who criticized Wallace's *Contributions to the Theory of Natural Selection.*

COLENSO, JOHN WILLIAM (1814–1893) English cleric who was the controversial first Anglican bishop of Natal, South Africa, and admired by Wallace for his liberal views on race and race relations.

COLLEY, THOMAS (?) English deacon under John William Colenso who, as a spiritualist, was joined by Wallace in defense of the medium Francis Ward Monck.

COMBE, GEORGE (1788–1858) Scottish phrenologist whose *Constitution of Man* influenced Wallace.

COUES, ELLIOT (1842–1899) American ornithologist, frontier historian, and theosophist who was a friend of Wallace.

CROOKES, WILLIAM (1832–1919) English physicist and scientific spiritualist who discovered thallium, invented the cathode-ray tube, and was a friend of Wallace.

CUVIER, GEORGES (1769–1832) French vertebrate anatomist, paleontologist, and geologist who was professor of natural history at the Collège de France and professor of comparative anatomy at the Musée d'Histoire Naturelle, Paris, and whose theory of the earth's formation, known as catastrophism, was countered by Charles Lyell.

DANA, JAMES DWIGHT (1813–1895) American geologist and zoologist who was professor of geology at Yale University.

DARWIN, CHARLES (1809–1882) English naturalist who was celebrated during the nineteenth century as the co-discoverer, with Wallace, of the theory of natural selection.

DARWIN, EMMA (1808–1896) Youngest daughter of Bessy and Josiah Wedgwood who was the first cousin and wife of Charles Darwin.

DARWIN, ERASMUS ALVEY (1804–1881) Older brother of Charles Darwin.

DARWIN, FRANCIS (1848–1925) English botanist who was the son of Charles and Emma Darwin and edited his father's letters,

DARWIN, GEORGE HOWARD (1845–1912) English physicist and professor of astronomy and experimental philosophy at Cambridge University who was the son of Charles and Emma Darwin.

DARWIN, HENRIETTA EMMA (later Litchfield) (1843–1927) Daughter of Charles and Emma Darwin who edited her mother's letters.

DAUBENY, CHARLES GILES BRIDLE (1795–1867) English professor of chemistry at Oxford University who was an early supporter of the theory of natural selection.

DE MORGAN, AUGUSTUS (1806–1871) English mathematician who was an innovator in the field of logic and a spiritualist investigator.

DOHRN, FELIX ANTON (1840–1909) German naturalist who founded the first important marine station.

DONKIN, HORATIO (?) English physician who, with his friend of Edwin Ray Lankester, tried to embarrass Wallace by charging the medium Henry Slade with fraud.

DOUBLEDAY, EDWARD (1811–1849) English curator of the entomological collections at the British Museum who advised Henry Bates and Wallace before their journey to Amazonia.

EASTWOOD, ALICE (1859–1953) Canadian-born botanist and friend of John Muir who accompanied Wallace on a botanical expedition in the Colorado Rockies during his visit to the United States.

EDWARDS, WILLIAM HENRY (1822–1909) American entomologist who advised Henry Bates and Wallace before their journey to Amazonia.

ELLIOTSON, JOHN (1791–1868) Scottish physician who introduced the use of the stethoscope in British medical practice and whose experiments in mesmerism intrigued Wallace.

ERNST, CARL WILHELM (?) American editor and journalist for the *Beacon* who was a spiritualist and an admirer of Wallace.

EVANS, MARY ANN (George Eliot) (1819–1880) English novelist who was the wife of George Lewes.

FAWCETT, HENRY E. (1833–1884) English economist and radical member of Parliament from Brighton who was postmaster general under William Gladstone and professor of political economy at Cambridge University.

FORBES, EDWARD (1815–1854) Scottish naturalist and professor of botany at King's College, London, whose creationist theory of polarity provoked Wallace to publish his Sarawak Law.

FOSTER, MICHAEL (1836–1907) English physiologist.

FOX, KATIE (?) American medium whose psychic experiences on March 31, 1848, in Rochester, New York, launched the modern spiritualist movement.

GALL, JOSEPH (1758–1828) German physician who was the founder of phrenology.

GALTON, FRANCIS (1822–1911) English statistician who was an early spiritualist investigator, an advocate of eugenics, and a cousin of Charles Darwin.

GASCOYNE-CECIL, ROBERT ARTHUR TALBOT (ninth earl of Salisbury) (1830–1903) English politician and leader of the Conservative Party (1885–1902) who was president of the British Association for the Advancement of Science in 1894.

GEACH, FREDERICK (?) English mining engineer who befriended Wallace in Timor.

GEIKIE, ARCHIBALD (1835–1924) Scottish professor of geology at the University of Edinburgh who was a friend of Wallace.

GEORGE, HENRY (1839–1897) American economist and politician whose *Progress and Poverty* reflected Wallace's radical political views.

GILMAN, DANIEL COIT (1831–1908) American educator and first president of Johns Hopkins University who sponsored Wallace during his visit to the United States.

GLADSTONE, WILLIAM EWART (1809–1901) English politician and classical scholar who, during the second of his four terms as Liberal prime minister (1880–1885), approved Wallace's civil pension.

GOULD, JOHN (1804–1881) English ornithologist, author, and artist who identified and classified some of Wallace's bird specimens from the Malay Archipelago.

GRAY, ASA (1810–1888) American botanist and professor of natural history at Harvard University who was an early supporter of the Darwin–Wallace theory of natural selection.

Gray, George Edward (1812–1898) English ornithologist who identified and classified many of Wallace's bird specimens from the Malay Archipelago.

Gray, John Edward (1800–1875) English curator of the zoological section of the British Museum who identified and classified many of Wallace's mammalian specimens from the Malay Archipelago.

Gulick, John Thomas (1832–1923) American missionary and naturalist who was a friend of George John Romanes, an opponent of the theory of natural selection, and an early proponent of the concept of geographic speciation (genetic drift).

Hall, Spencer Timothy (1812–1885) English phreno-mesmerist whose lectures in Leicester influenced Wallace.

Hampden, John (?) English member of the Flat Earth Society who challenged Wallace to prove that the earth was a sphere.

Haughton, Samuel (1821–1897) Irish clergyman, mathematician, and geologist at Trinity College, University of Dublin, who initially opposed the theory of natural selection.

Herschell, Farrer (1837–1899) English politician who was lord chancellor under William Gladstone and headed the Royal Commission on Vaccination.

Home, Daniel Dunglas (1833–1886) Scottish spiritualist medium.

Hooker, Joseph Dalton (1817–1911) English botanist who followed his father, William Hooker, as director of the Royal Botanic Gardens, Kew (1865–1885), and, as a close friend of Charles Darwin, facilitated the presentation of Darwin's and Wallace's papers at the Linnean Society in 1858.

Hooker, William Jackson (1785–1865) English botanist who was the first director of the Royal Botanic Gardens, Kew (1841–1865), during Henry Bates's and Wallace's travels in Amazonia.

Horsfield, Thomas (1773–1859) American naturalist who served in the East Indies under the Dutch and British, was curator of the East India Museum in London, and advised Henry Bates and Wallace about collecting in the tropics.

Humboldt, Friedrich Wilhelm Heinrich Alexander von (1769–1859) German geologist, botanist, engineer, and explorer whose *Personal Narrative* inspired Charles Darwin, Wallace, and all later scientific travelers.

Hunt, James (1832–1870) English nonpracticing physician who, as the first president of the Anthropological Society of London, strongly opposed the Darwin–Wallace theory of natural selection.

Hutton, Richard Holt (1826–1897) English editor of the *Spectator*.

Huxley, Thomas (1825–1895) English comparative anatomist, paleontologist, and educator who was famous for his role as "Darwin's bulldog."

James, William (1842–1910) American philosopher, psychologist, and spiritualist investigator who attended séances with Wallace during his visit to the United States.

Jenner, Edward (1749–1823) English physician who discovered the modern technique of vaccination.

Keeler, P. L. O. A. (?) American spiritualist medium.

Kingsley, Charles (1819–1875) English parson, novelist, and Christian socialist who was an admirer of Wallace's *Malay Archipelago*.

Lamarck, Jean-Baptiste-Pierre-Antoine de Monet de (1744–1829) French naturalist and evolutionist whose theory of the inheritance of acquired characteristics profoundly affected evolutionary thinking.

Lane-Fox, Augustus Henry (General Pitt-Rivers) (1827–1900) English army officer and "father of British archaeology" who supported Wallace's spiritualist investigations.

Lankester, Edwin Ray (1847–1923) English professor of zoology at University College, London, who was a friend and onetime adversary of Wallace.

Lesson, René-Primavère (1794–1849) French naturalist aboard the *Coquille*, which sailed to the Malay Archipelago.

Lewes, George (1817–1878) English critic, editor, playwright, and the husband of George Eliot.

Lindley, John (1799–1865) English botanist, specializing in orchids, who was the assistant librarian to botanist Joseph Bank, first professor of botany at the University of London, and author of *The Elements of Botany*, which taught Wallace the science of taxonomy.

Linnaeus, Carolus (Carl von Linné) (1707–1778) Swedish botanist who is considered the father of modern taxonomy.

Lockyer, Joseph Norman (1836–1920) English astronomer who was the first editor of *Nature*.

Lowell, Percival (1855–1916) American astronomer whose belief in intelligent life on Mars was countered by Wallace.

Lubbock, John (Lord Avebury) (1834–1913) English archaeologist, entomologist, and politician who was an early supporter of the Darwin–Wallace theory of natural selection.

Lunn, Henry Simpson (1859–1939) English humanitarian and religious leader who invited Wallace to give a series of lectures in Davos, Switzerland.

Lyell, Charles (1797–1875) Scottish geologist whose theory of the earth's formation, known as uniformitarianism, discredited the theory of catastrophism; whose *Principles of Geology* was influential; and who served as a mentor to Charles Darwin and Wallace.

Lyell, Mary Elizabeth Horner (1808–1873) Wife of Charles Lyell.

Macleay, William Sharp (1792–1865) English entomologist who developed anti-evolutionary ideas while working in the laboratories of Georges Cuvier in the 1810s and whose quinarian theory was popular in the 1830s and 1840s.

Macmillan, Alexander (1818–1896) Scottish liberal-minded publisher who founded Macmillan, which published *Nature* and most of Wallace's books.

Malthus, Thomas (1766–1834) English philosopher whose *Essay on the Principle of Population* influenced both Charles Darwin and Wallace.

Markham, Clements Robert (1830–1916) English geographer and writer who was president of Royal Geographical Society (1893–1905) and an authority on the civilization of the Incas.

MARSH, OTHNIEL CHARLES (1831–1899) American professor of vertebrate paleontology at Yale University.

MARTIUS, CARL FRIEDRICH PHILIPP VON (1794–1868) Austrian botanist and explorer who preceded Wallace in Brazil.

MASKELYNE, JOHN NEVIL (1839–1917) English magician who specialized in exposing the tricks of spiritualist mediums.

MELDOLA, RAPHAEL (1849–1915) English chemist who developed synthetic dyes, studied protective coloration, and was a friend of Wallace.

MEYER, ADOLPH BERNHARD (1840–1911) German explorer-naturalist who published the German translation of Darwin's and Wallace's papers that were presented at the Linnean Society in 1858.

MILL, JOHN STUART (1806–1877) English philosopher and economist who was an admirer of Wallace's *Malay Archipelago*.

MITTEN, WILLIAM (1819–1906) English chemist and amateur botanist, specializing in mosses, who was Wallace's father-in-law.

MIVART, ST. GEORGE JACKSON (1827–1900) British anatomist and biologist who opposed the Darwin–Wallace theory of natural selection.

MONCK, FRANCIS WARD (?) English spiritualist medium whose reputation Wallace defended.

MONGREDIEN, AUGUSTUS (1807–1888) English author, botanist, and advocate of free trade whose ideas on political economy influenced Wallace.

MONTAGU, MARY WORTLEY (1689–1762) English author and poet who, as the wife of the ambassador to Turkey, was an early advocate of variolation to prevent smallpox.

MUIR, JOHN (1838–1914) Scottish-born American naturalist and explorer who was an early leader of the environmental movement in the United States.

MURCHISON, RODERICK IMPEY (1792–1872) Scottish geologist who, as president of the Royal Geographical Society, helped Wallace get to the Malay Archipelago.

NATTERER, JOHANN (1788–1843) Austrian zoologist who accompanied Archduchess Leopoldina to Brazil, where he fathered an illegitimate daughter whom Wallace met in Guia.

NEWMAN, EDWARD (1801–1876) British entomologist who founded the Entomological Society of London and *Zoologist*.

NEWTON, ALFRED (1829–1907) Swiss-born English naturalist and professor of zoology and comparative anatomy at Cambridge University whom Wallace consulted while researching *The Geographical Distribution of Animals*.

NORDHOFF, CHARLES (1830–1901) German-born American journalist whom Wallace met in Washington, D.C.

OLCOTT, HENRY STEEL (1832–1907) American Buddhist reformer and theosophist who, with Madame Blavatsky, founded the Theosophical Society, an offshoot of spiritualism.

OWEN, JAMES JEROME (?–1895) American journalist who founded the *San Jose Daily Mercury* and was a spiritualist acquaintance of Wallace.

Owen, Richard (1804–1892) British comparative anatomist and paleontologist who was superintendent of the natural history departments at the British Museum (1856–1884).

Owen, Robert (1771–1858) Welsh social reformer who is considered the father of British socialism.

Paget, James (1814–1899) English surgeon who served on the Royal Commission on Vaccination.

Paine, Thomas (1737–1809) English political writer, deist, and freethinker whose ideas profoundly affected Wallace's religious views.

Pearson, Karl (1857–1921) English mathematician who was a protégé and biographer of Francis Galton.

Podmore, Edmund (1847–1888) English psychologist who coauthored a study of psychic phenomena.

Poulton, Edward Bagnall (1856–1943) English zoologist who was a friend and protégé of Wallace and extended his theory of protective coloration.

Powell, John Wesley (1834–1902) American Civil War hero and explorer who was director of the United States Geological Survey when Wallace visited the United States.

Prichard, James Cowles (1786–1848) English physician and ethnologist whose views on the human species influenced Wallace.

Ramsay, Andrew Crombie (1814–1891) Scottish geologist who nominated Wallace for a Royal Medal.

Riley, James Whitcomb (1849–1916) American poet and novelist who perpetrated the "Leonanie" hoax, which convinced Wallace that Edgar Allan Poe was communicating from the spirit world.

Romanes, George John (1848–1894) Canadian-born English naturalist who formulated the theory of physiological selection.

Rothenstein, William (1872–1945) English artist and writer who was commissioned to draw a portrait of Wallace.

Ryle, Herbert E. (?) English dean of Westminster Abbey who presided over the memorial service for Wallace in 1915.

Saunders, William Wilson (1809–1879) English insurance broker, botanist, and entomologist who was a patron of Wallace while he was in the Malay Archipelago.

Schiaparelli, Giovanni (1835–1910) Italian astronomer who observed Mars and its markings, which he called *canali* and others took as signs of life on the planet.

Schomburgk, Robert Hermann (1804–1865) German-born English explorer who discovered the source of curare while in British Guiana.

Sclater, Philip Lutley (1829–1913) English attorney and eminent ornithologist who founded *Ibis*, who was president of the Zoological Society of London, and whose concept of the zoogeographic distribution of animals influenced Wallace.

Sharpey, William (1802–1880) Scottish professor of anatomy and physiology who was an opponent of spiritualism.

Shaw, Norton (?–1868) English secretary of the Royal Geographical Society while Wallace was in the Malay Archipelago.

Silk, George (1823–?) English secretary to the archdeacon of Kensington who was a lifelong friend of Wallace.

Sims, Thomas (?) English professional pioneering photographer who was the husband of Wallace's sister, Fanny.

Slade, Henry (1835–1905) American spiritualist medium.

Slater, Matthew B. (?) English executor of Richard Spruce's estate who helped Wallace publish Spruce's *Notes of a Botanist on the Amazon & Andes.*

Sonnerat, Pierre (1745–1814) French naturalist and explorer who preceded Wallace in the Malay Archipelago.

Spencer, Herbert (1820–1903) English social and political philosopher who was a disciple of Jean-Baptiste Lamarck.

Spix, Johannes Baptist von (1781–1826) Austrian naturalist and explorer who preceded Wallace in Brazil.

Spruce, Richard (1817–1893) English botanist who was famous for obtaining cinchona seeds, which facilitated the production of quinine, for the British while traveling in the Amazon and Andes and was a friend of Wallace.

Spurzheim, Johann Gaspar (1776–1832) German physician who was a pupil of Joseph Gall and the mentor of George Combe.

St. John, Spenser (1825–1907) English diplomat who was secretary to and biographer of James Brooke and befriended Wallace while he was in the Malay Archipelago.

Stanford, Leland (1824–1893) American railroad magnate, United States senator, founder of Stanford University, and spiritualist investigator who befriended Wallace during his visit to the United States.

Stebbing, Thomas Roscoe Rede (1835–1926) English Anglican cleric and marine biologist.

Stevens, Samuel (?) English entomologist who served as agent for Wallace during his travels in Amazonia and the Malay Archipelago.

Stokes, George Gabriel (1819–1903) Irish mathematician who was an opponent of spiritualism.

Strang, William (1859–1921) English painter, etcher, engraver, and illustrator who was commissioned to create a portrait of Wallace.

Strutt, John William (Lord Rayleigh) (1842–1919) English physicist who discovered argon, was professor of natural philosophy at the Royal Institution and president of the Royal Society (1905–1908), and helped obtain the Order of Merit for Wallace.

Swainson, William (1789–1855) English zoologist, naturalist, and illustrator who traveled to Pernambuco, Brazil, in 1816; who collected fish, birds, insects, and plants; and whose *Treatise on the Geography and Classification of Animals* profoundly influenced Wallace.

Thistelton-Dyer, William Turner (1843–1928) English botanist who was a curator at the Royal Botanic Gardens, Kew, and a friend of Wallace.

Thomson, J. Arthur (1861–1933) English biologist and professor of natural history at the University of Aberdeen.

Thomson, William (Lord Kelvin) (1824–1907) Scottish mathematician and physicist whose calculations of the age of the earth implied that evolution by natural selection was impossible.

Turner, Herbert Hall (1861–1930) British astronomer and seismologist who reviewed Wallace's *Man's Place in the Universe.*

Tylor, Edward Burnett (1832–1917) English anthropologist.

Tyndall, John (1820–1893) Irish physicist who opposed Wallace's spiritualist investigations.

Wakefield, Bandusia (?) American botanist.

Waterton, Charles (1782–1865) English naturalist, explorer, and author who traveled in Guiana.

Wedgwood, Hensleigh (1803–1891) English philologist and police magistrate who was the brother of Emma Darwin.

Weir, John Jenner (1822–1894) English naturalist whose experiments on mimicry and protective coloration lent support to Charles Darwin's and Wallace's theories.

Weismann, August Friederich Leopold (1834–1914) German experimental biologist who disproved the theory of the inheritance of acquired characteristics and was a pioneer in the field of genetics.

Westermarck, Edward Alexander (1862–1939) Finnish anthropologist and philosopher who was mentored by Wallace.

Wilberforce, Samuel (1805–1873) English cleric who was the bishop of Oxford and Thomas Huxley's famous antagonist at the Oxford meeting of the British Association for the Advancement of Science in 1860

Wright, Edward Percival (1834–1910) Irish professor of botany at Trinity College, University of Dublin.

Select Bibliography

PRIMARY SOURCES

BOOKS AND ARTICLES

"Alfred Russel Wallace" [obituary]. *Lancet*, November 15, 1913, 1410.

"Alfred Russel Wallace" [obituary]. *New York Times*, November 8, 1913.

Allen, Grant. Review of *Darwinism*, by Alfred Russel Wallace. *Academy* 36 (1889): 41–42.

"Anniversary Meeting, President's Address, November 30, 1868." *Proceedings of the Royal Society* 17 (1869): 148.

Archer, T. C. Note to the Editor. *Hooker's Journal of Botany* 7 (1855): 213–14.

Bates, Henry Walter. "Contributions to an Insect Fauna of the Amazon Valley." *Transactions of the Linnean Society* 23 (1862): 495–566.

———. "Hints to Travelers." *Journal of the Royal Geographical Society* 34 (1864): 306–16.

———. *The Naturalist on the River Amazons*. London: Murray, 1892.

Bell, Thomas. "Presidential Address to the Members of the Linnean Society." *Journal of the Proceedings of the Linnean Society of London (Zoology)* 4 (1859): viii–ix.

Bellamy, Edward. *Looking Backward*. 1888. Reprint, New York: Penguin, 1994.

Carpenter, William Benjamin. "Epidemic Delusions." *Popular Science Monthly* 2 (1872–1873): 15–36.

———. "On Fallacies of Testimony Respecting the Supernatural." *Popular Science Monthly* 8 (1875–1876): 570–86.

———. "Mesmerism, Odylism, Table-turning, and Spiritualism." Parts 1 and 2. *Popular Science Monthly* 11 (1877): 12–25, 161–73.

———. *Mesmerism, Spiritualism, &c Historically & Scientifically Considered, Being Two Lectures Delivered at the London Institution, with Preface and Appendix*. New York: Appleton, 1877.

———. "The Curiosities of Credulity." *Athenaeum*, December 22, 1877, 814.

Chambers, Robert. *Vestiges of the Natural History of Creation*. Edited, with an introduction, by James A. Secord. 1844. Reprint, Chicago: University of Chicago Press, 1994.

This bibliography is by no means comprehensive and includes only those works from which I quote or to which I allude. For a more complete listing of Wallace's works, see Charles H. Smith, ed., *Alfred Russel Wallace: An Anthology of His Shorter Writings* (Oxford: Oxford University Press, 1991).

Claparède, R. E. "Critique of A. R. Wallace's *Contributions to the Theory of Natural Selection* in 'Scientific Serials.'" *Nature*, August 11, 1870, 306.
Clark, Hyde. Letter to the Editor. *Nature*, August 25, 1881, 380–81.
Colenso, John William. "On the Efforts of Missionaries Among Savages." *Anthropological Review and Journal of the Anthropological Society of London* 3 (1865): ccxlviii–cclxxii.
Combe, George. *The Constitution of Man Considered in Relation to External Objects.* 1834. Reprint, Delmar, N.Y.: Scholars' Facsimiles & Reprints, 1974.
Crookes, William. "Spiritualism Viewed by the Light of Modern Science." *Quarterly Journal of Science*, July 1870, 316–20.
———. "Some Further Experiments on Psychic Force." *Quarterly Journal of Science*, October 1871, 470–93.
Darwin, Charles. *Voyage of the Beagle (Journal of Researches into the Geology and Natural History of the Various Countries Visited by the HMS Beagle).* 1839. Reprint, New York: Penguin, 1989.
———. *The Origin of Species by Means of Natural Selection or the Preservation of Favored Races in the Struggle for Life.* 1859. Reprint of 6th ed., New York: Modern Library, n.d.
———. *The Descent of Man, and Selection in Relation to Sex.* 2 vols. 1871. Reprint, Princeton, N.J.: Princeton University Press, 1981.
———. *The Life and Letters of Charles Darwin.* 2 vols. Edited by Francis Darwin. 1887. Reprint, New York: Appleton, 1904.
———. *The Variation of Animals and Plants Under Domestication.* 2 vols. 2nd ed. London: Murray, 1888.
———. *The Autobiography of Charles Darwin and Selected Letters.* Edited by Francis Darwin. 1892. Reprint, New York: Dover, 1958.
———. *More Letters of Charles Darwin.* 2 vols. Edited by Francis Darwin. London: Murray, 1903.
———. *The Autobiography of Charles Darwin.* Edited by Nora Barlow. 1958. Reprint, New York: Norton, 1993.
———. *The Correspondence of Charles Darwin.* 9 vols. Edited by Frederick Burkhardt and Sydney Smith. Cambridge: Cambridge University Press, 1985–1994.
Dawson, Warren R. *The Huxley Papers: A Descriptive Catalogue of the Correspondence, Manuscripts and Miscellaneous Papers of the Rt. Hon. Thomas Henry Huxley . . . preserved in the Imperial College of Science and Technology, London.* London: Macmillan, 1946.
"The Death of Dr. Russel Wallace: A Veteran of Science." *Times* (London), November 8, 1913.
Dialectical Society. *Report on Spiritualism of the Committee of the London Dialectical Society Together with the Evidence, Oral and Written, and a Selection from the Correspondence.* London: Burns, 1873.
Dohrn, Anton. "Contributions to the Theory of Natural Selection." *Academy*, February 15, 1871, 138–40.
———. "Contributions to the Theory of Natural Selection." *Academy*, March 1, 1871, 159–60.

"Dr. A. R. Wallace, Scientist, Dead." *New York Times*, November 8, 1913.

Edwards, William H. *A Voyage up the River Amazon: Including a Residence at Pará*. New York: Appleton, 1847.

Forbes, Edward. "On the Manifestation of Polarity in the Distribution of Organized Beings in Time." *Proceedings of the Royal Institution of London* 57 (1854): 332–37.

Galton, Francis. *The Life, Letters and Labours of Francis Galton*. 2 vols. Edited by Karl Pearson. Cambridge: Cambridge University Press, 1924.

Geikie, Archibald. "Sir Charles Lyell, Bart., F.R.S." [obituary]. *Nature*, March 4, 1875, 341–42.

———. Review of *Island Life*, by Alfred Russel Wallace. Part 1. *Nature*, February 17, 1881, 357–59.

———. Review of *Island Life*, by Alfred Russel Wallace. Part 2. *Nature*, February 24, 1881, 390–93.

George, Henry. *Progress and Poverty*. 1882. Reprint, New York: Robert Schalkenback Foundation, 1997.

Gould, John. *Birds of New Guinea and the Adjacent Papuan Islands: Including Many New Species Recently Discovered in Australia*. 5 vols. London: Southeran, 1875–1888.

Gray, Asa. Review of *Origin of Species*, by Charles Darwin. *Annals and Magazine of Natural History*, 3rd ser., 6, no. 35 (1860): 373–86.

Gray, George Robert. "List of Birds by Mr. Wallace at the Molucca Islands, with Descriptions of New Species, etc." *Proceedings of the Zoological Society of London* (1860): 341.

Gray, John Edward. "List of Species of Mammalia Sent from the Aru Islands by Mr. A. R. Wallace to the British Museum." *Annals and Magazine of Natural History*, 3rd ser., 2, no. 9 (1858): 222–23.

———. "Additional Observations on the Genus *Cuscus*." *Annals and Magazine of Natural History*, 3rd ser., 9, no. 51 (1862): 244–51.

Gulick, John T. "Divergent Evolution Through Cumulative Segregation." *Journal of the Proceedings of the Linnean Society of London (Zoology)* 20 (1888): 189–274, 312–80.

Guppy, Henry F. J. "Notes on the Capabilities of the Negro for Civilisation." *Anthropological Review and Journal of the Anthropological Society of London* 2 (1864): ccxiii–ccxvi.

Haughton, Samuel. "On the Form of the Cells made by various Wasps and by the Honey Bee; with an Appendix on the 'Origin of Species.'" *Annals and Magazine of Natural History*, 3rd ser., 11, no. 77 (1863): 415–29.

Hooker, Joseph Dalton. *The Life and Letters of Joseph Dalton Hooker*. 2 vols. Edited by Leonard Huxley. New York: Appleton, 1918.

Humboldt, Alexander von. *Personal Narrative of a Journey to the Equinoctial Regions of the New Continent*. Translated, with an introduction, by Jason Wilson. 1814–1825. Reprint (abridged), New York: Penguin, 1995.

Hunt, James. "Opening Address to the Anthropological Society of London on 24 February 1863." *Anthropological Review* 1, no. 1 (1863): 1–20.

———. "On the Application of the Principle of Natural Selection to Anthropology: In Reply to Views Advocated by Some of Mr. Darwin's Disciples." *Anthropological Review* 4, no. 15 (1866): 320–40.

———. "On Physio-Anthropology, Its Aim and Method." *Anthropological Review and Journal of the Anthropological Society of London* 5 (1867): cclvi–vii.

Huxley, Thomas Henry. *Evidences as to Man's Place in Nature.* 1863. Reprint, London: Watts, 1921.

———. "A Liberal Education" (1868). In *Science and Education.* 1901. Reprint, New York: Citadel Press, 1964.

———. "On the Classification and Distribution of the Alectoromorphae and Heteromorphae." *Proceedings of the Zoological Society of London* (1868): 294–319.

———. *Darwinian Essays.* 1871. Reprint, New York: AMS Press, 1970.

———. *Life and Letters of Thomas Henry Huxley.* 2 vols. Edited by Leonard Huxley. New York: Appleton, 1913.

Jacob, Gertrude. *The Raja of Sarawak.* 2 vols. London: Macmillan, 1876.

James, William. "The Final Impressions of a 'Psychical Researcher'" (1909). In *Memories and Studies.* New York: Longman, Green, 1911.

Lamarck, Jean-Baptiste. *Zoological Philosophy.* 1809. Reprint, Chicago: University of Chicago Press, 1984.

Lankester, Edwin Ray. "Dr. Carpenter, C.B., F.R.S." [obituary]. *Nature*, November 26, 1885, 83–85.

———. "Charles Robert Darwin." In *Biographical Dictionary and Synopsis of Books, Ancient and Modern*, edited by Charles Warner, 4385–93. New York: International Society, 1896.

Lawrence, William. *Lectures on Comparative Anatomy, Physiology, Zoology and the Natural History of Man.* London: Bell & Daldy, 1866.

Lindley, John. *The Elements of Botany, Structural and Physiological; A Fifth Edition of the Outline of the First Principles of Botany.* London: Bradbury & Evans, 1847.

Litchfield, Henrietta [Darwin], ed. *Emma Darwin: A Century of Family Letters, 1792–1896.* 2 vols. New York: Appleton, 1915.

Lowe, Robert. "Reciprocity and Free Trade." *Nineteenth Century* 5 (1879): 992–1002.

Lyell, Charles. *Principles of Geology.* 3 vols. 1830–1833. Reprint, Chicago: University of Chicago Press, 1991.

———. *Principles of Geology, or the Modern Changes of the Earth and Its Inhabitants.* 11th ed. New York: Appleton, 1872.

———. *The Geological Evidences of the Antiquity of Man, with an Outline of Glacial and Post-Tertiary Geology.* 4th ed. 1873. Reprint, New York: AMS Press, 1973.

———. *Life, Letters, and Journals of Sir Charles Lyell, Bart.* 2 vols. Edited by K. Lyell. London: Murray, 1881.

———. *Sir Charles Lyell's Scientific Journals on the Species Question.* Edited by Leonard G. Wilson. New Haven, Conn.: Yale University Press, 1970.

Macmillan, Alexander. *Letters of Alexander Macmillan.* Edited by George Macmillan. Glasgow: University Press, 1908.

Malthus, Thomas. *An Essay on the Principle of Population, as it Affects the Future Improvement of Society with Remarks on the Speculations of Mr. Godwin, M.*

Condorcet, and Other Writers. Edited, with an introduction, by Antony Flew. 1798 and 1830. Reprint, New York: Penguin, 1985.
Markham, Clements R. *The Fifty Years' Work of the Royal Geographical Society*. London: Murray, 1881.
Meyer, Adolph Bernhard. Letter to the Editor. *Nature*, August 29, 1895, 415.
Mill, John Stuart. "Public Responsibility and the Ballot." *Reader*, April 29, 1865, 474–75.
Moseley, H. N. Review of *The Continuity of the Germ-Plasma Considered on the Basis of a Theory of Heredity*, by August Weismann. *Nature*, December 17, 1885, 154.
———. "Dr. August Weismann on the Importance of Sexual Reproduction for the Theory of Natural Selection." *Nature*, October 28, 1886, 629–32.
Mozley, H. N. "Compulsory Vaccination in the British Isles." *Vaccination Inquirer and Health Review* 7 (1885): 79–80.
Murchison, Roderick Impey. "Presidential Address." *Journal of the Royal Geographical Society* 29 (1859): clxxvii–viii.
———. "Presidential Address." *Journal of the Royal Geographical Society* 39 (1869): clxii–xv.
Newman, Edward. "President's Address to the Entomological Society." *Zoologist* 12 (1854): 4223–24.
———. "The Death of a Species." *Zoologist* 27 (1869): 1529–42.
Obituary of James Hunt. *Anthropological Review and Journal of the Anthropological Society of London* 8 (1870): lxxix–xxxiii.
Obituary of Alfred Russel Wallace. *Light: A Journal of Psychical, Occult, and Mystical Research*, November 15, 1913, 545.
Owen, Richard. "On the Anthropoid Apes, and Their Relations to Man." *Proceedings of the Royal Institution of Great Britain* 2 (1854–1858): 35–36.
Owen, Robert. *"A New View of Society" and Other Writings*. Edited by Gregory Claes. New York: Penguin, 1991.
Paine, Thomas. *The Age of Reason*. 1794. Reprint, New York: Buccaneer Books, 1976.
Prichard, James Cowles. *Researches into the Physical History of Mankind*. 4 vols. London: Houlston and Stoneman, 1851.
"The Promotion of Dr. A. R. Wallace." *Light: A Journal of Psychical, Occult, and Mystical Research*, November 15, 1913, 546–47.
P. S. "Wallace on Natural Selection." *Nature*, November 3, 1870, 8–9.
Romanes, George John. "Permanent Variation of Colour in Fish." *Nature*, June 5, 1873, 101.
———. "Perception and Instinct in the Lower Animals." *Nature*, August 7, 1873, 282–83.
———. "A Speculation Regarding the Senses." *Nature*, February 12, 1880, 348.
———. *The Life and Letters of George John Romanes*. Edited by Ethyl Romanes. London: Longman, Green, 1908.
Roosevelt, Theodore. "The Search for Truth." *Outlook*, December 2, 1911, 819–26.
Rothenstein, William. *The Portrait Drawings of William Rothenstein*. London: Chapman & Hall, 1926.

———. *Men and Memories: Recollections of William Rothenstein.* Vol. 2, *1900–22.* London: Faber and Faber, 1932.

St. John, Spenser. *The Life of Sir James Brooke, Rajah of Sarawak.* 1879. Reprint, Kuala Lumpur: Oxford University Press, 1994.

Salisbury, Lord [Robert Arthur Talbot Gascoyne-Cecil]. "Presidential Address, the British Association, Oxford, August 8 [1894]." *Nature,* August 9, 1894, 338–43.

Schomburgk, Robert. "On the 'Urari,' the Arrow Poison of the Indians of Guiana; with a Description of the Plant from which It Is Extracted." *Annals and Magazine of Natural History,* 1st ser., 7, no. 45 (1841): 407–27.

"Scientific Fallacy." *Medium and Daybreak,* March 29, 1872, 109.

Sclater, Philip Lutley. "The Geographical Distribution of Birds." *Journal of the Proceedings of the Linnean Society of London (Zoology)* 2 (1858): 130.

———. "Note on Wallace's Standard-wing, *Semioptera wallacii.*" *Ibis,* 1st ser., 2 (1860): 26–28.

"A Search After Truth" [obituary of Alfred Russel Wallace]. *Times* (London), November 8, 1913.

Snow, John. "On the Mode of the Communication of Cholera." In *Snow on Cholera.* 1854. Reprint of 2nd ed., New York: Hafner, 1965.

Spencer, Herbert. *First Principles.* New York: Collier, 1902.

Sprecher, Robert. "The Darwin–Bates Letters: Correspondence Between Two Nineteenth-Century Travellers and Naturalists." *Annals of Science* 25, no. 1 (1969): 1–47.

———. "The Darwin–Bates Letters: Correspondence Between Two Nineteenth-Century Travellers and Naturalists." *Annals of Science* 25, no. 2 (1969): 95–125.

Spruce, Richard. "On *Leopoldinia Piassaba,* Wallace." *Journal of the Proceedings of the Linnean Society of London (Botany)* 4 (1859): 58–60.

———. *Notes of a Botanist on the Amazon & Andes . . . During the Years 1849–1864.* 2 vols. Edited by Alfred Russel Wallace. London: Macmillan, 1908.

Stern, Bernhard J. "Letters of Alfred Russel Wallace to Lester F. Ward." *Scientific Monthly* 40 (1935): 375–79.

Stevens, Samuel. "Advertisement to Collectors." *Annals and Magazine of Natural History,* 2nd ser., 3, no. 13 (1849): 74–75.

———. "Directions for Collecting and Preserving Specimens of Natural History in Tropical Climates." 1855. Linnean Society of London Collection.

Swainson, William. *On the Natural History and Classification of Quadrupeds.* London: Longman, Rees, 1835.

———. *A Treatise on the Geography and Classification of Animals.* London: Longman, Rees, 1835.

Swedenborg, Emanuel. *Heaven and Hell.* Translated by G. F. Dole. 1758. Reprint, West Chester, Pa.: Swedenborg Foundation, 1994.

Tebb, William. "Sanitation, Not Vaccination the True Protection Against Small-Pox." October 12, 1881. [available at: www.whalete.to/vaccine/tebbl.html]

Thistelton-Dyer, William. "Mr. Romanes's Paradox." *Nature,* November 1, 1888, 7–8.

Thomson, J. Arthur. "A Great Naturalist." *Nature,* December 14, 1905, 145–46.

Turner, Herbert Hall. "Man's Place in the Universe: A Reply to Dr. Wallace." *Fortnightly Review*, April 1, 1903, 600–602.

"The Vaccination Acts." *Hansard's Parliamentary Debates*, 3rd ser., 334, no. 15 (April 5, 1889): 1741–1742.

Wallace, Alfred Russel. "On the Umbrella Bird (*Cephalopterus ornatus*) 'Ueramimbe,' Lingoa Geral." *Proceedings of the Zoological Society of London* (1850): 206–7.

———. *A Narrative of Travels on the Amazon and Rio Negro.* 1853. Reprint, New York: Haskell House, 1969.

———. "On the Rio Negro." *Journal of the Royal Geographical Society* 23 (1853): 212–17.

———. *Palm Trees of the Amazon and Their Uses.* 1853. Reprint, Lawrence, Kan.: Coronado, 1971.

———. Notebook: Fishes of the Amazon and Rio Negro, Descriptions. Natural History Museum, London.

———. "On the Habits of the Butterflies of the Amazon Valley." *Transactions of the Entomological Society of London*, n.s., 2, pt. 7 (1854): 253–64.

———. "On the Insects Used for Food by the Indians of the Amazon." *Transactions of the Entomological Society of London*, n.s., 2, pt. 8 (1854): 241–44.

———. "On Some Fishes Allied to *Gymnotus*." *Annals and Magazine of Natural History*, 2nd ser., 14, no. 83 (1854): 398–99.

———. "On the Monkeys of the Amazon." *Annals and Magazine of Natural History*, 2nd ser., 14, no. 84 (1854): 451–54.

———. "On the Law Which Has Regulated the Introduction of New Species." *Annals and Magazine of Natural History*, 2nd ser., 16, no. 93 (1855): 184–96.

———. "Observations on the Zoology of Borneo." *Zoologist* 14 (1856): 5113–17.

———. "Some Account of an Infant Orang-Utan." *Annals and Magazine of Natural History*, 2nd ser., 17, no. 101 (1856): 386–90.

———. "On the Orang-Utan or Mias of Borneo." *Annals and Magazine of Natural History*, 2nd ser., 17, no. 102 (1856): 471–76.

———. "On the Habits of the Orang-Utan in Borneo." *Annals and Magazine of Natural History*, 2nd ser., 17, no. 103 (1856): 26–32.

———. "Attempts at a Natural Arrangement of Birds." *Annals and Magazine of Natural History*, 2nd ser., 18, no. 105 (1856): 193–216.

———. "A New Kind of Baby." *Chambers's Journal*, November 22, 1856, 325–27.

———. "On the Great Bird of Paradise." *Annals and Magazine of Natural History*, 2nd ser., 20, no. 120 (1857): 411–16.

———. "On the Natural History of the Aru Islands." *Annals and Magazine of Natural History*, 2nd ser., 20, supplement (1857): 473–85.

———. "Note on the Theory of Permanent and Geographical Varieties." *Zoologist* 16 (1858): 5887–89.

———. "On the Entomology of the Aru Islands." *Zoologist* 16 (1858): 5889–94.

———. "On the Habits and Transformations of a Species of *Ornithoptera*, Allied to *O. Priamus*, Inhabiting the Aru Islands, near New Guinea." *Transactions of the Entomological Society of London*, n.s., 4 (1858): 272–73.

———. "On the Tendency of Varieties to Depart Indefinitely from the Original Type; Instability of Varieties Supposed to Prove the Permanent Distinctness of Species." *Journal of the Proceedings of the Linnean Society of London (Zoology)* 3 (1858): 53–62.

———. "On the Zoological Geography of the Malay Archipelago." *Journal of the Proceedings of the Linnean Society of London (Zoology)* 4 (1859): 172–84.

———. "Notes of a Voyage to New Guinea." *Journal of the Royal Geographical Society* 30 (1860): 172–77.

———. "Notes on *Semioptera Wallacii*, Gray, from a Letter Addressed to John Gould, Esq., FRS, dated 30 September 1859." *Proceedings of the Zoological Society of London* (1860): 61.

———. "The Ornithology of Northern Celebes." *Ibis*, 1st ser., 2 (1860): 140–47.

———. "On the Ornithology of Ceram and Wagiou." *Ibis*, 1st ser., 3 (1861): 283–91.

———. "Narrative of Search After Birds of Paradise." *Proceedings of the Zoological Society of London* (1862): 153–59.

———. "On the Physical Geography of the Malay Archipelago." *Journal of the Royal Geographical Society* 33 (1863): 217–34.

———. "Remarks on the Rev. S. Haughton's Paper on the Bee's Cells, and on the Origin of Species." *Annals and Magazine of Natural History*, 3rd ser., 12, no. 82 (1863): 303–9.

———. "The Origin of Human Races and the Antiquity of Man Deduced from the Theory of 'Natural Selection.'" *Anthropological Review and Journal of the Anthropological Society of London* 2 (1864): clviii–xx.

———. "On the Progress of Civilisation in Northern Celebes." *Reader*, October 15, 1864, 485–86.

———. "The Scientific Aspect of the Supernatural" (1866). In *Miracles and Modern Spiritualism*. 1896. Reprint, New York: Arno Press, 1975.

———. "Mimicry, and Other Protective Resemblances Among Animals" (1867). In *Contributions to the Theory of Natural Selection*. 1870. Reprint, New York: AMS Press, 1973.

———. "Creation by Law." *Quarterly Journal of Science*, October 1867, 471–88.

———. "A Theory of Birds' Nests; Shewing the Relation of Certain Differences of Colour in Birds to Their Mode of Nidification" (1868). In *Contributions to the Theory of Natural Selection*. 1870. Reprint, New York: AMS Press, 1973.

———. *The Malay Archipelago*. 1869. Reprint, Oxford: Oxford University Press, 1986.

———. "Museums for the People" (1869). In *Studies Scientific and Social*. Vol. 2. London: Macmillan, 1900.

———. "Sir Charles Lyell on Geological Climates and the Origin of Species." *Quarterly Review* 126 (1869): 359–94.

———. "The Limits of Natural Selection as Applied to Man." In *Contributions to the Theory of Natural Selection*. 1870. Reprint, New York: AMS Press, 1973.

———. "Government Aid to Science" [letter to the editor]. *Nature*, January 13, 1870, 288–89.

———. Review of *Hereditary Genius*, by Francis Galton. *Nature*, March 17, 1870, 501–3.

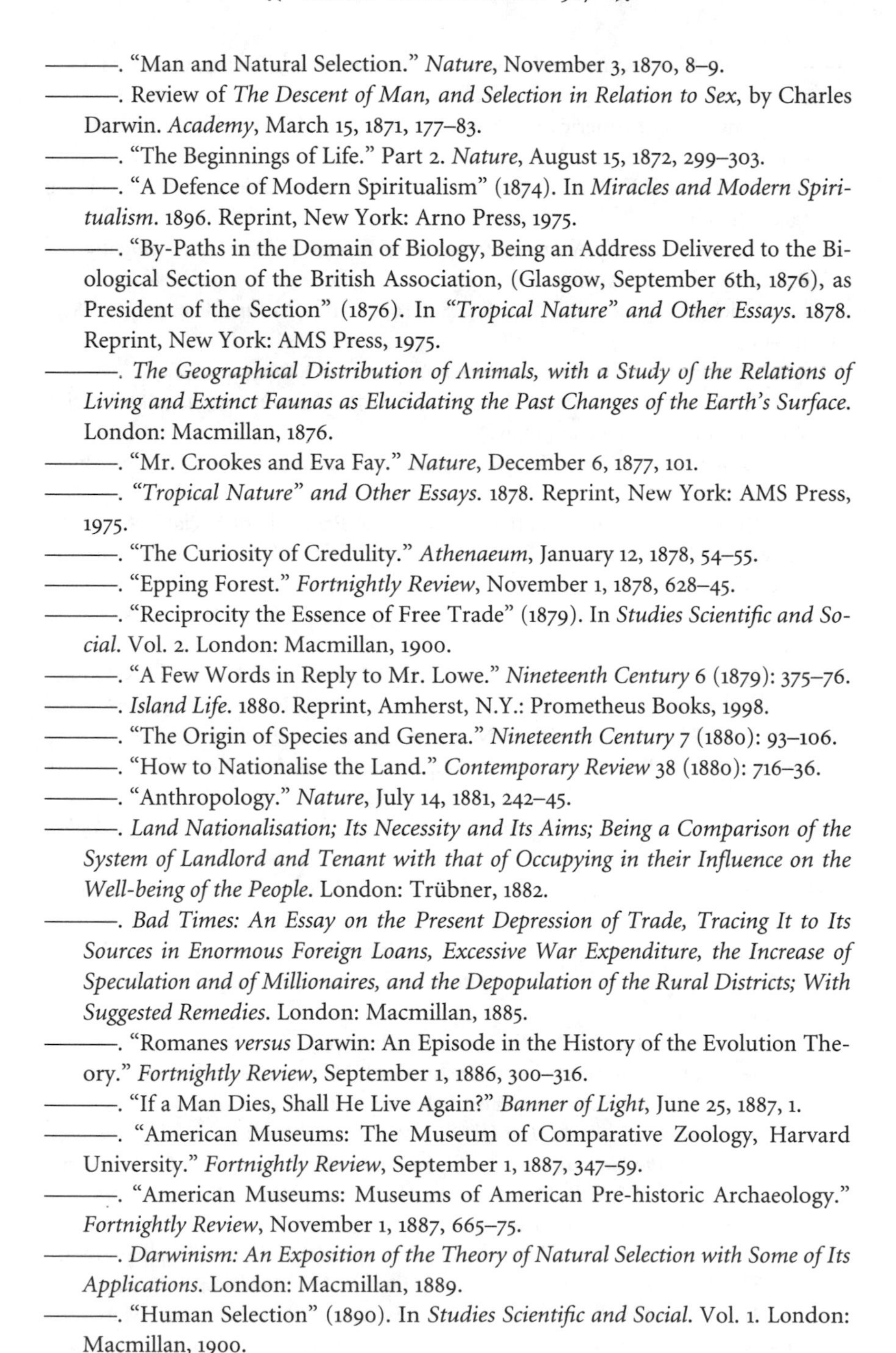

———. "Man and Natural Selection." *Nature*, November 3, 1870, 8–9.
———. Review of *The Descent of Man, and Selection in Relation to Sex*, by Charles Darwin. *Academy*, March 15, 1871, 177–83.
———. "The Beginnings of Life." Part 2. *Nature*, August 15, 1872, 299–303.
———. "A Defence of Modern Spiritualism" (1874). In *Miracles and Modern Spiritualism*. 1896. Reprint, New York: Arno Press, 1975.
———. "By-Paths in the Domain of Biology, Being an Address Delivered to the Biological Section of the British Association, (Glasgow, September 6th, 1876), as President of the Section" (1876). In *"Tropical Nature" and Other Essays*. 1878. Reprint, New York: AMS Press, 1975.
———. *The Geographical Distribution of Animals, with a Study of the Relations of Living and Extinct Faunas as Elucidating the Past Changes of the Earth's Surface*. London: Macmillan, 1876.
———. "Mr. Crookes and Eva Fay." *Nature*, December 6, 1877, 101.
———. *"Tropical Nature" and Other Essays*. 1878. Reprint, New York: AMS Press, 1975.
———. "The Curiosity of Credulity." *Athenaeum*, January 12, 1878, 54–55.
———. "Epping Forest." *Fortnightly Review*, November 1, 1878, 628–45.
———. "Reciprocity the Essence of Free Trade" (1879). In *Studies Scientific and Social*. Vol. 2. London: Macmillan, 1900.
———. "A Few Words in Reply to Mr. Lowe." *Nineteenth Century* 6 (1879): 375–76.
———. *Island Life*. 1880. Reprint, Amherst, N.Y.: Prometheus Books, 1998.
———. "The Origin of Species and Genera." *Nineteenth Century* 7 (1880): 93–106.
———. "How to Nationalise the Land." *Contemporary Review* 38 (1880): 716–36.
———. "Anthropology." *Nature*, July 14, 1881, 242–45.
———. *Land Nationalisation; Its Necessity and Its Aims; Being a Comparison of the System of Landlord and Tenant with that of Occupying in their Influence on the Well-being of the People*. London: Trübner, 1882.
———. *Bad Times: An Essay on the Present Depression of Trade, Tracing It to Its Sources in Enormous Foreign Loans, Excessive War Expenditure, the Increase of Speculation and of Millionaires, and the Depopulation of the Rural Districts; With Suggested Remedies*. London: Macmillan, 1885.
———. "Romanes *versus* Darwin: An Episode in the History of the Evolution Theory." *Fortnightly Review*, September 1, 1886, 300–316.
———. "If a Man Dies, Shall He Live Again?" *Banner of Light*, June 25, 1887, 1.
———. "American Museums: The Museum of Comparative Zoology, Harvard University." *Fortnightly Review*, September 1, 1887, 347–59.
———. "American Museums: Museums of American Pre-historic Archaeology." *Fortnightly Review*, November 1, 1887, 665–75.
———. *Darwinism: An Exposition of the Theory of Natural Selection with Some of Its Applications*. London: Macmillan, 1889.
———. "Human Selection" (1890). In *Studies Scientific and Social*. Vol. 1. London: Macmillan, 1900.
———. *Natural Selection and Tropical Nature: Essays on Descriptive and Theoretical Biology*. London: Macmillan, 1891.

———. "English and American Flowers." Part 1. *Fortnightly Review*, October 1, 1891, 530–32.
———. "English and American Flowers." Part 2. *Fortnightly Review*, December 1, 1891, 804–10.
———. "Inaccessible Valleys" (1893). In *Studies Scientific and Social.* Vol. 1. London: Macmillan, 1900.
———. "Are Individually Acquired Characters Inherited?" Part 2. *Fortnightly Review*, May 1, 1893, 655–68.
———. "Richard Spruce, Ph.D., F.R.G.S." [obituary]. *Nature*, February 1, 1894, 318–19.
———. "Another Substitute for Darwinism." *Nature*, October 4, 1894, 541.
———. "The Method of Organic Evolution" (1895). In *Studies Scientific and Social.* Vol. 1. London: Macmillan, 1900.
———. *Miracles and Modern Spiritualism.* 1896. Reprint, New York: Arno Press, 1975.
———. "Justice, not Charity, as the Fundamental Principle of Social Reform: An Appeal to My Readers" (1898). In *Studies Scientific and Social.* Vol. 2. London: Macmillan, 1900.
———. *The Wonderful Century: Its Successes and Its Failures.* New York: Dodd, Mead, 1899.
———. "The Dawn of a Great Discovery: My Relations with Darwin in Reference to the Theory of Natural Selection." *Black and White*, January 17, 1903, 78.
———. *Man's Place in the Universe.* New York: McClure, Phillips, 1904.
———. *My Life; A Record of Events and Opinions.* 2 vols. New York: Dodd, Mead, 1905.
———. "Fishes of the Rio Negro and Its Tributaries." *Proceedings of the Zoological Society of London* (1905): 189.
———. "Personal Suffrage, a Rational System of Representation and Election." *Fortnightly Review*, January 1, 1907, 3–9.
———. *Is Mars Habitable? A Critical Examination of Professor Percival Lowell's Book "Mars and its Canals," with an Alternative Explanation.* London: Macmillan, 1907.
———. "The World of Life: As Visualised and Interpreted by Darwinism" [excerpt from lecture presented at the Royal Institution]. *Nature*, January 28, 1909, 384–86.
———. "Flying Machines." *Daily News*, February 6, 1909, 4.
———. Acceptance speech. In *The Darwin–Wallace Celebration held on Thursday, 1st July 1908 by the Linnean Society of London.* London: Longman, Green, 1909.
———. *The World of Life: A Manifestation of Creative Power, Directive Mind, and Ultimate Purpose.* London: Bell, 1911.
———. *Social Environment and Moral Progress.* New York: Funk & Wagnall, 1913.
———. *The Revolt of Democracy.* New York: Funk & Wagnall, 1914.
———. *Alfred Russel Wallace: Letters and Reminiscences.* Edited by James Marchant. New York: Harper, 1916.
———. *Alfred Russel Wallace: An Anthology of His Shorter Writings.* Edited by Charles H. Smith. Oxford: Oxford University Press, 1991.

Walsh, J. J. "Experimental Proof of the Rotundity of our Earth." *Field, the Country Gentleman's Newspaper*, March 5, 1870, 199.
Waterton, Charles. *Wanderings in South America, the North-West of the United States and the Antilles in the Years 1812, 1816, 1820 & 1824*. New York: Sturgis & Walton, 1909.
Westermarck, Edward A. *The History of Human Marriage*. 3 vols. 5th ed. New York: Allerton, 1922.
Wikman, K. Rob. "Letters from Edward B. Tylor and Alfred Russel Wallace to Edward Westermarck." *Acta Acadamie Aboensis, Humanoria XIII* 7 (1940): 13–19.
Wright, E. Percival. "Tropical Nature." *Nature*, June 6, 1878, 140–41.

PERIODICALS

Annals and Magazine of Natural History
Anthropological Review and Journal of the Anthropological Society of London
Athenaeum
Daily News (London)
Fraser's Magazine
Hooker's Journal of Botany
Ibis
Journal of the Proceedings of the Linnean Society of London
Lancet
Land Nationalisation Tracts
Light: A Journal of Psychical, Occult, and Mystical Research
Medium and Daybreak a Weekly Journal Devoted to the History, Phenomena, Philosophy, and Teachings of Spiritualism
Nature
Proceedings of the Entomological Society of London
Proceedings of the Zoological Society of London
Quarterly Journal of Science
Science
Spectator
Spiritualist Newspaper
Times (London)
Vaccination Inquirer and Health Review
Zoologist

WALLACE CORRESPONDENCE AND RELATED MATERIAL

British Library, London

More than a thousand letters in Wallace archive: MS 46414–46442 (1856–1912)
Wallace–Alexander Macmillan correspondence: MS 42138A and MS 42138B (1875–1913)
Original manuscripts of many works
Henry Bates's pocketbook from Brazil

Cambridge University Library

Miscellaneous letters, including sixty from Wallace to Alfred Newton

Imperial College of Science, Technology, and Medicine, London

Thomas Henry Huxley Papers

Libraries and Guildhall Art Gallery, London

"Lloyds Loss and Casualty Book for 1852" [brief entry on the loss of the *Helen*]: MS 14932/12

Linnean Society of London

Miscellaneous letters (1854–1913)
Four Malay journals
Notebook ("Species Journal,"1855–1859): MS 180
North American journal
Part of Wallace's personal library, with many annotated texts

Manchester Archives and Local Studies, Manchester Central Library

Wallace–Matthew Slater correspondence (1886–1908)
Miscellaneous letters

Natural History Museum, London

Wallace's notebooks (Zoology Library)
Henry Bates's notebooks (Entomology Library)

Nature

Approximately seventy letters from Wallace to the editor

Oxford University Museum of Natural History

Miscellaneous letters

Royal Botanic Gardens, Kew

Richard Spruce Correspondence
Miscellaneous letters

Royal Geographical Society, London

Miscellaneous letters (1853–1908)

Royal Society of London

Miscellaneous letters

Thurrock Library, Grays

Material on The Dell, Grays

University College London

Letters to Sir Francis Galton

University of Edinburgh

Wallace's spiritualist library (acquired from Oxford University)

Wallace, Richard and A. J. R.

Letters from Wallace to William Wallace
Miscellaneous letters

Windsor Castle Archives

Material on the Order of Merit awarded to Wallace

Zoological Society of London

Miscellaneous letters

American Philosophical Society, Philadelphia

Carroll, P. Thomas, ed. *An Annotated Calendar of the Letters of Charles Darwin in the Library of the American Philosophical Society*. Wilmington, Del.: Scholarly Resources, 1976. [Darwin–George Romanes correspondence]

Catlett, J. Stephen, ed. *A New Guide to the Collections in the Library of the American Philosophical Society*. Philadelphia: American Philosophical Society, 1987. [Selected letters]

Daniel C. Gilman Papers, Milton S. Eisenhower Library,
Johns Hopkins University, Baltimore

Miscellaneous letters

Darwin Collection of the Dittrick Medical History Center,
Case Western Reserve University, Cleveland

Letters from Darwin and Wallace
Miscellaneous letters

Ernst Mayr Library, Harvard University, Cambridge, Mass.

Miscellaneous letters

Ferdinand Hamburger Archives, Johns Hopkins University Library, Baltimore

Miscellaneous letters

Houghton Library, Harvard University, Cambridge, Mass.

Letters from Wallace

Special Collections and Archives, Knox College Library, Galesburg, Ill.

Letter from Wallace to Charles Kingsley, May 7, 1869
Letter from Wallace to B. O. Flower, October 1, 1891
Letter from Wallace to George Seymour, April 12, 1903

SECONDARY SOURCES

Abreu, João Capistrano de. *Chapters of Brazil's Colonial History, 1500–1800*. Oxford: Oxford University Press, 1997.

Adler, Jeremy. "The Dueling Diagnoses of Darwin." *Journal of the American Medical Association*, April 23–30, 1997, 1275.

Aspects of Britain: Honours and Titles. London: Her Majesty's Stationery Office, 1996.

Barloon, Thomas J., and Russell Noyes, Jr. "Charles Darwin and Panic Disorder." *Journal of the American Medical Association*, January 8, 1997, 138–41.

Barquet, Nicolau, and Pere Domingo. "Smallpox: The Triumph over the Most Terrible of the Ministers of Death." *Annals of Internal Medicine*, October 15, 1997, 635–42.

Beddall, Barbara G. "Wallace, Darwin, and the Theory of Natural Selection: A Study in the Development of Ideas and Attitudes." *Journal of the History of Biology* 1 (1968): 261–323.

———. "Wallace, Darwin and Edward Blyth: Further Notes on the Development of Evolution Theory." *Journal of the History of Biology* 5 (1972): 153–58.

———. "Darwin and Divergence: The Wallace Connection." *Journal of the History of Biology* 21 (1988): 1–68.

———. "Wallace's Annotated Copy of Darwin's *Origin of Species*." *Journal of the History of Biology* 21 (1988): 265–89.

Beer, Max. *A History of British Socialism*. 2 vols. London: Bell, 1929.

Beerbohm, Max. Introduction to *The Portrait Drawings of William Rothenstein*. London: Chapman & Hall, 1916.

Birx, H. James. Introduction to *Island Life*, by Alfred Russel Wallace. 1880. Reprint, Amherst, N.Y.: Prometheus Books, 1998.

Bonner, John Tyler, and Robert M. May. Introduction to *The Descent of Man, and Selection in Relation to Sex*, by Charles Darwin. 2 vols. 1871. Reprint, Princeton, N.J.: Princeton University Press, 1981.

Borges, Jorge Luis. "Testimony to the Invisible." In *Testimony to the Invisible: Essays on Swedenborg*, edited by James F. Lawrence. West Chester, Pa.: Chrysalis Books, 1995.

Bowler, Peter J. *The Eclipse of Darwinism*. Baltimore: Johns Hopkins University Press, 1983.

Brackman, Arnold. *A Delicate Arrangement: The Strange Case of Charles Darwin and Alfred Russel Wallace*. New York: Times Books, 1980.

Brooks, John Langdon. *Just Before the Origin: Alfred Russel Wallace's Theory of Evolution*. New York: Columbia University Press, 1984.

Browne, Janet. *The Secular Ark: Studies in the History of Biogeography*. New Haven, Conn.: Yale University Press, 1983.

———. Introduction to *Voyage of the Beagle (Journal of Researches into the Geology and Natural History of the Various Countries Visited by the HMS Beagle)*, by Charles Darwin. 1839. Reprint, New York: Penguin, 1989.

———. "Biogeography and Empire." In *Cultures of Natural History*, edited by N. Jardine, J. A. Secord, and E. C. Spary. Cambridge: Cambridge University Press, 1997.

Burchfield, Joe D. *Lord Kelvin and the Age of the Earth*. Chicago: University of Chicago Press, 1990.

Burckhardt, Richard. Introduction to *Zoological Philosophy*, by Jean-Baptiste Lamarck. 1809. Reprint, Chicago: University of Chicago Press, 1984.

Burlington House. *A History of the British Association*. London: Lund, Humphries, 1922.

Burnett, John. *A History of the Cost of Living*. New York: Penguin, 1966.

Camerini, Jane. "Evolution, Biogeography, and Maps: An Early History of Wallace's Line." *Isis* 84 (1993): 700–727.

Clemens, Florence. "Conrad's Favorite Bedside Book." *South Atlantic Quarterly* 38 (1939): 305–15.

Clodd, Edward. "Memoir of the Author." Introduction to *The Naturalist on the River Amazons*, by Henry Walter Bates. London: Murray, 1892.

———. *Memories*. London: Watts, 1926.

———. *Pioneers of Evolution from Thales to Huxley*. Freeport, N.Y.: Books for Libraries Press, 1907.

Coates, B., and K. D. Bishop. *A Guide to the Birds of Wallacea*. Alderley, Queensland, Australia: Dove, 1997.

Cook, Gordon. *Manson's Tropical Diseases*. Philadelphia: Saunders, 1996.

Cooter, Roger. *The Cultural Meaning of Popular Science*. Cambridge: Cambridge University Press, 1984.

Cope, Ralph. "'I Will Gladly Do My Best': How Charles Darwin Obtained a Civil Pension for Alfred Russel Wallace." *Isis* 83 (1992): 3–26.

Cronin, Helena. *The Ant and the Peacock*. Cambridge: Cambridge University Press, 1994.

Curle, Richard. *The Last Twelve Years of Joseph Conrad*. London: Sampson, Low, Marston, 1929.

Cutright, Paul Russell, and Michael J. Brodhead. *Elliott Coues, Naturalist and Frontier Historian*. Chicago: University of Illinois Press, 1981.

D'Albe, E. E. Fournier. *The Life of Sir William Crookes*. London: Fisher Unwin, 1923.

Darlington, Philip J. *Zoogeography: The Geographical Distribution of Animals*. New York: Wiley, 1957.

de Beer, Gavin. *Charles Darwin and Alfred Russel Wallace: Evolution by Natural Selection*. Cambridge: Cambridge University Press, 1958.

Desmond, Adrian. *Archetypes and Ancestors*. Chicago: University of Chicago Press, 1982.

———. *Huxley*. Reading, Mass.: Addison-Wesley, 1997.

Desmond, Adrian, and James Moore. *Darwin*. New York: Norton, 1994.

Dick, Steven. *The Biological Universe: The Twentieth Century Extraterrestrial Life Debate and the Limits of Science*. Cambridge: Cambridge University Press, 1998.

Doyle, Arthur Conan. *The History of Spiritualism*. 2 vols. 1926. Reprint, New York: Arno Press, 1975.

Eiseley, Loren. *Darwin's Century*. New York: Anchor Books, 1958.

Ellegard, Alvar. *Darwin and the General Reader*. Chicago: University of Chicago Press, 1990.

"Extracts from the Report of the Royal Commission on Vaccination." *Vaccination Inquirer and Health Review* 12 (1891): 164–68.

Fichman, Martin. "Wallace: Zoogeography and the Problem of Land Bridges." *Journal of the History of Biology* 10 (1979): 45–63.

Freud, Sigmund. *Introductory Lectures on Psychoanalysis*. Translated by James Strachey. New York: Norton, 1977.

Fromm, Erich. Foreword to *Looking Backward*, by Edward Bellamy. New York: Buccaneer Books, 1994.

Gage, A. T. *A History of the Linnean Society*. London: Linnean Society, 1938.

Gardiner, B. G. "Picture Quiz: William Benjamin Carpenter." *Linnean* 16 (2000): 7–13.

George, Wilma. *Biologist-Philosopher: A Study of the Life & Writings of Alfred Russel Wallace*. London: Abelard-Schuman, 1964.

———. "Wallace and His Line." In *Wallace's Line and Plate Tectonics*, edited by T. C. Whitmore. Oxford: Clarendon Press, 1981.

Gillispie, Charles Coulton, ed. *Dictionary of Scientific Biography*. 14 vols. New York: Scribner, 1981.

Gould, Stephen Jay. "Wallace's Fatal Flaw." *Natural History*, January 1980, 26–39.

Gratzer, Walter. *A Bedside Nature: Genius and Eccentricity in Science, 1869–1953*. New York: Freeman, 1997.

Greenblatt, Samuel H. "Phrenology in the Science and Culture of the 19th Century." *Neurosurgery* 37 (1995): 790–803.

A History of the British Association. London: Lund, Humphries, 1931.

Himmelfarb, Gertrude. *Poverty and Compassion*. New York: Vintage, 1992.

———. *Darwin and the Darwinian Revolution*. Chicago: Dee, 1996.

Hoppen, K. Theodore. *The Mid-Victorian Generation, 1846–1886*. Oxford: Oxford University Press, 1998.

Hubbard, Elbert. *Little Journeys to Homes of Great Scientists: Wallace*. New York: Roycrofters, 1905.

Jardine, N., J. A. Secord, and E. C. Spary, eds. *Cultures of Natural History*. Cambridge: Cambridge University Press, 1997.

Jellinek, E. H. "Dr. H. C. Bastian, Scientific Jekyll and Hyde." *Lancet*, December 23, 2000, 2180–83.

Jung, Carl. "On Spiritualistic Phenomena." In *Psychology of the Occult*, edited and translated by R. F. C. Hull. Bollingen series 20. Princeton, N.J.: Princeton University Press, 1981.

———. "The Psychological Foundations of Belief in Spirits." In *Psychology of the Occult*, edited and translated by R. F. C. Hull. Bollingen series 20. Princeton, N.J.: Princeton University Press, 1981.

Kottler, Malcolm J. "Alfred Russel Wallace, the Origin of Man, and Spiritualism." *Isis* 65 (1976): 146–92.

———. "Charles Darwin and Alfred Russel Wallace: Two Decades of Debate over Natural Selection." In *The Darwinian Heritage*, edited by David Kohn. Princeton, N.J.: Princeton University Press, 1985.

Kuhn, Thomas S. *The Structure of Scientific Revolutions*. Chicago: University of Chicago Press, 1996.

Larsen, Anne. "Equipment for the Field." In *Cultures of Natural History*, edited by N. Jardine, J. A. Secord, and E. C. Spary. Cambridge: Cambridge University Press, 1997.

Lawrence, James F., ed. *Testimony to the Invisible: Essays on Swedenborg*. West Chester, Pa.: Chrysalis Books, 1995.

Lodge, Oliver. Foreword to *The Life of Sir William Crookes*, by E. E. Fournier D'Albe. London: Fisher Unwin, 1923.

Lucas, J. R. "Wilberforce and Huxley: A Legendary Encounter." *Historical Journal* 22 (1979): 313–30.

Lurie, Edward. *Louis Agassiz: A Life in Science*. Chicago: University of Chicago Press, 1960.

MacCleod, Roy M. "The X Club: A Social Network of Science in Late-Victorian England." *Notes and Records of the Royal Society of London* 24 (1970): 305–22.

———. "Of Medals and Men: A Reward System in Victorian Science, 1826–1914." *Notes and Records of the Royal Society of London* 26 (1971): 81–108.

Marsden, John. "On Human Credulousness." November 27, 1997. [available at: www.linnean.org]

Martine, Roddy. *Scottish Clan and Family Names: Their Arms, Origins, and Tartans*. Edinburgh: Mainstream, 1998.

Mayr, Ernst. "Wallace's Line in the Light of Recent Zoogeographic Studies." *Quarterly Review of Biology* 19 (1944): 1–14.

———. Introduction to *The Origin of Species*, by Charles Darwin. Facsimile of 1st ed. Cambridge, Mass.: Harvard University Press, 1975.

———. *Evolution and the Diversity of Life: Selected Essays*. Cambridge, Mass.: Harvard University Press, Belknap Press, 1997.

———. *The Growth of Biological Thought: Diversity, Evolution, and Inheritance*. Cambridge, Mass.: Harvard University Press, Belknap Press, 1982.

McIntyre, Loren. "Humboldt's Way." *National Geographic*, September 1985, 318–50.

McKie, Douglas. "The Origins and Foundation of the Royal Society of London." *Notes and Records of the Royal Society of London* 15 (1960): 1–37.

McKinney, H. L. Introduction to *Palm Trees of the Amazon and Their Uses*, by Alfred Russel Wallace. 1853. Reprint, Lawrence, Kan.: Coronado, 1971.

———. *Wallace and Natural Selection*. New Haven, Conn.: Yale University Press, 1972.

Middlemas, Keith. *The Life and Times of Edward VII*. London: Weidenfield & Nicolson, 1972.

Miller, Sally M., ed. *John Muir: Life and Work*. Albuquerque: University of New Mexico Press, 1996.

Milner, Richard. "Darwin for the Prosecution, Wallace for the Defense. Part 1: How Two Great Naturalists Put the Supernatural on Trial." *North Country Naturalist* 2 (1990): 19–36.

———. "Darwin for the Prosecution, Wallace for the Defense. Part 2: Spirit of a Dead Controversy." *North Country Naturalist* 2 (1990): 37–50.

Moon, H. P. *Henry Walter Bates, FRS (1825–1892): Explorer, Scientist and Darwinian*. Leicester: Leicestershire Museums, Art Galleries and Record Service, 1976.

Moore, James. "Wallace's Malthusian Moment: The Common Context Revisited." In *Victorian Science in Context*, edited by Bernard Lightman. Chicago: University of Chicago Press, 1997.

Murray, James. *Genetic Diversity and Natural Selection*. New York: Hafner, 1972.

Nowak, Ronald. *Walker's Primates of the World*. Baltimore: Johns Hopkins University Press, 1999.

Oppenheim, Janet. *The Other World: Spiritualism and Psychical Research in England, 1850–1914*. Cambridge: Cambridge University Press, 1988.

Porter, Andrew. *The Oxford History of the British Empire*. Vol. 3, *The Nineteenth Century*. New York: Oxford University Press, 1999.

Porter, Duncan. "With Humboldt, Wallace and Spruce at San Carlos de Rio." In *Richard Spruce, Botanist and Explorer*, edited by M. R. D. Seaward and S. M. D. Fitzgerald. Kew: Royal Botanic Gardens, 1996.

Poulton, Edward Bagnall. "Alfred Russel Wallace (1823–1913)." *Proceedings of the Royal Society of London* 95 (1924): i–xxxv.

Raby, Peter. *Alfred Russel Wallace: A Life*. Princeton, N.J.: Princeton University Press, 2001.

Rann, Ernest. "Dr. Alfred Russel Wallace at Home." *Pall Mall Magazine*, March 1909, 275–84.

Reid, Anthony. *Southeast Asia in the Early Modern Era*. Ithaca, N.Y.: Cornell University Press, 1993.

Ricklefs, M. C. *A History of Modern Indonesia Since c. 1300*. Palo Alto, Calif.: Stanford University Press, 1993.

Ridgely, Robert S., and Guy Tudor. *The Birds of South America: The Suboscine Passerines*. Austin: University of Texas Press, 1994.

Rudwick, Martin. Introduction to *Principles of Geology*, by Charles Lyell. Vol. 1. 1830. Reprint, Chicago: University of Chicago Press, 1991.

Schwartz, Joel S. "Darwin, Wallace, and the *Descent of Man*." *Journal of the History of Biology* 17 (1984): 271–89.

———. "Darwin, Wallace, and Huxley and *Vestiges of the Natural History of Creation*." *Journal of the History of Biology* 23 (1990): 127–53.

Severin, Tim. *Spice Island Voyage*. New York: Carrol & Graf, 1997.

Simon, Linda. *A Genuine Reality: A Life of William James*. New York: Harcourt, Brace, 1998.

Stevens, P. F. "George Bentham and the Darwin/Wallace Papers of 1858: More Myths Surrounding the Origin and Acceptance of Evolutionary Ideas." *Linnean* 11 (1995): 14–16.

Stocking, George W., Jr. *Victorian Anthropology*. New York: Free Press, 1987.

Stroud, Patricia T. *The Emperor of Nature*. Philadelphia: University of Pennsylvania Press, 2000.

Swales, J. D. "The Leicester Anti-Vaccination Movement." *Lancet*, October 24, 1992, 1019–21.

Trevelyan, George M. *British History in the Nineteenth Century, 1782–1901*. London: Longman, Green, 1934.

———. *A Shortened History of England*. New York: Viking, 1988.

Van Allen, Elizabeth J. *James Whitcomb Riley: A Life*. Bloomington: Indiana University Press, 1999.

Whitmore, T. C. Introduction to *Biogeographical Evolution of the Malay Archipelago*, edited by T. C. Whitmore. Oxford: Clarendon Press, 1987.

Williamson, Edwin. *The Penguin History of Latin America*. New York: Penguin, 1992.

Index